ESSENTIALS OF STATISTICS IN AGRICULTURAL SCIENCES

ESSENTIALS OF STATISTICS IN AGRICULTURAL SCIENCES

Edited by

Pradeep Mishra, PhD

Fozia Homa, PhD

AAP | APPLE ACADEMIC PRESS

Apple Academic Press Inc.
3333 Mistwell Crescent
Oakville, ON L6L 0A2
Canada

Apple Academic Press Inc.
1265 Goldenrod Circle NE
Palm Bay, Florida 32905
USA

Exclusive worldwide distribution by CRC Press, a member of Taylor & Francis Group

International Standard Book Number-13: 978-1-77188-752-6 (Hardcover)
International Standard Book Number-13: 978-0-42942-576-9 (eBook)

Library and Archives Canada Cataloguing in Publication

Title: Essentials of statistics in agricultural sciences / edited by Pradeep Mishra, PhD, Fozia Homa, PhD.

Names: Mishra, Pradeep (Professor of statistics), editor. | Homa, Fozia, editor.

Description: Includes bibliographical references and index.

Identifiers: Canadiana (print) 20190088230 | Canadiana (ebook) 2019008832X | ISBN 9781771887526 (hardcover) | ISBN 9780429425769 (ebook)

Subjects: LCSH: Agriculture—Statistical methods. | LCSH: Agriculture—Statistics.

Classification: LCC S566.55 .E87 2019 | DDC 630.2195—dc23 | 630.72/7—dc23

Library of Congress Cataloging-in-Publication Data

Names: Mishra, Pradeep (Professor of statistics), editor. | Homa, Fozia, editor.

Title: Essentials of statistics in agricultural sciences / [edited by] Pradeep Mishra, PhD, Fozia Homa, PhD.

Description: Oakville, ON ; Palm Bay, Florida : Apple Academic Press, 2019. |
Includes bibliographical references and index

Identifiers: LCCN 2019014411 (print) | LCCN 2019016618 (ebook) | ISBN 9780429425769 (ebook) |
ISBN 9781771887526 (hardcover : alk. paper)

Subjects: LCSH: Agriculture--Statistical methods. | Agriculture--Statistics

Classification: LCC S566.55 (ebook) | LCC S566.55 .E77 2019 (print) | DDC 338.1072/7--dc23

LC record available at https://lccn.loc.gov/2019014411

Apple Academic Press also publishes its books in a variety of electronic formats. Some content that appears in print may not be available in electronic format. For information about Apple Academic Press products, visit our website at **www.appleacademicpress.com** and the CRC Press website at **www.crcpress.com**

About the Editors

Pradeep Mishra, PhD

Assistant Professor, Department of Mathematics and Statistics, College of Agriculture, Jawaharlal Nehru Agricultural University, Jabalpur, Madhya Pradesh, India

Pradeep Mishra, PhD, is an Assistant Professor of Statistics at the College of Agriculture, Powarkheda at the Jawaharlal Nehru Krishi Vishwa Vidyalaya, Madhya Pradesh, India. During his PhD research work, he was selected for an INSPIRE Fellowship by the Department of Science and Technology of the Government of India. Dr. Mishra has served as a data management specialist at a private multinational company for several years. His area of specialization is agriculture statistics, time series, and design of experiment. He has published more than 68 research articles in different international and national journals. Also, he won several awards, including an award for young scientists in at an international conference (GRISAAS–2017); best doctoral degree in 2018 at an international conference on advances in agriculture and allied science technologies for sustainable development, in Hyderabad; a second best paper award from the Society of Economics & Development in PAU, 2018, among others. Dr. Mishra earned his BSc in Agriculture from the College of Agriculture, Bilaspur, affiliated with the Indira Gandhi Krishi Vishwavidyalaya, Raipur (C.G.); his MSc in Agricultural Statistics from the College of Agriculture, Jawaharlal Nehru Agricultural University; and his PhD in Agriculture Statistics, specializing in modeling and forecasting of foods crops in India and their yield sustainability.

Fozia Homa, PhD
Assistant Professor and Scientist, Department of Statistics, Mathematics and Computer Application, Bihar Agricultural University, Sabour, India

Fozia Homa, PhD, is an Assistant Professor and Scientist in the Department of Statistics, Mathematics, and Computer Application at Bihar Agricultural University, Sabour, India, and is the author or co-author of several journal articles. Dr. Fozia has received numerous awards in recognition of her research and teaching achievements from several organizations of national and international repute. She was also awarded with a SP Dhall Distinguished Publication Award in Statistics (2015) by the Society for Advancement of Human and Nature, Himachal Pradesh, India; Young Scientist Award (2015) by the Venus International Foundation, Chennai, India; and Best Young Researcher Award (2015) by GRABS Educational Trust, Chennai, India. She has been an active member of the organizing committees of several national and international seminars, conferences, and summits. Dr. Fozia Homa acquired her BSc (Statistics Hons) and MSc (Statistics) degrees from Banaras Hindu University, Varanasi, Utter Pradesh, India, and her PhD (Applied Statistics) with specialization in sampling techniques from the Indian Institute of Technology (Indian School of Mines), Dhanbad, Jharkhand, India. She has received several grants from various funding agencies to carry out her research projects. Her areas of specialization include sample surveys, population studies, and mathematical modeling.

Contents

Contributors

G. F. Ahmed
Assistant Professor, Computer Application, College of Agriculture, JNKVV, Powarkheda, (M.P.) – 461110, India

B. S. Dhekale
Assistant Professor, Sher-e-Kashmir University of Agriculture Sciences and Technology, Kashmir, India, E-mail: bhagyashreedhekale@yahoo.com

Shweta Dixit
Research Assistant Professor, School of Management, SRM Institute of Science and Technology, Chennai, Tamil Nadu – 603203, India

Fozia Homa
Assistant Professor-Cum-Scientist, Department of Statistics, Mathematics & Computer Application, Bihar Agricultural University, Sabour, Bhagalpur, Bihar, 813210, India

Mukti Khetan
Assistant Professor, Department of Statistics, Sambalpur University, Sambalpur, Odisha, 768019, India

G. R. Manjunatha
Scientist - B (Agricultural Statistics), Central Sericultural Research & Training Institute, Central Silk Board, MoT, Government of India, Berhampore, West Bengal, India

Pradeep Mishra
Assistant Professor, Department of Mathematics and Statistics, College of Agriculture, JNKVV, Powarkheda, (M.P.) – 461110, India, E-mail: pradeepjnkvv@gmail.com

D. Ramesh
Assistant Professor, Department of Statistics and Computer Applications, ANGRAU, Agricultural College, Bapatla, Andhra Pradesh, India, E-mail: dasyam.ramesh32@gmail.com

R. B. Singh
Professor, Department of Mathematics and Statistics, College of Agriculture, JNKVV, Jabalpur (M.P.), 482004, India

Supriya
Assistant Professor, College of Agriculture Campus, Kotwa, Azamgarh, NDUAT, Faizabad, India

A. K. Tailor
Technical Officer (Statistics), National Horticultural Research and Development Foundation, Regional Research Station, Nashik – 422003, (M.H.), India

G. K. Vani
Assistant Professor, Agricultural Economics & F.M., College of Agriculture, JNKVV, Jabalpur, (M.P.), 482004, India

Prashant Verma
Senior Research Scholar, Department of Statistics, Banaras Hindu University, Varanasi, Uttar Pradesh, 221005, India

K. P. Vishwajith
Department of Agriculture Statistics, Bidhan Chandra Krishi Vishwavidyalaya, Nadia – 741252, India

Abbreviations

ACF	autocorrelation function
ADF	Augmented Dickey-Fuller test
AIC	Akaike's information criterion
ARIMA	autoregressive integrated moving average
BD	binomial distribution
BIC	Bayesian information criterion
BLUEs	best linear unbiased estimators
BSS	sum of squares due to blocks
CC	contingency coefficient
CD	critical difference
CF	correlation factor
CFc	correction factor for checks
CFg	correction factor for genotypes
CLRM	classical linear regression model
CRD	completely randomized design
CRLB	Cramer Rao lower bound
CSS	column sum of squares
CT	central tendency
DES	Department of Economic and Statistics
ESS	error sum of squares
GARCH	generalized ARCH
LM	Lagrange multiplier
LOS	level of significance
LSD	Latin square design
MAE	mean absolute error
MAPE	mean absolute percentage error
ME	mean error
MLE	maximum likelihood estimator
MPE	mean percentage error
MSE	mean sum of square for error
MSS	mean sum of squares
MVB	minimum variance bound
MVUE	minimum variance unbiased estimator
NP Lemma	Neyman-Pearson Lemma

OFAT	one factor at a time
PCA	principal component
PD	poison distribution
RBD	randomized block design
RCBD	randomized completely block design
RMSE	root mean square error
RSS	row sum of squares
SE	standard errors
SELF	square error loss function
SRS	simple random sampling
SRSWOR	simple random sampling without replacement
SRSWR	simple random sampling with replacement
SS	sum of square
TPM	transition probability matrix
TRT	tick (√) on treatments
TSS	total sum of squares
UMVUE	Uniform Minimum Variance Unbiased Estimator
VIF	variance inflation factor
VIP	auxiliary regressions or variance inflation factor

Preface

Because of statistics' usefulness and varied range of applications, one can hardly find a branch where statistics has not been used. With the advent of computing facilities, the use of statistical theories has increased tremendously, including in agriculture. Along with other fields of science, statistics is being extensively used in agriculture, animal, fishery, dairy, and other fields in explaining various basic as well as applied problems. The selection of the appropriate statistical technique and its proper use for a given problem is mostly warranted for getting a meaningful explanation of the problems under consideration.

Students, teachers, researchers, and practitioners from agriculture, animal, fishery, dairy, and allied fields deal with flora, fauna, soil, air, water, nutrients, etc., along with socioeconomic and behavioral aspects for all-around development. Understanding of the theory and essence of both the theory of statistics and the specific field of study is a must.

In spite of knowing the utility of statistics, a large section of students, teachers, researchers, and practitioners in the fields of agriculture, animal, fishery, dairy sciences do not have much mathematical orientation and are scared of using statistics. Being mathematical science, having uncertainty and variability in its focal points of interest to reach to these huge users remains a challenging task to the statisticians, particularly the agricultural statisticians. Statistics needs to be presented to these users in their own terms/manners and language. In order to have a proper understanding of the problem on hand, correct and efficient handling of the data is needed. A good level of understanding of statistics is essential for planning, recording of information, analyzing, and interpreting data. It has been observed in many cases that many users lack a good comprehension of biostatistics and do not feel comfortable while making simple statistics-based decisions. Millions of students, readers, and researchers in-and-outside India need to have a good understanding of the subject of statistics to explain their problems in a better way. An attempt has been made in this book to present statistics in such a way that the students and researchers from agriculture and allied fields find it easy to handle and use in addressing many real-life problems in their own fields.

This book presents the subject in such a way so that readers do not require any prior knowledge about the subject. The ultimate aim of the book is to be a self-instructional textbook, which can be helpful to users in solving their problems using statistical tools and available software. Statistical theories have been discussed with the help of examples from real-life situations in agriculture and allied fields, followed by worked-out examples. Attempts have been made to familiarize users with the problems with examples on each topic in a lucid manner. Each chapter is followed by a number of problems and questions, which will help students in gaining confidence on solving those problems.

This book is comprised of seven chapters. The first chapter addresses and explains the subject of statistics, its usefulness, and applications with particular reference to agriculture and allied fields. A brief narration on statistics, highlighting its use, scope, steps in statistical procedure and limitations along with examples have been provided in Chapter 1. In the second chapter, extensive discussion has been made on principles of the design of field experiments: replication, randomization, local control, layout, uniformity trial, steps in designing field experiments. In this chapter, an elaborate discussion has been made on various field designs, starting with a completely randomized design, a randomized block design, and a Latin square design to combine analysis of various designs. In the third chapter, statistical inferences including different statistical tests have been discussed at length, while sampling methods have been discussed in Chapter 4. Time series analysis and its different aspects have been presented in Chapter 5. Chapters 6 and 7 are the most useful chapters from a practical utility point of view because these two chapters discuss the analysis of the problems concerning the previous chapters with the help of computer software. A good number of books and articles from different national and international journals have been consulted during the preparation of this book, which are recorded in the reference section. Interested readers will find more material from these references. The need of the students/teachers/researchers/practitioners in agriculture and allied fields have been the prime consideration during the preparation of this book.

Acknowledgments

Endeavor and hardship are very common aspects behind every achievement. But we would like to add something equally important and indispensable in this respect; i.e., direct or indirect help, cooperation, and inspiration of many, which paved the way to reach the destination.

At the onset of this acknowledgment, we ascribe all glory to the Gracious "Almighty" from whom all good things come, who has showered his choicest blessings and benevolent graces upon me in life and bestowing us with the blessing to complete this book.

To every faithful and holistic endeavor, there remains the contribution and help, and we wish to express our sincere thanks to Jawaharlal Nehru Krishi Vishwa Vidyalaya and Bihar Agricultural University for providing us the facilities to perform such a project and for supporting us with all the facilities and infrastructure for the completion of the project. We convey heartfelt thanks to Dr. P. K. Sahu, Head, and Professor, Department of Agricultural Statistics, Faculty of Agriculture, Bidhan Chandra Krishi Viswavidyalaya, and Dr. Mohammed Wasim Siddiqui, Scientist-cum-Assistant Professor, Bihar Agricultural University, without whose guidance we would never have turned this matter into a book.

With a profound and unfading sense of gratitude, we express our heartfelt respect and thanks to the co-authors of the chapters, colleagues, and other research team members for their support and encouragements for helping us to accomplish this venture. Their stimulating influence made the work complete with perfection.

We would like to thank Mr. Ashish Kumar and Mr. Rakesh Kumar of Apple Academic Press for their continuous supports to complete the project.

In the long run, we owe our best regards and deepest sense of gratitude to beloved parents and family members for their blessings, love, inspiration, and sacrifices and for inspiring our continuous zeal during the entire period of this venture, which made it possible for us to complete this endeavor.

CHAPTER 1

The Basics of Statistics

PRADEEP MISHRA[1], B. S. DHEKALE[2], R. B. SINGH[3], K. P. VISHWAJITH[4], and G. K. VANI[5]

[1]*Assistant Professor (Statistics), College of Agriculture, JNKVV, Powarkheda, (M.P.), 461110, India, E-mail: pradeepjnkvv@gmail.com*

[2]*Assistant Professor, Sher-e-Kashmir University of Agriculture Sciences and Technology, Kashmir, India*

[3]*Professor, Department of Mathematics and Statistics, College of Agriculture, JNKVV, Jabalpur (M.P.), 482004, India*

[4]*Department of Agriculture Statistics, Bidhan Chandra Krishi Vishwavidyalaya, Nadia, 741252, India*

[5]*Assistant Professor (Agricultural Economics & F.M.), College of Agriculture, JNKVV, Jabalpur (M.P.), 482004, India*

1.1 INTRODUCTION

Statistics plays a very important role in our daily life. Statistics help us to take in decision-making as well as provide reliable results. In making a conclusion, data is playing a key role in statistics, which give our results a proper path towards to decision making. So data handling is very important. If our data handling is not proper, definitely it gives the wrong picture for our results. Before we move to the statistical part, it is necessary to understand about data handling. Every process before statistical analysis like data collection, source, and conditions, the data entry and data checking is important.

(1) **Data collection:** When a researcher is going to collect data, it is self-important to keep all the important things in mind; because our decision-making or data analysis help us to understand the data. For example, if we are collecting any crop yield data, the yield may be affected by insect count, disease or environmental factors. So this kind of important information helps us in our final decision making.

(2) **Data source:** Data source is very important because of its effect on the quality of data. If data collected directly by the researcher, then it comes under primary data collection. When data is collected by some other person or agency, other than the researcher/current user then for researcher or current user it is secondary data. Also, when the researcher is using secondary data, the researcher should identify or check the accuracy of the data.

(3) **Data entry:** When a researcher is using primary and secondary data information and entering into computer or notebook, it needs to be careful that data goes into correct form until the last destination. It is very important because of 'any discrepancy in the data' effects on the results.

1.2 DEFINITION OF STATISTICS

Statistics is a science, which include the application of quantitative principles to the collection, data preparation, analysis, and presentation of numerical data. The practice of statistics utilizes data from some population in order to describe it meaningfully, to draw conclusions from it, and make informed decisions.

1.3 SCOPE AND LIMITATION OF STATISTICS

Every subject has its own scope and limitation. Statistics has a vast scope. The reason behind is that every field and the final conclusion drawn on the basis of data and statistical tools. Statistics is used in various disciplines including agriculture, medical, business forecasting, population studies, etc. The limitation with statistics is that without any data, it cannot help researcher in drawing any inference. Suppose we want to measure any qualitative characters (honesty, intelligence or work performance, etc.), then without any data, statistics is unable to do that.

1.4 DIAGRAMS AND GRAPHS

1.4.1 STATISTICS IN AGRICULTURE

Statistics, also called as science of probability, is the subject which has application in every industry and every other field including literature. Agriculture is dependent on whether phenomenon which in turn is probabilistic in nature. The agriculture and components within it are highly probabilistic in nature. The seed which germinates very well at certain soil, climate and whether condition may not germinate elsewhere under some other soil, climate and whether condition. Thus, to before one can sell seed, it needs to be statistically validated whether the seed will germinate in the market with varied climate and soil where it is proposed to be sold. Only when the seller has high degree of confidence based on statistical results from scientific experiments, he/she would sell the seed in market. There are more pertinent examples from field. The question of whether income insurance for farmer would serve the purpose is based on degree of correlation between crop production over the years and prices received in the market. The statistical analysis says that high degree of negative correlation means that when production is low then prices prevailing in the market are higher than average price but vice-versa being true. It is always a constant eagerness in everyone's mind how much different inputs in agricultural production affects yield, the answer lies with crop modeling performed using regression analysis. Even greater use is found with plant genetics and breeding department wherein clustering of phenotypes and genotypes is required to find the distant genotypes and phenotypes for better heterosis (dendrogram). The breeder is also interested in finding whether the phenotype combination observed is in line what Mendelian laws can be done using chi-square test. The list of application of statistical methods is long in agriculture. This book surveys the basic concepts as well as techniques and methods of statistics useful to students of agriculture and allied sciences in simple ways.

1.4.2 STATISTICS, DATA AND INFORMATION

Statistics works on data collected and compiled by various agencies/individuals. Data is defined as "facts and figures from which conclusions can be drawn." The data can be of various sorts, for example:

1. The daily weight of plants in a nursery measured by scientist.
2. The presence and absence recorded of a pigment in the plants in nursery.
3. The nursery temperature measured every hour for a week period.

The other types of data can be color bands obtained in chromatography, DNA sequence, rank of bull's semen quality etc. Once data have been collected and processed, then it is ready to be organized into information.

Thus, information is defined as "data which have been recorded, classified, organized, related, or interpreted within a framework so that meaning can emerge out of it." Thus, for the above three bullet point which provide example of data, corresponding information is as following:

1. 30 plants have weight ranging from 40 to 60 grams.
2. Out of 30 plants, only 10 have pigments on them.
3. Hourly temperature inside the nursery rose above 30°C only for one day in a week.

Information is commonly presented using methods of statistics. Statistical methods include descriptive statistics such as average, correlation and regression analysis, analysis of variance, etc.

For the three items provided in information section, the statistical counterpart is provided below:

1. Average plant weight for plants in nursery was 44 grams.
2. One out of every three plants in nursery has pigment.
3. Average daily temperature inside nursery was 22°C for the week.

In statistical analysis, diagrams and graphs helps to visualize the data and therefore, provide better insight into what data tells the user. It is also important because a graph/diagram tells same thing in less space than would be the case with tables. These are also important in finding patterns hidden in the data. Thus, graphs and diagrams are useful in preliminarily analysis of the data.

Data: 5, 9, 5, 4, 2, 3, 5, 2, 2, 1, 7, 2, 8, 1, 4
Frequency Distribution:

Number	1	2	3	4	5	7	8	9
Frequency	2	4	1	2	3	1	1	1

1.4.3 BAR DIAGRAM

Bar diagram plots the frequency of the given number against the number itself (Figure 1.1). Frequency is plotted on the y-axis with numbers on the x-axis. A perpendicular is dropped from the point to the respective x-axis point. This graph is one-dimensional because the only height of the bar is considered here, not the width.

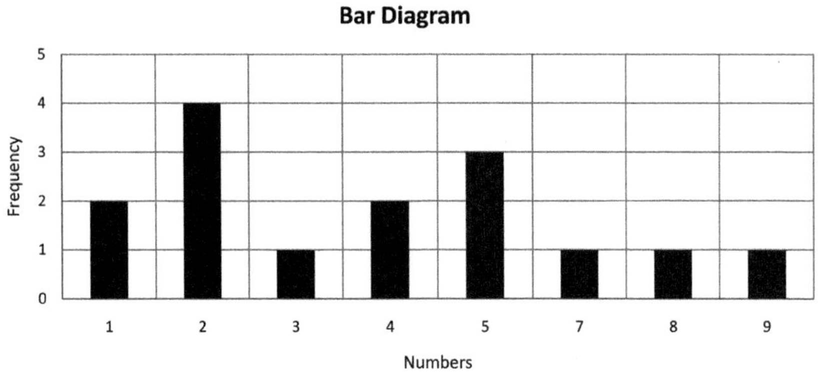

FIGURE 1.1 Bar diagram.

1.4.4 HISTOGRAM

A histogram is the plotting of class frequency against the class interval (Figure 1.2). Frequency is plotted on Y-axis with a corresponding class interval on X-axis. This diagram is two-dimensional because here not only the height of the bar is considered like a bar diagram but also the width of the bar, i.e., class interval.

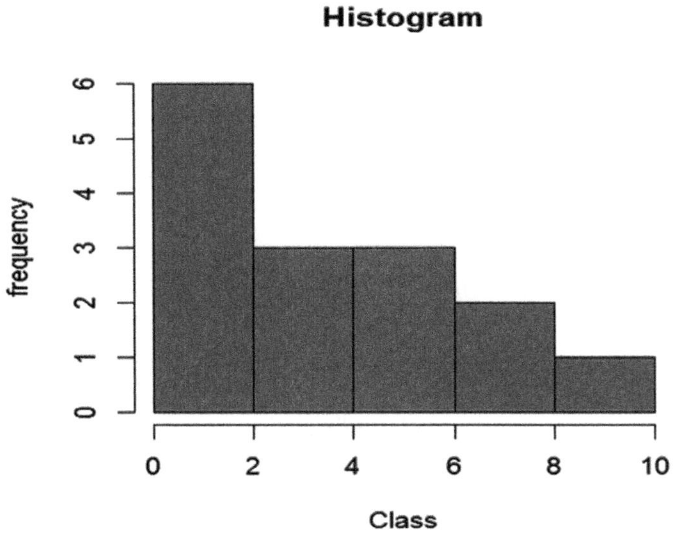

FIGURE 1.2 Histogram.

1.4.5 FREQUENCY POLYGON

A frequency polygon is when a straight-line joins the midpoints of each rectangle of the histogram (Figure 1.3). It is not possible to plot histograms of several distributions on the same axis to make comparisons, but it can be done with frequency polygon.

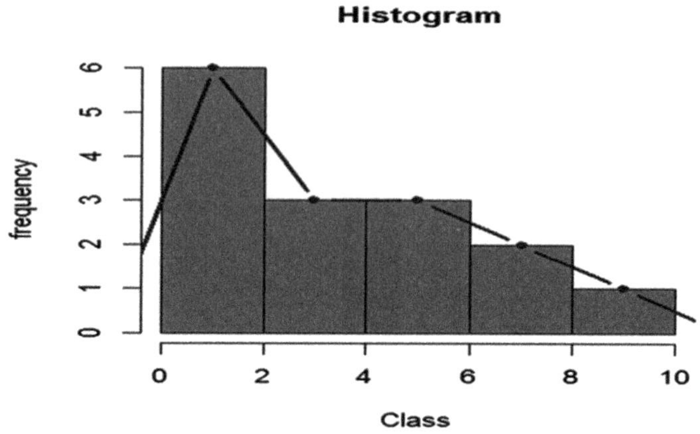

FIGURE 1.3 Frequency polygon.

1.4.6 SMOOTH FREQUENCY CURVE

It is a smooth curve imposed over the frequency polygon by a free hand method. The purpose is to remove any accidental fluctuation that might be present in the data.

1.4.7 OGIVE CURVE (CUMULATIVE FREQUENCY CURVE)

This curve plots class limits (lower or upper) against its cumulative frequency to depict the number of observations above or below that limit (Figure 1.4). The less than Ogive curve provides information on a number of observations falling below the upper-class limit in the sample distribution. While, the more than Ogive curve provides information on a number of observations above the lower-class limit in the sample distribution. The median is a location where less than Ogive curve intersects more than Ogive curve (Table 1.1).

TABLE 1.1 Ogive Curve (Cumulative Frequency Curve)

Frequency Distribution		Less than Ogive		More than Ogive	
Class	**Frequency**	**Number**	**Frequency**	**Number**	**Frequency**
0–1	0	1	0	1	15
1–3	6	3	6	3	9
3–5	3	5	9	5	6
5–7	3	7	12	7	3
7–9	3	9	15	9	0

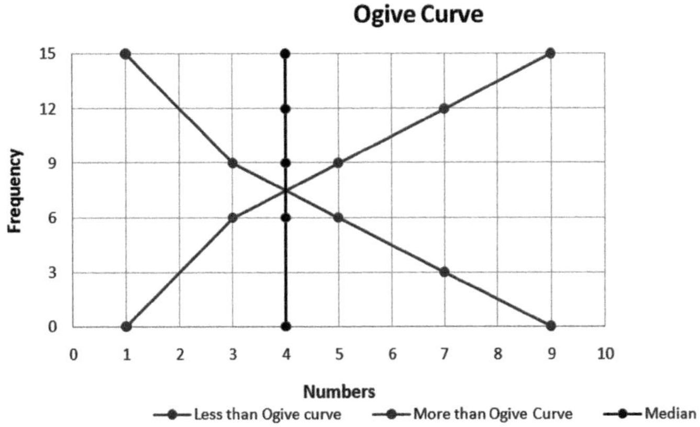

FIGURE 1.4 Ogive curve.

1.4.8 Q-Q PLOT

It is the scatter plot of theoretical quintiles against sample quintiles. Theoretical quintiles are obtained from the statistical distribution for which fit of the data is to be tested. Apart from plotting the sample quintile versus theoretical quintile, a line of equality is drawn so that departure from theoretical distribution can be visualized. Usually, Q-Q plot is used to check normality of variables from survey data (Figure 1.5).

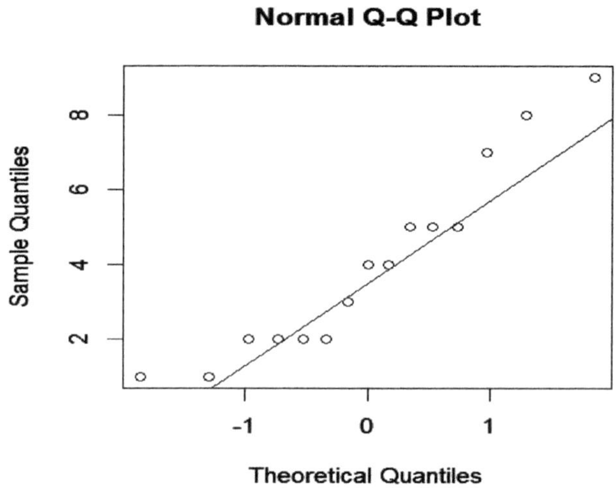

FIGURE 1.5 Q-Q plot.

1.4.9 BOXPLOT

Boxplot is a graphical representation of various positional measures of dispersions in a single plot (Figure 1.6). This plot provides graphical details of all three quartile values along with maximum and minimum values in a convenient and logical manner. The median is shown by the dark black line within the box, and the upper and lower limits of the inside box provide third and first quartile values. Above and below this inside box are some dotted lines that run up to maximum and minimum values of the distribution.

For this dataset, the median is four with first and third quartile values being two and five as is evident from the box plot. The maximum value in the distribution is 15 while the minimum is one as can be observed from box plot.

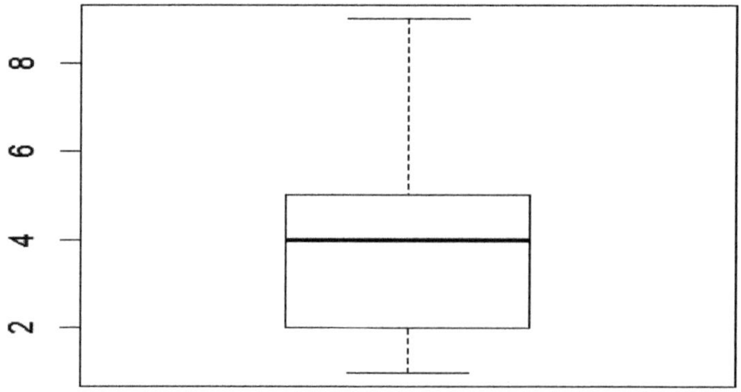

FIGURE 1.6 Boxplot.

1.4.10 P-P PLOT

This graph provides the plot of the empirical distribution of sample data against the best fitting von Mises distribution function (Figure 1.7). The best fitting von Mises distribution function is obtained with *maximum likelihood estimation* procedure. This plot is also called as von Mises Probability-Probability plot, and the "Probability–Probability plot" phrase has been popularly quoted as P-P plot.

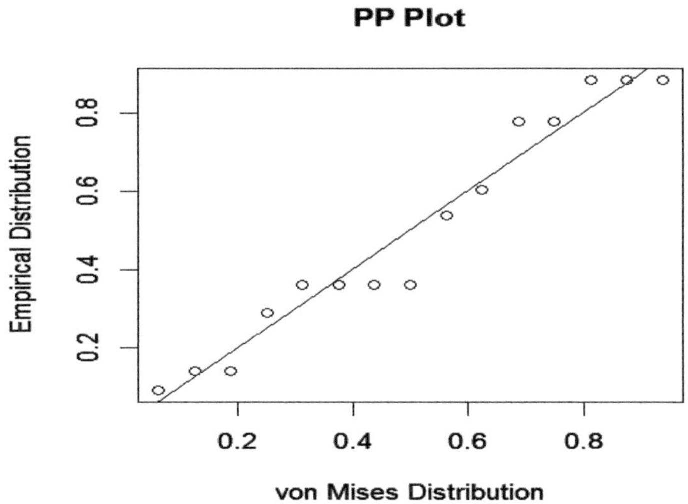

FIGURE 1.7 P-P plot.

1.4.11 DESCRIPTIVE STATISTICS

Before any statistical analysis of data, it is important to understand the nature of data. If we read the nature of data or having good information about that, then it helps the researcher to understand that why statistical analysis results are coming like that? Descriptive statistics tell about what is our data, and it doesn't tell us right or wrong about. To examine the nature and describe the basic features of each series, different descriptive statistics are used. In other words, we use descriptive statistics to simplify our data and obtain information about datasets. With inferential statistics, one tries to reach conclusions that extend beyond the immediate data alone. A descriptive summary is used to present quantitative measures in a manageable or sensible way. In a research study, we may have lots of measures. Each descriptive statistic reduces lots of data into an easier summary. Descriptive statistics can broadly be categorized into measures of central tendency (CT) and measures of dispersion. Among these, arithmetic mean, standard deviation/error, skewness, kurtosis, etc., are widely used to describe the given data for their obvious merits over other hosts of measures.

In statistics, no doubt data handling is important but is essential for the smooth processing of data by software. In the process of handling data, data cleaning and tabulation has a separate space carved out for them. Data tabulation is important in arriving at first filtered view of the data.

In this, the essential step is data classification which means sorting of data according to various characteristics/variables present in the data. This means sorting data based on geographical, chronological, and demographic or some other characteristics associated with data. This involves separating the observations that belong to different characteristics. The following table provides the classification of research data on number of farmers under various 'land holding' size of the farmer concerned.

Land holding of farmer (ha)	**No. of farmers**
Below 1	10
1–2	12
2–4	14
4–10	13
10 and above	11
Total	60

Here, the data for 60 farmers is classified into the class intervals. The land holding of farmer is divided into five classes. Each class has upper and lower limits which are in turn called as lower and upper class limits, respectively. The class interval is the other name for class width which means difference between upper and lower class limit. The class intervals appropriate to the data can be determined using following formula:

$$i = \frac{\text{Range}}{K}$$

where, i is the class interval, *range* is the difference between highest and lowest value taken by the variable in the data and K is the appropriate number of the classes. Here appropriate number of classes can be determined using Sturges' rule:

$$K = 1 + 3.322 \log_{10} N$$

where N is the number of observations in the data.

When the class interval includes the upper class limit in the interval then it is referred to as inclusive class interval, for example: 200–300, 301–400, 401–500. Here, '300' is included in the class interval and hence the next class interval starts from the next integer value 301. In other type of class interval which is referred as exclusive class interval, upper limit is not included and hence the upper class limit of the first interval becomes the lower class limit of the next class interval.

Tabulation is the entering data in rows and columns in a systematic manner. Tabulation needs classification and data cleaning to precede itself. A table summarizes the data in a meaningful and systematic manner to be called as statistical table. In a report, each table is supposed to have been numbered, provided a suitable title. The title of the table needs to be provided such that it reflects on what table needs to tell the reader. Apart from title and number, a table has caption, stub body, head note and footnote. The caption refers to the heading provided to the column. It must explain what is contained in column. The stub refers to row headings/designations. The stubs are usually wider that captions when tables are to be printed in portrait orientation (page layout). Body of the table contains numbers. This is the most vital part in the table. Head notes and footnotes are provided above and below the table for making the reader/user of the table grasp things from the table as desired.

1.4.11.1 MEAN OR AVERAGE

Mean is the sum of all numbers in the distribution divided by the count of all numbers, e.g., the arithmetic mean 'x' of 'n' observation $x_1, x_2, x_3, \ldots, x_n$ is given by $\bar{x} = \frac{1}{n}\sum_{i=1}^{n} x_i$. In the case of grouped data, $\bar{x} = \frac{1}{\sum_{i=1}^{n} f_i}\sum_{i=1}^{n} f_i x_i$, where x_i's are the mid-values of the classes and f_i's are the respective frequencies. Mean is most widely used for its simplicity in calculation and explanation. When a researcher is using mean or average, the following things need to be understood.

(a) Mean calculation is always done by all observations. So if any abnormal or extreme value is present in the data set, then it may mislead the results. So, it is good to check all individual values before the calculation of mean or average. In the case of extreme items, the mean gives a distorted picture of the series.

(b) It is easy for the interpretation and calculation.
(c) In the absence of any single observation, the mean calculation is not possible.

Example: The following table gives the height of 8 plants. Find out the average height of all plants.

Plants numbers	1	2	3	4	5	6	7	8
Heights (in cm)	44	21	32	28	34	29	48	30

Solution: Here, *N* or number of plants are 8.

So, find the average heights of all plant

$$n\overline{X} = \frac{266}{8} = 33.22cm$$

After calculation Mean or average heights of plants is 33.22 cm. This average comes from all eight plants individuals' value.

Example: Find out the average height of students for frequency distribution.

Frequency	0–10	10–20	20–30	30–40	40–50
Height (cm)	38	45	60	52	63

Solution:

Heights	No. of students (F)	Mid-point (X)	(FX)
0–10	38	5	190
10–20	45	15	675
20–30	60	25	1500
30–40	52	35	1820
40–50	63	45	2835
Total	258		7020

$$\overline{X} = \frac{1}{N}\sum FX$$

$$\overline{X} = \frac{1}{258} * 7020$$

$$\overline{X} = \frac{7020}{258} = 27.20\ cm$$

Here the point is, if the value and X are more in numbers, then the calculation of mean using method is time taking as well as it seems a little bit complicated calculation.

1.4.11.2 MEDIAN

Median of a distribution is the value of the variable which divides the distribution into two equal parts. It is the value which exceeds and is exceeded by the same number of observation; i.e., it is the value such that the number of observations above it is equal to the number of observations below it. Thus, the median is a positional average.

In the case of ungrouped data, if the number of observations is odd then the median is the value of the middle-most observation; after the values have been arranged in ascending or descending order of magnitude. In case of an even number of observations, there are two middle terms and median is obtained by taking the arithmetic mean of the values of the middle terms.

In case of continuous frequency distribution, the class corresponding to the cumulative frequency just greater than N/2 is called median class, and the value of the median is obtained by the following formula:

$$Median = l + \frac{h}{f}\left(\frac{N}{2} - C\right)$$

where l is the lower limit of the median class, f is the frequency of median class, h is the width of the median class, and C is the cumulative frequency of the class preceding the median class and $N = \sum f$

(a) It is easy for interpretation and calculation.

(b) Mean calculation is always done by all observations. So, the median is not affected by extreme values or abnormal data.

Example: Calculate the median of these five scores: 10, 30, 27, 29, 12.

Step 1: Sequence the scores in ascending order.

10, 12, 27, 29, 30

Step 2: The number of total scores equals 5, an odd number; the third number (27) is the raw score at the exact middle and becomes the median.

Example: Calculate the median of these six scores: 5, 6, 8, 50, 10, 70.

Step 1. Sequence the scores in ascending order 5, 6, 8, 10, 50, 70

Step 2. The number of scores equals 6, an even number; compute the average of the values for the third and fourth "middle" scores. The arithmetic mean of these raw scores of 8 and 10 equals the score of 9.

***Example*:** Calculate the median for the following frequency distribution:

Salary in (in $)	100–120	120–140	140–160	160–180	180–200	220–240	240–260
No. of employees	15	30	55	12	9	5	4

Solution:

Salary in (in $)	No. of employees	C.F.
100–120	15	15
120–140	30	45
140–160	55	100
160–180	12	112
180–200	9	121
220–240	5	126
240–260	4	130

where, $(n/2) = (130/2) = 65$. Hence the median class is 140 and 160, respectively. Since the median class is 140–160 and the lower limit of this class is 140, the frequency of the median class, 140–160 is 55.

$$Median = l + \frac{h}{f}\left(\frac{N}{2} - C\right)$$

$$\frac{N}{2} = \frac{130}{2} = 65, h = 20, f = 30, C = 55$$

$$Median = 140 + \frac{20}{30}\left(\frac{130}{2} - 55\right)$$

$$Median = 140 + \frac{20}{30}(65 - 55)$$

$$Median = 140 + \frac{20}{3} = 146.66$$

Median for the frequency distribution of number of employees is 146.66.

1.4.11.3 MODE

Mode is the value which occurs most frequently or predominant in a set of observations and around which the other items of the set cluster densely. In other words, the mode is the value of the variable which is predominant in the series.

In case of continuous frequency distribution, the mode is given by the equation:

$$M = L + \frac{h(f_1 - f_0)}{2f_1 - f_0 - f_2}$$

where *L* is the lower limit, *h* is the magnitude, and f_1 is the frequency of modal class, and f_0 and f_2 are frequencies of the classes proceeding and succeeding the modal class, respectively.

(a) It is easy for interpretation and calculation. The mode is mostly used in business forecasting.
(b) Mean calculation is always done by all observations. So the mode is not affected by extreme values or abnormal data.

Example: Find the mode for following frequency distribution:

Salary in (in $)	100–120	120–140	140–160	160–180	180–200	220–240	240–260
No. of employees	15	30	55	12	9	5	4

Solution:

Salary in (in $)	**No. of employees**	**C.F.**
100–120	15	15
120–140	30	F_0
		45
140–160	55	F_1
		100
160–180	12	F_2
		112
180–200	9	121
220–240	5	126
240–260	4	130

The model class is the class against the maximum frequency. This example the maximum frequency is 55, and the model class is 140–160.

$$M = L + \frac{h(f_1 - f_0)}{2f_1 - f_0 - f_2}$$

$$Mode = 140 + \left(\frac{20(55-30)}{2*55-30-12}\right)$$

$$140 + \left(\frac{20*25}{110-42}\right)$$

$$140 + \left(\frac{500}{68}\right)$$

$$140 + 7.35$$

$$147.35$$

It means that most frequent frequency distribution is 147.35; also from the above table, anyone can see that between the frequency of 140–160 no. of employee salary are maximum.

1.4.11.4 RELATIONSHIP OF MEAN, MEDIAN, AND MODE

No clear relationship between the arithmetic mean, median, and mode could be found. But for a moderately skewed (a dispersion property) uni-modal distribution the approximate relation holds well: Mean – Mode = 3 (Mean – Median). This can be utilized for an approximate value of whichever one of the three if the other two are given.

1.4.11.5 RANGE

A range is a difference between maximum and minimum observations of the distribution. If A and B are the maximum and values in a distribution, respectively, then its range is given by:

$$\text{Range} = X_{Max} - X_{Min} = A - B$$

***Example*:** Number of defective bulbs in different batches of a particular brand of electric bulb each of 1000 units are given as follows. Find out the range of defective bulbs manufactured by the company.

Batch	1	2	3	4	4	5	6	7	8	9	10	11
Defective bulb	12	14	4	5	6	3	9	6	7	8	11	8

From the above data, it is clear that the variable defective bulb per thousand bulbs (X) has the maximum value 14 (X_{Max}) and minimum (X_{Min}) value 3. Therefore, the range of the variable defective bulb per thousand lot is $(X_{Max} - X_{Min}) = 14 - 3 = 11$.

1.5 DISPERSION

Dispersion is the measure to extend to which individual items vary.

It is our common experience that there are certain varieties of a particular crop, which are very responsive to doses of inputs, and others are not. If these input responsive varieties are provided with a high dose of nutrient they come up with very good yields, on the other hand, if these varieties are put under input stressed condition then their performance will be very poor.

Like the measures of CT, there is a need to have measures for dispersion also. In fact, different measures of dispersions are available in the theory of statistics. Before going details into the discussion of different measures of dispersion, let us try to examine the characteristics of a good measure of dispersion. By and large, a good measure of dispersion should have the following characteristics:

a) It should be defined clearly.
b) It should be convincing and easy to understand.
c) It should be easy for calculation and mathematical treatment.
d) It should be based on all observations.
e) Good evaluate of dispersion should be least affected by the sampling fluctuations.
f) It should not be affected badly by the extreme observation.

There should not be any ambiguity in defining a measure; it should be clear and rigid in the definition. Unless a measure is convincing, i.e., easily understood and applied by the researcher, it is of least importance. Further uses of a measure, it should be put easily under mathematical treatments. A good measure should try to take care of all the observations in order to reflect the true nature of the data.

1.5.1 STANDARD DEVIATION

Standard deviation (often abbreviated as "StdDev" or "SD") provides an indication of how far the individual responses to a question vary or "deviate" from the mean. SD tells the researcher how spread out the responses are they concentrated around the mean, or scattered far and wide? SD generally does not indicate "right or wrong" or "better or worse" – a lower SD is not necessarily more desirable. It is used purely as a descriptive statistic. It describes the distribution in relation to the mean. The square of the standard deviation is known as the variance.

The stand deviation is defined as the square root of the mean of the squared deviation of individual values from their mean

$$S.D. = \sqrt{\frac{(x-\overline{x})^2}{n}}$$

Or

$$S.D. = \sqrt{\frac{(x-\overline{x})^2}{n-1}}$$

Example: Calculate the Standard deviation for below data:

7,1,1,3,3,3,1,5,6,2

Solution:

$$\overline{x} = \frac{7+2+1+3+4+3+2+5+6+2}{10} = 3.50$$

$$Deviation, (x-\overline{x}) = (7-3.5), (2-3.5), (1-3.5), (3-3.5), (4-3.5), (3-3.5), (2-3.5), (5-3.5), (6-3.5), (2-3.5)$$

$$s = \frac{\sqrt{(3.5)^2 + (1.5)^2 + (-2.5)^2 + (-0.5)^2 + (0.5)^2} + (-0.5)^2 + (-1.5)^2 + (1.5)^2 + (2.5)^2 + (-1.5)^2}{10-1}$$

$$s = \sqrt{\frac{12.25+2.25+6.25+0.25+0.25+2.25+2.25+6.25+2.25}{9}} = \sqrt{\frac{34.25}{9}}$$

$$s = \sqrt{3.80} = 1.94$$

Another method is:

$$s = \sqrt{\frac{\sum x^2 - \frac{(\sum x)^2}{n}}{n-1}}$$

$$s = \sqrt{\frac{(7^2+2^2+1^2+3^2+4^2+3^2+2^2+5^2+6^2+2^2) - \frac{(3.5)^2}{10}}{10-1}}$$

$$s = \sqrt{\frac{157 - \frac{1232}{10}}{9}} = \sqrt{\frac{157 - 123.2}{9}} = \sqrt{\frac{33.8}{9}}$$

$$s = \sqrt{3.75} = 1.94$$

So variance $s^2 = 3.76$.

Interpretation: For more standing about standard deviation some more examples with the interpretation given below.

T. No.	Trial–1	Trial–2
1	5	7
2	2	1
3	2	1
4	2	3
5	2	3
6	2	3
7	2	1
8	3	5
9	4	6
10	3	2
Mean	2.70	3.20
SD	1.06	2.15

At first glance (looking at the means only) it would seem that trial–2 was rated higher than trial–1. Looking at the mean alone tells only part of the story, yet all too often; this is what researchers focus on. But seeing the standard deviation, in Experiment–1 individual response is less deviated from the mean as compare trial–2.

Another example, we will see a different kind of scenario:

Treatments	Trial–1	Trial–2
1	3	1
2	3	2
3	3	2
4	3	3
5	3	3
6	3	3
7	3	3

Treatments	Trial–1	Trial–2
8	3	4
9	3	4
10	3	5
Mean	3.0	3.0
StdDev	0.00	1.15

- In the example ("Trial–1"), the Standard Deviation is zero due to all treatments were exactly the same mean value. The individual observation did not deviate at all from the average or mean.
- In "Trial–2," even though the group means is the same (3.0) as Trial–1, the Standard Deviation is higher. The Standard Deviation of 1.15 shows that the individual responses, or deviated from average, were a little over 1 point away from the mean.

1.5.2 *STANDARD ERROR*

The Standard Error ("Std Err" or "SE") is the sign of the consistency of the mean. A small SE is a sign that the sample mean is a more precise sign of the actual population mean. Larger sample size will normally result in a smaller SE (while SD is not directly affected by sample size). Keeping the view of the all above points, we can say that if the SD of this distribution helps us to understand how far a sample mean is from the true population mean, then we can use this to understand how accurate any individual sample mean in relation to the true mean. That is the essence of the Standard Error.

$$S.E.(\overline{x}) = \frac{\sigma}{\sqrt{n}}$$

Example: Calculate the Standard Error for below data:

7,1,1,3,3,3,1,5,6,2

Solution:

$$\overline{x} = \frac{7+2+1+3+4+3+2+5+6+2}{10} = 3.50$$

$$Deviation, (x-\overline{x}) = (7-3.5),(2-3.5),(1-3.5),(3-3.5),(4-3.5),(3-3.5),(2-3.5),(5-3.5),(6-3.5),(2-3.5)$$

$$s = \frac{\sqrt{(3.5)^2+(1.5)^2+(-2.5)^2+(-0.5)^2+(0.5)^2+(-0.5)^2+(-1.5)^2+(1.5)^2+(2.5)^2+(-1.5)^2}}{10-1}$$

$$s = \sqrt{\frac{12.25+2.25+6.25+0.25+0.25+2.25+2.25+6.25+2.25}{9}} = \sqrt{\frac{34.25}{9}}$$

$$s = \sqrt{3.80} = 1.94$$

So,

$$S.E.(\overline{x}) = \frac{\sigma}{\sqrt{n}} = \frac{1.94}{\sqrt{10}} = 0.613$$

Interpretation:

Treatment	Trial 1	Trial 2	Trial 3
1	3	3	3
2	3	3	4
3	3	3	2
4	3	3	5
5	4	3	7
6	4	3	3
7	3	3	4
8	3	3	2
9	3	3	8
10	3	3	5
Mean	3.2	3	4.44
Standard Error	0.13	0	0.69
Standard deviation	0.421	0	2.068

In (Trial–1) SE of 0.13, being relatively small, gives us and sign that our mean is relatively close to the true mean of the population. But (Trial–2) all samples have the same number, and due to this, the standard error is zero.

But (Trial–3) there is so much variation in respondent, so standard error is more as compared to (Trial–1) and (Trial–2). Also from the standard deviation value, anyone can see that standard deviation is higher in Trial–3, because count numbers are much more deviated from the mean.

1.5.3 SKEWNESS

It refers to asymmetry in the frequency distribution's shape. It tells about the magnitude and direction of asymmetry. The larger the distance between mean and mode, the larger the asymmetry.

$$S_k = \frac{\sqrt{\beta_1}(\beta_2 + 3)}{2(5\beta_2 + 6\beta_1 - 9)}$$

Based on skewness, distributions can be classified as follows:

1. **Positively skewed distributions:** In this type of distributions, mean is greater than mode and median lies in between mean and mode. This distribution is also called as a right-tailed distribution on account of the tail of the distribution resting on the right-hand side. In such distributions, frequencies are distributed over a wide range of values on a high curve (Figure 1.8).

FIGURE 1.8 Positively skewed distributions.

2. **Negatively skewed distributions:** This distribution is quite the opposite to the positively skewed distribution in respect of characteristics; mean is less than mode with the median being in between the two. This is also called left tailed distribution on account of a greater portion of frequencies spread on the right-hand side and tail portion resting on left-hand side (Figure 1.9).

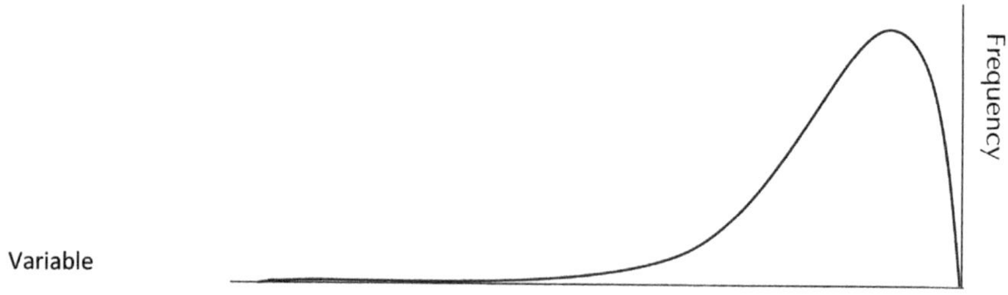

FIGURE 1.9 Negatively skewed distributions.

3. **Symmetric distributions:** This type of distribution has zero skewness, and the shape of the distribution is a bell-shaped but normal curve. In this case, the sum of positive deviations from median equals the sum of negative deviations from the median.

1.5.3.1 MEASURES OF SKEWNESS

The measures of skewness are of two types, namely, absolute, and relative measures of skewness. Absolute measures of skewness are expressed in the unit of the distributions while relative measures are unitless. Absolute measures of skewness are defined as the absolute difference between mean and mode.

$$Skewness = Mean - Mode$$

This measure can have values from negative to positive including zero. This measure is unsatisfactory on two counts: first, it is not unitless which makes a comparison of two different series with different units and second, two distributions can have the same magnitude of skewness although distributions vary greatly.

Relative measures of skewness are the following:

I. Karl Pearson's coefficient of skewness
II. The Bowley's coefficient of skewness
III. The Kelly's coefficient of skewness
IV. Measures of skewness based on moments

I. Karl Pearson's coefficient of skewness

This measure is also called Pearson's coefficient of skewness. The measure is the ratio of the absolute difference between mean and mode to standard deviation.

$$Skewness = \frac{Mean - Mode}{Standard\ Deviation}$$

This measure has no fix limits to its values, but its values usually range from −1 to +1.

Sometimes mode is , and then this measure cannot be calculated. In that case, the mode can be substituted by the following relationship:

$$Mode = 3\ Median - 2\ Mean$$

Upon substitution and simplification, this measure has another formula:

$$Coefficient\ of\ Skewness = \frac{3(Mean - Median)}{Standard\ Deviation}$$

II. The Bowley's coefficient of skewness

The earlier measure of skewness was based on measures of CT while this measure is based on a measure of dispersion. This measure is better if the distribution has missing values on extreme or distribution is open-ended and where extreme values/outliers are present. The formula is provided below:

$$Coefficient\ of\ skewness = \frac{Q_3 + Q_1 - 2Median}{Q_3 - Q_1}$$

This coefficient has its value ranges between ±1 with a coefficient equal to zero means zero skewness.

III. The Kelly's coefficient of skewness

This skewness coefficient is based on percentile and decile like Bowley's use of quartiles.

$$Coefficient\ of\ skewness = P_{50} - \frac{(P_{90} - P_{10})}{2}$$

$$Coefficient\ of\ skewness = D_5 - \frac{(D_9 - D_1)}{2}$$

IV. Measures of skewness based on moments

This skewness measure uses third (μ_3) and second (μ_2) moments about mean to create a coefficient of skewness. This is the coefficient that is mostly used in statistical analysis. This measure is denoted by β_1.

$$\beta_1 = \frac{\mu_3^2}{\mu_2^3}$$

In a symmetric distribution, $\beta_1 = 0$ while for positively and negatively skewed distributions β_1 is greater and less than zero, respectively.

1.5.3.2 KURTOSIS

Kurtosis is a Greek word, meaning "bulginess." In statistical parlance, it refers to the degree of flatness in the region about the mode of the frequency curve. This degree of flatness is measured by a coefficient, named as β_2.

$$\beta_2 = \frac{\mu_4}{\mu_2^2}$$

and its deviation γ_2 is given by $\beta_2 = \frac{\mu_4}{\mu_2^2}, \gamma_2 = \beta_2 - 3$

Skewness and kurtosis are the two opposite phenomena of the frequency distribution. If skewness refers to horizontal property, then kurtosis refers to the perpendicular property of the distribution.

If β_2 is greater than, less than and equal to three, then the curve is appropriately named as leptokurtic, platykurtic, and mesokurtic, respectively.

X	$X-\grave{X}$	$(X-\grave{X})^3$	$(X-\grave{X})^4$
5	1	1	1
9	5	125	625
5	1	1	1
4	0	0	0
2	–2	–8	16
3	–1	–1	1
5	1	1	1
2	–2	–8	16
2	–2	–8	16
1	–3	–27	81
7	3	27	81
2	–2	–8	16
8	4	64	256
1	–3	–27	81
4	0	0	0

Note: $\grave{X}$: *Mean*

Mean	4.0	Absolute measure of skewness	2.00
Mode	2.0	Pearson Coefficient of skewness	0.80
Median	4.0	Bowley's coefficient of skewness	–0.33
1st quartile	2.0	Kelly's coefficient of skewness	–0.70
3rd quartile	5.0	β_1 (moment based coefficient of skewness)	0.38
10th percentile	1.0	β_2 (measure of kurtosis)	2.31
50th percentile	4.0		
90th percentile	8.4		

Standard deviation	2.5
μ_3	8.8
μ_2	5.9
μ_4	79.5

1.5.4 COEFFICIENT OF VARIATION

The coefficient of variation measures the variability inconsistency between numbers of independent observation in the series. The coefficient of variation eliminates the unit of measurement of the standard deviation of a series of numbers by dividing it by the mean of these numbers.

$$C.V. = \frac{\sigma}{\overline{X}} * 100$$

where σ = standard deviation, and $\overline{X} = Mean$.

Examples help to understand the coefficient of variance and its importance in statistics. The coefficient of variation helps to understand the variability in data. If data is having less coefficient of variation (CV), so data has less variability and if CV is high, means data has more variability.

***Example*:**

Consider the distribution of the yields (per plot) of two maize varieties. For the first variety, the means and standard deviations are 70 kg and 15 kg, respectively. For the second variety, mean, and standard deviation are 40 kg and 8 kg, respectively. Then, for the first variety

$$C.V. = \frac{15}{70} * 100 = 10.5\%$$

$$Second variety$$

$$C.V. = \frac{8}{40} * 100 = 20.0\%$$

It is apparent that the variability in the first variety is less as compared to that in the second variety. But in terms of standard deviation, the interpretation could be reversed.

***Example*:**

Treatments	Trial–1	Trial–2	Trial–3
1	75	127	65
2	43	74	71
3	54	73	56
4	43	63	68
5	51	77	48
6	55	80	41
7	41	65	53
Mean	51.71	79.86	57.43
SE	4.43	8.19	4.18
SD	11.73	21.67	11.06
CV	22.68	27.13	19.26

- CV value indicates that the variation or instability or variability is less in the trial–1 (19.26). Highest variability found in the trial–2 (27.13).
- Also, if the standard error value is less, then CV value will also be less.
- High standard error (Experiment–2 SE = 8.19) explains the high variation, and hence CV is also more (27.13).

The scatter plot helps us to understand the coefficient of variation (Figure 1.10).

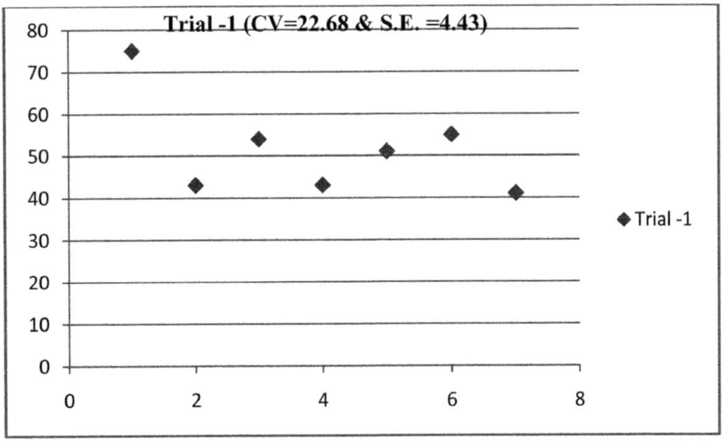

FIGURE 1.10 *(Continued)*

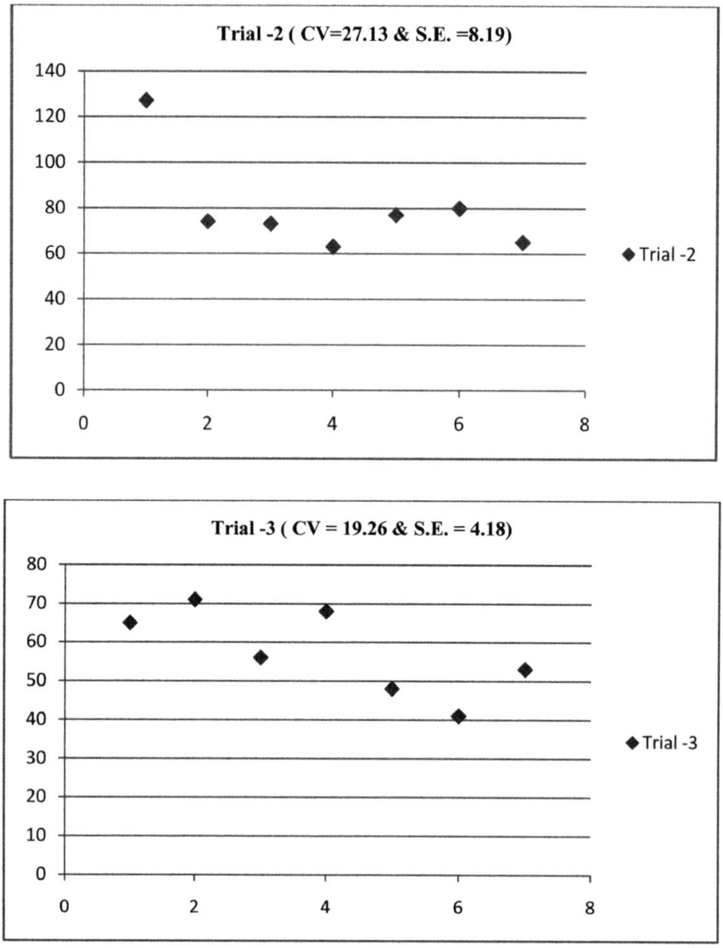

FIGURE 1.10 Scatter plot.

1.6 CORRELATION

Correlation is a method of an extent or degree of mutual dependence between two variables. In the study of two variables jointly, many times an investigator is interested to know the degree or extent of dependence between them. Actually, one wants to know whether the relation between the two variables is of high, moderate or low degree. If the two variables are said to independent, if a change in one variable has no effect on change in the other variable.

1.6.1 CORRELATION COEFFICIENT

To measure the degree, direction, and strong relationship between two or more variable correlation coefficient are used. There is no direct and simple interpretation for r itself. The relationship between variables is interpreted by the square of the correlation coefficient r^2, which is called the coefficient of determination. The coefficient of determination shows the amount of inconsistency or change in one of the variables is accounted for by the variability of the second variable.

r = correlation coefficient between two variables such as x and y.

n = number of observation.

The strength of the correlation is not dependent on the direction or the sign. Thus, $r = 0.90$ and $r = -0.90$ are equal in the degree of association of the measured variables. A positive correlation coefficient indicates that an increase in the first variable would correspond to an increase in the second variable, thus implying a direct relationship between the variables. A negative correlation indicates an inverse relationship whereas one variable increases the second variable decreases.

1.6.2 METHODS OF DETERMINING CORRELATION

1.6.2.1 GRAPHICAL METHOD

The extent of the relation between two variables can roughly be judged by plotting the pairs of observations as points on graph paper. These points are spread in different patterns and as such these are called scatter diagrams. Larger the number of points in a straight line, greater is the degree of relationship among them. In Figure 1.11 scatter diagrams, 1st shows that there is a perfect positive linear relationship between X and Y, i.e., the variable is proportional to Y and vice-versa, in this situation, the line flows from lower left side to upper right side. All the points are lies on the line.

In the second diagram, divulge the same phenomenon but in the opposite direction, i.e., if X increases, then Y decreases. In this case, the line runs from upper left to right bottom side. In the diagram, if delineated high positive and negative correlation respectively as most of the points lie near the straight lines or lie on them. In the diagram, if the same phenomena as previous expect that in these figures, the points lie farther from the lines indicating a low degree of correlation between the variables. If hardly any line can be drawn about which all the points concentrate. It means there is no correlation between the variables.

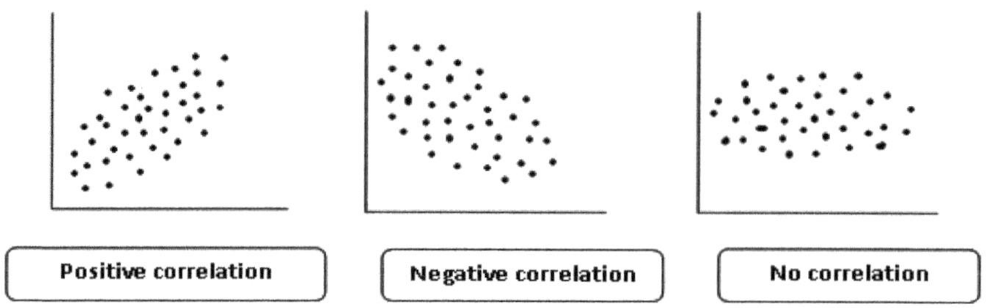

FIGURE 1.11 Scatter diagrams.

1.6.2.2 MATHEMATICAL MEASURE

A graph provides a rough idea about the type and extent of correlation between two variables. But more exactly the correlation can be measured numerically by calculating the coefficient of correlation. This is known as Person's coefficient of correlation and the formula for it was developed by Karl Pearson. This is based on three assumptions.

1. The variable *X* and *Y* are distributed normally.
2. The relationship among *X* and *Y* is linear.
3. There is a reason and effect relationship between *X* and *Y*.

$$r_{xy} = \frac{Cov(x, y)}{S_x . S_y}$$

From bivariate population, if there are *n* pairs of values of the variables *X* and *Y* as (x_1,y_1), (x_2,y_2), (x_3,y_3),.......... (x_n,y_n). Then, the formula is given as:

$$r_{xy} = \frac{\frac{1}{n}\sum_{i=1}^{n} x_i y_i - \overline{xy}}{\sqrt{\left(\frac{1}{n}\sum_{i=1}^{n} x_i^2 - \overline{x}^2\right)\left(\frac{1}{n}\sum_{i=1}^{n} y_i^2 - \overline{y}^2\right)}}$$

for i = 1, 2., *n*, and where r_{xy} is the correlation between *X* and *Y*, but mostly suffix XY is committed due to inconvenience.

The alternative formula is:

$$r = \frac{\sum_i x_i y_i - \frac{\left(\sum_i x_i\right)\left(\sum_i y_i\right)}{n}}{\sqrt{\left\{\sum_i x_i^2 - \frac{\left(\sum_i x_i\right)^2}{n}\right\}\left\{\sum_i y_i^2 - \frac{\left(\sum_i y_i\right)^2}{n}\right\}}}$$

1.6.2.3 TEST OF SIGNIFICANCE OF SIMPLE CORRELATION COEFFICIENT

Null hypothesis: ρ = 0, where ρ is the population correlation coefficient.

Test Statistic: $\frac{r}{\sqrt{1-r^2}}\sqrt{n-2}$

Conclusion: If t(Cal) > t(Tab) with (n–2) d. f. at the chosen level of significance, the null hypothesis is rejected. That is, there may be a significant correlation between the two variates. Otherwise, the null hypothesis is accepted.

1.6.2.4 PROPERTIES OF CORRELATION COEFFICIENT

1. The value of the coefficient of correlation lies between –1 to +1.
2. The value of *r* indicates, high, moderate, low positive or negative and nil degree of correlation as per the values of *r* given in the table below:

Degree of correlation	Positive corr. coeff.	Negative corr. coeff.
Prefect	r = +1	r = –1
High	$0.75 \leq r \leq 1$	$-1 \leq r \leq -0.75$
Moderate	$0.25 \leq r \leq 0.75$	$-0.75 \leq r \leq -0.25$
Low	$0 \leq r \leq 0.25$	$-0.25 \leq r \leq 0$
No correlation	0	0

3. It has no units, as it is a pure number.
4. The correlation coefficient is independent of the change of origin and scale in value between any two variables, but not in sign. If a constant value 'a' is added or subtracted from each value of *x* and 'b' from each value of *y* and also each value of *x* is divided (multiplied) by a constant 'c' and *y* by 'd,' the value of correlation coefficient calculated from coded value is same as that of original value. It means coding of data does not affect the value of r.

5. If X and Y are independent, then the correlation coefficient between them is zero, but the converse is need not be true.
6. Correlation coefficient between x and y is the same as the correlation coefficient between y and x.
7. Two independent variables are uncorrelated, but the converse may not be true.

1.6.2.5 LIMITATIONS OF THE CORRELATION COEFFICIENT

1. Correlation coefficient assumes a linear relationship between two variables. If there exists a non-linear relationship between the variables, then the correlation coefficient is no use. Thus, a low correlation coefficient indicates only a poor linear relationship between the variables, and it does not rule out the existence of a non-linear relationship.
2. High correlation coefficient does not mean high direct associations between the variables under multiple variables consideration problem. The high correlation coefficient between the two variables may be due to the influence of a third and/or fourth variable influencing both the variables. Unless and otherwise the effect of the third and/or fourth variable on the two variables are eliminated (percent correlation coefficient) the correlation between the variables may be misleading.
3. If the data are not homogeneous to some extent correlation coefficient may give rise to a misleading conclusion.
4. Given any two series of values for two variables, one can work out the correlation coefficient between the variables. Thus, one can work out the correlation coefficient between the population of n number of states in India and the production of wheat in n number of States of the United States of America. It is clear that one can get a correlation coefficient value between these two variables (viz. State wise population and production of wheat) but there is no logical basis for such type of correlation coefficient, one is not sure about the significance of such type of correlation coefficient. This type of correlation is known as nonsense correlation or spurious correlation.

1.6.2.6 APPLICATION OF CORRELATION COEFFICIENT

For example, the crop is affected by a number of factors like treatment applied, soil nutrient, rainfall, temperature, and so on. If we want to find out the correlation between the grain yield and the corresponding stem borer incidence in paddy, for example, the above factors will act as disturbing factors. Unless these contaminating factors are controlled or held constant the correlation between grain yield and stem borer incidence in paddy will not reveal the true nature of the relationship between them.

***Example*:**

The following data gives the estimated area and production of a particular crop in Gujarat State, India for last 10 years. Find the correlation between area and production.

Area (X)	2071	2401	2105	2244	2439	2556	2539	2412	2447	2910
Production (Y)	762	943	858	872	1037	860	1043	933	1042	1063

***Solution*:**

S. No.	X	Y	X2	Y2	XY
1	2071	762	4289041	580644	1578102
2	2401	943	5764801	889249	2264143
3	2105	858	4431025	736164	1806090
4	2244	872	5035536	760384	1956768
5	2439	1037	5948721	1075369	2529243
6	2556	860	6533136	739600	2198160
7	2539	1043	6446521	1087849	2648177
8	2412	933	5817744	870489	2250396
9	2447	1042	5987809	1085764	2549774
10	2910	1063	8468100	1129969	3093330
Total	24124	9413	58722434	8955481	22874183

$$r_{xy} = \frac{\sum_i X_i Y_i - \frac{(\sum_i X_i)(\sum_i Y_i)}{n}}{\sqrt{\left(\sum X_i^2 - \frac{(\sum X_i)^2}{n}\right)\left(\sum Y_i^2 - \frac{(\sum Y_i)^2}{n}\right)}}$$

$$r_{XY} = \frac{22874183 - \frac{24124 * 9413}{10}}{\sqrt{\left(58722434 - \frac{(24124)^2}{10}\right)\left(8955481 - \frac{(9413)^2}{10}\right)}}$$

$$r_{XY} = 0.74$$

1.6.3 RANK CORRELATION

At many occasions, units are not measures for certain characteristics but are ranked according to some criteria. In such situations, a certain number of units are ranked according to two criteria, or the units are ranked for a single criterion by two different judges or investigators independently. For example, the students of the class are ranked according to their marks in physics and mathematics, index of production and index of export, etc. in this situation it is worked out whether there is a correlation between ranks under two criteria. The other situation is in which some contestants in a beauty competition are ranked by two judges. In this situation, it is to measure what extent of agreement is there between the ranks awarded by two judges? A rank correlation was given by spearman and generally called as Spearman's rank correlation, denoted by r_s.

Let there are n units which are ranked as per two criteria, and then the rank correlation is given by:

$r_s = 1 - \frac{6\sum_i d_i^2}{n(n^2-1)}$, where d is the difference between ranks allocated.

The range of r_s is, $-1 \le r_s \le 1$.

***Example*:**

Nine participants in a singing contest were ranked by two judges as follows:

Participants	1	2	3	4	5	6	7	8	9
Judge-A	3	1	4	7	8	9	2	6	5
Judge-B	4	2	3	6	5	8	1	7	9

Find the rank correlation between the ranks given by the judge.

***Solution*:**

For the given number of participants, $n = 9$,

Difference (d_i)	–1	–1	1	1	3	1	1	–1	–4
d_i^2	1	1	1	1	9	1	1	1	16

$$\sum d_i^2 = 32, then$$

$$r_s = 1 - \frac{6\sum_i d_i^2}{n(n^2-1)} = 1 - \frac{6*32}{9(9^2-1)} = 1 - 0.267 = 0.733$$

The value of r_s is greater than 0.5. Hence, it is concluded that there is a fair degree of agreement between the ranks awarded by two judges.

1.7 REGRESSION

Regression is determined of the average association between two or more variables in terms of the original units of the data. If two variables are associated, unknown value of one of the variables can be estimated by using the known value of the other variable. Therefore, the estimated value may not be equal to the actually observed value, but it will be close to the actual value. The property of the tendency of actual value to lie close to the estimated value is called regression. In wider usage, regression is the theory of evaluation of the unknown value of a variable with the help of known values of the variables. The regression theory was first initiated by Sir Francis Galton in the field of genetics.

When data on two variables are known, by assuming one of the variables to be dependent on the other, we fit a linear equation to the data by the method of least square. The linear equation is called a regression equation.

For a bivariate data on x and y, the regression equation obtained with the assumptions that x is dependent on y is called regression of x on y. The regression of x on y is:

$$x-\bar{x}=b_{xy}\left(y-\bar{y}\right)$$

The regression equation obtained with the assumption that y is dependent on x is called regression of y on x. The regression of y on x is:

$$y-\bar{y}=b_{yx}\left(x-\bar{x}\right)$$

Here, the constants b_{xy} and b_{yx} are the regression coefficients. They are:

$$b_{xy}=\frac{\sum xy-\frac{\sum x\sum y}{n}}{\sum y^2-\left(\frac{\sum y}{n}\right)^2}$$

$$b_{yx}=\frac{\sum xy-\frac{\sum x\sum y}{n}}{\sum x^2-\left(\frac{\sum x}{n}\right)^2}$$

The regression equation of x on y is used for the inference of x values, and the regression equation of y on x is used for the inference of y values.

Graphical demonstration of the regression equation is called regression lines.

1.7.1 PROPERTIES OF REGRESSION COEFFICIENTS

Regression coefficients are the coefficient of the independent variables in the regression equations.

1. The regression coefficient b_{xy} is the change occurring in x for a unit change in y. The regression coefficient b_{yx} is the change occurring in y for a unit change in x.
2. The regression coefficients are independent of the origins of measurement of the variables. But, they are dependent on the scale.
3. The geometric mean of the regression coefficients b_{yx} and b_{xy} is equal to the coefficient of correlation.
4. The regression coefficient cannot be of opposite signs. If r is positive, both the regression coefficient will be positive. If r is negative, both the regression coefficient will be negative. If r is zero, both the regression coefficient will be zero.
5. Since coefficients of correlation, numerically, cannot be greater than 1, the product of the regression coefficient cannot be greater than 1.
6. If a one regression coefficient is greater than unity, then another one must be less than unity.
7. The arithmetic mean of two regression coefficients is never less than the correlation coefficient, provided the correlation coefficient is positive.
8. Two correlation between observed y and the predicted y_c is equal to $|r|$ where r is the correlation coefficient between y and x.

1.7.2 PROPERTIES OF REGRESSION LINE

1. Two regression line intersects at $(\bar{x}, \bar{y})$.
2. The regression line has a positive slope, if the variables are positively correlated. They have a negative slope, if the variables are negatively correlated.
3. If there is a perfect correlation, the regression lines coincide (there will be only one regression line).

1.8 COMPARISON BETWEEN THE CORRELATION COEFFICIENT AND REGRESSION COEFFICIENT

The comparison between the correlation coefficient and the regression coefficient is shown in the following table.

Correlation coefficient	**Regression coefficient**
A correlation coefficient (r_{XY}) measures the degree of linear association between any two given variables	Regression coefficient (b_{XY}, b_{YX}) measures the change in the dependent variable due to per unit change in the independent variable when the relation is linear
Correlation coefficient do not consider the dependency between the variables	One variable is dependent on the other variable
Correlation coefficient is a unit free measure	Regression coefficient (b_{XY}, b_{YX}) has the unit depending upon the units of the variables under consideration
Correlation coefficient is independent of the change in origin and scale	Regression coefficient (b_{XY}, b_{YX}) does not depend on a change in origin but depend on a change in scale
$-1 \leq r_{XY} \leq +1$	(b_{XY}, b_{YX}) does not have any limit

Example:

The following data give the area and production of a particular crop in India for the last eight years. Calculate the linear relationship between the area and the production for given data. Estimate the production when the area is 150 mha.

Area	164	176	178	184	175	167	173	180
Production	168	174	175	181	173	166	173	179

Solution:

Let x and y, respectively, denote the area and production. Then, the value of y corresponding to $x = 150$ has to be estimated. For this, the regression of y on x should be found, and the estimation should be made.

X	**y**	**u = x – 170**	**v = y – 170**	**u^2**	**Uv**
164	168	–6	–2	36	12
176	174	6	4	36	24
178	175	8	5	64	40
184	181	14	11	196	154
175	173	5	3	25	15
167	166	–3	–4	9	12

x	y	u = $x-170$	v = $y-170$	u^2	Uv
173	173	3	3	9	9
180	179	10	9	100	90
		37	29	475	356

$$x' = a + \frac{\sum u}{n} = 170 + \frac{37}{8} = 174.63 \text{ and } y' = b + \frac{\sum v}{n} = 170 + \frac{29}{8} = 173.63$$

Since the regression coefficients are independent of origins, the required regression coefficient is

$$b_{yx} = \frac{\sum xy - \frac{\sum x \sum y}{n}}{\sum x^2 - \left(\frac{\sum x}{n}\right)^2} = \frac{8*356 - 37*29}{8*475 - 37^2} = 0.7302$$

Thus, regression of y on x is: $y - \bar{y} = b_{yx}\left(x - \bar{x}\right)$;
On substitution we get:
y–173.63 = 0.7302*(x–174.63);
y = 0.7302x + 46.12.
Thus, the estimate of production is 155.65 m tons.

1.9 TESTING OF HYPOTHESIS

A hypothesis is a declaration or conjecture about the parameter(s) of population distribution(s), or hypothesis is a statement or assumption that is yet to be proved.

Parameter: Population constant is known as parameter. For example, the mean and variance of the population.

Statistic: Sample constant is called a statistic. It is a random variable. For example, the Student 't' test is a statistic.

1.9.1 HYPOTHESIS

A hypothesis is a general statement or assumption about the population which needed to prove. A statistical hypothesis is a statement about the probability distribution of population characteristics which are to be verified on the basis of sample information. On the basis of the amount of information provided by a hypothesis, a statistical hypothesis can be classified as simple or composite.

(i) **A simple hypothesis** is a statistical hypothesis which specifies all the parameters of the probability distribution of the random variable. For example, if we say that the yield of wheat variety in India is distributed normally with mean yield being 3.0.t/ha is a simple hypothesis.

(ii) **A composite hypothesis** is a statistical hypothesis which does not completely specifies all the parameters of the probability distribution of the random variable. For example, (i) yield of wheat variety in India is distributed normally with mean yield being 3.0 t/ha, or (ii) the yield of wheat varieties in India is distributed as normal with the standard deviation in yield being 0.56 t/ha. Then both these statements are composite in nature. Because we need information on mean and standard deviation from both the hypothesis to specify the distribution of yield.

1.9.2 TYPES OF HYPOTHESIS

Null hypothesis: H_o

A hypothesis which is to be actually tested for acceptance or rejection is termed as the null hypothesis. Consider, for example, the hypothesis may be set as 'wheat variety A will give the same yield per hectare as that of variety B' or there is no difference between the average yields of paddy varieties A and B. Such hypothesis of no difference is called a null hypothesis.

For example, there is no significant difference between the yields of wheat varieties A and B (or) they give the same yield per unit area. Symbolically, H_o: $\mu 1 = \mu 2$.

Alternative hypothesis: H_a

The statement about the population parameter(s) alternative to the null hypothesis is known as the alternate hypothesis. If H_o is accepted, what hypothesis is to be rejected and vice versa. An Alternative hypothesis is denoted by H_1 or H_a.

For example, there is a significant difference between the yields of wheat varieties A and B. Symbolically, H_1: $\mu 1 \neq \mu 2$.

If the statement is that A gives significantly less yield than B (or) A gives significantly more yield than B. Symbolically,

H_1: $\mu 1 < \mu 2$ (one-sided alternative-left tailed)

H_1: $\mu 1 > \mu 2$ (one-sided alternative-right tailed)

	Charge him	Release him
Butler did it	Correct	Error
Butler did not do it	Error	Correct

1.9.3 TWO TYPES OF ERROR

After applying a test, a decision is taken about the acceptance or rejection of null hypothesis vis-a-vis the Alternative hypothesis. There is always some possibility of committing an error in taking a decision about the hypothesis. This error is of two types.

1. Type-I error: Rejecting the null hypothesis (H_0), when it is true.
2. Type-II error: Accept the null hypothesis (H_a), when it is false.

	Ho is	
	True	**False**
Do not reject H_0	Correct decision	Type II error
Reject H_0	Type I error	Correct decision

Suppose new medicine is invented and it is given to few patients for curing a particular disease, but it is claimed that medicine has no or adverse effect then it is discontinued for use. Then we are committing type I error. But if the medicine is having an adverse or bad effect and claimed that it is good medicine and use for curing disease. Then we are committing type II error. If we look into both the errors, we can find out that type II error is more serious than type I error. So while testing any hypothesis, we are trying to minimize the type II error by committing some risks for type I error.

1.9.4 CRITICAL REGION

In any test, the area under the probability density curve is divided into two regions, viz., the region of acceptance and the region of rejection (Figure 1.12). A statistic is used to test the hypothesis H_0 and involves the choice of a region on the sampling distribution of the statistic. If the value of test statistic lies in this region of rejection. H_0 will be rejected and vice versa. The critical region is always on the tail of the distribution curve. It may be on both the tails or on one tail, depending upon the Alternative hypothesis. Moreover, the area of the critical region is equal to the level of significance.

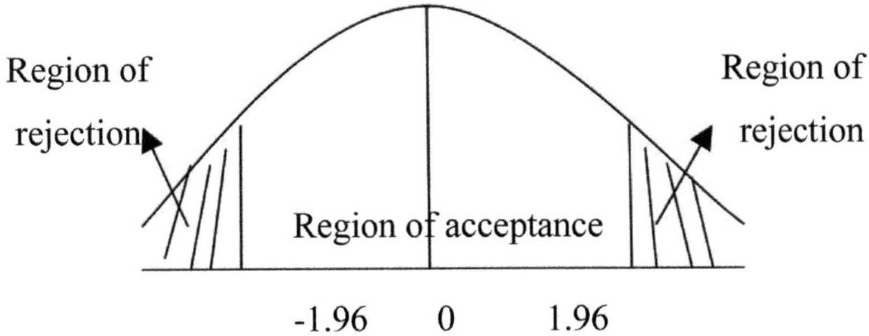

FIGURE 1.12 Area under probability density curve.

1.9.5 *LEVEL OF SIGNIFICANCE*

It is the probability of type-I error which we are ready to tolerate in making a decision about H_0. In other words, it is the probability of type-I error which is allowable. The level of significance is denoted by α and most commonly used values like 0.05 or 0.01 level for moderate and high precision, respectively.

1.9.6 *ONE TAIL TEST*

While testing the hypothesis, If the Alternative hypothesis is of the type $\mu \neq 0; \mu_1 \neq \mu_2$ etc., where the critical region lies on only one tail of the probability density curve. In this situation, the test is called a one-tail test (Figure 1.13).

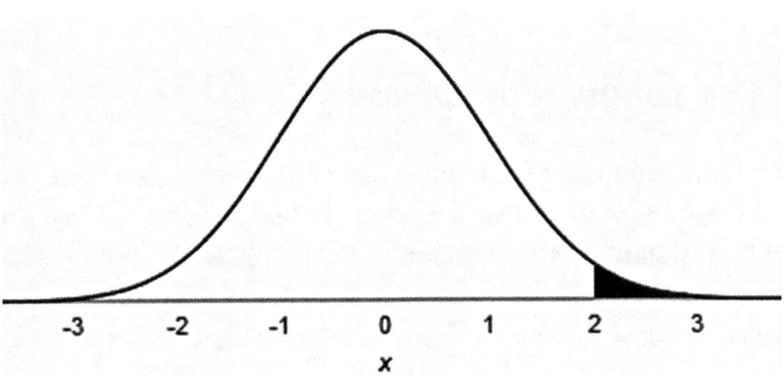

FIGURE 1.13 One tail test.

1.9.7 TWO TAIL TEST

If the Alternative hypothesis is of type $\mu \neq 0; \mu_1 \neq \mu_2$ etc., the critical region lies on both the tails. In this situation, the test is called a two-tailed test (Figure 1.14).

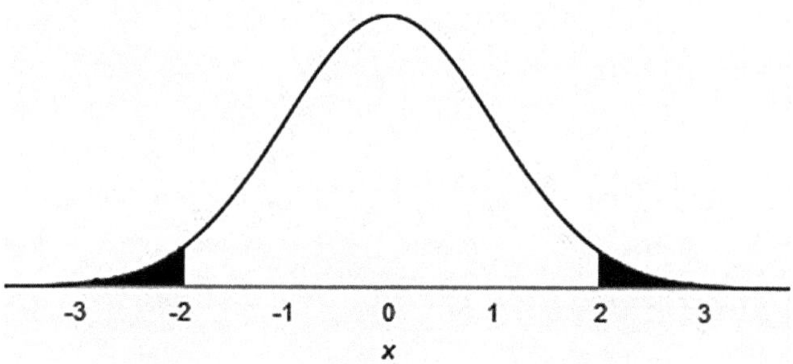

FIGURE 1.14 Two tail test.

1.9.8 NORMALITY TESTS FOR STATISTICAL ANALYSIS

In statistical analysis, for getting more precise results, data sets always need to be checked by the assumption of normality for many statistical procedures, namely parametric tests, because their validity depends on it. Following statistical tests and methods assume that data is normally distributed:

The assumptions of normality test are:

(1) Curve should be symmetric and bell-shaped.
(2) Parent population should be normal.
(3) Mean, median, and mode of the distribution are coinciding.

1.9.9 SYMMETRICAL DISTRIBUTIONS

A distribution is said to be symmetrical, when the mode, median, and mean are all in the middle of the distribution. Following graph shows a larger retirement age dataset with a distribution which is symmetrical. The mode, median, and mean all equal.

***Example*:**

Figure 1.15 shows the retirement age distribution, demonstrate the symmetrical distribution: 54, 59, 57, 55, 60, 58, 57, 58, 61, 59, 60 (Figure 1.16).

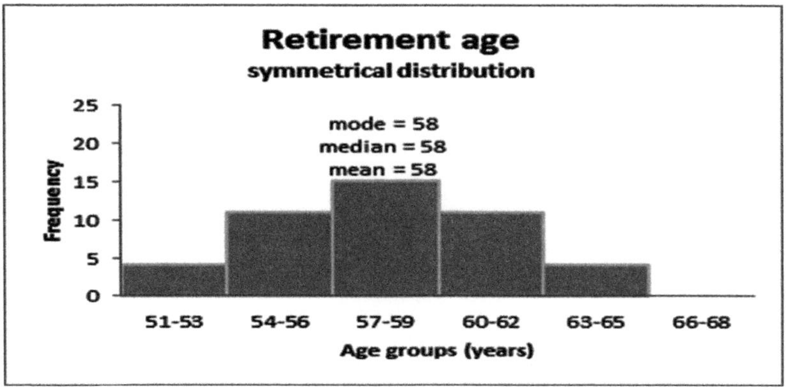

FIGURE 1.15 Retirement age distribution.

The following data shows the retirement age distribution in a tabular format.

Distances from the mean ordinates in terms of ± σ	Area under the normal curve
Z = ±0.745	50% = 0.50
Z = ±1.00	68.26% = 0.6826
Z = ±1.96	95% = 0.95
Z = ±2.0	95.44% = 0.9544
Z = ±2.58	99% = 0.99
Z = ±3.0	99.73% = 0.9973

1.9.10 CONFIDENCE INTERVAL

A confidence interval calculated for a measure of treatment effect shows the range within which the true treatment effect is likely to lie. The probability of Type I error = 1 – Confidence Level. For example, assigning a confidence level of 95% means you're only giving yourself 1–0.95 = 0.05, or 5% chance of making a type I error, that is, a 5% window of making the mistake of rejecting the null hypothesis.

It is good to know the standard deviation because we can say that any value is:

- **Likely** to be within one standard deviation (it should be 68 out of 100).

- **Very likely** to be within two standard deviations (it should be 95 out of 100).
- **Almost certainly** within three standard deviations (it should be 997 out of 1000).

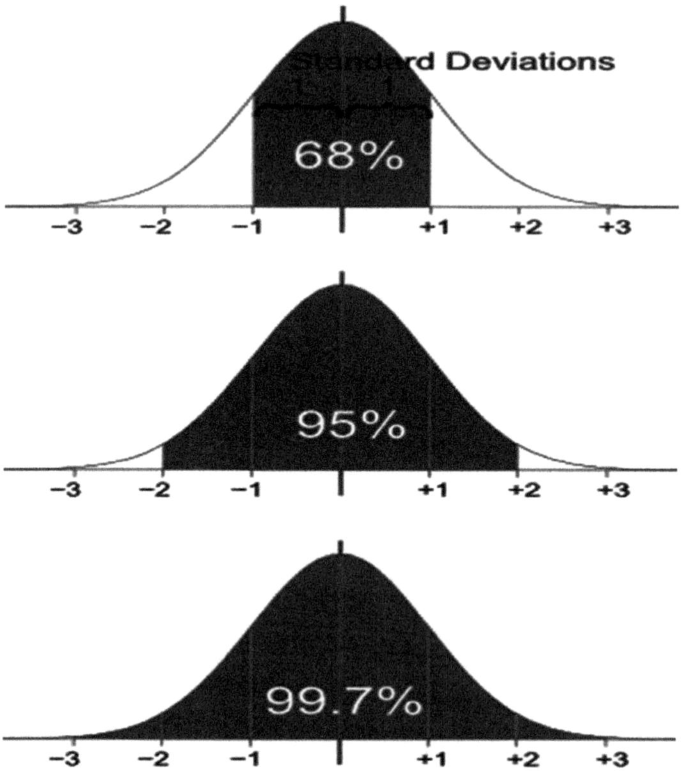

FIGURE 1.16 Area under normal curve (68%, 95%, and 99.7%).

1.9.11 DEGREES OF FREEDOM

It is the number of independent observations in a set or degrees of freedom is the numbers of observations minus the number of certain restriction have been placed on data.

Suppose, ten color balls are bought for distribution among ten students. If we distribute the balls starting from the 1st student, ball will be randomly distributed among all 10 balls, 2nd student from remaining 9 balls and so on. Then, 9th student will select from last two balls, but for the last student, he can't make choice. He has to take the last one left over ball. Hence out of 10 students, 9 students selected balls according to their choice.

1.9.12 STEPS IN TESTING OF HYPOTHESIS

The process of testing a hypothesis involves following steps.

1. Formulation of null and Alternative hypothesis.
2. Specify the level of significance. Generally, 5% or 1% is preferred.
3. Selection of test statistic and its computation.
4. Finding out the critical value from tables using the level of significance, sampling distribution and its degrees of freedom.
5. Determination of the significance of the test statistic.
6. Decision about the null hypothesis based on the significance of the test statistic.
7. Writing the conclusion in such a way that it answers the question on hand.

1.9.13 LARGE SAMPLE THEORY

A study of sampling distribution of statistic for large sample is known as large sample theory. Generally, a sample of size greater than 30 ($n \geq 30$) is known as large sample. If the sample drawn is large enough ($n \geq 30$), then the Central Limit Theorem applies, and the distribution of sample means is approximately normal. For large samples, the sampling distributions of statistic are normal (Z test).

1.9.14 SMALL SAMPLE THEORY

A study of sampling distributions for small samples is known as small sample theory. When the sample size n is less than 30 ($n < 30$), it is known as small sample. For small samples, the sampling distributions are t, F, and χ^2 distribution.

1.9.15 LARGE SAMPLE TEST

1.9.15.1 Z-TEST (STANDARD NORMAL DEVIATE TEST)

Z-test is generally carried out when the sample size is large (>30) or small sample with known population SD.

If a random sample of size n is drawn from an arbitrary population with mean $\overline{x}$ and variance σ^2 and any statistic be 'Z' with mean E(t) and variance V(t) then Z is asymptotically normally distributed with mean E(Z) and variance v(Z) as $n \to \infty$. The standard normal variety is

$$Z = \frac{z - E(z)}{\sqrt{V(z)}} \sim N(0,1)$$

A. One sample test: Testing equality of means for sample and population

In a sample of large size n, we examine whether the sample would have come from a population having a specified mean

(i) Case 1: When population S.D. is known.

Assumptions:

1. Population follows normal distribution.
2. Sampling procedure used is drawn at random process.

Condition:

1. Population S.D. is known.
2. Size of the sample may be small or large.

Null hypothesis (H_0): $\mu = \mu_o$

Alternative hypothesis (H_1): $\mu \neq \mu_o$ or $\mu > \mu_o$ or $\mu < \mu_o$

Test statistic: $Z = \left| \frac{\bar{x} - \mu}{\sigma / \sqrt{n}} \right| \sim N(0,1)$

***Example*:**

The average number of guava fruits per tree in a particular region was known from a considerable experience as 620 with a standard deviation 5.0. A sample of 30 trees gives an average number of fruits 500 per tree. Test whether the average number of fruits per tree selected in the sample is agreement with the average production in that region?

***Solution*:**

Null hypothesis: $\mu = \mu_o = 620$

$$Z = \frac{|500 - 620|}{5/\sqrt{30}} = 131.45$$

Conclusion: Z (calculated) > Z (tabulated), 1.96 at 5 percent level of significance. Therefore, it can be concluded that there is significant difference between sample mean and population mean with respect to average performance.

(ii) Case 2: If the S.D. in the population is not known still we can use the standard normal deviated test.

Assumption:

1. Population follows normal distribution.
2. Sampling procedure used is drawn at random process.

Conditions:

1. σ is not known.
2. Size of the sample is large (>30).

Null hypothesis: $\mu = \mu_o$
Alternative hypothesis (H_1): $\mu \neq \mu_o$ or $\mu > \mu_{o}$ or $\mu < \mu_o$

Test statistic: $Z = \frac{|500-620|}{5/\sqrt{30}} = 131.45$

where $S = \sqrt{\frac{1}{n}\sum_{i=1}^{n}(x_i - x')^2}$, where x' is the sample mean and n is the size of the sample.

Example:
The average daily milk production of a Jersey cow was given as 12 kgs. The distribution of daily milk yield in farm as follows:

Daily milk yield (kgs)	6–8	8–10	10–12	12–14	14–16	16–18
No. of cows	10	21	35	42	16	6

Test whether the performance of dairy farm was in agreement with the record.

Solution:

Null hypothesis: $\mu = \mu_o = 12$.
Using the mean formula we have, $x' = 11.78$, and using the Standard deviation formula we have $S = 2.49$.

Therefore, $Z = \frac{|11.78-12|}{2.49/\sqrt{130}} = 1.01$.

Conclusion: The calculated Z is less than the table Z, 1.96 at 5 percent level of significance. Therefore, the null hypothesis is accepted. That is there is no significant difference between the average daily milk yield of dairy farm and the previous record.

B. Two sample case

Testing equality of means for two independent samples

Given two sets of sample data of large size n_1 and n_2 from variables. We may examine whether the two samples come from the populations having the same mean. We may proceed as follows:

(i) Case 1: When common population S.D. is known.

Assumptions:

1. Population follows normal distribution.
2. Sampling procedure used is drawn at random process.

Conditions:

1. σ is known.
2. Size of sample may be small or large.

Null hypothesis: $\mu_1 = \mu_2$, where μ_1, μ_2 are the population means for 1st and 2nd populations, respectively.

Alternative hypothesis (H_1): $\mu_1 \neq \mu_2$ or $\mu_1 > \mu_2$ or $\mu_1 < \mu_2$

Test statistics: $Z = \frac{|\bar{x}_1 - \bar{x}_2|}{\sqrt{\sigma^2\left(\frac{1}{n1} + \frac{1}{n2}\right)}}$ where $\bar{x}_1$ and $\bar{x}_2$ are the means of 1st and 2nd samples respectively of size n_1 and n_2.

Conclusion: If Z(calculated) $\geq$ Z(tabulated), the null hypothesis is rejected. There is significant difference between two sample means. In other words, the two samples have come from two different populations having two different means. Otherwise, the null hypothesis is accepted.

(ii) Case 2: In this case, common population S.D. is not known.

Assumptions:

1. Populations are normal.
2. Samples are drawn independently and at random.

Conditions:

1. σ is not known.
2. Sizes of samples are large.

Null hypothesis: $\mu_1 = \mu_2$

Alternative hypothesis (H_1): $\mu_1 \neq \mu_2$ or $\mu_1 > \mu_2$ or $\mu_1 < \mu_2$

Test statistic: $Z =$ where $S_1^2 = \frac{1}{n_1}\sum(x_1 - \bar{x}_1)^2$ and $S_2^2 = S_2^2 = \frac{1}{n_2}\sum(x_2 - x_2')^2$

And $\bar{x}_1$, $\bar{x}_2$ are the means of 1st and 2nd samples with sizes n_1 and n_2 respectively.

Conclusion: If Z (calculated) $\geq Z$ (tabulated) at chosen level of significance, the null hypothesis is rejected. Otherwise, it is accepted.

Example:

A random sample of 90 duck farms of Indian runner breed gave an average production of 240 eggs per bird/year with S.D. of 18 eggs. Another random sample of 60 poultry farms of Khaki Campbell breed gave average eggs of 195 eggs per bird/year with a S.D. of 15 eggs. Distinguish between two breeds of birds with respect to their egg production.

Null hypothesis: $\mu_1 = \mu_2$;

$$Z = \frac{|240 - 195|}{\sqrt{\frac{18^2}{90} + \frac{15^2}{60^2}}} = 16.61$$

Conclusion: Z (calculated) > Z (tabulated), 1.96 at 5 percent of level of significance. Hence there is significant difference between two varieties of birds with respect to egg production.

C. Test for sampling from attributes

Sometimes there is need to have the tests of hypothesis for proportion of individual (or objects) having a particular characteristic. For example, to know whether the proportion of pest infected plants in the sample is in conformity with the proportion in the entire field (or population).

Here the number of plants in the sample is having only two outcomes infected or free which are distributed by binomial distribution with success p and failure q. For the binomial distribution, the first and second moment of the number of successes are 'np' and 'npq,' respectively.

Mean of proportion of successes = P.

S.E. of the proportion of successes = $\sqrt{PQ/n}$

1. One sample test:

Assumptions:

1. Population is normal.
2. Sample is drawn at random without replacement if it is a finite population.

Conditions:

1. P is known in the population.
2. Size of sample is large.

Null hypothesis: $P = P_0$

Alternative hypothesis (H_1): $P \neq P_0$

Test statistic: $Z = \frac{\left|\frac{X}{n} - P_o\right|}{\sqrt{\frac{P_o Q_o}{n}}}$, where P_o is the proportion in the population and $Q_0 = 1 - P_0$.

Conclusion: if *Z* (calculated) < *Z* (tabulated) at chosen level of significance, the null hypothesis is accepted. That is, there is no significant difference between the proportion in the sample and population. In other words, the sample may belongs to the given population. Otherwise, the null hypothesis is rejected.

Example:

It was estimated by division of plant pathology that 5 percent of wheat plants of particular variety is attacked by Karnal burnt diseases. A sample of 700 plants of the same variety of wheat crops was observed and found that 50 plants were infected with Karnal bunt. Test whether the sample results were in conformity with the population.

Null hypothesis: $P = P_0$

$$Z = \frac{\left|{}^{50}\!/_{700} - 0.05\right|}{\sqrt{{}^{0.05*0.95}\!/_{700}}} = 2.61$$

Conclusion: Here *Z* (calculated) < *Z* (tabulated), 1.96 at 5 percent level of significance, the null hypothesis is rejected. Therefore, there is significant difference between the proportion of diseased plants in the sample and the population.

2. Two sample test:

(i) Case 1: P is known: There are two population of individual (or objects) having the same proportions, P of possessing a particular character. Let $P_1 = X_1/n_1$ and $P_2 = X_2/n_2$ are the two proportions of individuals possessing the same attributes in the samples of sizes n_1 and n_2, respectively. It is to test whether the proportions in the samples are significantly different from each other or not.

Assumptions:

1. Populations are normal.
2. Samples are drawn independently and at random.

Conditions:

1. P is known.
2. Sizes of the samples are large.

Null hypothesis: $P_1 = P_2$

Alternative hypothesis (H_1): $P_1 \neq P_2$

Test statistic: $Z = \dfrac{|P_1 - P_2|}{\sqrt{P_0 Q_0 \left(\dfrac{1}{n_1} + \dfrac{1}{n_2}\right)}}$

Conclusion: If Z (calculated) < Z (tabulated) at chosen level of significance, the null hypothesis is accepted, i.e., there is no significant difference between the two proportions in the samples. In other words, their populations are having the same proportions, P_0.

Example:

In a farmer's survey, it was found that 54% of the farmer accepted new technique of rice cultivation. On conducting a survey in two village panchayats, 540 farmers accepted out of 1500 in the 1st panchayat and 400 out of 1000 in the 2nd panchayat. Test whether the difference between two panchayat is significant.

Null hypothesis: $P_1 = P_2 = P$ where P_1 and P_2 are the proportions in the 1st and 2nd populations, respectively.

$P_0 = 0.54$, $Q_0 = 1–0.54 = 0.46$, $P_1 = 540/1500 = 0.36$, $P_2 = 400/1000 = 0.40$.

$$Z = \frac{|0.36 - 0.40|}{\sqrt{0.54 * 0.46\left(\dfrac{1}{1500} + \dfrac{1}{1000}\right)}} = 1.97$$

Conclusion: Here Z (calculated) > Z (tabulated), 1.96 at 5 percent level of significance. There is significant difference between the two panchayats with regards to proportions of farmers accepting the improved rice cultivation technique.

(ii) Case 2: P is not known.

When the proportions in the populations are same and not known, then it is estimated from the proportions of the two samples.

If P_1 and P_2 are the proportions having an attributes in the two samples of sizes n_1 and n_2 respectively then P, its proportion having the same attribute in the population is estimated by taking the weighted average of P_1 and P_2.

$$P = \frac{n_1 P_1 + n_2 P_2}{n_1 + n_2} \text{ and } Q = 1\text{-}P,$$

Test statistic: $Z = \frac{|P_1 - P_2|}{\sqrt{PQ\left(\frac{1}{n_1} + \frac{1}{n_2}\right)}}$

Conclusion: If Z (calculated) $\leq Z$ (tabulated) at chosen level of significance, then the null hypothesis is accepted. Otherwise, the null hypothesis is rejected.

Example:

From two large samples of 600 and 700 carrots, 25% and 30% are found to be rotten. Can we conclude that the proportions of damaged carrots are equal in both the lots?

Null hypothesis: $P = P_0$

$$P_0 = \frac{n_1 p_1 + n_2 p_2}{n_1 + n_2} = 0.27,\ Q_0 = 1 - 0.27 = 0.73,\ P_1 = 0.30,\ P_2 = 0.25.$$

$$Z = \frac{(0.3 - 0.25)}{\sqrt{0.27x0.73(\frac{1}{500} + \frac{1}{700})}} = 1.92$$

Conclusion: Z (calculated) $< Z$ (tabulated), 1.96 at 5 percent level of significant. Therefore the null hypothesis is accepted. i.e., conclude that proportions of damaged carrots are equal in both the lots.

3. Student's *t*-Test

In case of small samples drawn from a normal population, the ratio of difference between sample and population means to its estimated standard error follows a distribution known as t-distribution, where

$$t = \frac{|\bar{x} - \mu|}{s/\sqrt{n}} \text{ where } s^2 = \frac{1}{n-1}\sum(x - \bar{x})^2$$

This follows a distribution with (n–1) degrees of freedom which can be written as $t_{(n-1)}$d.f. This test was given by Sir William Gossest and Prof. R.A Fisher. Sir William Gossest published his discovery in 1905 and published in his student name. Later this test is extended by Prof. R.A Fisher, known as 't' test or Student 't' test. 't' test is carried out when the sample size is small, i.e., $n < 30$.

A. One sample t-test: Equality of mean of sample and population

Assumptions:

1. Population is follows normal.
2. Sample is drawn at random.

Conditions:

1. σ is not known.
2. Size of sample is small.

Null hypothesis: $\mu = \mu_0$

Test statistic: $t = \frac{|\bar{x} - \mu 0|}{s/\sqrt{n}}$, where $s^2 = \frac{1}{n-1}\sum(x - x')^2$ and n is the sample size.

Conclusion: If t (calculated) < t (tabulated) with (n–1) d.f at chosen level of significance, the null hypothesis is accepted. That is, there is no significant difference between sample mean, and population mean. Otherwise, null hypothesis is rejected.

Example: The weights of rats (gms) given particular feed treatment follows normal distribution, a random sample of 10 rats were selected and weights were recorded as 96, 100, 102, 99, 104, 105, 99, 98, 100, and 101. Discus in the light of the above data the mean weight of rats in the population is 100.

Null hypothesis: $\mu = \mu_0 = 100$

x	d_i	d_i^2
96	–4	16
100	0	0
102	2	4
99	–1	1
104	4	16

x	d_i	d_i^2
105	5	25
99	–1	1
98	–2	4
100	0	0
101	1	1

$d_i = (x_i - A)$ where A = 100

$$\bar{x} = A + \sum \frac{d_i}{n} = 100 + \frac{4}{10} = 100.4$$

$$s = \sqrt{\frac{1}{n-1}\left[\sum d_i^2 - \frac{\left(\sum d_i\right)^2}{n}\right]}$$

$$= \sqrt{\frac{1}{9}\left[68 - \frac{16}{10}\right]} = 2.72$$

$$t = \frac{|100.4 - 100|}{2.72/\sqrt{10}} = 0.46$$

Conclusion: t (Calculated) < t (tabulated), (2.262) with 9 d.f. at 5 percent level of significance. Therefore, the null hypothesis is accepted. In other words, the sample may belong to the population whose mean weight is 100 gm.

B. Two sample t-test: Test for equality of two means

Case 1: Test for equality of two means (Independent Samples) variances are equal

Assumptions:

1. Populations are normal.
2. Samples are drawn independently and at random.

Conditions:

1. S.D. of the populations are same and not known.
2. Sample size for both samples is small, not necessarily equal

Null hypothesis: $\mu_1 = \mu_2$ where μ_1, μ_2 are the means of 1st and 2nd populations, respectively.

Test statistic: $t = \dfrac{|\bar{x}_1 - \bar{x}_2|}{\sqrt{S_n^2\left(\dfrac{1}{n_1} + \dfrac{1}{n_2}\right)}}$, where $s_n^2 = \dfrac{(n_1 - 1)S_1^2 + (n_2 - 1)S_2^2}{n_1 + n_2 - 2}$ and

$$s_1^2 = \frac{1}{n_1 - 1}\sum(x_{1i} - \bar{x}_1)^2 \text{ and } s_2^2 = \frac{1}{n_2 - 1}\sum(x_{2i} - \bar{x}_2)^2$$

Conclusion: If t (calculated) $\leq t$ (tabulated) with $(n_1+n_{2-}2)$ d.f at given level of significance, the null hypothesis is accepted. That is there is no significant difference between the two samples mean. Otherwise, the null hypothesis is rejected.

Example: Two types of diets were administered to two groups of ewe lambs. After a month following increases in weight (kg) were recorded. Test whether there is any significant difference between the two diets with respect to increase in weight.

Diet A	4	3	2	2	1	0	5	6	3
Diet B	5	4	4	2	3	2	6	1	

Null hypothesis: $\mu_1 = \mu_2$

X_1	X_2	X_1^2	X_2^2
4	5	16	25
3	4	9	16
2	4	4	16
2	2	4	4
1	3	1	9
0	2	0	4
5	6	25	36
6	1	36	1
3		9	

$$\bar{x}_1 = 2.89, \bar{x}_2 = 3.38, S^2 = 3.25.$$

$$t = \frac{2.89 - 3.38|}{\sqrt{3.25\left(\frac{1}{9} + \frac{1}{8}\right)}} = 0.56$$

Conclusion: t (calculated) < t (tabulated), (2.131) with 15 d.f. at 5 percent level of significance. Therefore, the null hypothesis is accepted. That is, there is no significant difference between the two diets with respect to increases in weight.

Case 2: Test for equality of two means (Independent Samples) variances are unequal

Assumptions:

1. Population follows normal distribution.
2. Sampling procedure used is drawn at random process.

Conditions:

- S.D. of the populations are not equal.
- Sample size for both samples is small and equal

Null hypothesis: $\mu_1 = \mu_2$ where $\mu_{1,}$ μ_2 are the means of 1st and 2nd populations respectively.

Test statistic: $t = \frac{|\bar{x}_1 - \bar{x}_2|}{\sqrt{\left(\frac{s_1^2}{n_1} + \frac{s_2^2}{n_2}\right)}}$

where $s_1^2 = \frac{1}{n_1 - 1}\sum(x_{1i} - \bar{x}_1)^2$ and $s_2^2 = \frac{1}{n_2 - 1}\sum(x_{2i} - \bar{x}_2)^2$

Conclusion: If t (calculated) $\leq t$ (tabulated) with (n_1+n_2-2) d.f at given level of significance, the null hypothesis is accepted. That is there is no significant difference between the two samples mean. Otherwise, the null hypothesis is rejected.

Case 3: Test for equality of two means (Independent Samples) variances are unequal and n1≠n2

Assumptions:

1. Population follows normal distribution.
2. Sampling procedure used is drawn at random process.

Conditions:

1. S.D. of the populations are not equal.
2. Sample size for both samples is small and equal

Test statistic: $t = \frac{\bar{x}_1 - \bar{x}_2 |}{\sqrt{\left(\frac{s_1^2}{n_1} + \frac{s_2^2}{n_2}\right)}}$

This 't' statistic follows neither t nor normal distribution, but it follows Behrens–Fisher d distribution. The Behrens–Fisher test is one of the laborious one. A simple alternative method has been suggested by Cochran & Cox. In this method, the critical value of *t* is altered as t_w (i.e.,) weighted t.

$$t_w = \frac{t_1\left(\frac{s_1^2}{n_1}\right)+t_2\left(\frac{s_2^2}{n_2}\right)}{\frac{s_1^2}{n_1}+\frac{s_2^2}{n_2}}$$

where t_1 is the critical value for *t* with (n_1-1) d.f. at a specified level of significance and t_2 is the critical value for *t* with (n_2-1) d.f. at a specified level of significance.

C. Paired t-test:

In the t-test for difference between two means, the two samples were independent of each other. When two small samples of equal size are drawn from two population, and the samples are dependent on each other than the paired t-test is used in preference to independent t-test.

In some other situations, two observations may be taken on the same experimental unit. For example, the soil properties before and after the application of micronutrient fertilizer may be observed on number of plots. This will result in paired observation. In such situations, we apply paired t-test. Or the same patients for the comparison of two drugs with some time interval; rats from the same litter for comparison of two diets; branches of same plant for comparison of the nitrogen uptake, etc., are some of the situations where paired-t can be used. In the paired t-test the testing of the difference between two treatments means was made more efficient by keeping all the other experimental conditions same.

Assumptions:

1. Population follows normal distribution.
2. Sampling procedure used is drawn at random process.

Conditions:

1. Samples are related with each other.
2. Sizes of the samples are small and equal.
3. S.D's in the population are equal and not known.

Null hypothesis: $\mu_1 = \mu_2$

Test statistic: $t = \frac{|\bar{d}|}{\sqrt{S_d^2/n}}$, where $d_i = (X_{1i} - X_{2i})$, $d' = \frac{\sum d_i}{n}$, n is the sample size and

$$S_d^2 = \frac{1}{n-1}\left[\sum d_i^2 - \frac{\left(\sum d_i\right)^2}{n}\right]$$

Conclusion: If t (calculated) < t (tabulated) with (n–1) d.f at 5 percent level of significance, the null hypothesis is accepted. That is, there is no significant difference between the means of the two samples. In other words, the two samples may belong to the same population. Otherwise, the null hypothesis is rejected.

Example: In an experiment the plots where divided into two equal parts and given soil treatment A and B to each part. Sorghum crop was cultivated on this plots and yield (kg/plot) was recorded and given below. Test the effectiveness of soil treatments on sorghum yield.

Soil treatment A	49	53	51	52	47	50	52	53
Soil treatment B	52	55	52	53	50	54	54	53

Solution:

Null hypothesis:

H_0: $\mu_1 = \mu_2$, there is no significant difference between the effects of the two soil treatments

Alternate hypothesis:

H_0: $\mu_1 \neq \mu_2$, there is significant difference between the effects of the two soil treatments

Level of significance = 5%

Test statistic

$$t = \frac{|\bar{d}|}{\sqrt{S_d^2/n}}, \text{ where } d_i = (X_{1i} - X_{2i}),\ \bar{d} = \frac{\sum d_i}{n}$$

x	y	d = x-y	d²
49	52	–3	9
53	55	–2	4
51	52	–1	1
51	52	–1	1
47	50	–3	16
50	54	–4	16
52	54	–2	4
53	53	0	0
	Total	–16	44

$$\bar{d} = -2$$

$$S_d^2 = \frac{1}{n-1}\left[\sum d_i^2 - \frac{\left(\sum d_i\right)^2}{n}\right] = 1.71$$

$$t = \frac{|\bar{d}|}{\sqrt{S_d^2 / n}} = 2/1.71 = 4.32$$

Conclusion: t (calculated) > t (tabulated), (2.365) for 7 df at 5 percent level of significance. Therefore, the null hypothesis is accepted. That is, there is a significant difference between the two soil treatments with respect to sorghum yield.

D. F-Test (Variance Ratio Test)

In the variance ratio test, we have to test two variance of two population is equal or not; i.e., σ_1^2 and σ_2^2. In case of testing two means of populations, assume that the population variance is same, but always this assumption may not hold good. We may have to draw two samples from different population, where the variances are not same. In such a situation, we cannot use t-test directly for testing equality of two means. Therefore, we have to test whether these two variances are same or not. For testing equality of two variances, we use F-test.

Null hypothesis: $H_0 : \sigma_1^2 = \sigma_2^2$;

Alternative hypothesis: $H_1 : \sigma_1^2 \neq \sigma_2^2 ; \alpha = 0.05$

Test criterion,

$$F = \frac{Larger\ variance}{Smaller\ variance} = \frac{S_1^2}{S_2^2} \vee \frac{S_2^2}{S_1^2}$$

$$F = \frac{S_1^2}{S_2^2};if\ S_1^2 > S_2^2$$

$$F = \frac{S_2^2}{S_1^2};if\ S_2^2 > S_1^2$$

$$S_1^2 = \frac{1}{n_1 - 1}\sum\left(X_1 - \bar{X}_1\right)^2$$

$$S_1^2 = \frac{1}{n_2 - 1}\sum\left(X_2 - \bar{X}_2\right)^2$$

This is 'F' with (n_1-1) and (n_2-1) d.f

Conclusion: If $F_{(cal)} < F_{(table)}$ for 5 percent, the test is significant, reject the null hypothesis. Therefore, we conclude that both variances are not same. Otherwise, if $F_{(cal)} < F_{(table)}$, the test is not significant. We accept the null hypothesis, and we conclude that both variances are same.

Example: The marks obtain by Ajay and Sujay for final exams in different subjects are given below. Check whether the variances of marks both the subjects are same or not.

Subject-X:	25	30	10	12	40	45	50
Subject-Y	20	25	30	22	17		

Solution:

Subject-X:	**Subject-Y**	$X_i - X'$	$(X_i - X')^2$	$Y_i - Y'$	$(Y_i - Y')^2$
25	20	–5.29	27.94	–2.80	7.84
30	25	–0.29	0.08	2.20	4.84
10	30	–20.29	411.51	7.20	51.84
12	22	–18.29	334.37	–0.80	0.64
40	17	9.71	94.37	–5.80	33.64
45		14.71	216.51		
50		19.71	388.65		

$$H_0 : S_1^2 = S_2^2; H_1 : S_1^2 \neq S_2^2; \alpha = 0.05$$

Test criterion, $$F = \frac{S_1^2}{S_2^2} or \frac{S_2^2}{S_1^2}$$

$$\overline{x} = \frac{212}{7} = 30.28, \overline{y} = \frac{114}{5} = 22.80$$

$$S_1^2 = \frac{1}{7-1}(1473.48) = 245.57$$

$$S_2^2 = \frac{1}{5-1}(98.80) = 24.70$$

$$\text{Therefore, } F = S_1^2 / S_2^2 = \frac{245.57}{24.37} = 9.94$$

$$F_{(table)} \text{ at } \alpha = 0.05 \text{ is } 4.5$$

Conclusion: Therefore, $F_{(cal)}$ is more than $F_{(table)}$ for 5% level of significance, so we conclude that the two variances are not same.

1.9.15.2 NORMAL DISTRIBUTION

Normal distribution is an approximation to the Binomial and Poisson distribution, but unlike these two distributions, it is a continuous frequency distribution. It was discovered by Johann Carl Friedrich Gauss, a German mathematician, in 1809 and hence referred to it as "Gaussian Distribution." Since Laplace made significant contribution to it hence also known as "Laplace's second law." It has been as well known as "Law of error." The probability density function for normal distribution (Figure 1.17) is provided below:

$$f(x:\mu,\sigma) = \frac{1}{\sigma\sqrt{2\pi}} e^{-\frac{1}{2}\left(\frac{x-\mu}{\sigma}\right)^2}$$

where x is a continuous variable having possible range of $\pm\infty$ with mean μ and standard deviation σ. The μ and σ are the two parameters defining this distribution.

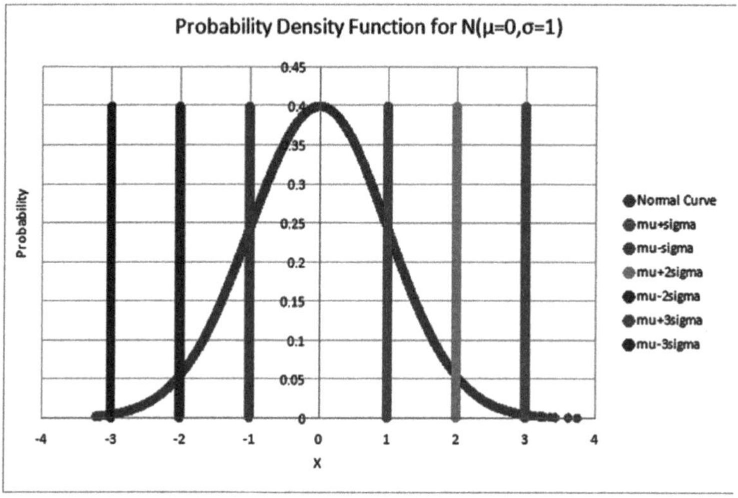

FIGURE 1.17 Probability density function for normal distribution.

1.9.15.2.1 Properties of Normal Distribution

1. The curve drawn for normal distribution is also known Normal curve is for continuous variables and is a smooth and perfectly symmetrical curve (zero skewness, $\beta_1 = 0$).
2. Normal curve is bell-shaped and extends on both sides up to infinity. It approaches the baseline as variable value approach infinity, but nevertheless, it does not meet the baseline (asymptotic to x-axis).
3. It is perfectly symmetrical about mean μ. It is mesokurtic curve, the one with $B_2 = 0$.
4. At x = μ, the maximum ordinate of the curve is reached, and the value of its ordinate at maximum is $\frac{1}{\sigma\sqrt{2\pi}}$.
5. The maximum ordinate point divides the normal curve into two equal and similar parts.
6. The normal distribution can be fully determined, and the normal curve can be completely specified if μ (mean) and σ (standard deviation/S.D.) are known.
7. The normal distribution takes values of variable from $-\infty$ to $+\infty$ but most of the values lies in range of $\mu \pm 3\sigma$. This range is known as effective range of normal curve.
8. For normal curve, following area properties holds good:

   $\mu \pm 0.67\sigma$ covers 50% area

$\mu \pm 1\sigma$ covers 68.27% area
$\mu \pm 1.96\sigma$ covers 95% area
$\mu \pm 2\sigma$ covers 95.45% area
$\mu \pm 2.58\sigma$ covers 99% area
$\mu \pm 3\sigma$ covers 99.73% area
$\mu \pm Q.D$ covers 50% area (Q.D. means Quartile Deviation)
$\mu \pm M.D.$ covers 57.51% area (M.D. means Mean Deviation)

9. For normal distribution, ratio of Q.D.: M.D.: S.D. is 10: 12: 15 and range is six times S.D.
10. All odd moments about mean are zero for normally distributed variable.

"All Normal curves are symmetrical but all symmetrical curves are not always normal."

1.9.15.2.2 Standard Normal Distribution/Unit Normal Distribution

The normal curve has two parameters, which decides its shape and size. There can be infinite combinations of these two parameters resulting in infinite variants of normal curve which will be difficult to compare one against another. To overcome this difficulty, it is essential to standardize the distribution to have zero mean and unit variance by change of origin and scale. Thus, the normal variable gets converted to normal variate with use of following formula:

$$z = \frac{x - \mu}{\sigma}$$

This standardization also helps in calculating the area under normal curve by use of Normal Table.

Now the Probability Density Function for Unit Normal Distribution is as following:

$$f(\mu, \sigma) = \frac{1}{\sigma\sqrt{2\pi}} e^{-\frac{1}{2}z^2}$$

Examples:

1. If marks in an exam for a sample of students has average marks of 50 out of 100 and a standard distribution of 12 then assuming that marks following normal distribution, find out probability of students getting between 50 and 80 marks out of 100.

Solution: Here, $\mu = 50$ and $\sigma = 12$ then standardization converts 80 to 2.5 on unit normal distribution.

Now find out the area under the standard normal curve for 2.5 in Normal Table provided in the Appendix which is 0.4938. This value is the probability of students getting marks between 50 and 80.

2. From the preceding question, what is probability of students getting more than 80 marks out of 100?

Solution: Now we know that the one half of standard normal curve below zero ($\mu = 50$) has area equal to 0.5 and the area from zero to 2.5 is 0.4938. Therefore, out of total area under curve, that is one, we deduct the sum of these two areas to arrive at probability of students getting more than 80 marks. This probability is = 1 – 0.5 – 0.4938 = 0.0062.

3. In continuation of preceding question, what is probability of students getting less than 80 marks?

Solution: This probability is 0.5 + 0.4938 = 0.9938.

1.9.15.3 BINOMIAL DISTRIBUTION

It is a discrete probability distribution expressing the probability of one set of binary outcome, i.e., success or failure.

The binomial distribution has following assumptions:

1. Each trial must result in one of two possible, mutually exclusive outcomes, either success or failure.
2. The probability of success remains constant from trial to trial.
3. Each trial is independent from previous trial.
4. The random experiment is performed repeatedly a finite number of times.

Let x be the number of success in n number of trials with probability of success p equals to $\frac{x}{n}$. Let q be the probability of failure and is equal to $1–p$.

The probability mass function is provided as below:

$$P(X = x : n, p) = \binom{n}{x} p^x q^{n-x} \text{ for } x = 1, 2, ..., n$$

For this distribution, product of n and p, i.e., np is the mean of distribution and npq equals variance of this distribution. Here, mean is greater than variance.

The probability mass function satisfies all the three axioms of probability as following:

1. Probability has range from zero to one, $0 \leq P(X = x{:}n,p) \leq 1$.
2. The probability of an entire sample space is one, $\sum_{x=1}^{n} P(X = x : n, p) = 1$
3. For two mutually exclusive events, the probability of either of these events is given by $P(X = x_1 \cup X = x_2{:}n,p) = P(X = x_1{:}n,p) + P(X = x_2{:}n,p)$

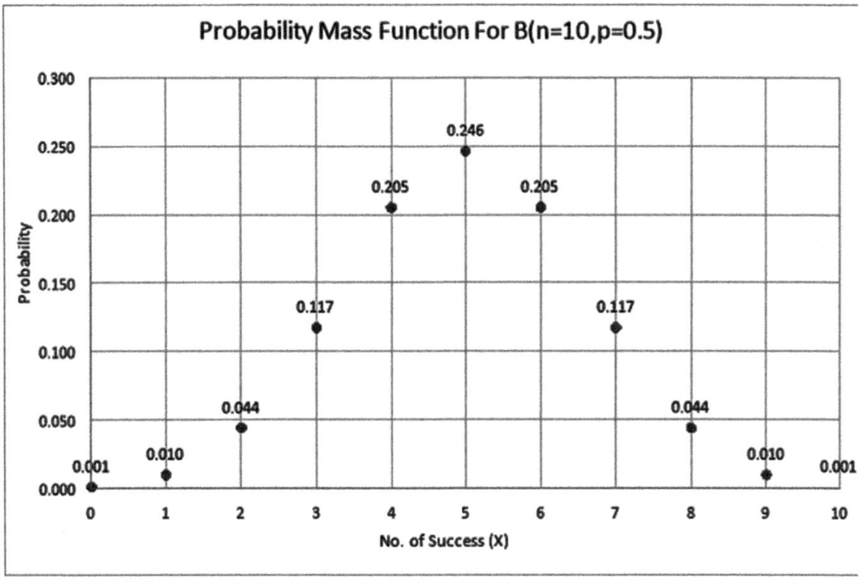

Example:

1. What is probability of getting 40 success out of 100 attempts if probability of success is 0.30?

Solution: Here, $n = 100$, $x = 40$ and $p = 0.3$. Then the probability can be found as following:

$$P(\mathrm{X} = 40 : \mathrm{n} = 100, \mathrm{p} = 0.3) = \binom{100}{40} 0.3^{40} 0.7^{100-40} = 0.00849$$

1.9.15.4 POISSON DISTRIBUTION

Poisson distribution discovered by French mathematician Simeon Denis Poisson (1781–1840) and hence named after him. Unlike Binomial distribution, here number of success from a trial are known but not number of failures. This distribution is a limiting form of Binomial distribution for $n \to \infty$ and $p \to 0$ with product of n and p remaining constant (λ). Thus, probability nearing zero means that this distribution is very well suited to rare events. Probability mass function for this distribution is provided as below:

$$P(X = x : \lambda) = e^{-\lambda} \frac{\lambda^x}{x!} \; for \; x = 1, 2, ..., \infty$$

where $e = 2.71828$.

This distribution has only one parameter, λ, which defines it.

Assumptions of Poisson Distribution:

1. The occurrences of the events are independent.
2. Theoretically, an infinite number of occurrences of the event must be possible in the interval.

This distribution has the property that its mean is equal to variance.

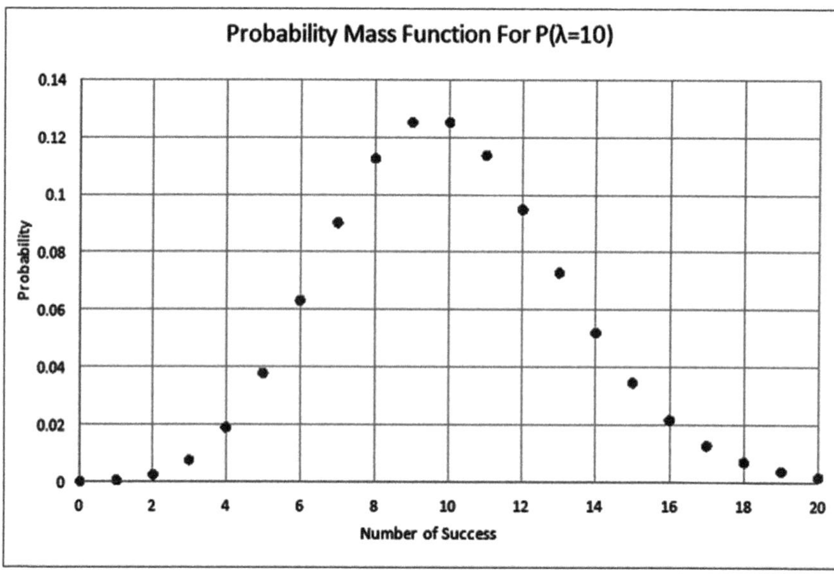

Examples:

1. If a highway in America has record of 10 accidents per day on an average. If it is given that number accidents follow Poisson distribution then find out probability of 50 deaths per day on this highway.

Solution: $P(X = 50 : \lambda = 10) = e^{-10} \frac{10^{50}}{50!} = 1.49273 \times 10^{-19}$

2. Find out probability of four earthquakes in a year if on an average this region suffers one earthquake a year.

Solution: $P(X = 4 : \lambda = 1) = e^{-1} \frac{1^4}{4!} = 1.53 \times 10^2$

1.10 PROBABILITY

1.10.1 DEFINITIONS OF PROBABILITY

1. Classical or a priori probability: It is defined as "ratio of number of favorable outcomes to total number of equally likely outcomes in an experiment."

$$Probability = \frac{Number\ of\ favourable\ outcomes}{Total\ number\ of\ equally\ likely\ outcomes}$$

This definition is called as a priori because if orderly examples, like rolling of unbiased dice, tossing of a fair coin, etc., are used then probability of event can be stated in advance of conducting an experiment. This definition of probability has a problem that not all experiments have events which are equally likely.

2. Relative frequency theory of probability: It is defined as "ratio of number of favorable outcomes to total number of outcomes in an experiment when experiment is repeatedly conducted a large number of times uniformly."

$$Probability = \lim_{n \to \infty} \frac{Number\ of\ favourable\ outcomes}{Total\ number\ of\ outcomes}$$

This definition attaches asymptotic property to definition of probability, i.e., when experiment is conducted large number of times, then empirical probability approaches theoretical probability as defined in classical definition of probability. This is definition is also called as an Empirical approach to probability. This definition is superior to classical probability. This definition was given by Von Mises.

3. Axiomatic approach to probability: This definition was given by Russian Mathematician A.N. Kolmogorov in 1933 in his book "Foundations of Probability." This definition of probability states three axioms of probability which are as following:

I. Probability of an event which belongs to sample space ranges from zero to one.
II. Probability of entire sample space equals to one. If S denotes sample space and $P(S)$ denotes probability of sample space then $P(S) = 1$. This means probability of a sure event is one.

III. If two events are mutually exclusive, then the probability of occurrence of either event is sum of probability of each event.

Sample Space: The set of all possible outcomes of a random experiment. Denoted by S. It can be either finite or infinite. If two coins are tossed simultaneously then the sample space of outcomes $S = \{HH, HT, TT, TH\}$ where H refers to Head and T refers to Tail.

Experiment: It is the process of making an observation or taking a measurement. Tossing two coins simultaneously is an experiment.

Trial/Event: It is an outcome of the experiment. The possible outcomes of tossing of coin are head and tail.

Independent Events: Two events are said to be independent when occurrence of one event does not prohibit the possibility of occurrence of other event in any way. If a coin and dice are tossed and rolled simultaneously then outcomes of tossing coin and rolling dice are independent.

Mutually Exclusive Events: Two events are said to be mutually exclusive events when occurrence of either event does not allow possibility of other event to occur in any way. Upon tossing a coin, head, and tail are mutually exclusive events.

Exhaustive Events: When total number of events includes all possible outcomes of a random experiment then events are said to be exhaustive. Upon tossing coin, exhaustive list of events is head and tail. Similarly, exhaustive list of events for tossing two coins simultaneously is $S = \{HH, HT, TT, TH\}$

Equally Likely Events: Those events which have equal likelihood of occurring, i.e., one event does not occur more often than others. Upon tossing an unbiased coin, getting head or tail is equally likely.

Complementary Events: If there exists two events which are mutually exclusive and exhaustive in nature then they are said to be complementary event to one another.

Dependent Events: Two events are said to be dependent events when occurrence of one event is dependent/conditional on occurrence of other event. If there are two events, reaching railways station and catching a train then later event is dependent on the former event.

Let there be two events, A and B. Let Probability of occurrence of events A and B are denoted by $P(A)$ and $P(B)$.

Simultaneous occurrence of two events is shown by *AB*. The inverted U symbol symbolizes simultaneous occurrence of two events, for example, $A \cap B$ and this is read as A intersection B.

Mutually exclusive events are shown by use of word "or" and by symbol $\cup$, English capital letter U which is read as Union. So for two mutually exclusive events *A* and *B*, we can write $A \cup B$ or $A \vee B$.

1.10.2 ADDITION THEOREM

For Mutually Exclusive Events

$$P(A \vee B) = P(A \cup B) = P(A) + P(B)$$

Probability of occurrence either event is sum of probability of occurrence of both events when both are mutually exclusive events.

For events which are not mutually exclusive

$$P(A \cup B) = P(A) + P(B) - P(A \cap B)$$

When the two events are not mutually exclusive events, then the probability of occurrence of either event is sum of probability of occurrence of both events less probability of simultaneous occurrence of these two events.

1.10.3 MULTIPLICATION THEOREM

For Independent events

$$P(AB) = P(A \cap B) = P(A).P(B)$$

Probability of simultaneous occurrence of two independent events is product of occurrence of each event independently.

Probability of simultaneous occurrence of two mutually exclusive events must be zero as shown below:

$$P(A \cap B) = P(AB) = P(A).P(B) = 0$$

This simplifies to mean that either of event must have zero probability of occurrence.

For dependent events

$$P(A \cap B) = P(AB) = P(B|A).P(A)$$

where $P(B \vee A)$ is the probability of occurrence of event B given event A has already occurred.

1.10.4 PERMUTATION AND COMBINATION

Permutation refers to number of ways in which items can be *arranged* while combination refers to the number of ways in which items can be *selected or chosen* from a given set of items.

$$\text{Permutation is } n_{Pr} = \frac{n!}{(n-r)!} = n(n-1)(n-2)\ldots(n-r+1)$$

$$\text{Combination is } n_{cr} = \frac{n!}{r!\times(n-r)!} = \frac{n_{Pr}}{r!} = \binom{n}{k} = \binom{n}{n-r}$$

Example: What is the chance that 53 Sundays occur in a leap year selected at random?

Solution: The leap year has one extra day over 365 days, i.e., 366 days a year. There are $\frac{366}{7}52$ complete weeks and two days extra $\left[\left(\frac{366}{7}-52\right)*7=2\right]$. These two days can be any combination of two consecutive days in a week, which is sample space in this case as shown below:

$$S = \left\{\begin{matrix} Monday-Tuesday, Tuesday-Wednesday, \\ Wednesday-Thursday, Thursday-Friday, \\ Friday-Saturday, Saturday-Sunday, Sunday-Monday \end{matrix}\right\}$$

This sample space has two events which contain Sundays, Saturday–Sunday, and Sunday–Monday. Thus, number of favorable outcomes is two, and the total number of outcomes is seven. Therefore, the probability of obtaining 53 Sundays in a randomly selected leap year is apriori fixed at $\frac{2}{7}$.

Example: Non-leap year: What is the chance that 53 Sundays occur in a non-leap year randomly?

Solution: In a non-leap year, there are 52 complete weeks and one day extra. This one-day can be any day out of seven days in a week. Thus, probability of having 53 Sundays in a randomly selected non-leap year turns out to be $\frac{1}{7}$.

Example: Let's assume there are 15 balls in an urn, out of which three are white, four are red, five are yellow and three are black. Let two balls be drawn at random from urn. Find out probability of getting either one red and one black ball or two yellow balls when balls are replaced upon being drawn from urn.

Solution: Since this example involves selection of items from a fixed set; therefore we need to find out combination of various items that can be formed in selection.

So when out of 15 balls only two are chosen then there can be $\binom{15}{2} = \frac{15!}{2!\times 13!} = \frac{15\times 14}{2\times 1} = 105$ combinations of various color balls.

Now one red ball out of five black ball can be drawn in $\binom{5}{1} = 5$ ways, and one black ball out of three black ball can be drawn in $\binom{3}{1} = 3$ ways. Thus, there are 5 × 3 = 15 ways in which two balls of red and black color each can be drawn from urn.

The probability of getting one red and one black is $\frac{15}{105} = \frac{1}{7} = \frac{3}{21}$.

Now two balls can be drawn out of five yellow balls in $\binom{5}{2} = 10$ ways.

Thus, probability of drawing two yellow balls out of 15 balls is $\frac{10}{105} = \frac{2}{21}$.

Now the two events are mutually exclusive, and hence the probability of getting either of the event happen is $\frac{3}{21} + \frac{2}{21} = \frac{5}{21} = 0.2381$.

Example: If an unbiased dice is rolled once then find the probability of getting either four or five.

Solution: There are six faces of dice, so the sample space is $S = \{1, 2, 3, 4, 5, 6\}$.

Total number of events in sample space are six. Probability of getting a four is $\frac{1}{6}$.

The probability of getting five is $\frac{1}{6}$.

Since upon rolling a dice, outcome four and five are mutually exclusive, hence the probability of getting either four or five is as follows:

$$P(4 \cup 5) = P(4) + P(5) = \frac{2}{6} = 0.333$$

Example: If a card is drawn from a pack of 52 cards then what is the probability of getting an ace or a card of diamond?

Solution: There are 52 cards in a pack of cards out of which 13 are of diamond, and 4 are ace. There is the only card which is both, diamond, and an ace, ace of the diamond.

Thus if the diamond is denoted by d and ace by a then probability of getting a card of diamond is $P(d)=\frac{13}{52}=\frac{1}{4}=0.25$, the probability of getting an ace is $P(a)=\frac{4}{52}=\frac{1}{13}=0.0769$ and probability of getting an ace of diamond is $P(d \cap a)=\frac{1}{52}$.

Now, $P(d \cup a)=P(d)+P(a)-P(d \cap a)=\frac{13}{52}+\frac{4}{52}-\frac{1}{52}=\frac{16}{52}=0.30769$

KEYWORDS

- **central tendency**
- **frequency polygon**
- **histogram**
- **statistics**

REFERENCES

Mishra, P., Joshi, R. P., Ramesh, D., Dhekale, B. S., & Sahu, P. K., (2018). "Modeling and forecasting of Sunn hemp in India." *International Journal of Current Microbiology and Applied Sciences, Special, 6*, 1284–1293.

Rangaswamy, R., (2010). *A Textbook of Agriculture Statistics* (2nd edn., pp. 27–39).

Sahu, P. K., & Das, A. K., (2004). *Agriculture and Applied Statistics: II* (pp. 426–428).

Sahu, P. K., & Mishra, P., (2013). "Modeling and forecasting production behavior and import-export of total spices in two most populous countries of the world." *Journal of Agriculture Research, 51*(4), 81–97.

Sahu, P. K., (2013). *Research Methodology: A Guide for Research in Agricultural Science, Social Science and Other Related Fields* (pp. 15–19).

Sharma, H. L., (2007). *Practical in Statistics* (pp. 36–45).

CHAPTER 2

Design of Experiments

D. RAMESH[1], G. R. MANJUNATHA[2], PRADEEP MISHRA[3], A. K. TAILOR[4], and B. S. DHEKALE[5]

[1]*Assistant Professor, Department of Statistics and Computer Applications, ANGRAU, Agricultural College, Bapatla, Andhra Pradesh, India, E-mail: dasyam.ramesh32@gmail.com*

[2]*Scientist-B (Agricultural Statistics), Central Sericultural Research and Training Institute, Central Silk Board, MoT, Government of India, Berhampore, West Bengal, India*

[3]*Assistant Professor (Statistics), College of Agriculture, JNKVV, Powarkheda, (M.P.) 461110, India*

[4]*Technical Officer (Statistics) National Horticultural Research and Development Foundation, Regional Research Station, Nashik–422 003, (M.H), India*

[5]*Assistant Professor, Sher-e-Kashmir University of Agriculture Sciences and Technology, Kashmir, India*

2.1 TERMINOLOGIES

Experimental designs is a plan of layout of an experiment according to which the treatments are allocated for the plots based on the theory of probability with a certain set of rules and procedure to obtain appropriate observations.

A statistical hypothesis is a statement about population parameters, i.e., theoretical models or probability or sampling distributions.

Degrees of freedom is the difference between the total number of items and the total number of constraints. If "n" is the total number of items and "p" the total number of constraints then the degrees of freedom (d.f.) is given by d.f. = n – p. In other words, the number of degrees of freedom generally refers to the number of independent observations in a sample set data.

Level of significance (LOS): The maximum probability at which we would be willing to risk a type-I error is known as LOS or the size of Type-I error. The LOS usually employed in the testing of hypothesis is at 5% and 1%. The LOS is always fixed in advance before collecting the sample information. LOS 5% means the results obtained will be true is 95% out of 100 cases, and the results may be wrong is 5 out of 100 cases. Or simply the probability of committing a type-I error and is denoted as α.

Blocks: In agricultural experiments, most of the times we divide the whole experimental unit (field) into relatively homogeneous sub-groups or strata. These strata, which are more uniform amongst themselves than the field as a whole, are known as blocks.

Experimental unit: Experimental unit is the object to which treatment is applied to record the observations.

Treatments: The objects of comparison in an experiment are defined as treatments.

2.2 INTRODUCTION

In any of the scientific study, planning in the form of designing experiment is one of the very important steps for the true way to examine the hypothesis of the study. Moreover, it helps to generate valid information (data) from an experiment. Therefore, designing experiments is an integrated component of each and every scientific study. The basic objectives of the experimental designs are generally to reduce the experimental error as low as possible and upsurge the precision of the experiment. The *experimental error* is a value of the observation in any experiment, vary considerably. These variations produced in observation may be due to known and also unknown sources. The variation produced by these sources is, of course, irrelevant and undesirable from the viewpoint of a good experiment, but they are unavoidable and inherent in the process of experiment. The variation produced by a set of unknown factors beyond the control of the experiment is called experimental error, which may be either random error or systematic error. For example, the nature of error of a faulty thermometer for instance which records always 1°F less or more is an instance of *systematic error*, whereas if it records the same body temperature sometimes more or less an unpredictable manner, then the nature of the error is *random*. In this situation, it is understood that any number of repeated measurements (replications) with the same thermometer would not overcome this systematic error (while estimating the average) whereas it is very likely that the random error would cancel out with repeated measurements (while estimating the average). For reducing the experimental error, we adopt certain techniques should be followed are: *randomization, replication,* and *local control (blocking/grouping)*. These are commonly known as Design of Experiments basic principles which are mainly for valid estimation as well as diminution of experimental errors as depicted in the following Fisher's diagram (Figure 2.1).

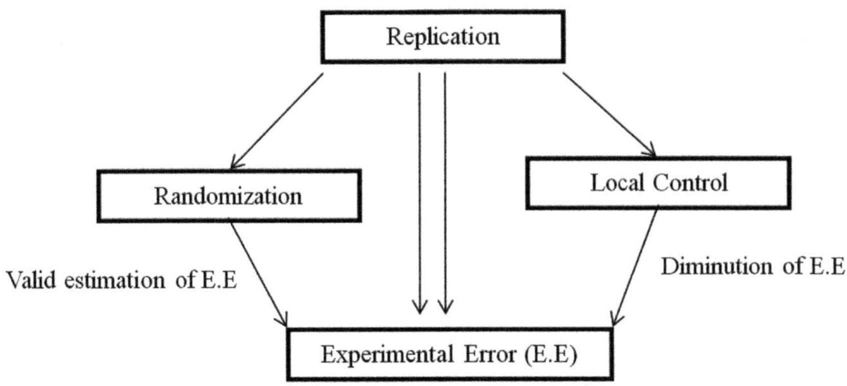

FIGURE 2.1 Fisher's diagram.

2.3 BASIC PRINCIPLES OF EXPERIMENTAL DESIGNS

2.3.1 REPLICATION

It is the repetition of treatments of the experiment under the identical condition, i.e., the experimental material on which all procedure should be the same. It keeps the random error to as low as possible.

- The number of replications to be decided in any experiments depends upon many factors like experimental material, number of treatments, etc.
- As for thumb rule, the number of replications should provide at least 12 degrees of freedom for experimental error.

2.3.1.1 ADVANTAGES

- Replication reduces the random error as low as possible.
- It provides the precision by reducing the standard error mean.
- It provides the (valid is for/by randomization only) estimate of experimental error.

2.3.2 RANDOMIZATION

Assigning the treatments randomly to experimental units (based on the law of probability) is technically known as randomization. It eliminates systematic error completely, further ensures that whatever error component persist in the observation is purely random in nature.

For example, assume the soil fertility variations in one direction, i.e., low to high as shown in Figure 2.2.

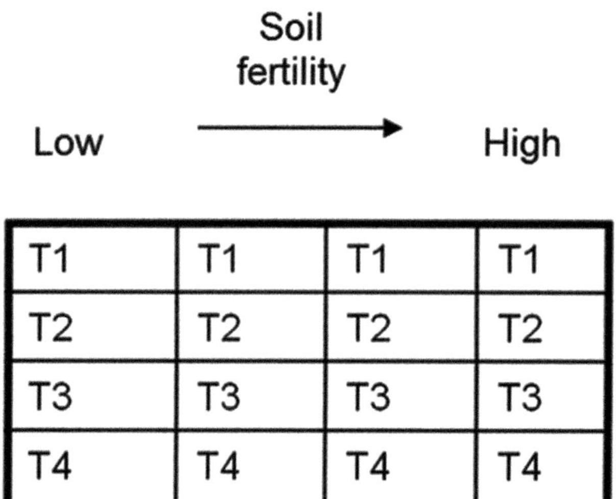

FIGURE 2.2 Allotment of treatments systematically when soil fertility gradient in one direction.

Here the treatments are allocated systematically. As a consequence of this, the variation caused due to soil fertility variation will be systematic in nature, and it cannot be eliminated, and thereby error cannot be estimated properly even with any number of replications. The nature of the error is both systematic and random.

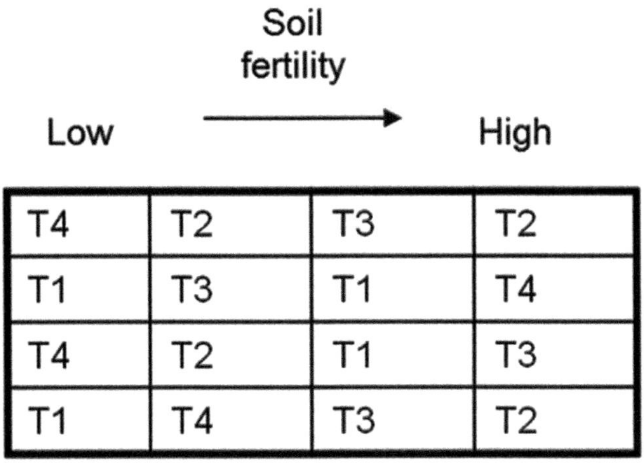

FIGURE 2.3 Allotment of treatments randomly when soil fertility gradient in one direction (appropriate method).

As a consequence of treatments is allocated randomly using random number table (Figure 2.3), the variation caused due to soil fertility variation will be systematic in nature is eliminated completely and whatever the error that persists is of random in nature/chance factor.

2.3.2.1 ADVANTAGES

- Randomization eliminates the systematic error.
- Ensures whatever the error component that still persists in the observation is purely random in nature.

2.3.3 TYPES OF RANDOMIZATION

- **Unrestricted randomization or complete randomization:** Allocation of treatment randomly to the experiment without any restrictions, e.g., CRD.
- **Restricted randomization**: Allocation of treatment randomly to the experiment with certain restriction either one-way blocking or two-way blocking, e.g., RBD, LSD.

2.3.4 LOCAL CONTROL

It is the statistical device used to make the experimental material into a homogenous one, as far as possible. It is another device to decrease or regulate the variation due to the extraneous factor and upsurge the precision of the experiment. This can be achieved by grouping, blocking, and balancing. It is also defined as "control of all factors except the one about which we are investigating (i.e., treatments)."

- ***Grouping*:** For example, the poultry birds in an experiment are heterogeneous w.r.t. their body weights, to make any treatment comparison, say two levels of feeding on their growth rate, the birds have to be grouped into homogenous groups w.r.t. their body weights. This type of local control (grouping) is to achieve homogeneity of experimental units by increasing the precision of an experiment as well as arriving at a valid conclusion.
- ***Blocking*:** In the field, experiments are generally heterogeneous w.r.t. soil fertility, then the field can be divided into the smaller homogenous blocks. Such blocks should be constructed perpendicular to soil fertility gradient so that the plots will be homogenous. This kind of homogeneity of plots makes a comparison of treatments easier.
- ***Balancing*:** It is used in a specific situation where it aims at having a uniform effect of other factors to all the plots. This is used in complex design like lattice design. Here more precision can be gained by balancing.

On the whole, randomization and local control try to eliminate systematic error and random error, respectively, whereas with replication tries to reduce both the errors in field experiments.

The readers may follow that the term 'treatment' is used in explaining the basic principles of design of experiments. It is better to give an idea about it by a simple definition.

2.3.5 TREATMENT

In experimental designs, treatment refers to any stimulus which is directly related to response or observation of an experiment and which is applied so as to observe the change of effect (positive or negative) of the response in an experimental situation or to compare its effect with other stimuli used in the same experiment. In practice, treatment may refer to any physical substance or a procedure or anything, which has been capable of the controlled application according to the requirement of the experimenter. For example, in the varietal experiment, varieties are the treatments; in irrigation trail, the different numbers of irrigations are the treatments, and in sowing experiment, the sowing methods are the treatments.

2.3.6 ANALYSIS OF VARIANCE

It is basically a technique of decomposing the total variation present in data into different meaningful components (known or Unknown), usually abbreviated as ANOVA. The difference of two samples mean value can be done by using t-test. Analysis of variance is the generalization of the comparison of several samples mean values. The idea of ANOVA was introduced by Sir R. A. Fisher, the noted British Statistician.

Consider an example: if three mulberry varieties are to be compared to find their varietal effect, this could be done with the help of field experiment, in which each variety is planted in 10 plots, and total will be 30 plots. The yield (Biomass) is harvested from each plot. Now we wish to know if there are any differences between these varieties with respect to biomass, then the solution to this problem is provided by the technique of ANOVA.

The total variation in any set of numerical data is due to a number of causes which may be classified as: (i) assignable causes, and (ii) chance causes. The variation due to assignable causes can be detected and measured whereas the variation due to chance causes is beyond the control of humans and cannot be traced separately.

Let us consider a factor *T* with *t* levels with equal replication *r* and their responses *Y*. The result may be tabulated as shown in the following table.

Treatments	**1**	**2**	**i**	**t**
	Y_{11}	Y_{21}	Y_{i1}	Y_{t1}
	Y_{12}	Y_{22}	Y_{i2}	Y_{t2}
	…	…	…	…
	Y_{1j}	Y_{2j}	Y_{ij}	Y_{tj}
	…	…	…	…
	Y_{1r}	Y_{2r}	Y_{ir}	Y_{tr}
Total	$Y_{1.}$	$Y_{2.}$	$Y_{i.}$	$Y_{t.}$
Mean	$\bar{Y}_{1}$	$\bar{Y}_{2}$	$\bar{Y}_{i}$	$\bar{Y}_{t}$

The response of the jth individual unit in the ith treatment may be represented by the equation

$Y_{ij} = \bar{Y}_i - e_{ij}$
$= i^{th}$ treatment mean – random error
$(Yij - \mu) = (\bar{Y}i - \mu) + (Y_{ij} - \bar{Y}_i)$
$\Sigma\Sigma(Y_{ij} - \mu)^2 = \Sigma\Sigma\,[(\bar{Y}_i - \mu) + (Y_{ij} - \bar{Y}_i)]^2$
$\Sigma\Sigma(Y_{ij} - \mu)^2 = \Sigma\Sigma\,(\bar{Y}_i - \mu)^2 + \Sigma\Sigma\,(Y_{ij} - \bar{Y}_i)^2$
$SS(Y) = SS(T) + SS(E)$
The bird view of ANOVA table is showed below:

Source of variation (*sv*)	**Degree of freedom (*df*)**	**Sum of squares (*SS*)**	**Mean sum square (*MSS*)**	**F Calculated (F cal.)**
Treatment	t–1	SS(T)	SS(T)/t–1 = A	A/B
.				
.				
.				
Error	Subtracting	SS(E)	SS(E)/edf = B	
Total	tr–1 or n–1			

If F calculated value is greater than F table value [In MS-excel, = FINV (0.05, df_1, df_2)] then we accept the Alternative hypothesis (H_1), i.e., $\tau_1 = \tau_2 \neq \tau_3 \ldots = \tau_t$ otherwise we reject the Alternative hypothesis (H_1), i.e., $\tau_1 = \tau_2 \neq \tau_3 \ldots = \tau_t$. Once it gets confirmed regarding F significance, then move for pair-wise treatment comparison by post-hoc tests like LSD, DMRT, Student Newman Kuel, Tukeys, etc.

$$lsd_{0.01} = t_\alpha \sqrt{EMS\left(\frac{1}{r_i} + \frac{1}{r_j}\right)}$$

2.3.7 ASSUMPTIONS OF ANOVA

- Various treatment and environmental effects are additive in nature.
- The experimental errors are independent.
- The experimental errors (e_{ij}) are distributed normally with mean zero and constant variance [$e_{ij} \sim$ NID $(0, \sigma_e^2)$].

2.3.8 DATA TRANSFORMATION

In general, the agricultural field experiments parameters like yield measurements, height, and girth measurements, etc. follow a normal distribution.

However, certain types of measurements in biological fields, like percentage infestation, germination percentage, insect population, etc., do not follow a normal distribution as such. When one or more assumptions fail we left with two options. One is that a new model can be developed to which the data may conform. The second is that certain corrections may be made in the data. The corrected data meet the assumptions of the ANOVA model. Developing a new model is not so easy, but it is easy to make the correction in the data. *The corrections are done by means of converting the data from their original form to a new form such conversion of data is known as the transformation of data.*

Among the transformations, *logarithmic, square root,* and *angular or arc-sign transformation* are widely used to make the data corrected for ANOVA.

a) **Logarithmic or log transformation:** The data follows multiplicative, i.e., where the variance is proportional to the mean can be corrected with the help of log transformation. Number of silkworm larvae per unit area, Number of insect per plants, number of plants per unit area, etc. are the typical examples where logarithmic transformation can be used effectively. The procedure is to take simply the logarithm of each and every observation and carry out the analysis of variance following the usual procedure with the transformed data. However, if in the data set small values (zero values) are recorded then instead of taking log(x) it will be better take log(x + 1).
b) **Square root transformation:** Square root transformation is used for percentage data which are derived from count data consisting of small whole numbers. This type of data generally follows a poison distribution (PD) in which the variance tends to be proportional to the mean. Data obtained from counting the rare events like number of diseased leaf per plant, number of accidents occurred in highway, the percentage of infestation (disease or pest) in a plot (either 0 to 30% or 70 to 100%), etc. are the examples where square root transformation can be useful before going for analysis of variance to draw a meaningful conclusion. If the data range either between 0 to 30% or between 70 to 100% requires sq. root transformation otherwise no need, i.e., between 30% to 70%. If most of the values in a data set are small (less than 10) coupled with the presence of 0 values, instead of using transformation, better to use as it is.
c) **Angular or Arc-sign transformation:** Angular transformation is applicable for data on proportions, i.e., data obtained from a

count and the data expressed as decimal fractions and percentages. Such type of data generally follows Binomial Distribution (BD). *If percentage data arose from count data like percentage egg hatching (which is originated from the ratio of non-egg hatch to the total number of eggs kept for hatching) should be transformed and not the percentage data such as percentage sericillin or percentage protein, which are not derived from the count data.*

Moreover, all the percentage data arising out of count data need not be subjected to arc-sin transformation before ANOVA. For percentage data ranging between either 0 to 30% or 70 to 100% but not both, the square root transformation should be used. For percentage data ranging between 30 to 70%, no transformation is required. Percentage data which overlaps the above two situations (either 0 to >30% or <70 to 100% or 0 to 100%) should only be put under *arc-sin* transformation.

Procedure in MSEXCEL for arc sign transformation as follows:

= degrees (asin*(sqrt*(*p*/100))),

where p is the percentage value, if p is zero (add by $1/4n$) and 100 (subtract by $1/4n$), where n is the number of counts on which percentage obtained before transformation.

Analysis of variance is taken up on the transformed data and inferences are made accordingly with the transformed means. However, in the result table, original means is preferred over the transformed means which helps to state the fact clearly.

2.3.4 BOX PLOT

The box plot is a graphical representation of data that shows a data set's lowest value, highest value, median value, and the size of the first and third quartile. The box plot is useful in analyzing small data sets that do not lend themselves easily to histograms. Because of the small size of a box plot, it is easy to display and compare several box plots in a small space. A box plot is a good alternative or complement to a histogram and is usually better for showing several simultaneous comparisons.

A box and whisker plot is a way of summarizing a set of data measured on an interval scale. It is often used in explanatory data analysis. This type of graph is used to show the shape of the distribution, its central value, and its variability. In a box and whisker plot, the ends of the box are the upper and lower quartiles, so the box spans the Interquartile range, and the median is marked by a vertical line inside the box the whiskers are the two lines outside the box that extend to the highest and lowest observations (Figure 2.4).

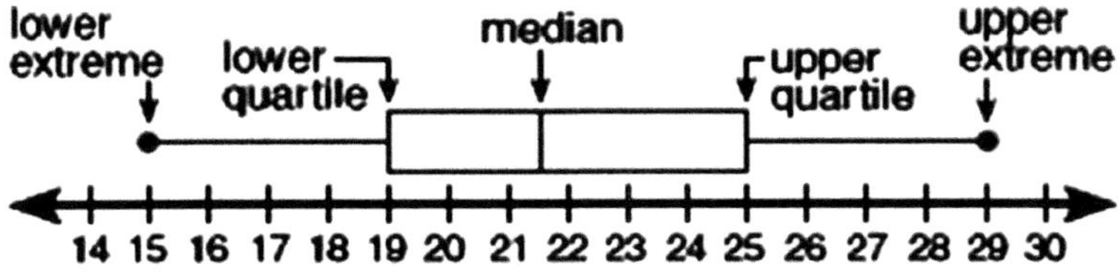

FIGURE 2.4 Box and whisker plot.

Box plots are a powerful display for comparing distributions. They provide a compact view of where the data are centered and how they are distributed over the range of the variable. It is used to compare experimental treatments by comparing box plots.

2.4 COMPLETELY RANDOMIZED DESIGN (CRD)

The CRD is the simplest of all the designs and also called the one way ANOVA. In this design, treatments are allocated at randomly, i.e., any place of the experimental material as experimental material is homogenous. In the case of field experiments, the whole field is divided into a required number of plots equal sizes, and then the treatments are randomized in these plots. In field experiments, there is generally large variation among experimental plots due to soil heterogeneity. Hence, CRD is not preferred in field experiments. In a controlled condition like a lab, pot culture, greenhouse experiments, and even poultry and animal science experiments, CRD is most useful as to achieve homogeneity of experimental materials is easy.

2.4.1 ADVANTAGES AND DISADVANTAGES OF CRD

1. Layout is very easy and completely flexible design.
2. It is most commonly used in controlled condition experiments such as in Microbiology, Chemistry, Physiology, etc. This design is useful in pot cultural experiments where the same type of soil is usually used.

3. Any number of replications and treatments can be used. The number of replications may vary from treatment to treatment.
4. It yields a maximum number of degrees of freedom for the estimation of error than the other designs which is done through the increase of replications.
5. Analysis of data is simplest compared to other designs.
6. Missing values will not create a problem in analyzing data.

2.4.2 LIMITATIONS OF CRD

1. Less accurate than other designs.
2. Since randomization is unrestricted error component includes all the variations among experimental units.
3. It is suitable for a small number of treatments.
4. It is difficult to find homogenous experimental units in all perspective.
5. It cannot be applicable for field condition.

2.4.3 ANALYSIS OF CRD

This design deals with levels of one factor. The data collected from this design are one-way classified and the analysis of which is followed through the model given below:

$$y_{ij} = \mu + t_i + e_{ij}$$

where y_{ij} means the observation obtained in j^{th} replicate of ith treatment ($j = 1, 2,..., r_i$; $i = 1, 2,..., v$). μ is the general mean, t_i is the fixed effect of ith treatment, and e_{ij} is a random error, assumed to be normally and independently distributed with mean zero and common variance.

2.4.4 STEPS FOR ANALYSIS OF CRD WITH UNEQUAL AND EQUAL REPLICATIONS

Example 1: A trail of five *Rhizoctonia solani* isolates on PDA medium was conducted in a CRD. The results are presented in the below table.

R. solani isolates	Mycelial growth		
T1	29.0	28.0	29.0
T2	33.5	31.5	29.0
T3	26.5	30.0	
T4	48.5	46.5	49.0
T5	34.5	31.0	

Solution:

Step 1:

H_0: $\tau_1 = \tau_2 = \tau_3 = \tau_4 = \tau_5$ (Effect of all *R. solani* isolates on PDA medium are same)
vs. H_1: $\tau_1 = \tau_2 = \tau_3 \neq \tau_4 = \tau_5$ (Effect of all *R. solani* isolates on PDA medium are not same).

Step 2:

Calculate treatment total,

$T_1 = 86.0$, $T_2 = 94.0$, $T_3 = 56.5$, $T_4 = 144.0$, $T_5 = 65.5$ and Grand total (G) $= 446.0$

No. of replications for treatments T_1, T_2, T_3, T_4 and T_5 are $r_1 = 3$, $r_2 = 3$, $r_3 = 2$, $r_4 = 3$ and $r_5 = 2$, respectively, and $n = 13$.

Step 3:

First compute, $CF = \dfrac{(G)^2}{n} = \dfrac{(446)^2}{13} = 15301.23$

Total S.S = sum of squares of 13 observation – C
$= (29^2 + 28^2 + . + 34.5^2 + 31^2) - 15301.23$
$= 16090.50 - 15301.23 = 789.27$.

$$\text{Treatments S.S} = \frac{86^2}{3} + \frac{94^2}{3} + \ldots + \frac{65.5^2}{3} - \text{C.F}$$
$$= 2465.33 + 2945.33 + \ldots + 2145.12 - 15301.23$$
$$= 762.68$$

Error sum square = TSS – Tr.SS = 789.27–762.68 = 26.58

Step 4:

ANOVA table

Source of variation	**Degree of freedom**	**Sum of squares**	**Mean sum square**	**Computed F**	**Tabular F at 1%**
Treatment	5–1 = 4	762.69	190.67	57.38**	7.00
Error	13–5 = 8	26.58	3.32		
Total	13–1 = 12	789.27			

** Significant at 1% level.

Step 5:

Compare the calculated *F* value of the table *F* value at α and decide on the significance of the difference among treatments using the following rules:

1. If the calculated *F* value is larger than the tabular *F* value at 1% LOS (α), the variation due to treatments is said to be *significant*. Such a result is generally indicated by placing a double asterisk on the calculated *F* value in the analysis of variance.
2. If the calculated *F* value is smaller than or equal to the tabular *F* value at 1% LOS, the variation due to treatments is said to be *non-significant*. Such a result is indicated by placing *ns* on the calculated *F* value in the analysis of variance or by leaving the *F* value without any such marking. Note that a no significant *F* in the ANOVA indicates the failure of the experiment to detect any differences among treatments. It does not mean that all treatments are the same, because the failure to detect treatment differences based on the non-significant *F* test could be the result of either a very small or no difference among the treatments or due to large experimental error, or both.

For the above example, the computed F value of 57.38 is larger than the tabular F value of 7.00 at the 1% α. Hence, the treatment differences are said to be significant, i.e., accept H_1. *In other words, chances are less than 1 in 100 that all the observed differences among the five treatment means could be due to chance. It should be noted that such a significant F test verifies the existence of some differences among the treatments tested but does not specify the particular pair (or pairs) of treatments that differ significantly.*

Step 6:

Pair-wise comparison of treatments by post-hoc test:

The experimenter aims to compare the five isolates of *R. solani*, with respect to the mycelial growth on PDA medium. The steps involved in applying the *lsd* (least significant difference) test would be the following.

Compute the *lsd* value at a LOS. Because some treatments have three replications and others have two, three sets of *lsd* values must be computed.

For comparing two treatments each having three replications (r = 3), compute the *lsd* value as follows.

$$lsd_{.01} = t_{\alpha}\sqrt{2 \times \frac{MSE}{r}} = 3.35 \times \sqrt{\frac{2(3.32)}{3}} = 4.98mm$$

For comparing two treatments each having two replications (r = 2), compute the *lsd* value as follows.

$$lsd_{.01} = t_{\alpha}\sqrt{2 \times \frac{MSE}{r}} = 3.35 \times \sqrt{\frac{2(3.32)}{2}} = 6.10mm$$

where, MSE = 3.32 and t_{α} = 3.35(for 8 degrees of freedom at 1% α).

For comparing two treatments in which one treatment having two replications and another with three replications, the *lsd* value is,

$$lsd_{0.01} = t_{\alpha}\sqrt{EMS\left(\frac{1}{r_1}+\frac{1}{r_2}\right)} = 3.35\sqrt{3.32\left(\frac{1}{3}+\frac{1}{2}\right)} = 5.57mm$$

Calculate the coefficient of variation (*cv*) with the help of grand mean and mean sum of square for error (MSE), as follows:

$$cv = \frac{\sqrt{MSE}}{grandmean} \times 100 = [1.82/34.30] \times 100 = 5.31\%$$

The *cv* indicates the consistency of experiment and is a good index of the reliability of the experiment. It explains the degree of precision with which the treatments are compared. It is an expression of the overall experimental error as a percentage of the overall mean; thus, the higher the *cv*, the lower is the reliability of the experiment. The *cv* varies greatly with the type of experiment, the crop grown and the characters measured. An experienced experimenter, however, can make a reasonably good judgment on the acceptability of a particular *cv* value for a given type of experiment. Experimental results having a *cv* value of >30% are to be viewed with caution.

Treatment	RS 1 *(r = 3)*	RS 2 *(r = 3)*	RS 3 *(r = 2)*	RS 4 *(r = 3)*	RS 5 *(r = 2)*
RS 1	0.00	2.66	0.42	19.33**	4.05
(r = 3)		[4.98]	[5.57]	[4.98]	[5.57]
RS 2	—	0.00	3.08	16.67**	1.39
(r = 3)			[5.57]	[4.98]	[5.57]
RS 3	—	—	0.00	19.75**	4.47
(r = 2)				[5.57]	[6.10]
RS 4	—	—	—	0.00	15.28**
(r = 3)					[5.57]
RS 5	—	—	—	—	0.00
(r = 2)					

Value in [] is absolute difference of two treatment means and ** significant at 1% α.

Conclusion: The above table explained that RS4 is showing significantly superior w.r.t. mycelial growth on PDA as compared to others RSs and others are at par each other.

Before going further, one should decide replication numbers required in an experiment for achieving a reasonable level of reliability as mentioned here. As said earlier, make the error degrees of freedom around 12 is a thumb rule to fix replication numbers. The hint behind this rule is that critical values derived from some of the distributions like Student's *t* or *F* almost stabilize after 12 *df* thereby providing some extent of stability to the conclusions drawn from such experiments. If anybody looks at the tables of Snedecor's F, the value below 12 *df* is very high, and it varies. Thus, a small variation in the treatment effects will not be able to detect significance when smaller *df* for error. *On the other hand, the table values of Snedecor's F stabilize after 12 df. So in order to be able to capture small variations, the error df should be around 12.*

Example 2: Find out the yielding abilities of 5varieties of mustard wherein experiment was conducted in the greenhouse using a CRD with four pots per variety. Following table represents the seed yield of mustard (g/pot).

Varieties of mustard (g/pot)				
M_1	M_2	M_3	M_4	M_5
17	24	19	17	12
22	29	18	23	12
22	24	19	21	15
24	25	16	24	14

Solution:

Step 1:

$H_0: \tau_1 = \tau_2 = \tau_3 = \tau_4 = \tau_5$ (Varietal effect of all *mustard* are same)
vs. $H_1: \tau_1 = \tau_2 = \tau_3 \neq \tau_4 = \tau_5$ (Varietal effect of all *mustard* are not same)

Step 2:

Calculate treatment total,

$T_1 = 85.0$, $T_2 = 102.0$, $T_3 = 85.0$, $T_4 = 72.0$, $T_5 = 53.0$ and Grand total (G) = 397.0

No. of replications for treatments are same, i.e., $r = 4$ and $n = 5\ x\ 4 = 20$.

Step 3:

First compute, $CF = \frac{(G)^2}{n} = \frac{(397)^2}{20} = 7880.45$

Total S.S = sum of squares of 20 observation – C. F

$= (17^2 + 22^2 + . + 15^2 + 14^2) - 7880.45$

$= 8307 - 7880.45 = 416.55.$

$$\text{Treatments S.S} = \frac{85^2}{4} + \frac{102^2}{4} + \ldots + \frac{53^2}{4} - \text{C.F}$$

$$= 1806.25 + 2601 + \ldots + 702.25 - 7880.45$$

$$= 331.30$$

Error sum square = TSS – Tr.SS = 416.55–331.30 = 85.25

Step 4:

ANOVA table:

Sources of variation	DF	SS	MSS	F cal.	$F_{0.01}$
Varieties	4	331.30	82.825	14.573**	4.893
Error	15	85.25	5.683		
Total	19	416.55			

** Significance at 1% level.

Conclusion: Accept alternative hypothesis as calculated F is > Table or critical F. Thus, the next step is post hoc test for pair-wise comparison of treatments.

Step 5:

$$CD \text{ or lsd} = t_{0.01} \times \sqrt{\frac{2 \times MSE}{r}}$$

$$CD = 2.94 \times \sqrt{\frac{2 \times (5.683)}{4}}$$

$$CD = 4.96$$

Step 6: The arrangement of treatment means either in ascending or descending order.

V_2	25.50
V_1	21.25
V_3	21.25
V_4	18.00
V_5	13.25

Comparing the pair-wise treatment means with above CD or *lsd* value. If the difference value of pair-wise treatment means is greater than CD or *lsd* value then the particular pair of treatments significantly different from each other otherwise non-significant. Similarly do for all possible pairs of treatment [i.e., $^5C_2$ ways].

Conclusion: Variety 2 gives on par with varieties 1 and 3, and significantly higher yield than varieties 4 and 5; Varieties 1 and 3 are on par with variety 4, and significantly higher yield than variety 5.

2.5 RANDOMIZED COMPLETELY BLOCK DESIGN (RCBD OR RBD)

In completely randomized design (CRD) no question of adopting the principle of local control as the experimental units are already homogeneous. Generally, when experiments involve a large number of experimental units, CRD cannot ensure the precision of the estimates of treatment effects. In the field of agricultural experiments, generally, the experimental materials are heterogeneous. In these situations, the *p* local control measure is adopted, and the material of the experiment is grouped into homogeneous subgroups. The subgroup is commonly termed as block/replication. The blocks are formed with units having common characteristics which may influence the response under study. In the field of agricultural experiments, the soil fertility is a vital character that influences the crop responses. The uniformity trial is used to detect the soil fertility of a field. If the fertility gradient is found to run in one direction (say from North to South), then the blocks are formed in the opposite direction (from East to West), i.e., perpendicular to the fertility gradient.

2.5.1 ADVANTAGES OF RBD

1. It increases the precision of an experiment.
2. Adoption of local control in RBD reduces the experimental error.
3. RBD is more efficient than CRD as the amount of information obtained in RBD is more as compared to CRD.
4. It is flexible to include any number of replications and treatments in the design, if large numbers of homogeneous units are available.
5. Statistical analysis is simple and less time-consuming.
6. Missing of one or two values can be tackled easily.

2.5.2 DISADVANTAGES OF RBD

When the number of treatments is increased, the block size will also increase. If the block size is large, it may be difficult to maintain homogeneity within blocks. Consequently, the experimental error will be increased. Hence, RBD may not be suitable for a large number of treatments.

2.5.3 ANALYSIS OF RBD

This design deals with levels of two factors such as blocks and treatments. The data collected from this design are two-way classified and the analysis of which is followed under the following model:

$$y_{ij} = \mu + t_i + b_j + e_{ij}$$

where y_{ij} means the observation obtained from i^{th} treatment in j^{th} block or replicate ($j = 1, 2,\ldots, r$; $i = 1, 2,\ldots, v$). μ is the general mean, t_i is the fixed or true effect of ith treatment, b_j is the true effect of the jth block and e_{ij} is a random error, assumed to be normally and independently distributed with mean zero and common variance.

Example: A trial was conducted with a RBD layout to study the comparative performance of Maize under rainfed condition. The re-arranged data are given in the following table.

Variety	Grain Yield of Maize (Kg/plot)			
	I	II	III	IV
T1	17.1	19.2	20.4	22.9
T2	29.5	25.5	30.2	24.5
T3	19.1	18.5	19.5	16.5
T4	30.2	34.5	32.6	35.9
T5	20.3	20.6	28.1	25.2

Solution:

Step 1:

$H_0: \tau_1 = \tau_2 = \tau_3 = \tau_4 = \tau_5$ (Varietal effect of all maize are same)
vs. $H_1: \tau_1 = \tau_2 = \tau_3 \neq \tau_4 = \tau_5$ (Varietal effect of all maize are not same)

Step 2:

Calculate treatment total and block total

$T_1 = 79.60$, $T_2 = 109.70$, $T_3 = 73.60$, $T_4 = 133.20$, $T_5 = 94.20$; $B_1 = 116.20$, $B_2 = 118.30$, $B_3 = 130.80$, $B_4 = 125.00$ and Grand total (G) = 490.30.

Total number of observations i.e., $n = 20$.

Step 3:

First compute, $CF = \frac{(G)^2}{n} = \frac{(490.30)^2}{20} = 12019.70$

Total S.S = sum of squares of 20 observation – C.F

$= (17.1^2 + 19.2^2 + . + 28.1^2 + 25.2^2) - 12019.70$

$= 689.23$

$$\text{Block or Repln. S.S} = \frac{116.2^2 + 118.3^2 + \ldots + 125.0^2}{5} - \text{C.F}$$

$$= 26.49$$

$$\text{Treatments S.S} = \frac{79.60^2 + 109.70^2 + \ldots + 94.20^2}{4} - \text{C.F}$$

$$= 581.07$$

Error sum square = TSS- RSS-Tr.SS = 108.16

Step 4:

ANOVA Table:

sv	*df*	*SS*	*MSS*	*F cal*	*F tab at 5% α*
Replication/ Block	r–1 = 3	26.49	8.83	1.22	3.28
Variety	t–1 = 4	581.07	145.27	20.14*	3.05
Error	(r–1) * (t–1) = 12	108.16	7.21		
Total	n–1 = 19	689.23			

* Significance at 5% α.

Conclusion: Accept Alternative hypothesis as calculated F under varietal effect is > Table or critical F. Thus, the next step is post hoc test for pair-wise comparison of varieties.

Step 5:

$$SE(D) = \sqrt{\frac{2EMS}{r}} = \sqrt{\frac{2(7.21)}{4}}$$

$$= 1.89$$

$$CD = t_\alpha \times SE(d)$$

$$= (2.179) \times (1.89) = 4.05$$

Step 6:

The arrangement of treatment means either in ascending or descending order.

Variety	Mean
T_4	33.30
T_2	27.43
T_5	23.55
T_1	19.90
T_3	18.40

Comparing the pair-wise treatment means with the above CD value calculated. If the difference value of pair-wise treatment means is greater than CD or *lsd* value then the particular pair of treatments significantly different from each other, otherwise non-significant. Similarly, do for all possible pair of treatments ($^5C_2$ ways).

Conclusion: The result revealed that Maize variety-T_4 exhibits significantly higher grain yield than all other varieties. The remaining varieties are all on par with each other, except variety-T_2.

2.6 LATIN SQUARE DESIGN (LSD)

As we know that the limitation of RBD, the design is not able to apply when the variability in the experimental material is in two directional which can be controlled through forming of rows and columns and then treatments are allotted to the individual plots. Each plot is received one treatment. For randomization, rows are randomized first, later columns. There is no randomization is possible within column and row. Such designs are generally known as *row-column designs/two-directional elimination of heterogeneity setting designs/three-way orthogonal designs*. The most common of such designs are Latin Square Designs (LSD). These designs have a number of replications equal to the number of rows = columns = treatments and vice-versa.

2.6.1 ADVANTAGES OF LSD

- It minimizes the two-way variation in the experimental material which is not possible in the CRD and RBD.
- Analysis of data is not complicated.
- It is more efficient than CRD and RBD as the experimental error is small due to its three-way grouping.
- Estimation of missing value is also simple.

2.6.2 DISADVANTAGES OF LSD

- It is not flexible like RBD.
- In the design, treatment numbers are equal to No. of rows and No. of rows are equal to No. of columns.

- LSDs for lesser order 2, 3 & 4 may not be efficient like CRD & RBD with same replication numbers.
- It is seldom to use when the treatment numbers are < 5 (because it does not provide adequate *df* for experimental error) and >12 (needs the more plot size which affects the homogeneity within row or plot).

2.6.3 ANALYSIS OF LSD

In LSD there are three factors such as p rows, q columns, and v treatments and each at v levels. The data collected from this design are three-way classified, and the analysis carried out under the following model:

$$y_{ijk} = \mu + r_i + c_j + t_k + e_{ijk}$$

where y_{ijk} means the observation obtained in i^{th} row, j^{th} column and k^{th} treatment; μ, r_i, c_j & t_k (i, j, k = 1, 2, 3,..., v) are the fixed effects denoting general mean, row, column, and treatment effect, respectively. The e_{ijk} is random error, assumed to be normally and independently distributed with mean zero and common variance.

Suppose the following data is pertaining to the yield data from an experiment on five varieties of wheat (P, Q, R, S & T), conducted in LSD. Analyze the following data and find the superior variety?

S–44.0	P–29.1	T–31.1	Q–42.0	R–47.2
T–26.2	Q–43.1	P–29.0	R–44.3	S–38.1
R–40.6	T–38.5	Q–43.1	S–45.8	P–29.2
P–35.8	R–36.1	S–51.7	T–33.7	Q–49.9
Q–49.3	S–34.6	R–46.1	P–31.3	T–29.4

From the above problem, initially, a hypothesis has to be formulated as follows.

H_0: (1) There is no significant difference between Varietal effects.
(2) There is no significant difference between row effects.
(3) There is no significant difference between column effects.

H_1: (1) There is a significant difference between Varietal effects.
(2) There is a significant difference between row effects.
(3) There is a significant difference between column effects.

Data has to be arranged in the following format.

	C1	C2	C3	C4	C5	Total	Mean
R1	44.0	29.1	31.1	42.0	47.2	193.4	38.68
R2	26.2	43.1	29.0	44.3	38.1	180.7	36.14
R3	40.6	38.5	43.1	45.8	29.2	197.2	39.44
R4	35.8	36.1	51.7	33.7	49.9	207.2	41.44
R5	49.3	34.6	46.1	31.3	29.4	190.7	38.14
Total	195.9	181.4	201.0	197.1	193.8	969.2	
Mean	39.18	36.28	40.2	39.42	38.76		

The stepwise procedure for calculation of the different sum of squares (SS) and ANOVA are as follows:

(1) Grand total = 44+26.2+…+49.9+29.4 = 969.20

(2) Correction Factor $CF = \frac{GT^2}{t \times t} = \frac{969.20^2}{5 \times 5} = 37573.946$

(3) Total Sum of Squares (TSS) = $[44^2+26.2^2+\ldots+49.9^2+29.4^2]$-CF = 1385.914

(4) Row Sum of Squares (RSS) $= \frac{1}{5}(193.4^2 + \ldots.190.7^2) - CF = 74.49$

(5) Column Sum of Squares (CSS) $= \frac{1}{5}(195.9^2 + \ldots.193.8^2) - CF = 44.18$

For obtaining treatment sum of square (TrSS), we need to prepare a table in below format.

	P	Q	R	S	T
	35.8	49.3	40.6	44.0	26.2
	29.1	43.1	36.1	34.6	38.5
	29.0	43.1	46.1	51.7	31.1
	31.3	42.0	44.3	45.8	33.7
	29.2	49.9	47.2	38.1	29.4
Total	154.4	227.4	214.3	214.2	158.9
Mean	30.88	45.48	42.86	42.84	31.78

(6) Treatment Sum of Squares (TrSS)

$$= \frac{1}{5}(154.4^2 + \ldots.158.9^2) - CF = 947.15$$

(7) Error Sum of Squares (ESS) = TSS – TrSS – CSS – RSS = 320.09

ANOVA table					
sv	*df*	*SS*	*MSS*	*F-cal*	*F-at (Prob. = 0.05)*
Row	4	74.50	18.63	0.70	3.26
Column	4	44.18	11.05	0.41	3.26
Treatment	4	947.15	236.79	8.88*	3.26
Error	12	320.09	26.67		
Total	24	1385.91			

From the ANOVA table, it is found that the effects of treatments were significant. So, our next task is to calculate the CD as follows.

Critical Difference (CD) for any two treatment means are

$$=\sqrt{\frac{2EMSS}{v}}\times t_{\alpha@5\%}=\sqrt{\frac{2\times 26.67}{5}}\times 2.178=7.117$$

For pair-wise comparison, arrange the treatments mean either ascending or descending order

Q	45.48
R	42.86
S	42.84
T	31.78
P	30.88

Conclusion: As mentioned earlier, by using the respective CD, treatment means are compared. It was found that treatments-Q, *R* & S were on par with each other. Similarly, treatments *T* and P also non-significant to each other. On the whole, treatment Q superior to the rest of the treatments.

2.7 FACTORIAL EXPERIMENT

In statistics, full factorial is an experiment which consists of two or more number of factors, each with describes possible values or levels. A full factorial design may also be known as "fully crossed design." Such these type experiments allow the investigator to study the effect of each factor as the response variable, as well as the effect of interactions between factors on the response variable for the vast majority of factorial experiments each factor has two levels only. If the factorial experiment would have four treatments combinations in total, then it is called 2^2 factorial designs.

If the number of combinations in a full factorial design is too high to logistically feasible, a fractional factorial design may be done. In the analysis of experimental results, the effect of each factor can be determined with the same accuracy as if only one factor had been variable at a time and the interaction effect between the factor can also be evaluated.

A factorial design was used in the 19th century by John Bennet Lawes and Joseph Herry Gilbert. The name factorial experiment was given by Yates in 1926. Before this, such experiments were called complex experiments, which were given by R. A. Fisher.

2.7.1 TYPES

There are two types of factorial experiment are, namely,

i. Symmetrical factorial; and
ii. Asymmetrical factorial.

i. *Symmetrical factorial* – where the levels of each factor are the same in the experiment is known as a symmetrical factorial experiment. Usually denoted by "l^f," where "l" indicates levels and "f" indicates factors.

Example: 2^2, 2^3, 3^2, 3^3.

ii. *Asymmetrical factorial*- where the levels of each factor are different in the experiment is known as an asymmetrical factorial experiment. Sometimes these are called as "mixed factorial experiments."

Usually for two factorial experiments it's represented as $m \times n$, where 'm' indicates levels of first factor and "n" indicates levels of the second factor; where for three-factorial experiments it is represented as $m \times n \times p$, where 'm' indicates levels of the first factor, 'n' indicates levels of second factor and 'p' indicates levels of the third factor.

2.7.2 STATISTICAL MODEL

The statistical model for a two-factor factorial: *a* level of *A, b* level of *B* and *r* replications in a completely RBD with one observation per experimental unit is,

$$Y_{ijk} = \mu + \alpha_i + \beta_j + (\alpha\beta)_{ij} + \varepsilon_{ijk}$$

For $i = 1, 2, 3 \ldots.a$
$j = 1, 2, 3,\ldots b$
$k = 1, 2, 3, \ldots r$
where μ is the true mean effect, α_i is true effect of the ith level of A and B_j is the true effect of the jth level of B and ε_{ijk} is the true effect of the kth experimental unit subject to (i,j) treatment combination.

2.7.3 EFFECT MEASURED

Two types of effects are commonly measured in factorial experiment namely (i) *main effects* (ii) *interaction effects.*

(i) The *main effect* of a factor may be defined as a measure of change in response due to change in the level of the factor averaged over all levels of all other factors in the factorial experiment. In other words, mean of simple effects and the simple effect is the difference between the responses for fixed level of other factors.
(ii) The *interaction effect,* in general, is an additional effect which is due to the combined effect of two or more factors. Also the interaction *AB* between two factors *A* and *B* is the same as *BA*. Furthermore, the interaction between two factors is known as *first-order interaction,* where for three factors, it is called *second-order interaction* and so on.

2.7.4 MERITS OF FACTORIAL EXPERIMENTS

1. It allows additional factors to be examined at no additional cost.
2. These are more efficient than one factor at a time (OFAT) experiments. It provides more information at similar or lower cost.
3. Factorial design increases the scope of the experiments, and it's inductive value, and it does so mainly by giving information not only on the main factors but on their interactions.
4. It takes less time than that of simple experiments.
5. In the factorial design, the effect of a factor is estimated at several levels of other factors and the conclusion held over a wide range of conditions.

2.7.5 *DEMERITS OF FACTORIAL EXPERIMENTS*

1. When there are a more number of factors involved in the experiments, then a large sized block is required to accommodate all the treatments.

2.7.6 *BASIC IDEAS AND NOTATIONS*

In order to develop extended notation to present the analysis of design in a concise form let us start for simplicity with a 2^2 factorial experiment.

2.7.6.1 2^2 *FACTORIAL EXPERIMENTS*

This is a very simplest example of a factorial experiment in which each factor at two levels and their numerical measures are always represented by capital letters and their treatment combinations represented by small letters.

Suppose we have two planting method of onion crop (P_1, P_2) and two irrigation frequency (i_1, i_2), where, P_1 – planting method of onion nursery on raised bed; and P_2 – planting method of onion nursery on flatbed.

Similarly, i_1 – frequency of irrigation at 16 days interval in flatbed and two splits in raised bed; and i_2 – frequency of irrigation at 8 days interval in flatbed and two splits in raised bed.

Then there will be four treatment combinations, i.e., P_1i_1, P_1i_2, P_2i_1, P_2i_2

The above treatment combinations can be displayed in the following table:

Plant Stand of Onion Nursery			
Planting method	**Irrigation frequency at 16 and 8 days interval, respectively**		**Response of irrigation frequency**
	i_1	i_2	
P_1	P_1i_1 (65)	P_1i_2 (85)	$(P_1i_1 - P_1i_2)$ –20
P_2	P_2i_1 (125)	P_2i_2 (150)	$(P_2i_1 - P_2i_2)$ –25
Response of irrigation frequency	$(P_2i_1 - P_1i_1)$ +60	$(P_2i_2 - P_1i_2)$ +65	

The main effect of (65) and (60) are called simple effect of planting method of onion nursery, and the effect of (20) and (25) are known as a simple effect of irrigation frequency. Now, if we take average of these two effects on relation to factors as planting methods and irrigation frequency than we can get main effect, i.e.,

$$\frac{65+60}{2}=62.50 \text{ main effect of planting methods of onion nursery}$$

$$\frac{(-25)+(-20)}{2}=-22.5 \text{ main effect of irrigation frequency}$$

It is revealed that the frequency of irrigation for planting method of onion nursery is decrease the plant stand in the nursery.

Symbiotically, the two simple effects in relation to planting method in presence of initial level of irrigation frequency $i_1 = P_2i_1 - P_1i_1 = P - (1)$ and in the presence of $i_2 = P_2i_2 - P_1i_2 = RI - I$

Therefore, the mean effect of the factor is

$$P=\frac{1}{2}\left[P-(1)+PI-I\right]$$

$$P=\frac{1}{2}\left[P-(1)*(i+(1))\right]$$

Similarly, two simple effect in relation to irrigation frequency in the presence of initial and final level of variety will be $i - (1)$ and $Pi - P$ these are called the simple effect of irrigation frequency taking mean of these two simple effect to irrigation frequency. Thus

$$I=\frac{1}{2}\left[i-(1)+Pi-P\right]$$

$$I=\frac{1}{2}\left[(P+(1))*(I-(1))\right]$$

It may be noted that the simple effect of irrigation frequency at two level of planting method P_1 & P_2 are not same. Similarly, two simple effect of planting methods at different irrigation level are also not same. This shows that the effect of one factor changes the level of another factor. It revealed that the factors are not independent. The dependence of one factor as other factor in their responses is known as irrigation levels.

The measure of interaction is the half of the difference between two simple effects.

Example:

The interaction between planting method is

$$\frac{60-65}{2}=-2.5$$

Similarly, the interaction can be found by simple effect of irrigation frequency is

$$\frac{-25+20}{2}=-2.5$$

The interaction may be negative or positive; the interaction between two factors is known as two-factor interaction or first-order interaction.

2.7.6.2 SS DUE TO FACTORIAL EFFECT

In fact, factorial effect, main effect, and interaction are orthogonal contrast. Its SS can be obtained by 2^2*r, where, *r* is the number of replications/blocks each carrying 1 degree of freedom; hence,

1. SS due to main effect of $P=\frac{(P)^2}{2^2 * r}$

   where *P* is the factorial effect of total of the factor *P* equal to *(Pi)-(i)+(P)-(1)*
2. SS due to main effect of $I=\frac{(I)^2}{2^2 * r}$

   where *I* is the factorial effect of total of the factor *P* equal to *(Pi)-(P)+(i)-(1)*
3. Similarly SS due to interaction of $PI=\frac{(PI)^2}{2^2 * r}$

   where *PI* is the factorial effect of total of the factor *PI* equal to *(Pi)-(P)-(i)+(1)*

Now these SS may be tested by F Test as

$$F=\frac{MS\ due\ to\ factorial\ effect}{MSE}$$

where MSE is the error mean square of the analysis of variance

A Skelton of ANOVA table for 2^2 factorial experiments in RBD is given below:

Source of variance	***df***	**SS**	**MSS**	**F Cal.**	**F Tab.**
Block/Replications	r–1	SS (r)	MS (r)		
Main effect of A	1	$(A)^2/4r$	MS (A)	MS (A)/MSE	
Main effect of B	1	$(B)^2/4r$	MS (B)	MS (B)/MSE	
Interaction of (AB)	1	$(AB)^2/4r$	MS (AB)	MS (AB)/MSE	
Error	3(r–1)	By subtraction	MSE		
Total	4r–1	$\sum_{i}\sum_{j}\left(Y_{ij}-\bar{Y}.\right)^2$			

i. The standard error of a factorial effect total

$$\sqrt{4r\sigma_e^2}$$

ii. The standard error of a factorial effect mean

$$\sqrt{\sigma_e^2/r}$$

Thus, the estimates of SE can be obtained by replacing σ_e^2 by MSE in ANOVA table.

2.7.6.2 FACTORS OF 2^3 FACTORIAL EXPERIMENT

2^3 factorial experiment consisting of 3-factors (P, I, and PE), each at two levels, P, I, and PE will denote the second levels of the factor respectively. Here PE indicates about the percentage plant establishment. In this way, we will get 8 number of combination treatments.

2.7.6.3 ASYMMETRICAL FACTORIAL

In this type of experiment, all the factors are not the same number of levels. The simplest of the factorial experiments are with two factors say *P & Q*, where, P = 2 levels of P_0P_1; Q = 3 levels of $q_0q_1q_2$.

This type of experiment known as asymmetrical factorial (2x3) experiment. Thus there will be six number of treatments combinations.

$$P_0q_0,\ P_0q_1,\ P_0q_2,\ P_1q_0,\ P_1q_1, \text{and } P_1q_2$$

2.7.6.4 2X3 FACTORIAL DESIGN

Level of factor P	Level of factor Q			Total
	q_0	q_1	q_2	
P_0	T_{00}	T_{01}	T_{02}	P_0
P_1	T_{10}	T_{11}	T_{12}	P_1
Total	Q_0	Q_1	Q_1	GT

1. Correction factor $= \dfrac{(GT)^2}{N(6r)}$
2. TSS (correlated) will be calculate from individual observations
3. SS due to the factor $P = \dfrac{P_0^2 + P_1^2}{3r} - CF$
4. SS due to the factor $Q = \dfrac{Q_0^2 + Q_1^2 + Q_2^2}{2r} - CF$
5. SS due to the cells $= \dfrac{T_{00}^2 + T_{01}^2 + T_{02}^2 + T_{10}^2 + T_{11}^2 + T_{12}^2}{r} - CF$
6. SS due to interaction PQ = SS (Cells) – SS (P) – SS (Q)
7. Error SS = TSS (Corrected) – SS (P) – SS (Q) – SS (PQ)

2.7.6.5 AN EXAMPLE OF 2X3 ASYMMETRICAL FACTORIAL EXPERIMENT

We illustrate the asymmetrical factorial design step by step procedure for an analysis of variance of a two-factor experiment in RCBD design with an experiment involving three different planting stage in onion seed crop with two levels of neck cutting at RRS NHRDF Nashik with four replications. The list of the six factorial treatment combinations is shown in Table 2.1, and data in Table 2.2

Step 1: Denote the number of replications by *r*, the level of factor P (i.e., Different planting stage) by p and the level of factor *T* (i.e., neck cutting) by t.

TABLE 2.1 The 3x2 Factorial Combinations of Two Neck Cutting and Three Different Planting Stage Levels

Planting stage levels	Treatment combinations	
	T_1	T_2
P_1	P_1T_1	P_1T_2
P_2	P_2T_1	P_2T_2
P_3	P_3T_1	P_3T_2

TABLE 2.2 Umbel Size (cm) of Onion Seed Crop of at Two Level of Neck Cutting Tested with Three Different Planting Stage in a RCB Design

Treatments	Replications				Total
	R1	**R2**	**R3**	**R4**	
P1T1	6.26	6.28	5.88	6.5	24.92
P1T2	6.52	6.78	6.68	6.56	26.54
P2T1	6.56	6.34	6.42	5.76	25.08
P2T2	7.10	6.94	6.84	6.86	27.74
P3T1	5.78	5.30	6.00	6.14	23.22
P3T2	6.34	6.02	6.24	6.40	25.00
Total	**38.56**	**37.66**	38.06	38.22	152.5

Step 2: Compute the treatment total (T), replication total (R) and the grand totals (G) as shown in Table 2.2 and compute the Treatment SS, replications SS, Interaction SS and error SS, following the procedure

$$C.F. = \frac{(GT)^2}{rt} = \frac{(152.5)^2}{6*4} = 969.01$$

$$Total\ SS = \sum_{\substack{i=1\\j=1}}^{tr} X_{ij}^2 - C.F. = (6.26)^2 + (6.28)^2 + (5.88)^2 + (6.50)^2 + ... + (6.40)^2 - 969.01 = 38883.06$$

$$Treatment\ SS = \frac{\sum_{i=1}^{t} T_i^2}{r} - C.F. = \frac{(24.92)^2 + (26.54)^2 + (25.08)^2 + (27.74)^2 + (23.22)^2 + (25.00)^2}{4} - 969.01 = 4.15$$

$$Replication\ SS = \frac{\sum_{i=1}^{t} R_j^2}{t} - C.F. = \frac{(38.56)^2 + (37.66)^2 + (38.06)^2 + (38.22)^2}{6} - 969.01 = 0.07$$

$$Total\ SS(PT) = \frac{\sum_{i=1}^{t} Total_{ij}^2}{r} - C.F. = \frac{(24.92*24.92)+(26.54*26.54)+(25.08*25.08)+(27.74*27.74)+(23.22*23.22)+(25*25)}{4} - 969.01 = 3.00$$

$$SS\ due\ to(P) = \frac{\sum P_1 + \sum P_2 + \sum P_3}{rt} - C.F. = \frac{(51.46)+(52.82)+(48.22)}{8} - 969.01 = 1.40$$

$$SS\ due\ to(T) = \frac{\sum T_1 + \sum T_2}{rp} - C.F. = \frac{(73.22)+(79.28)}{12} - 969.01 = 1.53$$

$$Interaction(PT) = Total\ SS(PT) - SS\ due\ to(P) - SS\ due\ to(T) = 3.00 - 1.40 - 1.53 = 0.08$$

$$Sum\ of\ square\ error = Treatment\ SS - Replications\ SS - Total\ SS = 4.15 - 0.07 - 3.00 = 1.07$$

ANOVA Table

Source of variance	*df*	Sum of squares	Mean square	F Cal.	F. Tab
Replication	(4–1) = 3	0.07	0.0233	-	-
Treatment	(6–1) = 5	4.14	0.822	-	-
Treatment (P)	(3–1) = 2	1.39	0.695	9.75*	3.68
Treatment (T)	(2–1) = 1	1.53	1.53	21.37*	4.54
Interaction (PT)	(2–1) (3–1) = 2	0.08	0.04	0.55	3.68
Error	(4–1)(3*2–1) = 15	1.07	0.071	-	-
Total	4*3*2–1 = 23	3.00	-	-	-

*Significant.

Compute the coefficient of variation as:

$$cv = \frac{\sqrt{Error\ MS}}{Grand\ mean} x100 = \frac{\sqrt{0.071/15}}{152.5/24} x100 = 4.21$$

The result shows that non-significant interaction between P and T, indicating that the varietal difference was not significantly affected planting stage level applied of onion seed crop and that the planting effect did not differ significantly with various neck cutting level. Main effect both of P & *T* were found significant.

2.7.6.6 AN EXAMPLE OF 2^2 SYMMETRICAL FACTORIAL EXPERIMENT

An experiment was conducted on garlic varieties with the combinations of two levels of planting method, i.e., raise and flatbed on every two levels of different irrigation frequency with five replications at RRS NHRDF, Karnal during 1998–1999. The list of the four factorial treatment combinations is shown in Table 2.3, and data in Table 2.4

Step 1: Denote the number of replications by *r*, the level of factor P (i.e., different planting method) by p and the level of factor I (i.e., Irrigation frequency) by i.

TABLE 2.3 The 2x2 Factorial Combinations of Two Planting Method and Two Irrigation Frequency Levels

Planting stage levels	**Treatment combinations**	
	i_1	i_2
P_1	P_1i_1	P_1i_2
P_2	P_2i_1	P_2i_2

TABLE 2.4 Yield (q/ha) of Garlic Crop of at Two Level of Planting Method Tested with Two Different Irrigation Frequency in a RCB Design

Treatments	**Replications**					**Total**
	R1	**R2**	**R3**	**R4**	**R5**	
P1I1	95	97	96	100	96	484
P1I2	97	96	96	97	96	482
P2I1	46	48	47	49	47	237
P2I2	50	46	49	49	49	243
TOTAL	288	287	288	295	288	1446

Step 2: Compute the treatment total (*T*), replication total (*R*) and the grand totals (*G*) as shown in Table 2.2 and compute the Treatment *SS*, replications *SS*, Interaction *SS* and error *SS*, following the procedure

$$C.F.=\frac{(GT)^2}{rt}=\frac{(1446)^2}{4*5}=104545.80$$

$$Total\ SS=\sum_{\substack{i=1\\j=1}}^{tr}X_{ij}^2-C.F.=(95)^2+(97)^2+(96)^2+(100)^2+......+(49)^2-104545.80=11844.2$$

$$Treatment\ SS=\frac{\sum_{i=1}^{t}T_i^2}{r}-C.F.=\frac{(484)^2+(482)^2+(237)^2+(243)^2}{5}-104545.80=11813.8$$

$$Replication\ SS=\frac{\sum_{i=1}^{t}R_j^2}{t}-C.F.=\frac{(288)^2+(287)^2+(288)^2+(295)^2+(288)^2}{4}-104545.80=10.70$$

$$Total\ Tr.SS(PI)=\frac{\sum_{i=1}^{t}Total_{ij}^2}{r}-C.F.=\frac{(484*484)+(482*482)+(237*237)+(243*243)}{5}-104545.80=11813.80$$

$$SS\ due\ to(P)=\frac{\sum P_1+\sum P_2}{rt}-C.F.=\frac{(966)+(480)}{10}-104545.80=11809.80$$

$$SS\ due\ to(I)=\frac{\sum T_1+\sum T_2}{rp}-C.F.=\frac{(721)+(725)}{10}-104545.80=0.80$$

$$Interaction(PI) = Total\ Tr.SS(PI) - SS\ due\ to(P) - SS\ due\ to(I) = 11813.80 - 11809.80 - 0.80 = 3.20$$
$$Sum\ of\ square\ error = Treatment\ SS - Replications\ SS - Total\ SS = 11844.20 - 10.70 - 11813 = 19.70$$

ANOVA Table

Source of variance	*df*	Sum of squares	Mean square	F Cal.	F. Tab
Replication	4	10.70	2.68	-	-
Treatment	3	11813.80	3937.93	-	-
Treatment (P)	1	11809.80	11809.8	7193*	4.75
Treatment (I)	1	0.80	0.80	0.49	4.75
Interaction (PI)	1	11813.80	11813.80	1.95	4.75
Error	12	19.70	1.64	-	-
Total	19	11844.20	-	-	-

*Significant.

The result shows a non-significant interaction between P and I, indicating that there is no significance difference in interaction effect on the yield of garlic and that the planting effect differs nonsignificantly with irrigation frequency level. Main effect both of P were found significant, but irrigation frequency level found quite non -significant.

2.8 SPLIT PLOT DESIGN

In the agricultural experiments, some factors (e.g., irrigation, tillage, drainage, weeds management, etc.) are to be applied in larger experimental units where other factors can be applied in smaller experimental units comparatively. Further, some experimental materials may be rare while the other experimental materials may be available in greater quantity or when the levels of one (or more) treatment factors are easy to change, while the alteration of levels of other treatment factors are expensive, or may be time-consuming. One more point may be noted that although two or more different factors are to be tested in the experiment, one factor may require to be tested with higher precision than the other factor. In all these situations, a design called the split-plot design is suitable to adopt.

The following guidelines are suggested, to make such a choice:

2.8.1 DEGREE OF PRECISION

To achieve greater degree of precision for factor B than for factor A, assign factor B to the sub-plot and factor A to the main plot; e.g., generally plant breeder who plans to evaluate ten promising rice varieties with three levels of fertilization, would probably wish to have better precision for varietal comparison than for fertilizer response. Thus, he would designate variety as the subplot factor and fertilizer as the main plot factor. Or, an Agronomist would assign variety to main plot and fertilizer to sub-plot if he wants greater precision for fertilizer response than variety effect.

2.8.2 RELATIVE SIZE OF THE MAIN EFFECTS

If the main effect of one factor (A) is expected to be more larger and easier to detect than that of the another factor (B), factor A can be assigned to the main plot and factor B to the sub-plot. This increases the chance of identifying the difference among levels of factor B which has a smaller effect.

2.8.3 MANAGEMENT PRACTICES

The common type of situation when the split-plot design is routinely suggestive is the difficulties in the execution of other designs, i.e., practical execution of plans. The cultural practices required by a factor may order the use of large plots. For practical expediency, such a factor may be assigned to the main plot, e.g., in an experiment to evaluate water management and variety, it may be desirable to allocate water management to the main plot to minimize water movement between adjacent plots, facilitate the simulation of the water level required, and reduce border effects. Or, if ploughing is one of the factors of interest, then one cannot have different depths of ploughing in different plots scattered randomly apart.

2.8.4 LAYOUT AND RANDOMIZATION

The stepwise procedure for layout of a $m \times n$ factorial experiment conducted in Split-plot design with r replications is as given below

Step 1: Divide the whole area of experiment into r replications considering field condition and layout.

Step 2: Now divide each and every replication into m number of blocks (main plots) of equal in size and homogeneity.

Step 3: Afterwards, divide each and every main plot into *n* number of experimental units (subplots) of equal size.

Following proper layout, Randomization is taken up in two different steps: one for main plots and another for subplots.

Step 4: m levels of main plots of each and every replication are randomly allocated.

Step 5: Now *n* levels of subplot factor is randomly allocated to each and every main plot separately.

2.8.5 ANALYSIS

The analysis of variance for a split-plot design is divided into two components:
(a) main plot analysis; and
(b) subplot analysis.

2.8.6 STATISTICAL MODEL

For the above experiment with m levels of main plot factor, *n* levels of subplot factor in *r* replications, the model is as follows:

$$y_{ijk} = \mu + \gamma_i + \alpha_j + e_{ij} + \beta_k + \upsilon_{jk} + e_{ijk}$$

where, i = 1, 2, 3….r; j = 1, 2, 3….m; k = 1, 2, 3….n; μ = general effect.

α_j = additional effect due to j-th level of main plot factor A and $\sum_{j=1}^{m} \alpha_j = 0$

β_k = additional effect due to k-th level of sub plot factor B and $\sum_{k=1}^{n} \beta_k = 0$

υ_{jk} = interaction effect due to j-th level of main plot factor and k-th level of sub plot factor B

and $\sum_{j=1} \upsilon_{jk} = \sum_{k=1} \upsilon_{jk} = 0$ for all j and all k.

e_{ij} = error I = error associated with i-th replication and j-th level of main plot factor

and $e_{ij} \sim$ i.i.d. $N(0, \sigma_m^2)$

e_{ijk} = error II= error associated with i-th replication and j-th level of main plot factor, k-th level of sub plot factor

and $e_{ijk} \sim$ i.i.d. $N(0, \sigma_s^2)$

2.8.7 HYPOTHESIS TO BE TESTED

$$H_0 : \gamma_1 + \gamma_2 .... + \gamma_r = 0$$
$$\alpha_1 + \alpha_2 .... + \alpha_m = 0$$
$$\beta_1 + \beta_2 .... + \beta_n = 0$$
$$\upsilon_{11} + \upsilon_{12} .... + \upsilon_{jk} ...... + \upsilon_{mn} = 0$$

H_1: γ's are not all equal.
β's are not all equal.
β's are not all equal.
υ's are not all equal.

Source of Variation	DF	SS	MSS	F-ratio
Replication	r–1	RSS	RMS	RMS/ ErMS(I)
Main plot factor (A)	m–1	SS(A)	MS(A)	MS(A)/ ErMS(I)
Error I	(r–1)(m–1)	ErSS(I)	ErMS(I)	
Subplot factor (B)	(n–1)	SS(B)	MS(B)	MS(B)/ ErMS(II)
Interaction (AxB)	(m–1)(n–1)	SS(AB)	MS(AB)	MS(AB)/ ErMS(II)
Error II	m(n–1)(r–1)	ErSS(II)	ErMS(II)	
Total	mnr–1			

The steps for procedure for comparison of different SS and mean sum of squares (MSS) are given as follows:

1. Grand Mean = $\sum_{i=1}^{r} \sum_{j=1}^{m} \sum_{k=1}^{n} y_{ijk} = GT$
2. Correlation Factor (CF) = $\frac{GT^2}{mnr}$
3. Total SS = $\sum_{i=1}^{r} \sum_{j=1}^{m} \sum_{k=1}^{n} y^2_{ijk} - CF$
4. Work out the SS due to the main plot factor and the replication. For the purpose the following table is required to be framed:

	a_1	a_1	.	a_m	total
R_1					
R_2					
. . .					
R_r					
Total					

where $y_{ij()} = \sum_{k=1}^{n} y_{ijk}$

$$TSS\ (table\ I) = \frac{1}{n}\sum_{i}^{r}\ \sum_{j}^{m} y^2_{ij()} - CF$$

$$\text{Replication SS} = \frac{1}{mn}\sum_{i=1}^{r} y^2_{i()()} - CF$$

$$SS(A) = \frac{1}{nr}\sum_{j=1}^{m} y^2_{()j()} - CF$$

SS Error I = TSS (Table I) – RSS – SS (A)

5. Work out the SS due to the subplot factor and the interaction. For the purpose the following table is required:

	a_1	a_1	.	a_m	Total
b_1					
b_2					
. . .					
b_n					
Total					

$$TSS\ (table\ II) = \frac{1}{r}\sum_{j=1}^{m}\ \sum_{k=1}^{n} y^2_{()jk} - CF$$

$$SS(B) = \frac{1}{mr}\sum_{k=1}^{n} y^2_{()()k} - CF$$

SS Error II = TSS – RSS – SS (A) – SS Error I – SS (B) – SS (AB). MSS are calculated dividing SS by corresponding degrees of freedom.

6. F-ratio for replication and main plot factor is obtained by comparing the respective MSS against MSS due to error I. On the other hand, the F-ratio corresponding to subplot factor and interaction effects are worked out by comparing the irrespective MSS against MSS due to error II.
7. Calculated F-ratios are checked with tabulated value of F at required LOS and degrees of freedom.

8. If F-test is significant, next thing will be to estimate the Standard errors (SE) for different types of comparison as given below:

(a) SE for difference (SED) between two replication means and corresponding CD value will be

$$SED = \sqrt{\frac{2EMSS(I)}{mn}} CD = \sqrt{\frac{2EMSS(I)}{mn}} \times t_{\alpha/2;error(I)df}$$

(b) SED between two main plot treatment means and corresponding CD value will be

$$SED = \sqrt{\frac{2EMSS(I)}{rn}} CD = \sqrt{\frac{2EMSS(I)}{rn}} \times t_{\alpha/2;error(I)df}$$

(c) SED between two subplot treatment means and corresponding CD value will be

$$SED = \sqrt{\frac{2EMSS(II)}{rm}} CD = \sqrt{\frac{2EMSS(II)}{rm}} \times t_{\alpha/2;error(II)df}$$

(d) SED between two subplot treatment means at same level of main plot treatment means, and corresponding CD value will be

$$SED = \sqrt{\frac{2EMSS(II)}{r}} CD = \sqrt{\frac{2EMSS(II)}{r}} \times t_{\alpha/2;error(II)df}$$

(e) SED between two main plot treatment means at same or different level of subplot treatment, but the ratio of the treatment mean difference and the above SE does not follow *t* distribution. An approximate value of *t* is given by where t_1 and t_2 tabulate t-values at error(I) and error(II) *df*, respectively, at chosen LOS and corresponding CD value will be as follows:

$$SED = \sqrt{\frac{2[(n-1)ErMS(II) + ErMS(I)]}{rn}};$$

$$t = \frac{t_1 ErMS(I) + t_2 (n-1)ErMS(II)}{ErMS(I) + (n-1)ErMS(II)}; CD = SED \times t_{cal}$$

Note: Main plot error should be lesser than subplot error and similarly, *cv* of respective.

2.8.8 ADVANTAGES

(1) The main advantage of this design is that the requirement of larger area for certain factor for effective management purposes in practical field has been purposefully used to increase precision of other factor in the experiment without scarifying the amount of information required by the experiment.
(2) In this design, Error II is smaller than Error I, which results subplot effect and interaction effects are estimated precisely.
(3) It is the simplest of all incomplete block design where the technique of confounding can be employed.

2.8.9 DISADVANTAGES

(1) Here, the plot size and precision of the measurement of effects are not equal for the factors.
(2) Randomization and layout are somewhat complicated.
(3) If both factors require larger size plot size, then split-plot design is not suitable.

Example: A field experiment was conducted to identify the best method of ploughing in four varieties of Potato. Three different methods of ploughing were allocated randomly to three main plots in each replication individually, and in each main plot, four varieties were assigned randomly among four plots. Yields (t/ha) were recorded from the individual plots and given below. Analyze the data and draw your conclusions?

	Plough–1				**Plough–2**				**Plough–3**			
	V1	**V2**	**V3**	**V4**	**V1**	**V2**	**V3**	**V4**	**V1**	**V2**	**V3**	**V4**
Rep–1	8.7	9.1	7.8	7.2	9.5	12.6	11.2	9.8	7.5	9.5	8.2	7.9
Rep–2	8.6	9.2	7.9	7.3	9.4	12.5	11	9.6	7.6	9.8	8.4	8
Rep–3	8.5	9.3	8.2	7.4	9.6	12.3	10.9	10	7.4	9.7	8.5	8.1

From the above problem, initially, hypothesis has to be formulated as follows.

H_0: (1) There is no significant difference between main plot effects [ploughing]
(2) There is no significant difference between subplot effects [Varieties]

(3) There is no significant difference between interaction effects [P×V]

(4) There is no significant difference between replication effects

H_1: (1) There is a significant difference between main plot effects [ploughing]

(2) There is a significant difference between subplot effects [varieties]

(3) There is a significant difference between Interaction effects [P×V]

(4) There is a significant difference between replication effects

The step-by-step procedure for calculation of different SS and ANOVA are as follows:

Grand total = 8.7 + 9.1 + ...+8.5+8.1 = 328.20

$$\text{Correction Factor } CF = \frac{GT^2}{p \times v \times r} = \frac{328.20^2}{3 \times 4 \times 3} = 2992.09$$

Total Sum of Squares (TSS) = $[8.7^2+9.1^2+...+8.5^2+8.1^2]$–CF = 75.87

SS of Main plot and replication can be worked by using Table 2.5.

TABLE 2.5 Table of Totals of Ploughing *x* Replications

Ploughing	Replication			Total
	R1	**R2**	**R3**	
P1	32.8	33.0	33.4	99.2
P2	43.1	42.5	42.8	128.4
P3	33.1	33.8	33.7	100.6
Total	109.0	109.3	109.9	328.2
Average	36.33	36.43	36.63	

$$\text{Table 2.5 TSS } = \frac{1}{V}(32.8^2 + ...33.7^2) - CF = 45.37$$

$$\text{Replication SS } = \frac{1}{PV}(109^2 + ... + 109.9^2) - CF = 0.035$$

$$\text{Main Plot SS } = \frac{1}{Vr}(99.2^2 + ... + 100.6^2) - CF = 45.207$$

SS of Error-I = Table 2.5; TSS – Main Plot; SS – Replication; SS = 0.128; SS of subplot and Interaction can be worked by using Table 2.6.

TABLE 2.6 Table of Totals of Ploughing *x* Variety

Ploughing	Variety				Total	Average
	V1	V2	V3	V4		
P1	25.8	27.6	23.9	21.9	99.2	8.27
P2	28.5	37.4	33.1	29.4	128.4	10.7
P3	22.5	29	25.1	24	100.6	8.38
Total	76.8	94	82.1	75.3		
Average	8.53	10.44	9.12	8.37		

$$\text{Table 2.6 TSS} = \frac{1}{r}(25.8^2 + ... + 24^2) - CF = 75.39$$

$$\text{Variety SS} = \frac{1}{\text{Pr}}(76.8^2 + ... + 75.3^2) - CF = 23.99$$

P×V SS = Table 2.6 TSS – Ploughing SS – Variety SS.

SS of Error-II = TSS-Main Plot; SS – Replication; SS – Variety; SS-P×V SS – SS of Error-I = 0.31.

ANOVA table

SOV	DF	SS	MSS	F-cal	F-at (Prob. = 0.05)
Replication	2	0.035	0.018	0.545	6.94
Main plot	2	45.207	22.603	704.519	6.94
Error-I	4	0.128	0.032		
Subplot	3	23.992	7.997	464.366	3.16
Interaction	6	6.198	1.033	59.978	2.66
Error-II	18	0.31	0.017		
Total	35	75.87			

From the above table, it is found that effects of Ploughing, Variety, and their interactions were significant. So, our next task is to calculate CD as follows.

(i) Critical Difference (CD) for any two main plot treatment means

$$= \sqrt{\frac{2EMSS - I}{Vr}} \times t_{error-Idf\,@5\%LOS} = \sqrt{\frac{2*0.032}{4*3}} \times 2.776 = 0.2027$$

(ii) Critical Difference (CD) for any two Subplot treatment means

$$= \sqrt{\frac{2EMSS - II}{Pr}} \times t_{error-IIdf\,@5\%LOS} = \sqrt{\frac{2*0.017}{3*3}} \times 2.101 = 0.129$$

(iii) Standard error of Difference between two main plot treatment means at same or different level of subplot treatment

$$SED = \sqrt{\frac{2[(n-1)EMSS-II + EMSS-I}{Vr}} = \sqrt{\frac{2[(4-1)0.017+0.032]}{4*3}} = 0.1176$$

It follows t-distribution with,

$$t_{tab} = \frac{t_1 EMSS-I + t_2(n-1)EMSS-II}{EMSS-I+(n-1)EMSS-II} = \frac{(2.776*0.032)+(2.101*3*0.017)}{0.032+3*0.017} = 2.361$$

So, CD can be calculated as CD = SED*t-tab
= 0.1176*2.361 = 0.278

As mentioned in theory, by using respective CD, treatment means are compared. It was found that ploughing (P–2), Variety (V–2) and the interaction (P2V2) were significant among others.

2.9 STRIP PLOT DESIGN

In two factorial experiments, when both factors require larger plots, then split plot is not suitable, then the adaptable design is Strip plot design. In this design interaction effect between the two factors are measured with high precision than either of the two factors. In agricultural field experiments with factors like irrigation, different methods of mulching, ploughing, etc. require larger sized plots for the convenience of management. Here, the factors assigned to the horizontal row is called "horizontal factor" and the factors assigned to the vertical row is called "vertical factor." Both these factors always remain perpendicular to each other. Size and shape of both vertical and horizontal strips may not same.

2.9.1 RANDOMIZATION AND LAYOUT

Just like split-plot design, randomization, and allocation of the two factors is done in two steps. Initially, m horizontal factors are randomly allocated to m horizontal strips. Then, *n* levels of another factor are allotted to vertical strips. Here, the procedure is repeated for each and every replication with proper randomization. Then, the ultimate layout will be "$m \times n$" strip plot design with "r" replication.

2.9.2 ANALYSIS

The analysis of strip plot design is performed in three steps:

(a) analysis of horizontal factor effects;
(b) analysis of vertical factor effects; and
(c) analysis of interaction factor effects.

2.9.3 STATISTICAL MODEL

$$y_{ijk} = \mu + \gamma_i + \alpha_j + e_{ij} + \beta_k + e_{ik} + (\alpha\beta)_{jk} + e_{ijk}$$

where, i = 1, 2, 3….r; j = 1, 2, 3….m; k = 1, 2, 3….n;
μ = general effect

α_j = additional effect due to j-th level of vertical factor A and $\sum_{j=1}^{m} \alpha_j = 0$

β_k = additional effect due to k-th level of horizontal factor B and $\sum_{k=1}^{n} \beta_k = 0$

$(\alpha\beta)_{jk}$ = interaction effect due to j-th level of factor Aand k-th level of factor B and $\sum_{k=1} (\alpha\beta)_{jk} = 0$.

e_{ij} = error I = error associated with i-th replication and j-th level of vertical factor A and e_{ij} ~ i.i.d. $N(0,\sigma_1^2)$

e_{ik} = error II = error associated with i-th replication and k-th level of horizontal factor B and e_{ik} ~ i.i.d. $N(0,\sigma_2^2)$

e_{ijk} = error III= error associated with i-th replication , j-th level of vertical factor A and j-th level of main plot factor k-th level of horizontal factor B and e_{ijk} ~ i.i.d. $N(0,\sigma_3^2)$

2.9.4 HYPOTHESIS TO BE TESTED

$$H_0 : \gamma_1 = \gamma_2 .... = \gamma_r = 0$$
$$\alpha_1 = \alpha_2 .... = \alpha_m = 0$$
$$\beta_1 = \beta_2 .... = \beta_n = 0$$
$$(\alpha\beta)_{11} = (\alpha\beta)_{12} .... = (\alpha\beta)_{jk} ...... = (\alpha\beta)_{mn} = 0$$

H_1: γ's are not all equal
β's are not all equal
β's are not all equal
$(\alpha\beta)$'s are not all equal

Initially construct, two ways mean table of (a) replication × horizontal factor (b) replication × vertical factor (c) horizontal factor × vertical factor. The steps for procedure for comparison of different SS and MSS are given as follows:

$$\text{Grand Mean} = \sum_{i=1}^{r}\sum_{j=1}^{m}\sum_{k=1}^{n} y_{ijk} = GT$$

$$\text{Correlation Factor (CF)} = \frac{GT^2}{mnr}$$

$$\text{Total SS} = \sum_{i=1}^{r}\sum_{j=1}^{m}\sum_{k=1}^{n} y^2_{ijk} - CF$$

Work out the SS due to the vertical factors A and the replication. For the purpose the following table is required to be framed:

	a_1	a_1	...	a_m	Total
R_1					
R_2					
...					
R_r					
Total					

$$\text{where } y_{ij0} = \sum_{k=1}^{n} y_{ijk}$$

$$TSS\ (table\ I) = \frac{1}{n}\sum_{i}^{r}\sum_{j}^{m} y^2_{ij0} - CF$$

$$\text{Replication SS} = \frac{1}{mn}\sum_{i=1}^{r} y^2_{i00} - CF$$

$$SS(A) = \frac{1}{nr}\sum_{j=1}^{m} y^2_{0j0} - CF$$

SS Error I = TSS (Table I) - RSS - SS (A)

Work out the SS due to the horizontal factor B and the replication. For the purpose the following table is required to be framed:

	b_1	b_1	…	b_n	Total
R_1					
R_2					
…					
R_r					
Total					

$$TSS\ (table\ II) = \frac{1}{m}\sum_{i=1}^{r}\sum_{k=1}^{n} y^2_{i()k} - CF$$

$$SS(B) = \frac{1}{mr}\sum_{k=1}^{n} y^2_{()()k} - CF$$

SS Error II = TSS (table II) – RSS – SS (B)

Workout the SS due to the vertical factor A and horizontal factor B interaction. For the purpose the following table is required to be prepared:

	b_1	b_1	…	b_n	Total
a_1					
a_2					
…					
a_m					
Total					

$$TSS\ (table\ III) = \frac{1}{r}\sum_{j=1}^{m}\sum_{k=1}^{n} y^2_{()jk} - CF$$

$SS(AB) = TSS(table - III) - SS(A) - SS(B)$

SS Error II = TSS (table III)- RSS - SS (B)

SS Error III = TSS (table III)- RSS - SS (A) - SS (B) - SS (AB)-ErSS I-ErSS II

MSS are calculated dividing SS by corresponding degrees of freedom.

SOV	DF	SS	MSS	F-ratio
Replication	r–1	RSS	RMS	RMS/ ErMS(I)
Vertical plot factor (A)	m–1	SS(A)	MS(A)	MS(A)/ ErMS(I)
Error I	(r–1)(m–1)	ErSS(I)	ErMS(I)	
Horizontal plot factor (B)	(n–1)	SS(B)	MS(B)	MS(B)/ ErMS(II)
Error II	(r–1)(n–1)	ErSS(II)	ErMS(II)	
Interaction (AxB)	(m–1)(n–1)	SS(AB)	MS(AB)	MS(AB)/ ErMS(II)
Error III	(m–1) (n–1)(r–1)	ErSS(III)	ErMS(III)	
Total	mnr–1			

F- ration for replication and vertical factor are obtained by comparing the respective MSS against MSS due to error I. On the other hand, the F-ratio corresponding to the horizontal factor and interaction between horizontal and vertical are worked out by comparing respective MSS against MSS due to error II and error III, respectively.

Calculated F-ratios are compared with tabulated value of F at appropriate LOS and degrees of freedom.

If F-test is significant, next task will be to estimate the SE for different types of comparison as given below:

SED between two replication means and corresponding CD value will be:

$$SED = \sqrt{\frac{2EMSS - I}{mn}} CD = \sqrt{\frac{2EMSS - I}{mn}} \times t_{\alpha/2;error-Idf}$$

SED between two vertical treatment means and corresponding CD value will be:

$$SED = \sqrt{\frac{2EMSS - I}{rn}} CD = \sqrt{\frac{2EMSS - I}{rn}} \times t_{\alpha/2;error-Idf}$$

SED between two horizontal treatment means and corresponding CD value will be:

$$SED = \sqrt{\frac{2EMSS - II}{rm}} CD = \sqrt{\frac{2EMSS - II}{rm}} \times t_{\alpha/2;error-IIdf}$$

SED between two vertical treatment means at same level of horizontal treatment means, and corresponding CD value will be:

$$SED = \sqrt{\frac{2[(n-1)ErMSIII + ErMSI]}{rn}}$$

SED between two horizontal treatment means at same or of vertical treatment:

$$SED = \sqrt{\frac{2[(m-1)ErMSIII + ErMSII]}{rm}}$$

But the ratio of the treatment mean difference and the above SE of (d), (e) does not follow *t* distribution. An approximate value of *t* is given by where t_1 = tabulated values at error I; t_{11} = tabulated values at error II and t_{111} = tabulated values at error III *df,* respectively, at chosen LOS and corresponding CD value will be as follows:

$$\frac{\{(n-1)ErMSIII \times t_{111}\} + \{t_1 \times ErMSI\}}{ErMSI + (n-1)ErMSIII} : and; \frac{\{(m-1)ErMSIII \times t_{111}\} + \{t_{11} \times ErMSII\}}{ErMSII + (m-1)ErMSIII};$$

2.9.5 ADVANTAGES AND DISADVANTAGES

The major advantage of this design is that it can accommodate two different factors which require larger plot sizes for feasibility of management. Here, Interaction factor are estimated more precisely than other factors.

Most disadvantage of this design is the complicated randomization, layout, and complicated statistical analysis

Example: A field experiment was conducted to find the efficacy of three different sources of organic manure (M1-M2-M3) and four irrigation schedules (I1-I2-I3-I4) was conducted in strip plot design with three replications? Below are yield (t/ha) for different treatments? Identify the best organic manure, irrigation schedule along with their combination?

Horizontal Plot	Vertical Plot											
	M–1				M–2				M–3			
	I–1	I–2	I–3	I–4	I–1	I–2	I–3	I–4	I–1	I–2	I–3	I–4
Rep–1	11.2	10.2	14.5	12.3	11.8	10.9	16.2	13.5	11	9.5	12.5	10.6
Rep–2	11.3	10.5	14.6	12.5	11.9	10.8	16.5	13.8	10.9	9.8	12	10.8
Rep–3	11.5	10.4	14.2	12.8	11.6	10.7	16.6	13.9	10.5	9.7	12.4	10.7

From the above problem, initially, hypothesis has to be formulated as follows.

H_0: (1) There is no significant difference between Vertical plot effects [Manures-M]

(2) There is no significant difference between Horizontal plot effects [Irrigation schedule-I]

(3) There is no significant difference between Interaction effects [M×I]

(4) There is no significant difference between replication effects

The stepwise procedure for calculation of different SS and ANOVA are as follows:

Grand total = 11.2+11.3+…+10.8+10.7 = 434.6

Correction Factor $CF = \frac{GT^2}{h \times v \times r} = \frac{434.6^2}{4 \times 3 \times 3} = 5246.48$

Total Sum of Squares (TSS) = $[11.2^2+11.3^2+\ldots+10.8^2+10.7^2]$-CF = 127.112

SS of Vertical factor and replication can be worked by using Table 2.7.

TABLE 2.7 Table of Totals of Manure × Replications

Ploughing	Manure			Total	Average
	M1	M2	M3		
R1	48.2	52.4	43.6	144.2	12.02
R2	48.9	53	43.5	145.4	12.12
R3	48.9	52.8	43.3	145	12.08
Total	146	158.2	130.4		

Table 2.7 $TSS = \frac{1}{H}(146^2 + \ldots + 130.4^2) - CF = 32.50$

Replication $SS = \frac{1}{HV}(144.2^2 + \ldots. + 145^2) - CF = 0.062$

Vertical factor $SS = \frac{1}{Hr}(146^2 + ... + 130.4^2) - CF = 32.36$

SS of Error-I = Table 2.7 TSS – Main Plot SS – Replication SS = 0.078
SS of Horizontal strip can be worked by using Table 2.8.

TABLE 2.8 Table of Totals of Ploughing × Variety

	Irrigation Schedule			
	I–1	**I–2**	**I–3**	**I–4**
R1	34	30.6	43.2	36.4
R2	34.1	31.1	43.1	37.1
R3	33.6	30.8	43.2	37.4
Total	101.7	92.5	129.5	110.9
Average	11.3	10.28	14.39	12.32

Table 2.8 $TSS = \frac{1}{r}(34^2 + ... + 37.4^2) - CF = 83.42$

Horizontal factor $SS = \frac{1}{Vr}(101.7^2 + ... + 110.9^2) - CF = 83.21$

SS of Error-II = TSS (Table 2.8) – Replication SS – Horizontal factor SS = 0.14

SS of Interaction effect can be worked by using Table 2.9.

TABLE 2.9 Table of Totals of Vertical Factor × Horizontal Factor

	M1	**M2**	**M3**
I–1	34	35.3	32.4
I–2	31.1	32.4	29
I–3	43.3	49.3	36.9
I–4	37.6	41.2	32.1

Interaction factor $SS = \frac{1}{r}(34^2 + ... + 32.1^2) - CF - SS(V) - SS(H) = 10.64$

SS of Error-III = TSS – RSS – SS (H) – SS (V) – SS (V*H) – SS of Error-I SS of Error-II = 0.61

After calculation of SS, ANOVA tables has to be prepared as shown below.

ANOVA table

SOV	DF	SS	MSS	F-cal	F-at (Prob. = 0.05)
Replication	2	0.06	0.03	1.6	6.94
Vertical factor	2	32.36	16.18	832.17	6.94
Error-I	4	0.08	0.02		
Horizontal factor	3	83.21	27.74	1155.71	4.76
Error-II	6	0.14	0.02		
Interaction	6	10.64	1.77	34.78	3.00
Error-III	12	0.61	0.05		
Total	**35**	**127.11**			

From the above table, it is found that effects of Vertical, Horizontal, and their interactions were found to be significant. So, our next task is to calculate CD as follows.

Critical Difference (CD) for any two vertical factor means

$$=\sqrt{\frac{2EMSS-I}{Hr}}\times t_{error-Idf\,@5\%LOS}=\sqrt{\frac{2*0.019}{4*3}}\times 2.776=0.1562$$

Critical Difference (CD) for any two horizontal factor means

$$=\sqrt{\frac{2EMSS-II}{Vr}}\times t_{error-IIdf\,@5\%LOS}=\sqrt{\frac{2*0.024}{3*3}}\times 2.447=0.179$$

Standard error of difference between two vertical factor means at same level of horizontal factor.

$$SED=\sqrt{\frac{2[(n-1)EMSS-III+EMSS-I}{Hr}}=\sqrt{\frac{2[(4-1)0.051+0.019]}{4*3}}=0.169$$

It follows t-distribution with,

$$t_{tab}=\frac{t_1EMSS-I+t_3(n-1)EMSS-III}{EMSS-I+(n-1)EMSS-III}=\frac{(2.776*0.019)+(2.179*3*0.051)}{0.019+3*0.051}=2.245$$

(Here, $t_1=t_{0.025,4}=2.776; t_2=t_{0.025,6}=2.447; t_3=t_{0.025,4}=2.179;$)

So, CD can be calculated as CD = SED*t-tab
= 0.169*2.245 = 0.379

Standard error of difference between two horizontal factor means at same level of vertical factor

$$SED = \sqrt{\frac{2[(m-1)EMSS - III + EMSS - II}{Vr}} = \sqrt{\frac{2[(3-1)0.051 + 0.024]}{3*3}} = 0.167$$

It follows t-distribution with,

$$t_{tab} = \frac{t_2 EMSS - II + t_3(m-1)EMSS - III}{EMSS - I + (m-1)EMSS - III} = \frac{(2.447*0.024) + (2.179*2*0.051)}{0.051 + 2*0.024} = 2.23$$

(Here, $t_1 = t_{0.025,4} = 2.776; t_2 = t_{0.025,6} = 2.447; t_3 = t_{0.025,4} = 2.179;$)

So, CD can be calculated as CD = SED*t-tab = 0.167*2.23 = 0.372

As mentioned in theory, by using respective CD, treatment means are compared.

2.10 AUGMENTED DESIGNS

In many practical situations, experimenters are to conduct experiments, where it becomes very difficult to replicate the treatments due to scarcity of experimental materials, which is often experienced in breeding experiments.

Sometimes a large number of germplasms are required to be evaluated at same time, and then it becomes very difficult to accommodate a large number of germplasm (say 100) in CRD setup or RBD setup.

Similarly, sometimes experimenter wants to compare not only the factorial treatments among themselves but also against a treatment called control (which may be standard package of practice; water/zero levels of all factors under consideration). There should be modification required for regular factorial design

To deal the above situations, experimental designs called *Augmented designs* have been developed. Federer has given augmented designs for the first two situations; suitable adjustment is possible in analysis of variance to handle the problem of comparing the treatments with control treatment.

Just by example, if one is interested to know the efficacy of three doses of a particular growth hormone along with four doses of fertilizer in a crop improvement trail. Then experiment conducted in factorial RBD with 3*4 = 12 treatment combinations. If experimenter wants to compare treatment effects against a treatment "control," i.e., no growth hormone and fertilizer, then factors are to be increased; finally, treatment combinations will be 4*5 = 20 per replication. In this process there we need eight more experimental units in each replication. Alternate methods is to use augmented design where treatment combinations will be 3*4+1 = 13 treatments.

Augmented design is framed on any standard design, either by complete or incomplete block design. Here block size may vary, but homogeneity among the experimental units within a block should be ensured.

2.10.1 AUGMENTED COMPLETELY RANDOMIZED DESIGN

The basic experimental design CRD is used here. Let "v" number of test genotypes which are not repeated because of scarcity and "c" number of checks which are repeated "r" times. Then the total plots required are $N = v + rc$.

Here the whole experimental area divided into N homogeneous units. N number of plots are randomly allotted to 'v' and 'c' genotypes. Then there shall be v + c = e entries.

The analysis of variance table as follows.

SOV	DF	SS	MS	F-ratio
Entries (e)	e–1	*eSS*	*eMS*	*eMS/ErMS*
Checks (c)	c–1	*cSS*	*cMS*	*cMS/ErMS*
Genotypes (g)	v–1	*gSS*	*gMS*	*gMS/ErMS*
Check vs. Genotype	1	*cgSS*	*cgMS*	*cgMS/ErMS*
Error	c(r–1)	*ErSS*	*ErMS*	
Total	N–1	*TSS*		

Grand total = $G = \sum_{i=1}^{v} g_i + \sum_{k=1}^{r}\sum_{j=1}^{c} c_{kj}$ where g_i is the value corresponding to ith genotype, and c is the value corresponding to jth check and kth replicate.

Correction factor $CF = \frac{G^2}{N}$

Total Sum of squares $(TSS) = \sum_{i=1}^{v} g_i^2 + \sum_{k=1}^{r}\sum_{j=1}^{c} c_{kj}^2 - CF$

Sum of square due to entries $(eSS) = \sum_{i=1}^{v} g_i^2 + \frac{1}{r}\sum_{j=1}^{c} c_j^2 - CF$

Correction factor for Checks $(CFc) = \frac{1}{c \times r} (\sum_{k=1}^{r} \sum_{j=1}^{c} c_{kj})^2$

Sum of square due to checks $(cSS) = \frac{1}{r} \sum_{j=1}^{c} c_j^2 - CF_c$

Correction factor for genotypes $(CFg) = \frac{1}{v} (\sum_{i=1}^{v} g_i)^2$

Sum of square due to entries $(eSS) = \sum_{i=1}^{v} g_i^2 + \frac{1}{r} \sum_{j=1}^{c} c_j^2 - CF$

Sum of square due to genotypes $(gSS) = \sum_{i=1}^{v} g_i^2 - CF_g$

Sum of square check vs. genotypes $(cgSS) = eSS - cSS - gSS$
Error sum of squares $(ErSS) = TSS - eSS$
SE of mean difference for:

(a) Between two checks: $\sqrt{\frac{2ErSS}{r}}$

(b) Between a check and a genotype $\sqrt{ErMS(1 + \frac{1}{r})}$

2.10.2 AUGMENTED RANDOMIZED BLOCK DESIGN

Let us discuss an augmented design in randomized complete blocks.

Let there be c = number of checks.
r = number of blocks
n_k = number of test genotypes in k[th] block
n = total number of test genotypes.

2.10.3 RANDOMIZATION AND LAYOUT

1. k[th] block is divided into $(n_k + c)$ number of plots.
2. Randomly allocate the 'c' number of checks among $(n_k + c)$ number of plots in k[th] block.
3. Randomly allocate the total $\sum_{k=1}^{r} n_k = n$ test genotypes with 'n_k' test genotypes in the 'k[th]' block.

The structure of ANOVA is as follows:

SOV	DF	SS	MS	F-ratio
Blocks	r–1	*BSS*	*BMS*	*BMS/ErMS*
Entries (e)	e–1	*eSS*	*eMS*	*eMS/ErMS*
Checks (c)	c–1	*cSS*	*cMS*	*cMS/ErMS*
Genotypes (g)	n–1	*gSS*	*gMS*	*gMS/ErMS*
Check vs. Genotype	1	*cgSS*	*cgMS*	*cgMS/ErMS*
Error	(c–1)(r–1)	*ErSS*	*ErMS*	
Total	N–1	*TSS*		

As only the checks are replicated but not the test genotypes, before calculating the different components of ANOVA, there is to adjust the effects of test genotypes.

(i) Block effects $r_k = \frac{1}{c}\left(\sum_{j=1}^{c} c_{jk} - \bar{c}\right)$; it may be noted that sum of block effects must be equal to zero, i.e., $\sum_{k=1}^{r} r_k = 0$

(ii) Mean effect (m) = $\frac{1}{e}\left[G - (r-1)\bar{c} - \sum_{k=1}^{r} n_k r_k\right]$

(iii) Check effect $(c_j) = \bar{c}_j - m$ where $\bar{c}_j = \frac{1}{r}\sum_{k=1}^{r} c_{jk}$

(iv) Adjust genotypic response $(g^{I}_{i}) = g_i - r_{jk}$

Grand total (G) = $\sum_{i=1}^{n} g_i + \sum_{k=1}^{r}\sum_{j=1}^{c} c_{kj}$

Correction factor (CF) = $\frac{G^2}{N}$

Total sum of squares (TSS) $= \sum_{i=1}^{n} g_i^2 + \sum_{k=1}^{r}\sum_{j=1}^{c} c_{kj}^2 - CF$

Sum of squares due to blocks (BSS) $= \sum_{k=1}\left(\frac{R^2{}_k}{n_k + c}\right) - CF$

Sum of squares due to entries (Ess) $= (m \times G) + \sum_{k=1}^{r} r_k R_k + \sum_{j=1}^{c} c_j\left(\sum_{k=1}^{r} c_{jk}\right) + \sum_{i=1}^{n} g_i g^{n}{}_i - \sum_{k=1}^{r} \frac{R^2{}_k}{(n_k + c)}$

Correction factor for checks (CFc) $= \frac{1}{c \times r} + \left(\sum_{k=1}^{r}\sum_{j=1}^{c} c_{kj}\right)^2$

Sum of squares due to checks (cSS) $= \frac{1}{r}\sum_{j=1}^{c}\left(\sum_{k=1}^{r} c_{kj}\right)^2 - CF_c$

Correction factor for genotypes (CFg) $= \frac{1}{n}\left(\sum_{i=1}^{n} g_i\right)^2$

Sum of squares due to genotypes (gSS) = $eSS - cSS - gSS$

Sum of squares due to check vs. genotypes (cgSS) = $eSS - cSS - gSS$

Error Sum of squares (ErSS) $= \sum_{i=1}^{n} g_i^2 + CF_g$

The SE for testing the different varietal means are given below:

(i) For two checks means $= \sqrt{\frac{2ErMS}{r}}$

(ii) For two test genotype means in the same blocks $= \sqrt{2ErMS}$

(iii) For any two entries mean in the same block $= \sqrt{ErMS\left(1+\frac{1}{c}\right)}$

(iv) For means between a check and a test genotypes $= \sqrt{ErMS\left(1+\frac{1}{c}+\frac{1}{r}+\frac{1}{cr}\right)}$

Note: While adopting this design, it should be kept in mind that these designs are used in situations where no other design fit best for the purpose. Other guidelines are

1. Each and every block should be kept to higher possible homogeneous condition.
2. The number of test genotypes should be same in every block to facilitate statistical analysis.

Example: An experiment was conducted to evaluate germplasm trial; in this, sixteen test genotypes were tested in five blocks along with three number of standard check varieties. Given below is the layout and responses of total 19 [(Test genotypes(16)+ Checks(3)] genotypes. Analyze the data to examine: (a) whether the test genotypes are superior over the checks, (b) which of the test varieties is superior, and (c) which of the test genotypes is superior. Note that in the given layout g stands for test genotypes and c stands for check varieties.

Block 1	Block 2	Block 3	Block 4	Block 5
g16,30	c3,14	g15,8	g5,20	c1,7
c1,10	g3,4	c2,6	c2,6	g1,7
g2,6	c1,8	c1,6	g8,18	c2,4
g4, 15	g13,40	g7,11	c3,17	g10,16
c2,3	g6,25	g12,6	g14,12	c3,18
c3,15	c2,5	g11,18	c1,9	
g9,24		c3,16		

Solution: From above, it is clear that the above experiment has been conducted in the augmented randomized block design (RBD) with three checks each being replicated in each of the five blocks and there are sixteen test genotypes randomly allocated among the five blocks.

Thus, c = number of checks = 3.

V = number of test genotypes in different blocks = 4,3,4,3 and 2 respectively in block 1, block 2, block 3, block 3, block 4 and block 5.

r = number of blocks = 5 & n = number of test genotypes = 16.

Only the checks are replicated for five times but not the test genotypes, so we are to adjust the effects of test genotypes as follows:

We make the following table, and following quantities are worked out:

Checks	R1	R2	R3	R4	R5	Total	Mean
C1	10	8	6	9	7	40.00	8.00
C2	3	5	6	6	4	24.00	4.80
C3	15	14	16	17	18	80.00	16.00
Total	28.00	27.00	28.00	32.00	29.00	144.00	28.80
Genotypes	30	4	8	20	7		
	6	40	11	18	16		
	15	25	6	12			
	24		18				
Total	75.00	69.00	43.00	50.00	23.00	260.00	
Rep Total	103.00	96.00	71.00	82.00	52.00	404.00	

i) Block effects $r_k = \frac{1}{c}\left(\sum_{j=1}^{c} c_{jk} - \bar{c}\right) = \frac{1}{3}\left(\sum_{j=1}^{3} c_{jk} - \bar{c}\right)$, where

$$\bar{c} = \frac{1}{r}\sum_{j=1}^{c}\sum_{k=1}^{r} c_{jk} = \bar{c} = \frac{1}{5}\sum_{j=1}^{3}\sum_{k=1}^{5} c_{jk} = \frac{1}{5}\times 144.0 = 28.80$$

$$r_1 = \frac{1}{c}\left(\sum_{j=1}^{c} c_{j1} - \bar{c}\right) = r_1 = \frac{1}{3}\left(\sum_{j=1}^{3} c_{j1} - \bar{c}\right) = \frac{1}{3}(28.0 - 28.8) = -0.267$$

$$r_2 = \frac{1}{c}\left(\sum_{j=1}^{3} c_{j2} - \bar{c}\right) = \frac{1}{3}\left(\sum_{j=1}^{3} c_{j2} - \bar{c}\right) = \frac{1}{3}(27.0 - 28.8) = -0.600$$

$$r_3 = \frac{1}{c}\left(\sum_{j=1}^{3} c_{j3} - \bar{c}\right) = \frac{1}{3}\left(\sum_{j=1}^{3} c_{j3} - \bar{c}\right) = \frac{1}{3}(28.0 - 28.8) = -0.267$$

$$r_4 = \frac{1}{c}\left(\sum_{j=1}^{3} c_{j4} - \bar{c}\right) = \frac{1}{3}\left(\sum_{j=1}^{3} c_{j4} - \bar{c}\right) = \frac{1}{3}(32.0 - 28.8) = 1.067$$

$$r_5 = \frac{1}{c}\left(\sum_{j=1}^{3} c_{j5} - \bar{c}\right) = \frac{1}{3}\left(\sum_{j=1}^{3} c_{j5} - \bar{c}\right) = \frac{1}{3}(29.0 - 28.8) = 0.067$$

It may be noted that sum of the block effects must be equals to zero, i.e.,

$$\sum_{k=1}^{r} r_k = -0.267 - 0.600 - 0.267 + 1.067 + 0.067 = 0$$

ii) Mean effect (m) = $\frac{1}{e}\left[G - (r-1)\bar{c} - \sum_{k=1}^{r} n_k r_k\right]$

$$= \frac{1}{19}\left[404.0 - (5-1)28.8 - \{4\times(-0.267) + 3\times(-0.60) + 4\times(-0.267) + 3\times(1.067) + 2\times\times(0.067)\}\right]$$

$$= \frac{1}{19}[404.0 - 115.8 - 0.061] = 15.1367$$

Check effects (Cj) = $\bar{c} - m, (j = 1,2,3)$. Thus we have

$$c_1 = \overline{c_1} - m = 8.0 - 15.1367 = -7.1367$$

$$c_2 = \overline{c_2} - m = 4.8 - 15.1367 = -10.3367$$

$$c_3 = \overline{c_3} - m = 16.0 - 15.1367 = 0.8633$$

iii) Adjustment for genotypic responses $(g_i') = g_i - r_{ik}$, where r_{ik} is the response of the ith test genotype and r_{ik} is the block effect of the block in which i[th] genotype occurs.

Corresponding effects of the ith test genotype is obtained by subtracting the grand mean (m) from the above-adjusted effect of genotype, (g_i') i.e., $(g_i'') = g_i' - m$

Genotypes	Genotypic response (r_{ik})	Block (k)	Block effect (r_{ik})	Adjusted response $(g_i') = g_i - r_{ik}$	Adjusted genotype effects $(g_i'') = g_i' - m$
g1	7	5	0.067	6.933	–8.204
g2	6	1	–0.267	6.267	–8.870
g3	4	2	–0.600	4.600	–10.537
g4	15	1	–0.267	15.267	0.130
g5	20	4	1.067	18.933	3.769
g6	15	2	–0.600	25.600	10.463
g7	11	3	–0.267	11.267	–3.870
g8	18	4	1.067	16.933	1.796
g9	24	1	–0.267	24.267	9.130
g10	16	5	0.067	15.933	0.796
g11	18	3	–0.267	18.267	3.130
g12	6	3	–0.267	6.267	–8.870
g13	40	2	–0.600	40.600	25.463
g14	12	4	1.067	10.933	–4.204
g15	8	3	–0.267	8.267	–6.870
g16	30	1	–0.267	30.267	15.130

Grand total (G) = $\sum_{i=1}^{n} g_i + \sum_{k=1}^{r}\sum_{j=1}^{c} c_{kj} = 260 + 144 = 404$, where is the value corresponding to i[th] (i = 1,2,…,16) genotype and c_{kj} is the value corresponding to j[th] (j = 1,2,3) check and k[th](k = 1,2,…,5) replicate.

$$\text{Correction factor (CF)} = \frac{G^2}{N} = \frac{404^2}{31} = 5265.032$$

$$\text{Total sum of Square (TSS)} = \sum_{i=1}^{n} g_i^{\ 2} + \sum_{k=1}^{r}\sum_{j=1}^{c} c_{kj}^{\ 2} - CF$$

$$= \sum_{i=1}^{16} g_i^{\ 2} + \sum_{k=1}^{5}\sum_{j=1}^{3} c_{kj}^{\ 2} - CF$$

$$= (7^2 + 6^2 + ... + 8^2 + 30^2) + \left(10^2 + 8^2 + ... + 17^2 + 18^2\right) - 5265.032$$

$$= 2172.968.$$

Sum of square due to entries (eSS)

$$= (m \times G) + \sum_{k=1}^{r} r_k R_k + \sum_{j=1}^{c} c_j \left(\sum_{k=1}^{r} c_{jk}\right) + \sum_{i=1}^{n} g_i g_i^{\ n} - \sum_{k=1}^{r} \frac{R_k^{\ 2}}{c + v_k}$$

$$= [15.1367 \times 404] + \left[103 \times (-0.267) + 96 \times (-0.600) + 71(-0.267) + 82 \times 1.067 + 52 \times 0.067\right]$$

$$+ \left[40 \times (-7.1367) + 24 \times (-10.3367) + 80 \times (0.8633)\right]$$

$$+ \left[7 \times (-8.204) + 6 \times (-8.870) + 4 \times (-10.537) + ... + 30 \times (15.130)\right]$$

$$- \left[\frac{1}{(4+3)}\left(103^2\right) + \frac{1}{(3+3)}(96^2) + \frac{1}{(4+3)}(71^2) + \frac{1}{(3+3)}\left(82^2\right) + \frac{1}{(2+3)}(52^2)\right]$$

$$= 6115.227 + (-13.080) + (-464.485) + 1778.473 - 5433.038$$

$$= 1983.097$$

$$\text{Correction factor for checks (CFc)} = \frac{1}{c \times r}\left(\sum_{k=1}^{r}\sum_{j=1}^{c} c_{kj}\right)^2 = \frac{1}{3 \times 5}\left(144^2\right) = 1382.4$$

$$\text{Sum of squares due to checks (cSS)} = \frac{1}{r}\sum_{j=1}^{c}\left(\sum_{k=1}^{r} c_{kj}\right)^2 - CFc = \frac{1}{5}\left(40^2 + 24^2 + 80^2\right) - 1382.4 = 332.8$$

$$\text{Correction factor for genotypes (CFg)} = \frac{1}{n}\left(\sum_{i=1}^{n} g_i\right)^2 = \frac{1}{16} \times 260^2 = 4225.0$$

$$\text{Sum of squares due to genotypes (gSS)} = \sum_{i=1}^{n} g_i^{\ 2} - CFg$$

$$= \left(7^2 + 6^2 + 4^2 + ... + 8^2 + 30^2\right) - 4225.0$$

$$= 5696.0 - 4225.0$$

$$= 1471.0$$

Sum of squares for checks vs. genotypes (cgSS) = eSS-cSS-gSS

= 1983.097–332.8–1471.0

= 179.297

Error sum of squares (ErSS) = TSS-BSS-Ess
= 2172.968–168.006–1983.097
= 21.865

sv	*df*	*SS*	*MSS*	*Cal. F*	*Tab. F*
Blocks	4	168.006	42	15.37	3.84
Entries (e)	18	1983.097	110.2	40.31	3.18
Checks (c)	2	332.8	166.4	60.89	4.46
Genotypes (g)	15	1471.0	98.07	35.88	3.22
C *vs* G	1	179.297	179.3	65.6	5.32
Error	8	21.865	2.733		
Total	30	2172.968			

From above ANOVA table it is clear that the calculated F-values are greater than the corresponding tabulated F values at 5% LOS and respective degrees of freedom:

(i) There remain significant differences among the nineteen entries (check plus test genotypes).

(ii) There remain significant differences among the three checks.

(iii) There remain significant differences among the sixteen test genotypes.

So our next task will be to compare the checks among themselves, among the genotypes and among the check and the genotypes for that we need to calculate the CD values corresponding to the above comparisons.

Standard error of mean difference for:

a) Between two checks: $\sqrt{\frac{2ErMS}{r}} = \sqrt{\frac{2\times 2.733}{5}} = 1.046$

and corresponding CD (0.05) = SED x $t_{0.025,\text{error df}}$ = 1.046 x 2.306 = 2.412

b) Between any two entries in the same block:

$$\sqrt{ErMS\left(1+\frac{1}{r}\right)} = \sqrt{2.733\times\left(1+\frac{1}{5}\right)} = 1.811$$

and corresponding CD(0.05) = SEd x $t_{0.025,\text{error df}}$ = 1.811 x 2.306 = 4.176

c) Between two test genotypes: $\sqrt{2ErMS} = \sqrt{2\times 2.733} = 2.338$
and corresponding CD(0.05) = SEd x $t_{0.025,\text{error df}}$ = 2.338 x 2.306 = 5.391

d) Between means of a check and a test genotypes

$$= \sqrt{ErMS \times \left(1+\frac{1}{r}+\frac{1}{c}+\frac{1}{rc}\right)} = \sqrt{2.733 \times \left(1+\frac{1}{5}+\frac{1}{3}+\frac{1}{15}\right)} = 2.091$$

and corresponding CD(0.05) = SEd x $t_{0.025,\text{error df}}$ = 2.091 x 2.306 = 4.822

Among the checks		
	Check	**Mean**
	c3	16
	c1	8
	c2	4.8
CD (0.05)	2.412	

Among the genotypes

g13	g16	g6	g9	g5	g8	g11	g10	g4	g14	g7	g15	g1	g2	g12	g3
40	30	25	24	20	18	18	16	15	12	11	8	7	6	6	4

CD(0.05) = 5.391

Entries in the same block

	B1	B2	B3	B4	B5
Checks	10	8	6	9	7
	3	5	6	6	4
	15	14	16	17	18
Genotypes	30	4	8	20	7
	6	40	11	18	16
	15	25	6	12	
	24		18		

CD(0.05) = 4.176

Among the entries

Entry	g13	g16	g6	g9	g5	g11	g8	C3	g10	g4	g14	g7	C1	g15	g1	g2	g12	C2	g3
Mean	40	30	25	24	20	18	18	16	16	15	12	11	8	8	7	6	6	4.8	4

CD(0.05) = 4.822

Conclusion*:* From the above, we can conclude that:

i. C3 is the best check which having significantly higher response than any other checks.

ii. Among the test genotypes, g13 is the best one which having significantly higher response than any other genotypes followed by g16, g6, g9, and so on.
iii. Among the entries g16 in block 1, g13 in block 2, g11 in block 3, g5 in block 4 and g10 in block 5 are the best entries, respectively.
iv. Irrespective of this block g13 and g16 are found to be the best and the better performer comparing to the other test and check genotypes.

2.11 CONFOUNDING

When the number of factors and/or levels of the factors are increased, then the number of treatment combinations increase very rapidly, and it is not possible to accommodate all of these treatment combinations in a single homogeneous block. A new technique is therefore necessary for designing experiments with a large number of treatments. One such device is to take blocks of size less than the number of treatments and have more than one block per replication. The treatment combinations are then divided into as many groups as the number of blocks per replication. The different groups of treatments are allocated to the blocks.

This is the device of reducing the size of block taking one or more interaction contrasts identical with block contrasts are known as confounding. Generally, higher order interactions, that is, interactions with three or more factors are confounded, because their loss is immaterial. As an experimenter is particularly interested in main effects and two-factor interactions, so these should not be confounded as far as possible.

The designs for such confounded factorials can be also be called as incomplete block designs. However usual incomplete block designs for single factor experiments cannot be adopted as the contrasts of interest in two kinds of experiments are different. The treatment groups are first allocated at random to the different blocks. The treatments allotted to a block are then distributed at random to its different units.

When there are two or more replications, if the same set of interactions are confounded in all the replications, confounding is called *Complete Confounding*, and if different sets of interaction are confounded in different replications, confounding is called *Partial Confounding*. In complete confounding, all the information on confounded interactions are lost. Where in partial confounding, the confounded interactions can be recovered from those replications in which they are not confounded.

2.11.1 ADVANTAGES OF CONFOUNDING

It reduces the experimental error considerably by stratifying the experimental material into homogeneous subgroups or subsets. The removal of the variation among incomplete blocks (freed from treatments) within replicates results in smaller error mean square as compared with a RBD, thus making the comparisons among some treatment effects more precisely.

2.11.2 DISADVANTAGES OF CONFOUNDING

In confounding, the increased precision is obtained at cost of information (partial or complete) from certain relatively unimportant interactions.

The confounded contracts are replicated fewer times than other contrasts, and as such, there is loss of information on them, and they can be estimated with a lower degree of precision as the number of replications is reduced. Statistical analysis is become complex, especially, when some experimental units are missing.

2.11.3 CONFOUNDING IN 2^3 EXPERIMENT

Although 2^3 is a factorial with small number of treatment combinations but for illustration purpose, this example has been considered. Let the three factors be A, B, C each at two levels:

Factorial Effect Treatment Combination	A	B	C	AB	AC	BC	ABC
(1)	-	-	-	+	+	+	-
(a)	+	-	-	-	-	+	+
(b)	-	+	-	-	+	-	-
(ab)	+	+	-	+	-	-	-
(c)	-	-	+	+	-	-	+
(ac)	+	-	+	-	+	-	-
(bc)	-	+	+	-	-	+	-
(abc)	+	+	+	+	+	+	+

The various factorial effects are as follows:

A: (abc)+(ac)+(ab)+(a)-(bc)-(c)-(b)-(1)

B: (abc)+(bc)+(ab)+(b)-(ac)-(c)-(a)-(1)

C: (abc)+(ac)+(bc)+(c)-(ab)-(a)-(b)-(1)
AB: (abc)+(c)+(ab)+(1)-(bc)-(ac)-(b)-(a)
AC: (abc)+(ac)+(ab)+(a)-(bc)-(c)-(b)-(1)
BC: (abc)+(bc)+(a)+(1)-(ac)-(c)-(ab)-(b)
ABC: (abc)+(a)+(b)+(c)-(bc)-(ca)-(ab)-(1)

Let the highest order interaction "ABC" be confounded and we decide to use blocks of four units (plots) each per replicate.

Thus in order to confound the interaction, ABC with blocks all the treatment combinations with positive sign are allocated at random in one block and those with negative signs in the other block. Thus the following arrangement gives ABC confounded with blocks, and hence we drop information on ABC.

	Replication-I			
Block–1	(1)	(ab)	(ac)	(bc)
Block–2	(a)	(b)	(c)	(abc)

It can be observed that contrast estimating ABC is identical to the contrast estimating block effects. The other six factorial effects viz. A, B, C, AB, AC, and BC each contain two treatments in block-1 (or 2) with the positive signs and two with negative signs so that they are orthogonal with block totals, and hence these differences are not influenced among blocks and can thus be estimated and tested as usual without any difficulty. Whereas for confounded interaction, all the treatments in one group are with positive sign and in the other with negative signs.

Similarly, if AB is to be confounded, then the two blocks will consists of:

Block 1 (abc) (c) (ab) (1)

Block 2 (bc) (ac) (b) (a)

Here AB is confounded with block effects and cannot be estimated independently whereas all other effects A, B, C, AC, BC, and ABC can be estimated separately.

2.11.4 COMPLETE CONFOUNDING

If the same interaction "ABC" is confounded in all the other replications, then the interaction is said to be completely confounded.

2.11.5 PARTIAL CONFOUNDING

When an interaction is confounded in one replicate and not in another, the experiment is said to be partially confounded. Consider again 2^3 experiment with each replicate divided into two blocks of four units each. It is not necessary to confound the same interaction in all the replicates, and several factorial effects may be confounded in one single experiment.

Suppose, the following plan confounds the interaction ABC, AB, BC, and AC in replications I, II, III, and IV, respectively.

Rep-I		Rep-II		Rep-III		Rep-IV	
Block–1	**Block–2**	**Block–3**	**Block–4**	**Block–5**	**Block–6**	**Block–7**	**Block–8**
(abc)	(ab)	(abc)	(ac)	(abc)	(ab)	(abc)	(ab)
(a)	(ac)	(c)	(bc)	(bc)	(ac)	(ac)	(bc)
(b)	(bc)	(ab)	(a)	(a)	(b)	(b)	(a)
(c)	(1)	(1)	(b)	(1)	(c)	(1)	(c)

In the above arrangement, the main effects A, B, and C are orthogonal with block totals and are entirely free from the block effects. The interaction ABC is completely confounded with blocks in the replicate I, but in the other three replications the interaction ABC is orthogonal with blocks, and consequently, an estimate of ABC may be obtained from replicates II, III, and IV. Similarly, it is possible to recover information on the other confounded interactions namely AB (from I, III, IV), BC (from I, II, IV), and AC (from I, II, III). Since the partially confounded interactions are estimated from only a portion of the observations, they are determined with a lower degree of precision than the other effects.

KEYWORDS

- **binomial distribution**
- **completely randomized design**
- **Latin square design**
- **level of significance**
- **randomized completely block design**
- **total sum of squares**

REFERENCES

Chakrabarti, M. C., (1962). *Mathematics of Design and Analysis of Experiments*. Asia Publ. House.
Cochran, W. G., & Cox, D. R., (1957). *Experimental Designs* (2[nd]edn.). John Wiley.
Das, M. *N*., &Giri, *N*. C., (2011). *Design and Analysis of Experiments* (2[nd]edn.). New Age International.
Dean, A. M., & Voss, D., (1999). *Design and Analysis of Experiments*. Springer.
Design Resources Server, Indian Agricultural Statistics Research Institute (ICAR), New Delhi–110012, India. www.iasri.res.in/design.
Dey, A., (1986). *Theory of Block Designs*. Wiley Eastern.
John, J. A., & Quenouille, M. H., (1977).*Experiments: Design and Analysis*. Charles & Griffin.
Kempthorne, O., (1976). *Design and Analysis of Experiments*. John Wiley.
Montgomery, D. C., (2005). *Design and Analysis of Experiments*. John Wiley.
Raghavarao, D., (1971). *Construction and Combinatorial Problems in Design of Experiments*. John Wiley.
Sahu, P. K., & Das, A. K., (2014). *Agriculture and Applied Statistics-II* (2[nd]edn.). Kalyani Publishers.

CHAPTER 3

The Axioms of Statistical Inference: Estimation and Hypothesis Testing

MUKTI KHETAN[1], PRASHANT VERMA[2], SHWETA DIXIT[3], and F. HOMA[4]

[1]*Assistant Professor, Department of Statistics, Sambalpur University, Sambalpur, Odisha – 768019, India*

[2]*Senior Research Scholar, Department of Statistics, Banaras Hindu University, Varanasi, Uttar Pradesh – 221005, India*

[3]*Research Assistant Professor, School of Management, SRM Institute of Science and Technology, Chennai, Tamil Nadu – 603203, India*

[4]*Assistant Professor-cum-Scientist, Department of Statistics, Mathematics and Computer Application, Bihar Agricultural University, Sabour, Bhagalpur, Bihar – 813210, India*

3.1 INTRODUCTION

Statistical inference is basically drawing of valid statistical conclusions about the population characteristics on the basis of a sample drawn from the population in a systematic manner. Two important problems in statistical inference are: (i) estimation and (ii) testing of hypothesis.

In many situations, one may be interested in the value of an unknown population characteristic, which is known as estimation, whereas the testing of hypothesis is one which helps to collect a representative sample and examine it to see if our hypothesis, which is a statement about population holds true.

The theory of estimation was developed by Prof. R.A. Fisher in 1930 and is classified into two subclasses one is point estimation, and the other one is Interval estimation. The point estimation is used to provide an estimate of the population parameter, whereas interval estimation provides a probable range of the population parameter with a certain degree of accuracy.

3.2 POINT ESTIMATION

A point estimator is a sample statistic (any function, T_n of the random sample $x_1, x_2, \ldots, x_n$), which is used to estimate a population parameter. A particular numerical value of the point estimator is known as the estimate of the population parameter. For example, the sample mean $\bar{x}$ is a point estimate of a population mean μ and any numerical value of θ is known as the point estimate of μ. However, before using sample statistics as a point estimator, one should confirm whether the estimator demonstrates certain properties associated with good point estimators. The following are some of the properties that should be satisfied by an estimator.

(i) unbiasedness;
(ii) consistency;
(iii) efficiency; and
(iv) sufficiency.

3.2.1 *UNBIASEDNESS*

If the expected value of the estimator is equal to the population parameter being estimated, the estimator is said to be an unbiased estimator of the population parameter; otherwise, it is called a biased estimator. An estimator $T_n (x_1, x_{2,\ldots,} x_n)$ is said to be an unbiased estimator of population parameter θ if

$$E(T_n) = \theta$$

If $E(T_n) \neq \theta$, T_n is said to be a biased estimator.

If $E(T_n) > \theta$, T_n is said to be a positively biased estimator.

If $E(T_n) < \theta$, T_n is said to be a negatively biased estimator.

Here, the biasedness of the estimator can be explained by $b(\theta) = E(T_n) - \theta$, where b(θ) is the bias of the estimator.

3.2.1.1 *SOME NUMERICAL EXERCISE BASED ON UNBIASEDNESS*

Example: If *X* has a binomial distribution with parameter *n* and *p*, show that the sample proportion, $\frac{X}{n}$ is an unbiased estimator of *p*.

Solution:

Since $X \sim B(n, \theta)$
Therefore $E(X) = np$

$$E\left(\frac{X}{n}\right) = \frac{1}{n}E(X) = \frac{1}{n}\,np = p$$

And hence that $\frac{X}{n}$ is an unbiased estimator of p.

Example: If S^2 is the variance of a random sample from an infinite population with the finite variance σ^2, then $E(S^2) = \sigma^2$.

Solution:

$$E(S^2) = E\left[\frac{1}{n-1}\sum_{i=1}^{n}(X_i - \bar{X})^2\right]$$

$$= \frac{1}{n-1}E\left[\sum_{i=1}^{n}\{(X_i - \mu) - (\bar{X} - \mu)\}^2\right]$$

$$= \frac{1}{n-1}\left[\sum_{i=1}^{n}E\{(X_i - \mu)^2\} - nE\{(\bar{X} - \mu)^2\}\right]$$

$$= \frac{1}{n-1}\left[\sum_{i=1}^{n}\sigma^2 - n\frac{\sigma^2}{n}\right]\left[as\ E\{(X_i - \mu)^2\} = \sigma^2\ and\ E\{(\bar{X} - \mu)^2\} = \frac{\sigma^2}{n}\right]$$

$$E(S^2) = \frac{1}{n-1}\left[n\sigma^2 - n\frac{\sigma^2}{n}\right] = \sigma^2$$

Example: Suppose $x_1, x_2, \ldots, x_n$ is a random sample from a normal population $T = \frac{1}{n}\sum_{i=1}^{n}X_i^2$. Show that $T = \frac{1}{n}\sum_{i=1}^{n}X_i^2$ is an unbiased estimator of $\mu^2 + 1$.

Solution:

$$E(T) = E\left\{\frac{1}{n}\sum_{i=1}^{n} X_i^2\right\}$$

$$= \frac{1}{n}\sum_{i=1}^{n} E\left(X_i^2\right)$$

$$= \frac{1}{n}\sum_{i=1}^{n}\left[V(X) + \{E(X)\}^2\right]$$

$$= \frac{1}{n}\sum_{i=1}^{n}\left[1 + \mu^2\right]$$

$$= \frac{1}{n} n\left[1 + \mu^2\right]$$

$$E(T) = 1 + \mu^2$$

It is proved that $T = \frac{1}{n}\sum_{i=1}^{n} X_i^2$ is an unbiased estimator of $\mu^2 + 1$.

Example: Show that $\frac{\sum_{i=1}^{n} x_i\left(\sum_{i=1}^{n} x_i - 1\right)}{n(n-1)}$ is an unbiased estimator of θ^2, for the sample $x_1, x_2, \ldots, x_n$ drawn on X which takes the value 1 or 0 with respective probabilities è or $(1-$è$)$.

Solution:

Given that

$$P(X = x) = \begin{cases} 1-\theta & x = 0 \\ \theta & x = 1 \end{cases}$$

$$i.e.\ X \sim Bernoulli\ (1, \theta)$$

$$Then,\ \ T = \sum_{i=1}^{n} X_i \sim Binomial\ (n, \theta)$$

$$\rightarrow E(T) = n\theta$$

We have to prove that.

$$E\left[\frac{\sum_{i=1}^{n} x_i\left(\sum_{i=1}^{n} x_i - 1\right)}{n(n-1)}\right] = \theta^2$$

Now taking the left-hand side,

$$E\left[\frac{\sum_{i=1}^{n} x_i\left(\sum_{i=1}^{n} x_i - 1\right)}{n(n-1)}\right] = E\left[\frac{T(T-1)}{n(n-1)}\right]$$

$$= E\left[\frac{T^2 - T}{n(n-1)}\right]$$

$$= \left[\frac{E(T^2) - E(T)}{n(n-1)}\right]$$

$$= \left[\frac{\left\{V(T) - [E(T)]^2\right\} - E(T)}{n(n-1)}\right]$$

$$= \left[\frac{\left\{n\theta(1-\theta) - n^2\theta^2\right\} - n\theta}{n(n-1)}\right]$$

$$= \left[\frac{\left\{n\theta(1-\theta) - n^2\theta^2\right\} - n\theta}{n(n-1)}\right]$$

$$= \left[\frac{n\theta - n\theta^2 - n^2\theta^2 - n\theta}{n(n-1)}\right] = \frac{n(n-1)\theta^2}{n(n-1)} = \theta^2$$

3.2.2 CONSISTENCY

The next feature allied with good point estimators is consistency. A point estimator is quoted as a consistent estimator if the values of the point estimator tend to become adjacent to the population parameter as the sample size becomes substantially large. An estimator $T_n = T(x_1, x_{2,\ldots,} x_n)$ is said to be a consistent estimator of θ, if T_n converges to θ in probability, i.e., $T_n \xrightarrow{p} \theta$ as $n \to \infty$. In other words, T_n is a consistent estimator of θ if $\forall \varepsilon > 0, \eta > 0$, $\exists$ a positive integer $n \geq m\ (\varepsilon, \eta)$ such that

$$P\left[\left|T_n - \theta\right| < \varepsilon\right] \to 1 \qquad \text{as n} \to \infty$$

$$\Rightarrow P\left[\left|T_n - \theta\right| < \varepsilon\right] > 1 - \eta; \quad \forall \text{ n} \geq m$$

where m is some large value of *n*.

3.2.2.1 THEOREM 1: INVARIANCE PROPERTY OF CONSISTENT ESTIMATORS

If T_n is a consistent estimator of θ and $\psi\{\theta\}$ is a continuous function of θ, then $\psi(T_n)$ is a consistent estimator of $\psi(\theta)$.

Proof: Since, T_n is a consistent estimator of, θ i.e., $T_n \xrightarrow{p} \theta$ as $\forall \varepsilon > 0, \eta > 0$ i.e., $\forall \varepsilon > 0, \eta > 0$ $\exists$ a positive integer $n \geq m\ (\varepsilon, \eta)$ such that

$$P\left[\left|T_n - \theta\right| < \varepsilon\right] \to 1 - \eta; \quad \forall \text{ n} \geq m$$

Since $\psi(.)$ is a continues function for every positive ε_1; however, there exists a small positive integer ε_1 such that $\left|\psi(T_n) - \psi(\theta)\right| < \varepsilon_1$ whenever $\left|T_n - \theta\right| < \varepsilon$ i.e., $\left|T_n - \theta\right| < \varepsilon \to \left|\psi(T_n) - \psi(\theta)\right| < \varepsilon_1$ (1)

For two event A and B if $A \Rightarrow B$ then $A \subseteq B \Rightarrow P(A) \leq P(B)$ (2)

Using (2) in (1)

$$\left|T_n - \theta\right| < \varepsilon \Rightarrow \left|\psi(T_n) - \psi(\theta)\right| < \varepsilon_1$$

$$\left|T_n - \theta\right| < \varepsilon \subseteq \left|\psi(T_n) - \psi(\theta)\right| < \varepsilon_1$$

$$P\left[\left|T_n-\theta\right|<\varepsilon\right] \leq \mathrm{P}\left[\left|\psi(T_n)-\psi(\theta)\right|<\varepsilon_1\right]$$

$$\mathrm{P}\left[\left|\psi(T_n)-\psi(\theta)\right|<\varepsilon_1\right] \geq P\left[\left|T_n-\theta\right|<\varepsilon\right] \quad (3)$$

Since T_n is consistent estimator i.e., $P\left[\left|T_n-\theta\right|<\varepsilon\right] \rightarrow 1-\eta; \quad \forall\, n \geq m$
From Eq. (3)

$$\mathrm{P}\left[\left|\psi(T_n)-\psi(\theta)\right|<\varepsilon_1\right] \geq P\left[\left|T_n-\theta\right|<\varepsilon\right] \rightarrow 1-\eta \quad (4)$$

So from Eq. (4), we can say that $\psi(T_n)$ is consistent for $\psi(\theta)$.

3.2.2.2 THEOREM 2: SUFFICIENT CONDITIONS FOR CONSISTENCY

Let $\{T_n\}$ be a sequence of estimators such that for all $\theta \in \Theta$

(i) $E(T_n) \rightarrow \theta$ as $n \rightarrow \infty$ (ii) $V(T_n) \rightarrow 0$ as $n \rightarrow \infty$

Then T_n is consistent estimator of θ.

Remarks

1. If T is a consistent estimator of θ then, T^2 is also a consistent estimator of θ^2.
2. A consistent estimator is not necessarily unbiased.

Some Numerical Exercise Based on Consistency

Example: Prove that in sampling from $N(\mu,\sigma^2)$, the sample mean is consistent estimator of μ.

Solution:

We have to prove that $\bar{X} = \frac{1}{n}\sum_{i=1}^{n} X_i$ is consistent estimator of μ.

For this, we have to know two things for consistency

(i) $E(\bar{X}) = \mu$ or $\lim_{n\rightarrow\infty} E(\bar{X}) = \mu$

(ii) $V(\bar{X}) = 0$ or $\lim_{n\rightarrow\infty} V(\bar{X}) = 0$

Since $X \sim N(\mu,\sigma^2)$ then $\bar{X} \sim N(\mu,\frac{\sigma^2}{n})$

$$E(\bar{X}) = \mu \quad \text{or} \quad (5)$$

and $V\left(\bar{X}\right)=\frac{\sigma^2}{n}$

$$\lim_{n\to\infty} V\left(\bar{X}\right)=0 \tag{6}$$

From equation (5) and (6) implies that $\bar{X}$ is consistent estimator of μ.

Example: If X_1, X_2,…,X_n are random samples on a Bernoulli variate X taking the value 1 with probability p and value 0 with probability of p(1–p). Show that $\frac{1}{n}\sum_{i=1}^{n} X_i\left(1-\frac{1}{n}\sum_{i=1}^{n} X_i\right)$ is consistent estimator of p(1–p).

Solution:

$$X \sim Bernoulli(p) \text{ or } X \sim B(p)$$

Then $T=\sum_{i=1}^{n} X_i \sim B(n,p)$

$$E(T)=E(\sum_{i=1}^{n} X_i)=np$$

$$\Rightarrow E\left(\bar{X}\right)=E(\frac{1}{n}\sum_{i=1}^{n} X_i)=\frac{1}{n}E(\sum_{i=1}^{n} X_i)=\frac{1}{n}np=p \tag{7}$$

and

$$V(T)=V(\sum_{i=1}^{n} X_i)=np(1-p)$$

$$\Rightarrow V\left(\bar{X}\right)=V(\frac{1}{n}\sum_{i=1}^{n} X_i)=\frac{1}{n^2}V(\sum_{i=1}^{n} X_i)=\frac{1}{n^2}np(1-p)=\frac{p(1-p)}{n}$$

$$\lim_{n\to\infty} V(\bar{X})=0 \tag{8}$$

Equations (7) and (8) implies that $\bar{X}$ is consistent estimator for p.
We have to prove that $\bar{X}(1-\bar{X})$ is consistent estimator for p(1 – p)
Using Remark 1,
If $\bar{X}$ is consistent estimator for p,
Then $\bar{X}(\bar{X}-1)$ is consistent estimator for p(1 – p).

Example: For Poisson distribution with parameter θ, show that $\frac{1}{\bar{X}}$ is consistent estimator for $\frac{1}{\theta}$.

Solution:

$X \sim P(\theta)$

$$\Rightarrow E\left(\bar{X}\right) = E\left(\frac{1}{n}\sum_{i=1}^{n} X_i\right) = \frac{1}{n}\left[\sum_{i=1}^{n} E\left(X_i\right)\right] = \frac{1}{n} n\theta = \theta \tag{9}$$

and

$$V\left(\bar{X}\right) = V\left(\frac{1}{n}\sum_{i=1}^{n} X_i\right) = \frac{1}{n^2}\left[\sum_{i=1}^{n} V\left(X_i\right)\right] = \frac{1}{n^2} n\theta = \theta$$

$$\lim_{n\to\infty} V(\bar{X}) = 0 \tag{10}$$

Equations (9) and (10) implies that θ is a consistent estimator for θ.
Using Remark 1,
If θ is consistent estimator for θ
Then $\frac{1}{\bar{X}}$ is consistent estimator of $\frac{1}{\theta}$.

Example: Suppose $X_1, X_2, \ldots, X_n$ are sample values independently drawn from population with mean μ and variance σ^2. Consider the estimator $Y_n = \frac{X_1 + X_2 + \ldots + X_n}{n+1}$. Discuss whether they are unbiased and consistent for μ.

Solution:

Given that $X_i \sim N(\mu, \sigma^2)$ i=1,2,...,5

$E(X_i) = \mu$ i=1,2,...,n

$V(X_i) = \sigma^2$ i=1,2,...,n

$$Y_n = \frac{1}{n+1}\sum_{i=1}^{n}(X_i)$$

$$E(Y_n) = E\left[\frac{1}{n+1}\sum_{i=1}^{n}(X_i)\right] = \frac{1}{n+1}\sum_{i=1}^{n} E(X_i) = \frac{1}{n+1}\sum_{i=1}^{n}\mu = \frac{n}{n+1}\mu$$

$$E(Y_n) = \frac{1}{1+\frac{1}{n}}\mu$$

$$\lim_{n\to\infty} E(Y_n) = \mu \qquad (11)$$

Equation (11) implies that for large n, Y_n is an unbiased estimator for μ.

$$V(Y_n) = V\left[\frac{1}{n+1}\sum_{i=1}^{n}(X_i)\right] = \frac{1}{(n+1)^2}\sum_{i=1}^{n}V(X_i) = \frac{1}{n+1}\sum_{i=1}^{n}\mu = \frac{n}{(n+1)^2}\sigma^2$$

$$V(Y_n) = \frac{n}{n^2+1+2n}\sigma^2$$

$$\lim_{n\to\infty} V(Y_n) = 0$$

$$\lim_{n\to\infty} V(Y_n) = 0 \qquad (12)$$

Equations (11) and (12) implies that Y_n is a consistent estimator for μ.

3.2.3 EFFICIENCY

Suppose a simple random sample containing n elements is used to yield two unbiased, consistent point estimators of the same population parameter. In such situation, we would favor the point estimator with smaller variance. The point estimator with the smaller variance is considered to have greater relative efficiency than the other. Let T_1 and T_2 two consistent estimators with variances Var(T_1) and Var(T_2) respectively. If we have Var (T_1) < Var (T_2), then T_1 is more efficient than $T_2$and have a greater chance of being close to the population parameter. With this, we can assert that estimator T_1 preferable over the estimator T_2.

If T_1 is the estimator with variance V_1 and T_2 is any other estimator with variance V_2, then the efficiency E is defined as

$$E = \frac{V_1}{V_2}$$

If E < 1 then T_1 is better estimator than T_2.
If E > 1 then T_2 is better estimator than T_1.
If E = 1 then both estimators are equally efficient.

3.2.3.1 REMARK

If we have more than 2 estimators, then the one with lesser variance will be the best estimator among all.

3.2.3.2 SOME NUMERICAL EXERCISE BASED ON EFFICIENCY

Example: A random sample X_1, X_2,…,X_5 of size 5 is drawn from a normal population with unknown mean μ. Consider the following estimators to estimates. μ

$$T_1 = \frac{X_1 + X_2 + X_3 + X_4 + X_5}{5}$$

$$T_2 = \frac{X_1 + X_2}{2} + X_3 - X_4$$

$$T_3 = \frac{2X_1 + X_2 + \lambda X_3}{3}$$

If we consider λ in such a manner that T_3 is an unbiased estimator of μ. So find value of λ and T_1 and T_2 are unbiased for μ. Find out the estimator which is best among T_1, T_2, and T_3.

Solution:

Given that $X_i \sim N(\mu, \sigma^2)$ i=1,2,...,5

$$E(X_i) = \mu$$

$$V(X_i) = \sigma^2$$

$$E(T_1) = E\left[\frac{1}{5}\sum_{i=1}^{5}(X_i)\right] = \frac{1}{5}\sum_{i=1}^{5}E(X_i) = \frac{1}{5}\sum_{i=1}^{5}\mu = \frac{1}{5}5\mu = \mu$$

$$E(T_1) = \mu \tag{13}$$

Equation (13) implies that T_1 is an unbiased estimator for μ.

$$E(T_2) = E\left[\frac{X_1 - X_2}{2} + X_3 - X_4\right] = \left[\frac{E(X_1) - E(X_2)}{2} + E(X_3) - E(X_4)\right] = \left[\frac{\mu - \mu}{2} + \mu - \mu\right] = \mu$$

$$E(T_2) = \mu \tag{14}$$

Equation (14) implies that T_2 is an unbiased estimator for μ.
Given that T_3 is an unbiased estimator for μ
i.e., $E(T_3) = \mu$

$$E(T_3) = \frac{2E(X_1) + E(X_2) + \lambda E(X_3)}{3} = \mu$$

$$\Rightarrow \frac{2\mu + \mu + \lambda\mu}{3} = \mu$$

$$\Rightarrow \lambda = 0$$

So for value of $\lambda = 0$, T_3 is an unbiased estimator for μ.

$$V(T_1) = V\left[\frac{1}{5}\sum_{i=1}^{5}(X_i)\right] = \frac{1}{5^2}\sum_{i=1}^{5}V(X_i) = \frac{1}{25}\sum_{i=1}^{5}\sigma^2 = \frac{1}{25}5\sigma^2 = \frac{1}{5}\sigma^2 = 0.02\sigma^2$$

$$V(T_2) = V\left[\frac{X_1 + X_2}{2} + X_3 - X_4\right] = \left[\frac{V(X_1) + V(X_2)}{2^2} + V(X_3) + V(X_4)\right]$$

$$= \left[\frac{\sigma^2 + \sigma^2}{4} + \sigma^2 + \sigma^2\right] = \frac{5}{2}\sigma^2 = 2.5\sigma^2$$

$$V(T_3) = V\left[\frac{2X_1 + X_2 + \lambda X_3}{3}\right] = \left[\frac{2^2V(X_1) + V(X_2) + 0V(X_3)}{3^2}\right] = \left[\frac{4\sigma^2 + \sigma^2}{9}\right] = \frac{5}{9}\sigma^2 = 0.55\sigma^2$$

Among all variance, T_1 have least variance. So, T_1 is best estimator among T_1, T_2, and T_3.

Example: If X_1, X_2, X_3 is a random sample of size 3 from the population with mean σ^2 and variance σ^2. Define

$$T_1 = X_1 + X_2 - X_3$$
$$T_2 = 2X_1 + 3X_3 - 4X_2$$
$$T_3 = \frac{\lambda X_1 + X_2 + X_3}{3}$$

Find the value of λ if T_3 is an unbiased estimator of λ and which is the best among T_1, T_2, and T_3.

Solution:

Given that $X_i \sim N(\mu,\sigma^2)$ i=1,2,...,5

$$E(X_i)=\mu \quad \text{i=1,2,3}$$

$$V(X_i)=\sigma^2 \quad \text{i=1,2,3}$$

$$E(T_1)=E(X_1+X_2-X_3)=E(X_1)+E(X_2)-E(X_3)=\mu+\mu-\mu=\mu \tag{15}$$

Equation (15) implies that T_1 is an unbiased estimator for μ.

$$E(T_2)=E[2X_1+3X_3-4X_2]=2E(X_1)+3E(X_3)-4E(X_2)$$

$$E(T_2)=2\mu+3\mu-4\mu$$

$$E(T_2)=\mu \tag{16}$$

Equation (16) implies that T_2 is an unbiased estimator for μ.
Given that T_3 is an unbiased estimator for μ

i.e., $E(T_3)=\mu a$ (17)

$$E(T_3)=\frac{\lambda E(X_1)+E(X_2)+E(X_3)}{3}=\mu$$

$$\Rightarrow \frac{\lambda\mu+\mu+\mu}{3}=\mu$$

$$\Rightarrow \lambda=1$$

So for the value of $\lambda=1$, T_3 is an unbiased estimator for μ.

$$V(T_2)=V[2X_1+3X_3-4X_2]=4V(X_1)+9V(X_3)+16V(X_2)=17\sigma^2 \tag{18}$$

$$V(T_2)=V[2X_1+3X_3-4X_2]=4V(X_1)+9V(X_3)+16V(X_2)=17\sigma^2 \tag{19}$$

$$V(T_3)=V\left[\frac{X_1+X_3+X_2}{3}\right]=\frac{V(X_1)+V(X_3)+VE(X_2)}{9}=\frac{1}{3}\sigma^2 \tag{20}$$

Among all variance, T_3 have the least variance. So, T_3 is best estimator among T_1, T_2, and T_3.

Example: If X_1, X_2,…,X_n and X_4 be a random sample of a normal distribution $N(\mu,\sigma^2)$. Find the efficiency of $T=\frac{1}{7}[X_1+3X_2+2X_3+X_4]$ relative to $\bar{X}=\frac{1}{4}\left[\sum_{i=1}^{4}X_i\right]$.

Solution:

$$\text{Given that } X_i \sim N(\mu,\sigma^2) \qquad \text{i=1,2,...,4}$$

$$V(X_i)=\sigma^2 \qquad \text{i=1,2,3,4}$$

$$V(T)=V\left\{\frac{1}{7}[X_1+3X_2+2X_3+X_4]\right\}$$

$$V(T)=\frac{1}{49}[V(X_1)+9V(X_2)+4V(X_3)+V(X_4)]=\frac{15\sigma^2}{49} \qquad (21)$$

$$V(\bar{X})=V\left\{\frac{1}{4}\left[\sum_{i=1}^{4}X_i\right]\right\}=\frac{1}{16}\left[\sum_{i=1}^{4}V(X_i)\right]=\frac{1}{16}\left[\sum_{i=1}^{4}\sigma^2\right]=\frac{4}{16}\sigma^2=\frac{1}{4}\sigma^2 \qquad (22)$$

$$E=\frac{V(T)}{V(\bar{X})}=\frac{15}{49}\sigma^2\frac{4}{\sigma^2}=\frac{60}{49}$$

$$E=\frac{60}{49}>1$$

$$\Rightarrow V(T)>V(\bar{X}) \qquad (23)$$

Equation (23) implies that $\bar{X}$ is more efficient than T.

Example: Say X_1, …,X_n are sample values independently drawn from population mean μ and variance σ^2. Consider

$$T_1=\frac{1}{n+1}[X_1+X_2+...+X_n]$$

and

$$T_2=\frac{1}{n^2}[X_1+2X_2+3X_3...+nX_n].$$

Solution:

$$\text{Given that } X_i \sim N(\mu,\sigma^2) \qquad \text{i=1,2,...,n}$$

$$V(X_i)=\sigma^2 \qquad \text{i=1,2,...,n}$$

$$V(T_1)=V\left\{\frac{1}{n+1}[X_1+X_2+...+X_n]\right\}=\frac{1}{(n+1)^2}\{V(X_1)+V(X_2)+...+V(X_n)\}$$

$$V(T_1)=\frac{1}{(n+1)^2}\{\sigma^2+\sigma^2+...+\sigma^2\}=\frac{n\sigma^2}{(n+1)^2}$$

$$T_2=\frac{1}{n^2}\left[\sum_{i=1}^{n} iX_i\right]$$

$$V(T_2)=V\left\{\frac{1}{n^2}\left[\sum_{i=1}^{n} iX_i\right]\right\}=\frac{1}{n^4}\left[\sum_{i=1}^{n} i^2V(X_i)\right]=\frac{\sigma^2}{n^4}\left[\sum_{i=1}^{n} i^2\sigma^2\right]$$

$$E=\frac{V(T_1)}{V(T_2)}=\frac{n\sigma^2}{(n+1)^2}\frac{n^4}{\sigma^2 n(n+1)(2n+1)}=\frac{6n^4}{(n+1)^2(2n+1)}$$

$$E=\frac{V(T_1)}{V(T_2)}=\frac{n\sigma^2}{(n+1)^2}\frac{n^4}{\sigma^2 n(n+1)(2n+1)}=\frac{6n^4}{(n+1)^2(2n+1)}$$

For large *n,* i.e., $n \to \infty$

$$E=\frac{6}{(1)^2(2)}=6$$

$$\Rightarrow V(T_1)>V(T_2) \tag{24}$$

Equation (24) implies that T_2 is more efficient than T_1.

3.2.4 SUFFICIENCY

The last attribute of a good point estimator is sufficiency, and as the name suggests, a point estimator is said to be sufficient estimator for population parameter, if it contains all the information in sample regarding the population parameter. If T_n is an estimator of population parameter θ based on a sample $x_1, x_2, ..., x_n$ of size n and its density function is $f(x;\theta)$ such that the conditional distribution of $x_1, x_2, ..., x_n$ given T_n, is independent of the parameter θ, then T_n is sufficient estimator for a population parameter θ.

3.2.4.1 THEOREM 3: FACTORIZATION THEOREM

The Factorization is the theorem given by Neymann fisher's which gives the necessary and sufficient condition for a distribution to have a sufficient statistic.

$T_n = t(x_1, x_2, ..., x_n)$ is sufficient for population parameter θ if and only if the joint density function $L(x;\theta)$ of n sample values $x_1, x_2, ..., x_n$ can be expressed in the form.

$$L(x;\theta) = f(x_1, x_2, ..., x_n; \theta) = h(x).g_\theta(t(x_1, x_2, ..., x_n))$$

where $g_\theta(t(x_1, x_2, ..., x_n))$ depend on θ and $x_1, x_2, ..., x_n$ only through the value of $t(x_1, x_2, ..., x_n)$ and $h(x_1, x_2, ..., x_n)$ is independent of population parameter θ.

3.2.4.2 SOME NUMERICAL EXERCISE BASED ON SUFFICIENCY

Example: $X \sim Beroulli\ (1,\theta)$; find the sufficient estimator for θ.

Solution:

$$X \sim Beroulli\ (1,\theta)$$

$$f(x) = \begin{cases} \theta^x (1-\theta)^{1-x} & x=0,1 \\ 0 & \text{otherwise} \end{cases}$$

$$\prod_{i=1}^{n} f(x_i) = \prod_{i=1}^{n} \theta^x (1-\theta)^{1-x}$$

$$L(x;\theta)=\theta^{\sum_{i=1}^{n}x}(1-\theta)^{n-\sum_{i=1}^{n}x}$$

$$L(x;\theta)=1.\left(\theta^{\sum_{i=1}^{n}x}(1-\theta)^{n-\sum_{i=1}^{n}x}\right)$$

$$h(x)=1$$

$$g_\theta(t(x_1,x_2,...,x_n))=\sum_{i=1}^{n}x_i$$

$$L(x;\theta)=h(x).g_\theta(t(x_1,x_2,...,x_n))$$

So by Neymann fisher factorization theorem, $t(x_1,x_2,...,x_n)=\sum_{i=1}^{n}x_i$ is sufficient statistic for θ.

Example: $X \sim Poisson\ (\theta)$; find the sufficient estimator for θ.

Solution:

$$X \sim Poisson\ (\theta)$$

$$f(x)=\begin{cases}\frac{1}{\underline{|x}}\theta^x e^{-x\theta} & x=0,1,2,...\\ 0 & otherwise\end{cases}$$

$$L(x;\theta)=\prod_{i=1}^{n}f(x_i)=\prod_{i=1}^{n}\frac{1}{\underline{|x_i}}\theta^{x_i}e^{-x_i\theta}$$

$$L(x;\theta)=\frac{1}{\prod_{i=1}^{n}\underline{|x_i}}\theta^{\sum_{i=1}^{n}x}e^{-n\theta}$$

$$h(x)=\prod_{i=1}^{n}\underline{|x_i}$$

$$g_\theta(t(x_1,x_2,...,x_n))=\sum_{i=1}^{n}x_i$$

$$L(x;\theta)=h(x).g_\theta(t(x_1,x_2,...,x_n))$$

So by Neymann fisher factorization theorem, $t(x_1,x_2,...,x_n)=\sum_{i=1}^{n}x_i$ is sufficient statistic for θ.

Example: $X \sim Beta\ (\alpha,\beta)$; find the sufficient estimator for (α,β).

Solution:

$$X \sim Beta\ (\alpha,\beta)$$

$$f(x)=\begin{cases}\dfrac{1}{B(\alpha,\beta)}x^{\alpha-1}(1\text{-}x)^{\beta-1} & 0<x<1\\ 0 & \text{otherwise}\end{cases}$$

$$L(x;\alpha,\beta)=\prod_{i=1}^{n}f(x_i)=\prod_{i=1}^{n}\frac{1}{B(\alpha,\beta)}x_i^{\alpha-1}(1\text{-}x_i)^{\beta-1}$$

$$L(x;\alpha,\beta)=\frac{1}{\{B(\alpha,\beta)\}^n}\left(\prod_{i=1}^{n}x_i\right)^{\alpha-1}\left(\prod_{i=1}^{n}(1-x_i)\right)^{\beta-1}$$

$$L(x;\alpha,\beta)=\frac{1}{\left(\prod_{i=1}^{n}x_i\right)\left(\prod_{i=1}^{n}(1-x_i)\right)}\frac{\left(\prod_{i=1}^{n}x_i\right)^{\alpha}\left(\prod_{i=1}^{n}(1-x_i)\right)^{\beta}}{\{B(\alpha,\beta)\}^n}$$

$$h(x)=\left(\prod_{i=1}^{n}x_i\right)\left(\prod_{i=1}^{n}(1-x_i)\right)$$

$$g_{\alpha,\beta}(t(x_1,x_2,...,x_n))=\frac{\left(\prod_{i=1}^{n}x_i\right)^{\alpha}\left(\prod_{i=1}^{n}(1-x_i)\right)^{\beta}}{[B(\alpha,\beta)]^n}$$

$$L(x) = h(x).g_{\alpha,\beta}(t(x_1, x_2, ..., x_n))$$

So by Neymann fisher factorization theorem, $t(x_1, x_2, ..., x_n) = (\prod_{i=1}^{n} x_i, \prod_{i=1}^{n}(1 - x_i))$ is jointly sufficient statistic for (α, β).

Example: $X \sim Gamma\ (\alpha, \lambda)$; find the sufficient estimator for (α, λ).

Solution:

$$X \sim Gamma\ (\alpha, \lambda)$$

$$f(x) = \begin{cases} \frac{1}{\overline{|\alpha}} \lambda^{\alpha} x^{\alpha-1} e^{-\lambda x} & 0<x<\infty \\ 0 & \text{otherwise} \end{cases}$$

$$L(x; \alpha, \lambda) = \prod_{i=1}^{n} f(x_i) = \prod_{i=1}^{n} \frac{1}{\overline{|\alpha}} \lambda^{\alpha} x_i^{\alpha-1} e^{-\lambda x_i}$$

$$L(x; \alpha, \lambda) = \frac{1}{\{\overline{|\alpha}\}^n} \left(\prod_{i=1}^{n} x_i \right)^{\alpha-1} e^{-\lambda \sum_{i=1}^{n} x_i}$$

$$L(x; \alpha, \lambda) = \frac{1}{\left(\prod_{i=1}^{n} x_i \right)} \frac{\left(\prod_{i=1}^{n} x_i \right)^{\alpha} e^{-\lambda \sum_{i=1}^{n} x_i}}{\{\overline{|\alpha}\}^n}$$

$$h(x) = \left(\prod_{i=1}^{n} x_i \right)$$

$$g_{\alpha,\lambda}(t(x_1, x_2, ..., x_n)) = \frac{\left(\prod_{i=1}^{n} x_i \right)^{\alpha} e^{-\lambda \sum_{i=1}^{n} x_i}}{\{\overline{|\alpha}\}^n}$$

$$L(x;\alpha,\lambda) = h(x).(g_{\alpha,\lambda}(t(x_1, x_2,..., x_n))$$

So by Neymann fisher factorization theorem, $t(x_1, x_2,..., x_n) = (\prod_{i=1}^{n} x_i, \sum_{i=1}^{n} x_i)$ is jointly sufficient statistic for (α,λ).

3.2.4.3 REMARKS

1. Invariance Property of sufficient estimator: If T_n is a sufficient estimator for the population parameter θ and if $\psi(T_n)$ is one to one function of T_n, then $\psi(\theta)$ is sufficient for $\psi(\theta)$.
2. A sufficient estimator may or may not be an unbiased estimator of the population parameter.
3. A sufficient estimator is always a consistent estimator of the population parameter.
4. A sufficient estimator is the most efficient estimator of the population parameter.

3.3 CRAMER RAO INEQUALITY

3.3.1 THEOREM

Suppose $x_1, x_2...x_n$ is the random sample from $g(x;\theta)$ and $T(\underline{X}) = t(x_1, x_2...x_n)$ is an unbiased estimator for $\gamma(\theta)$ based on a random sample of size *n*. Cramer Rao inequality state under the following regularity condition:

(i) $\frac{\delta}{\delta\theta}\log g(x;\theta)$ exits for all $\underline{x}=(x_1, x_2,..., x_n)$ and $\theta \in \Theta$

(ii) $$\frac{\delta}{\delta\theta}\left[\int\int....\int\prod_{i=1}^{n} g(x_i;\theta)dx_1....dx_n\right] = \int\int....\int\frac{\delta}{\delta\theta}\prod_{i=1}^{n} g(x_i;\theta)dx_1....dx_n$$

(iii) $$\left[\begin{array}{l}\frac{\delta}{\delta\theta}\left[\int\int....\int t(x_1, x_2,..., x_n)\prod_{i=1}^{n} g(x_i;\theta)dx_1....dx_n\right] \\ = \int\int....\int t(x_1, x_2,..., x_n)\frac{\delta}{\delta\theta}\prod_{i=1}^{n} g(x_i;\theta)dx_1....dx_n\end{array}\right]$$

(iv) $0 < E\left(\frac{\delta}{\delta\theta}\log g(x;\theta)\right)^2 < \infty \quad \forall \theta \in \Theta$

If $T = t(x_1, x_2 ... x_n)$ is an unbiased estimator for $\gamma(\theta)$ then

$$Var(t) \geq \frac{\left[\gamma'(\theta)\right]^2}{E\left(\frac{\delta}{\delta\theta}\log g(\underline{x};\theta)\right)^2} = \frac{\left[\gamma'(\theta)\right]^2}{I(\theta)} \qquad \text{here } g(\underline{x};\theta) = \prod_{i=1}^{n} g(x_i;\theta)$$

where $\gamma'(\theta) = \frac{\delta}{\delta\theta}\gamma(\theta)$ and $I(\theta)$ is the fishers information of $\gamma(\theta)$.

"In other words, Cramer Rao Inequality provides a lower bound to the variance of the unbiased estimator of $\gamma(\theta)$."

"Sometimes Cramer Rao Inequality called Cramer Rao lower bound (CRLB)."

Proof: We have $T(\underline{X}) = T = t(x_1, x_2 ..., x_n)$ is an unbiased estimator of $\gamma(\theta)$. i.e.,

$$E\left(T(\underline{X})\right) = \gamma(\theta)$$

$$\gamma(\theta) = E\left(T(\underline{X})\right)$$

$$\gamma(\theta) = E\left(T(\underline{X})\right) = \int\int ....\int t(x_1, x_2, ..., x_n)\prod_{i=1}^{n} g(x_i;\theta)dx_1....dx_n \qquad (25)$$

Differentiate both sides with respect to θ from equation (25).

$$\frac{\delta}{\delta\theta}\gamma(\theta) = \frac{\delta}{\delta\theta}E\left(T(\underline{X})\right) = \frac{\delta}{\delta\theta}\left[\int\int ....\int t(x_1, x_2, ..., x_n)\prod_{i=1}^{n} g(x_i;\theta)dx_1....dx_n\right]$$

$$\gamma'(\theta) = \left[\int\int ....\int t(x_1, x_2, ..., x_n)\frac{\delta}{\delta\theta}\prod_{i=1}^{n} g(x_i;\theta)dx_1....dx_n\right]$$

(Using regularity condition (iii))

$$\gamma'(\theta) = \left[\int\int....\int t(x_1, x_2, ..., x_n) \frac{\frac{\delta}{\delta\theta}\prod_{i=1}^{n} g(x_i;\theta)}{\prod_{i=1}^{n} g(x_i;\theta)} \prod_{i=1}^{n} g(x_i;\theta) dx_1....dx_n \right]$$

$$= \int\int....\int \left\{ t(x_1, x_2, ..., x_n) \frac{\delta}{\delta\theta} \log \prod_{i=1}^{n} g(x_i;\theta) \right\} \prod_{i=1}^{n} g(x_i;\theta) dx_1....dx_n$$

$$\gamma'(\theta) = E\left[T(\underline{X}) \frac{\delta}{\delta\theta} \log \prod_{i=1}^{n} g(x_i;\theta) \right] \quad \text{where } T(\underline{X}) = t(x_1, x_{2,...,} x_n) \tag{26}$$

Since we have

$$\int\int....\int \prod_{i=1}^{n} g(x_i;\theta) dx_1....dx_n = 1$$

Differentiating both sides with respect to θ, we get

$$\frac{\delta}{\delta\theta} \int\int....\int \prod_{i=1}^{n} g(x_i;\theta) dx_1....dx_n = 0$$

$$\int\int....\int \frac{\delta}{\delta\theta} \prod_{i=1}^{n} g(x_i;\theta) dx_1....dx_n = 0$$

(Using regularity condition (ii))

$$\int\int....\int \frac{\frac{\delta}{\delta\theta}\prod_{i=1}^{n} g(x_i;\theta)}{\prod_{i=1}^{n} g(x_i;\theta)} \prod_{i=1}^{n} g(x_i;\theta) dx_1....dx_n = 0$$

$$\int\int....\int \left\{ \frac{\delta}{\delta\theta} \prod_{i=1}^{n} g(x_i;\theta) \right\} \prod_{i=1}^{n} g(x_i;\theta) dx_1....dx_n = 0 = 0$$

$$E\left[\frac{\delta}{\delta\theta} \log \prod_{i=1}^{n} g(x_i;\theta) \right] = 0 \tag{27}$$

Now,

$$Cov\left[T(\underline{X}), \frac{\delta}{\delta\theta}\log\prod_{i=1}^{n} g(x_i;\theta)\right] = \left\{\begin{array}{l} E\left[T(\underline{X}), \frac{\delta}{\delta\theta}\log\prod_{i=1}^{n} g(x_i;\theta)\right] \\ -E(T(\underline{X}))E\left(\frac{\delta}{\delta\theta}\log\prod_{i=1}^{n} g(x_i;\theta)\right) \end{array}\right\}$$

(Using regularity condition (iii))

$$E\left[T(\underline{X}), \frac{\delta}{\delta\theta}\log\prod_{i=1}^{n} g(x_i;\theta)\right] = E\left[T(\underline{X}), \frac{\delta}{\delta\theta}\log\prod_{i=1}^{n} g(x_i;\theta)\right] - E(T(\underline{X}))E\left(\frac{\delta}{\delta\theta}\log\prod_{i=1}^{n} g(x_i;\theta)\right)$$

$$E\left[T(\underline{X}), \frac{\delta}{\delta\theta}\log\prod_{i=1}^{n} g(x_i;\theta)\right] = E\left[T(\underline{X}), \frac{\delta}{\delta\theta}\log\prod_{i=1}^{n} g(x_i;\theta)\right]$$

(Using equation (27) $E\left(\frac{\delta}{\delta\theta}\log\prod_{i=1}^{n} g(x_i;\theta)\right) = 0$)
From Eq. (26)

$$\gamma'(\theta) = Cov\left[T(\underline{X}), \frac{\delta}{\delta\theta}\log\prod_{i=1}^{n} g(x_i;\theta)\right] \le \left[Var(T(\underline{X})), Var\left(\frac{\delta}{\delta\theta}\log\prod_{i=1}^{n} g(x_i;\theta)\right)\right]^{1/2}$$

Using result $Cov(X,Y) \le \left[Var(X), Var(Y)\right]^{1/2}$ (Chauchy Schwartz Inequality)

$$\left[\gamma'(\theta)\right]^2 \le \left[Var(T(\underline{X})), Var\left(\frac{\delta}{\delta\theta}\log\prod_{i=1}^{n} g(x_i;\theta)\right)\right] \le \left[Var(T(\underline{X})), E\left(\frac{\delta}{\delta\theta}\log\prod_{i=1}^{n} g(x_i;\theta)\right)^2\right]$$

Here,

$$Var\left(\frac{\delta}{\delta\theta}\log\prod_{i=1}^{n} g(x_i;\theta)\right) = E\left[\frac{\delta}{\delta\theta}\log\prod_{i=1}^{n} g(x_i;\theta)\right]^2 - \left[E\left(\frac{\delta}{\delta\theta}\log\prod_{i=1}^{n} g(x_i;\theta)\right)\right]^2$$

$$Var(T(\underline{X})) = Var(t) \ge \frac{\left[\gamma'(\theta)\right]^2}{E\left(\frac{\delta}{\delta\theta}\log\prod_{i=1}^{n} g(x_i;\theta)\right)^2}$$

3.3.2 REMARKS

An unbiased estimator $T(\underline{X})$ of $\gamma(\theta)$ is called a Minimum Variance Bound (MVB) estimator of the parameter θ if it acquires CRLB. If

$T(\underline{X}) = T = t(x_1, x_2 ..., x_n)$ is an unbiased estimator of parameter θ. i.e., $E(T(\underline{X})) = \gamma(\theta)$, i.e., $\gamma(\theta) = \theta \Rightarrow \gamma'(\theta) = 1$

So CRLB reduce to

$$Var(t) \geq \frac{1}{E\left(\frac{\delta}{\delta\theta}\log\prod_{i=1}^{n} g(x_i;\theta)\right)^2}$$

CRLB can be written as

$$Var(t) \geq \frac{[\gamma'(\theta)]^2}{-E\left(\frac{\delta^2}{\delta\theta^2}\log\prod_{i=1}^{n} g(x_i;\theta)\right)}$$

Condition for the equality sign in Cramer Rao Inequality:
We have

$$Var(t) \geq \frac{[\gamma'(\theta)]^2}{-E\left(\frac{\delta^2}{\delta\theta^2}\log\prod_{i=1}^{n} g(x_i;\theta)\right)}$$

$$\text{i.e., } [\gamma'(\theta)]^2 \leq Var(t)\left[-E\left(\frac{\delta^2}{\delta\theta^2}\log\prod_{i=1}^{n} g(x_i;\theta)\right)\right]$$

i.e., $$[\gamma'(\theta)]^2 \leq E(t-\gamma(\theta))^2\left[-E\left(\frac{\delta^2}{\delta\theta^2}\log\prod_{i=1}^{n} g(x_i;\theta)\right)\right] \quad (28)$$

The sign of equality will old in Cramer Rao inequality if and only if the sign of equality holds in Eq. (25). Sign of equality will hold in Eq. (25) by Chauchy Schwartz inequality, if, and only if the variables $(t-\gamma(\theta))$ and $\left(\frac{\delta}{\delta\theta}\log\prod_{i=1}^{n} g(x_i;\theta)\right)$ are proportional to each other.

A necessary and sufficient condition for an unbiased estimator t to attain the lower bound of its variance is given by

$$\frac{\delta}{\delta\theta}\log\prod_{i=1}^{n} g(x_i;\theta) = \frac{(t-\gamma(\theta))}{\lambda(\theta)} \tag{29}$$

where $\lambda(\theta)$ is a constant independent of $x_1, x_2, \ldots, x_n$ but may depend on θ. If the $\frac{\delta}{\delta\theta}\log\prod_{i=1}^{n} g(x_i;\theta)$ is expressible in the form (26), then

(i) t is an unbiased estimator of $\gamma(\theta)$.
(ii) MVB estimator (t) for $\gamma(\theta)$ exits.
(iii) $V(t) = \left|\gamma'(\theta).\lambda(\theta)\right|$.

3.3.3 SOME NUMERICAL EXERCISE BASED ON MVUE

Example: Obtain MVB estimator for μ in normal distribution $N(\mu,\sigma^2)$ where σ^2 is known.

Solution:

$$X \sim N(\mu,\sigma^2)$$

where σ^2 is known

$$f(x) = \begin{cases} \frac{1}{\sigma\sqrt{2\pi}} e^{-\frac{1}{2}\left(\frac{x-\mu}{\sigma}\right)^2} & -\infty < x < \infty, -\infty < \mu < \infty, \sigma^2 > 0 \\ 0 & \text{otherwise} \end{cases}$$

Likelihood function:

$$L(\mu) = \prod_{i=1}^{n} f(x_i) = \prod_{i=1}^{n} \frac{1}{\sigma\sqrt{2\pi}} e^{-\frac{1}{2}\left(\frac{x_i-\mu}{\sigma}\right)^2}$$

$$L(\mu) = \frac{1}{\sigma^n\left(\sqrt{2\pi}\right)^n} e^{-\frac{1}{2}\sum_{i=1}^{n}\left(\frac{x_i-\mu}{\sigma}\right)^2}$$

$$\log L(\mu) = \log\left\{\frac{1}{\sigma^n\left(\sqrt{2\pi}\right)^n} e^{-\frac{1}{2}\sum_{i=1}^{n}\left(\frac{x_i-\mu}{\sigma}\right)^2}\right\}$$

$$\log L(\mu) = \log\left(\frac{1}{\sigma\sqrt{2\pi}}\right)^n - \frac{1}{2}\sum_{i=1}^{n}\left(\frac{x_i-\mu}{\sigma}\right)^2$$

$$\log L(\mu) = n\log\left(\frac{1}{\sigma\sqrt{2\pi}}\right) - \frac{1}{2}\sum_{i=1}^{n}\left(\frac{x_i-\mu}{\sigma}\right)^2$$

$$\frac{\delta \log L(\mu)}{\delta\mu} = \frac{\delta}{\delta\mu}\left\{n\log\left(\frac{1}{\sigma\sqrt{2\pi}}\right) - \frac{1}{2}\sum_{i=1}^{n}\left(\frac{x_i-\mu}{\sigma}\right)^2\right\}$$

$$\frac{\delta \log L(\mu)}{\delta\mu} = -\frac{1}{\sigma}\sum_{i=1}^{n}\left(\frac{x_i-\mu}{\sigma}\right)$$

$$\frac{\delta \log L(\mu)}{\delta\mu} = \frac{\sum_{i=1}^{n} x_i - n\mu}{\sigma^2} = \frac{n\overline{x} - n\mu}{\sigma^2}$$

$$\frac{\delta \log L(\mu)}{\delta\mu} = \frac{\overline{x} - \mu}{\frac{\sigma^2}{n}} \tag{30}$$

Now we compare Eq. (29) with Eq. (30), we get $t = \overline{x}$, $\gamma(\mu) = \mu$ and $\lambda(\mu) = \frac{\sigma^2}{n}$.

So $\overline{x}$ is an unbiased estimator of $\overline{x}$ and variance of $\overline{x}$ is $V(\overline{x}) = \left|\gamma'(\mu).\lambda(\mu)\right| = \frac{\sigma^2}{n}$.

Example: If $X \sim Exp(\theta)$ then show that $\overline{x}$ is an MVB estimator for θ and has variance $\frac{\theta^2}{2}$.

Solution:

$$X \sim Exp(\theta)$$

$$f(x) = \begin{cases} \frac{1}{\theta} e^{-\frac{x}{\theta}} & 0 < x < \infty, \\ 0 & \text{otherwise} \end{cases}$$

Likelihood function:

$$L(\theta) = \prod_{i=1}^{n} f(x_i) = \prod_{i=1}^{n} \frac{1}{\theta} e^{-\frac{x_i}{\theta}}$$

$$L(\theta) = \frac{1}{\theta^n} e^{-\sum_{i=1}^{n} \frac{x_i}{\theta}}$$

$$\log L(\theta) = \log \left\{ \frac{1}{\theta^n} e^{-\sum_{i=1}^{n} \frac{x_i}{\theta}} \right\}$$

$$\log L(\theta) = -n \log \theta - \sum_{i=1}^{n} \frac{x_i}{\theta}$$

$$\frac{\delta \log L(\theta)}{\delta \theta} = \frac{\delta}{\delta \theta} \left\{ -n \log \theta - \sum_{i=1}^{n} \frac{x_i}{\theta} \right\}$$

$$\frac{\delta \log L(\theta)}{\delta \theta} = \frac{-n}{\theta} + \frac{1}{\theta^2} \sum_{i=1}^{n} x_i$$

$$\frac{\delta \log L(\theta)}{\delta \theta} = \frac{\sum_{i=1}^{n} x_i - n\theta}{\theta^2} = \frac{n\overline{x} - n\theta}{\theta^2}$$

$$\frac{\delta \log L(\theta)}{\delta \theta} = \frac{\overline{x} - \theta}{\frac{\theta^2}{n}} \tag{31}$$

Now we compare Eq. (29) with Eq. (31) we get $t=\bar{x}$, $\gamma(\theta)=\theta$ and $\lambda(\theta)=\frac{\theta^2}{n}$. So $\bar{x}$ is an unbiased estimator of θ and variance of $\bar{x}$ is $V(\bar{x})=\left|\gamma'(\theta).\lambda(\theta)\right|=\frac{\theta^2}{n}$.

3.4 COMPLETENESS

'Completeness' of a statistic, in the context of a given probability family, is a property of the estimator that guarantees that the unbiased estimator is unique which may be written as a function of a sufficient statistic; this estimator is then automatically the UMVUE. A statistic T is said to be complete if

$$E_\theta\left(g(T)\right)=0 \qquad \forall\theta$$

$$\Rightarrow g(T)=0 \qquad a.e.$$

Then we can say that $g(T)$ is a complete sufficient statistic for the parameter θ.

3.4.1. SOME NUMERICAL EXERCISE BASED ON COMPLETENESS

Example: If $X_1, X_2, ...X_n$ are random samples from Bernoulli distribution

$$f(x)=\begin{cases}\theta^x(1-\theta)^{1-x} & \text{x} = 0,\ 1\\ 0 & \text{otherwise}\end{cases}$$

Then show that $T=\sum_{i=1}^{n}X_i$ is complete sufficient statistic.

Solution: Likelihood function of $x_1, x_2, ...x_n$ is

$$L(\theta)=\prod_{i=1}^{n}f(x_i)=\prod_{i=1}^{n}\theta^{x_i}(1-\theta)^{1-x_i}=\underbrace{1}_{h(\underline{x})}\ \underbrace{\theta^{\sum_{i=1}^{n}x_i}(1-\theta)^{n-\sum_{i=1}^{n}x_i}}_{g_\theta(\sum_{i=1}^{n}x_i)}$$

So, by NFF, $T(\underline{X})=T=\sum_{i=1}^{n}X_i$ is sufficient statistics for θ.

Distribution of sufficient statistics $T(\underset{\sim}{X}) = T = \sum_{i=1}^{n} X_i$

Since sum of Bernoulli variate is a binomial variate, so

$$T(\underset{\sim}{X}) = \sum_{i=1}^{n} X_i \sim B(n, \theta)$$

$$P(T = t) = \begin{cases} \dfrac{n!}{t!(n-t)!} \theta^t (1-\theta)^{n-t} & \text{t =0,1,...,n} \\ 0 & \text{otherwise} \end{cases}$$

Take

$$E_\theta\left(g(T)\right) = 0$$

$$\Rightarrow \sum_{t=0}^{n} g(T) P(T = t) = 0$$

$$\Rightarrow \sum_{t=0}^{n} g(T) \frac{n!}{t!(n-t)!} \theta^t (1-\theta)^{n-t} \quad = 0$$

$$\Rightarrow \sum_{t=0}^{n} g(T) \frac{n!}{t!(n-t)!} \left(\frac{\theta}{1-\theta}\right)^t (1-\theta)^n \quad = 0$$

$$\Rightarrow \sum_{t=0}^{n} g(T) \frac{n!}{t!(n-t)!} \left(\frac{\theta}{1-\theta}\right)^t \quad = 0$$

$$\Rightarrow \sum_{t=0}^{n} g(T) \frac{n!}{t!(n-t)!} \rho^t \quad = 0 \tag{32}$$

where $\rho = \dfrac{\theta}{1-\theta}$

Equation (32) represents power series. Sum of finite power series is zero only when each coefficient of power series is zero, i.e.,

$$\Rightarrow g(T)\frac{n!}{t!(n-t)!}=0 \qquad \text{t=0,1...n}$$

$$\Rightarrow g(T)=0 \qquad \text{t=0,1...n}$$

$$T(\underline{X})=T=\sum^{n} X_i \quad \text{i.e.,}$$

So $T(\underline{X})=T=\sum_{i=1}^{n} X_i$ is complete sufficient statistic for θ.

3.5 UNIFORM MINIMUM VARIANCE UNBIASED ESTIMATOR (UMVUE)

An estimator $T(\underline{X})$ is said to be uniformly minimum variance unbiased estimator (UMVUE) for $\gamma(\theta)$ if

(i) $E\left(T(\underline{X})\right)=\gamma(\theta) \quad \forall\theta$

(ii) $V\left(T(\underline{X})\right)\leq V\left(T^{'}(\underline{X})\right) \quad \forall\theta$

where $T^{'}(\underline{X})$ is any other estimator such that $E\left(T^{'}(\underline{X})\right)=\gamma(\theta)$.

3.5.1. SOME NUMERICAL EXERCISE BASED ON UMVUE

Example: If $X_1, X_2, ...X_n$ are random samples from Bernoulli $(1,\theta)$ Find MVUE of $g(\theta)=\theta(1-\theta)$.

Solution:

Likelihood function of $x_1, x_2, ...x_n$ is

$$L(\theta)=\prod_{i=1}^{n} f(x_i)=\prod_{i=1}^{n}\theta^{x_i}(1-\theta)^{1-x_i}=\underset{h(\underline{x})}{\underset{\downarrow}{1}} \quad \underbrace{\theta^{\sum_{i=1}^{n} x_i}(1-\theta)^{n-\sum_{i=1}^{n} x_i}}_{g_\theta(\sum_{i=1}^{n} x_i)}$$

So, by Neymann Factorization Theorem, $T(\underline{X})=T=\sum_{i=1}^{n} X_i$ is sufficient statistics for θ.

Here $\gamma(\theta)=\theta(1-\theta)$

$$E_\theta\left(g(T)\right)=\gamma(\theta)$$

$$E_\theta\left(g(T)\right)=\theta(1-\theta)$$

$$\Rightarrow \sum_{t=0}^{n} g(T)P(T=t)=\theta(1-\theta)$$

$$\Rightarrow \sum_{t=0}^{n} g(T)\frac{n!}{t!(n-t)!}\theta^{t}(1-\theta)^{n-t} \quad =\theta(1-\theta)$$

$$\Rightarrow \sum_{t=0}^{n} g(T)\frac{n!}{t!(n-t)!}\left(\frac{\theta}{1-\theta}\right)^{t}(1-\theta)^{n} \quad =\theta(1-\theta)$$

$$\Rightarrow \sum_{t=0}^{n} g(T)\frac{n!}{t!(n-t)!}\left(\frac{\theta}{1-\theta}\right)^{t} \quad =\frac{\theta(1-\theta)}{(1-\theta)^{n}}$$

$$\Rightarrow \sum_{t=0}^{n} g(T)\frac{n!}{t!(n-t)!}\left(\frac{\theta}{1-\theta}\right)^{t} \quad =\theta(1-\theta)^{1-n}$$

$$\rho=\frac{\theta}{1-\theta} \quad \Rightarrow(1-\theta)\rho=\theta \quad \Rightarrow \theta=\frac{\rho}{1+\rho} \tag{33}$$

where $\rho=\dfrac{\theta}{1-\theta} \quad \Rightarrow(1-\theta)\rho=\theta \quad \Rightarrow \theta=\dfrac{\rho}{1+\rho}$

$$\Rightarrow \sum_{t=0}^{n} g(T)\frac{n!}{t!(n-t)!}\rho^{t} \quad =\rho(1+\rho)^{n-2}$$

$$\Rightarrow \sum_{t=0}^{n} g(T)\frac{n!}{t!(n-t)!}\rho^{t} \quad =\rho\sum_{t=0}^{n-2}\frac{(n-2)!}{t!(n-2-t)!}(\rho)^{t}=\sum_{t=0}^{n-2}\frac{(n-2)!}{t!(n-2-t)!}(\rho)^{t+1}$$

Let $t+1=u$

$$\Rightarrow \sum_{t=0}^{n} g(T)\frac{n!}{t!(n-t)!}\rho^{t} \quad = \sum_{u=1}^{n-1}\frac{(n-2)!}{u-1!(n-u-1)!}(\rho)^{u}$$

$$\Rightarrow g(0)\frac{n!}{0!(n-0)!}\rho^0+\sum_{t=1}^{n-1}g(T)\frac{n!}{t!(n-t)!}\rho^t+g(n)\frac{n!}{0!(n-n)!}\rho^n=\sum_{t=1}^{n-1}\frac{(n-2)!}{t-1!(n-1-t)!}(\rho)^t$$

Comparing coefficient of ρ we get

$$\Rightarrow g(0)\frac{n!}{0!(n-0)!}=0\rightarrow g(0)=0 \text{ for t=0}$$

$$\Rightarrow g(n)\frac{n!}{n!(n-n)!}=0\rightarrow g(n)=0 \text{ for t=n}$$

and $\Rightarrow$ $1\le t\le n-1$

$$\Rightarrow g(T)\frac{n!}{t!(n-t)!}=\frac{(n-2)!}{t-1!(n-1-t)!}$$

$$\Rightarrow g(T)=\frac{T(n-T)}{n(n-1)}$$

So UMVUE of $\theta(1-\theta)$ is

$$\Rightarrow f(T)=\begin{cases}0 & \text{t=0,n}\\ \dfrac{T(n-T)}{n(n-1)} & 1\le t\le n-1\end{cases}$$

3.6 RAO-BLACKWELL THEOREM

Theorem: Let $x_1,x_2...,x_n$ be a random sample from the distribution whose p.d.f is $f(x;\theta)$. Let the statistic $Z=z(x_1,x_2...,x_n)$ be an unbiased estimator for $\gamma(\theta)$. And let $T=t(x_1,x_2...,x_n)$ be sufficient statistic defined as $\phi(t)=E(Z/T=t)$ which independent of θ.

Then

(i) $E_\theta(\phi(T))=\gamma(\theta)$

(ii) $Var(\phi(T))\le Var(Z)$

Proof: Z is an unbiased estimator for $\gamma(\theta)$, i.e.,

$$E(U) = \gamma(\theta) \tag{34}$$

Now

$$\phi(T) = E\left(Z / T = t\right)$$

Taking expectation both sides, we get

$$E\left(\phi(T)\right) = E\left\{E\left(Z / T = t\right)\right\}$$

$$E\left(\phi(T)\right) = E\left(Z\right)$$

$$E\left(\phi(T)\right) = \gamma\left(\theta\right)$$

i.e., $\gamma(\theta)$ is an unbiased estimator for $\gamma(\theta)$ which is the first condition of Rao Blackwell theorem.

Now

$$V\left(Z\right) = E\left[Z - E(Z)\right]^2$$

$$= E\left[Z - \gamma(\theta)\right]^2$$

$$= E\left[\left(Z - \phi(T)\right) + \left(\phi(T) - \gamma(\theta)\right)\right]^2$$

$$= E\left[\left(Z - \phi(T)\right)^2 + \left(\phi(T) - \gamma(\theta)\right)^2 + 2\left(Z - \phi(T)\right)\left(\phi(T) - \gamma(\theta)\right)\right] \tag{35}$$

$$E\left[\left(Z - \phi(T)\right)\left(\phi(T) - \gamma(\theta)\right)\right] = \int\int \left(z - \phi(t)\right)\left(\phi(t) - \gamma(\theta)\right) f(z,t)\ dzdt$$

$$E\left[\left(Z - \phi(T)\right)\left(\phi(T) - \gamma(\theta)\right)\right] = \int\int \left(z - \phi(t)\right)\left(\phi(t) - \gamma(\theta)\right) f(t)\, f(z / t) dzdt$$

Because $f(z / t) = \dfrac{f(z,t)}{f(t)}$

Since

$$= \int (\phi(t) - \gamma(\theta)) \left\{ \int (z - \phi(t)) f(t) f(z/t) dz \right\} dt \tag{36}$$

Since

$$\int (z - \phi(t)) f(t) f(z/t) dz = E(Z - \phi(T)/T = t)$$

$$= E(Z/T = t) - \phi(t)$$

$$E[(Z - \phi(T))(\phi(T) - \gamma(\theta))] = \int\int (z - \phi(t))(\phi(t) - \gamma(\theta)) f(t) f(z/t) dz dt = 0 \tag{37}$$

Using Eq. (37) in Eq. (36) we get

$$E[(Z - \phi(T))(\phi(T) - \gamma(\theta))] = 0$$

Using Eq. (38) in Eq. (35) we get

$$\text{Var (Z)} = E\left[(Z - \phi(T))^2 + (\phi(T) - \gamma(\theta))^2\right]$$

$$\text{Var (Z)} = Var(\phi(T)) + E(Z - \phi(T))^2$$

$$\text{Var (Z)} \leq Var(\phi(T))$$

Which is the second condition of Rao Blackwell theorem?

3.7 METHODS OF ESTIMATION

Previously we have discussed the properties of a good estimator, and now we shall briefly discuss some commonly used methods for obtaining such estimators. The methods are as follows:

(i) Method of Maximum Likelihood Estimation
(ii) Method of Moments Estimation
(iii) Method of Minimum Variance Estimation
(iv) Method of Least Squares Estimation
(v) Bayes Estimation

3.7.1 METHOD OF MAXIMUM LIKELIHOOD ESTIMATION

The method of maximum likelihood is widely regarded as the best general method of finding estimators. The principle is given by Sir R. A. Fisher in 1921. In this approach, we find the value of θ that maximizes the likelihood function. In this approach, we get the maximum likelihood estimator (MLE) for the population parameter by differentiating the likelihood or log likelihood with respect to the parameter(s) and setting the derivative to zero. In other terms, the method of MLE is an estimator for the unknown population parameter $\theta = (\theta_1, \theta_2 ..., \theta_n)$, maximizes the likelihood function. The procedure for finding the MLE is as follows

1. Obtain the likelihood function of sample, i.e., $L(\theta)$ or $L(\theta; x_1, x_2 ... x_n) = \prod_{i=1}^{n} f(x_i)$
2. Take logarithmic of $L(\theta)$
3. Partially differentiate logarithmic of $L(\theta)$ with respect to its parameter, i.e., $\frac{\delta}{\delta\theta}\log L(\theta)$ and equate them to zero, i.e., $\frac{\delta}{\delta\theta}\log L(\theta) = 0$.
4. Obtain the value of the population parameter θ.
5. Find the second derivative of logarithmic of $L(\theta)$, i.e., $\frac{\delta^2}{\delta\theta^2}\log L(\theta)$.
6. If the second derivative is negative, then L is maximum at $\theta = \hat{\theta}$. Such that $\hat{\theta}$ is called MLE for the population parameter θ. i.e., $\frac{\delta^2}{\delta\theta^2}\log L(\theta) < 0 | \theta = \hat{\theta}$. Then $\hat{\theta}$ is MLE for the parameter θ.

3.7.1.1 PROPERTIES OF ESTIMATES OBTAINED BY MAXIMUM LIKELIHOOD ESTIMATOR

1. MLE's are not necessarily unbiased. For example, $X \sim N(\mu, \sigma^2)$, then $\bar{X}$ is MLE of μ and $s^2 = \frac{1}{n-1}\sum_{i=1}^{n}(x-\bar{x})^2$ is MLE for σ^2, but $\bar{X}$ is unbiased for μ and $s^2 = \frac{1}{n-1}\sum_{i=1}^{n}(x-\bar{x})^2$ is biased for θ.
2. MLE's are always consistent.
3. MLE's are most efficient.
4. If a sufficient estimator exists, it is a function of the MLE.
5. MLE's are not unique in general.

(vi) If T_n is the MLE of θ and $\psi(\theta)$ is one to one function of θ, then $\psi(T_n)$ is the MLE of $\psi(\theta)$ which is known as invariance property of MLE.

(vii)Asymptotic distribution of MLE's is normal, i.e.,

$$\hat{\theta}_{mle} \sim N\left(\theta_0, \frac{1}{E\left[\frac{\delta}{\delta\theta}\log L(\theta)\right]^2}\right)$$

where θ_0 is the true value of the parameter.

3.7.1.2 SOME NUMERICAL EXERCISE BASED ON MAXIMUM LIKELIHOOD ESTIMATOR

Example: If $X \sim N(\mu,\sigma^2)$ find the MLEs for μ when σ^2 is known.

Solution:

$X \sim N(\mu,\sigma^2)$ where is known.

$$f(x) = \begin{cases} \frac{1}{\sigma\sqrt{2\pi}} e^{-\frac{1}{2}\left(\frac{x-\mu}{\sigma}\right)^2} & -\infty < x < \infty, -\infty < \mu < \infty, \sigma^2 > 0 \\ 0 & \text{otherwise} \end{cases}$$

Likelihood function:

$$L(\mu) = \prod_{i=1}^{n} f(x_i) = \prod_{i=1}^{n} \frac{1}{\sigma\sqrt{2\pi}} e^{-\frac{1}{2}\left(\frac{x_i-\mu}{\sigma}\right)^2}$$

$$L(\mu) = \frac{1}{\sigma^n\left(\sqrt{2\pi}\right)^n} e^{-\frac{1}{2}\sum_{i=1}^{n}\left(\frac{x_i-\mu}{\sigma}\right)^2}$$

$$\log L(\mu) = \log\left\{\frac{1}{\sigma^n\left(\sqrt{2\pi}\right)^n} e^{-\frac{1}{2}\sum_{i=1}^{n}\left(\frac{x_i-\mu}{\sigma}\right)^2}\right\}$$

$$\log L(\mu) = \log\left(\frac{1}{\sigma\sqrt{2\pi}}\right)^n - \frac{1}{2}\sum_{i=1}^{n}\left(\frac{x_i-\mu}{\sigma}\right)^2$$

$$\log L(\mu) = n\log\left(\frac{1}{\sigma\sqrt{2\pi}}\right) - \frac{1}{2}\sum_{i=1}^{n}\left(\frac{x_i-\mu}{\sigma}\right)^2$$

$$\frac{\delta \log L(\mu)}{\delta\mu} = \frac{\delta}{\delta\mu}\left\{n\log\left(\frac{1}{\sigma\sqrt{2\pi}}\right) - \frac{1}{2}\sum_{i=1}^{n}\left(\frac{x_i-\mu}{\sigma}\right)^2\right\}$$

$$\frac{\delta \log L(\mu)}{\delta\mu} = \frac{1}{\sigma}\sum_{i=1}^{n}\left(\frac{x_i-\mu}{\sigma}\right)$$

$$\frac{\delta \log L(\mu)}{\delta\mu} = 0$$

$$\frac{1}{\sigma}\sum_{i=1}^{n}\left(\frac{x_i-\mu}{\sigma}\right) = 0 \Rightarrow \overline{x} = \hat{\mu}$$

$$\frac{\delta^2 \log L(\mu)}{\delta\mu^2} = -\frac{n}{\sigma^2} < 0 \tag{38}$$

So M.L.E of the μ is $\overline{x}$.

Example:
If $X \sim Exp(\theta)$ find the MLE of θ.

Solution:

$$X \sim Exp(\theta)$$

$$\left[\frac{\partial^2 \log L(\theta)}{\partial\theta^2}\right]_{[\theta=\overline{x}]} = \frac{-n}{\overline{x}^2} < 0$$

Likelihood function:

$$L(\theta) = \prod_{i=1}^{n} f(x_i) = \prod_{i=1}^{n} \frac{1}{\theta} e^{-\frac{x_i}{\theta}}$$

$$L(\theta) = \frac{1}{\theta^n} e^{-\sum_{i=1}^{n} \frac{x_i}{\theta}}$$

$$\log L(\theta) = \log \left\{ \frac{1}{\theta^n} e^{-\sum_{i=1}^{n} \frac{x_i}{\theta}} \right\}$$

$$\log L(\theta) = -n \log \theta - \sum_{i=1}^{n} \frac{x_i}{\theta}$$

$$\frac{\delta \log L(\theta)}{\delta \theta} = \frac{\delta}{\delta \theta} \left\{ -n \log \theta - \sum_{i=1}^{n} \frac{x_i}{\theta} \right\}$$

$$\frac{\delta \log L(\theta)}{\delta \theta} = \frac{-n}{\theta} + \frac{1}{\theta^2} \sum_{i=1}^{n} x_i$$

$$\frac{-n}{\theta} + \frac{1}{\theta^2} \sum_{i=1}^{n} x_i = 0$$

$$\sum_{i=1}^{n} x_i = n\theta$$

$$\hat{\theta} = \overline{x} \tag{39}$$

$$\frac{\delta^2 \log L(\theta)}{\delta \theta^2} = \frac{-n}{\theta} + \frac{n\overline{x}}{\theta^2}$$

$$\frac{\delta^2 \log L(\theta)}{\delta \theta^2} = \frac{n}{\theta^2} - \frac{2n\overline{x}}{\theta^3} = \frac{n}{\theta^3} [\theta - 2\overline{x}]$$

$$\frac{\delta^2 \log L(\theta)}{\delta\theta^2}\left|\left[\hat{\theta}=\overline{x}\right]=\frac{n}{\theta^3}[\overline{x}-2\overline{x}]\right.$$

$$\left[\frac{\partial^2 \log L(\theta)}{\partial\theta^2}\right]_{[\theta=\overline{x}]}=\frac{-n}{\overline{x}^2}<0$$

So from equation (39) $\overline{x}$ the MLE of θ.

3.7.2 METHOD OF MOMENT ESTIMATOR

The method of moment approach is the oldest and given by Karl Pearson in 1894. This approach runs onto equating population moments to corresponding sample moments and solve for the parameter(s). Let the density function $f(x;\theta_1,\theta_2,...,\theta_m)$ with m parameters $\theta_1,\theta_2,...,\theta_m$, μ_r' is the population moment and m_r' is the sample moment about origin thus the method of moment is

$$\mu_r' = m_r' \;; \quad r=1,2,...,m \tag{40}$$

By solving Eq. (40), we get method of moment estimator.

3.7.2.1 PROPERTIES OF ESTIMATES OBTAINED BY METHOD OF MOMENT ESTIMATOR

1. This method is easy to employ and usually provides consistent estimators for respective parameters.
2. Under general conditions, the method of moment estimates will have asymptotically normal distributions when n is extremely large.
3. In general, the method of moments estimators are less efficient than MLEs. In some cases, they will perform equally good.

3.7.2.2 SOME NUMERICAL EXERCISE BASED ON METHOD OF MOMENT ESTIMATOR

Example: $X \sim Gamma\,(\alpha,\lambda)$; find the estimate of (α,λ) by the method of moments.

Solution:

$$X \sim Gamma\ (\alpha, \lambda)$$

$$f(x) = \begin{cases} \frac{1}{\Gamma\alpha} \lambda^{\alpha} x^{\alpha-1} e^{-\lambda x} & 0<x<\infty \\ 0 & \text{otherwise} \end{cases}$$

$$\Rightarrow E(X) = \mu_1' = \frac{\alpha}{\lambda}, \qquad V(X) = \mu_2 = \frac{\alpha}{\lambda^2}$$

We know $\mu_2 = \mu_2' - \mu_1'^2$

$$\Rightarrow \mu_2' = \mu_2 + \mu_1'^2$$

$$\Rightarrow \mu_2' = \frac{\alpha}{\lambda^2} + \left(\frac{\alpha}{\lambda}\right)^2 = \frac{\alpha(1+\alpha)}{\lambda^2}$$

By method of moment estimator

$$\mu_1' = m_1' \text{ and } \mu_2' = m_2' \tag{41}$$

where $m_1' = \frac{1}{n}\sum_{i=1}^{n} x_i = \bar{x}$ and $m_2' = \frac{1}{n}\sum_{i=1}^{n} x_i^2$

So from Eq. (41)

$$\mu_1' = m_1' \Rightarrow \frac{\alpha}{\lambda} = \bar{x} \tag{42}$$

$$\mu_2' = m_2' \Rightarrow \frac{\alpha(1+\alpha)}{\lambda^2} = \frac{1}{n}\sum_{i=1}^{n} x_i^2 \tag{43}$$

From Eq. (42)

$$\left(\frac{\alpha}{\lambda}\right)^2 = \bar{x}^2 \tag{44}$$

Divide the Eqs. (42) and (44) we get

$$\frac{\frac{\alpha(1+\alpha)}{\lambda^2}}{\left(\frac{\alpha}{\lambda}\right)^2} = \frac{\frac{1}{n}\sum_{i=1}^{n} x_i^2}{\overline{x}^2}$$

$$\Rightarrow \frac{(1+\alpha)}{\alpha} = \frac{\frac{1}{n}\sum_{i=1}^{n} x_i^2}{\overline{x}^2}$$

$$\Rightarrow 1 + \frac{1}{\alpha} = \frac{\frac{1}{n}\sum_{i=1}^{n} x_i^2}{\overline{x}^2}$$

$$\Rightarrow \frac{1}{\alpha} = \frac{\frac{1}{n}\sum_{i=1}^{n} x_i^2}{\overline{x}^2} - 1$$

$$\Rightarrow \frac{1}{\alpha} = \frac{\frac{1}{n}\sum_{i=1}^{n} x_i^2 - \overline{x}^2}{\overline{x}^2}$$

$$\Rightarrow \frac{1}{\alpha} = \frac{\frac{1}{n}\left\{\sum_{i=1}^{n} x_i^2 - \overline{x}^2\right\}}{\overline{x}^2}$$

$$\Rightarrow \frac{1}{\alpha} = \frac{\frac{1}{n}\left\{\sum_{i=1}^{n} (x_i - \overline{x})^2\right\}}{\overline{x}^2}$$

$$\Rightarrow \frac{\overline{x}^2}{\frac{1}{n}\left\{\sum_{i=1}^{n} (x_i - \overline{x})^2\right\}} = \alpha \qquad (45)$$

$$\frac{\alpha}{\lambda}=\overline{x}\Rightarrow\frac{\dfrac{\overline{x}^2}{\frac{1}{n}\left\{\sum_{i=1}^{n}(x_i-\overline{x})^2\right\}}}{\lambda}=\overline{x}$$

$$\Rightarrow\frac{\overline{x}}{\frac{1}{n}\left\{\sum_{i=1}^{n}(x_i-\overline{x})^2\right\}}=\lambda \tag{46}$$

Equations (45) and (46) give the estimator (α,β) by method of moments.

Example: $X\sim Beta(\theta+1,1)$ find the method of moments for θ.

Solution:

$$X\sim Beta(\theta+1,1)$$

$$\Rightarrow E(X)=\mu_1^{'}=\frac{\theta+1}{\theta+1+1}=\frac{\theta+1}{\theta+2}$$

By method of moment estimator

$$\mu_1^{'}=m_1^{'} \tag{47}$$

where $m_1^{'}=\frac{1}{n}\sum_{i=1}^{n}x_i=\overline{x}$

So from Eq. (38)

$$\Rightarrow\frac{\theta+1}{\theta+2}=\overline{x} \tag{48}$$

$$\Rightarrow\theta+1=\overline{x}\theta+2\overline{x}$$

$$\Rightarrow\theta-\overline{x}\theta=2\overline{x}-1$$

$$\Rightarrow\theta(1-\overline{x})=2\overline{x}-1$$

$$\Rightarrow\hat{\theta}=\frac{2\overline{x}-1}{(1-\overline{x})} \tag{49}$$

Equation (49) gives the estimator $\hat{\theta}$ by the method of moments.

3.7.3 METHOD OF LEAST SQUARE

Let *X* and *Y* be two random variables and suppose that we wish to find the functional relationship between *X* and *Y* say Y = f(X).

We assume that both $E(Y^2)$ and E[f(X)] exists. Our objective is to find the function f(X). The principal of least square is that it consists in minimizing the sum of squares of residuals. So that

$E[Y\text{-}f(X)]^2$ is minimum.

Similarly, for the *n* observation, we can say that

$E[Y - f(x_1, x_2, ..., x_n)]^2$ is minimum

After minimizing $E[Y - f(x)]^2$ we get best linear fit for any given curve.

3.7.3.1. PROPERTIES OF ESTIMATES OBTAINED BY METHOD OF LEAST SQUARE

1. The least square estimators are unbiased in case of linear models.
2. The least square estimators do not possess the asymptotically normal property likewise method of moment estimators.
3. Under fairly general setup we obtain best linear unbiased estimators (BLUEs).

3.7.4 METHOD OF MINIMUM VARIANCE ESTIMATOR

If $L(\theta) = \prod f(x_i;\theta)$, is the likelihood function of a random sample of n-observations $x_1, x_2, ...x_n$ from a population with probability function $f(x_i;\theta)$. Then the problem is to find statistic $t = t(x_1, x_2, ...x_n)$, such that

$$E(T) = E(T(X)) = \int_{-\infty}^{\infty} ... \int_{-\infty}^{\infty} t(x_1, ..., x_n) L(\theta) dx_1 dx_2 ... dx_n = \gamma(\theta \qquad (50)$$

$$E(T) = \int_{-\infty}^{\infty} \int_{-\infty}^{\infty} .... \int_{-\infty}^{\infty} [t(x) - \gamma(\theta)] L(\theta) dx_1 dx_2 ... dx_n$$

And $V(T)=\int_{-\infty}^{\infty}\int_{-\infty}^{\infty}....\int_{-\infty}^{\infty}[t-E(t)]^2 L(\theta)dx_1dx_2...dx_n$ (51)

is minimum.

In other words, we have to minimize Eq. (51) subject to condition (50). Then we get minimum variance unbiased estimator.

3.7.4.1 PROPERTIES OF ESTIMATES OBTAINED BY METHOD OF LEAST SQUARE

1. The Minimum Variance Unbiased Estimator (MVUE) described above are unbiased as well as they have minimum variance.
2. Sometimes there may not exist any MVUE for a given scenario or set of data. This can happen in two ways: (a) When there is no unbiased estimators, (b) Even if we have unbiased estimators, none of them gives uniform unbiased estimators.

3.7.5. BAYES ESTIMATIONS

In the estimation process so far we have a density $f(x;\theta)$ for each θ from which a random sample is drawn, i.e., $x_1, x_2...x_n$. When we have a prior information in the form of density, say h(θ), known as prior density of θ and $f(x|\theta)$, known as conditional density of *x* for a given value of parameter θ. When both *X* and θ are random variable then the density $f(x;\theta)$ can be given by;

$$f(x;\theta)=f(x|\theta)h(\theta)$$

Using the Bayes' Probability rule here,

$$f(x;\theta)=\frac{f(x|\theta)h(\theta)}{\int f(x|\theta)h(\theta)d\theta}$$

Then the Bayesian point estimator of a parametric function τ(θ) of θ can be estimated by the following expression

$$E[\tau(\theta)]=\int\tau(\theta)f(\theta|x)d\theta$$

i.e.,

$$\hat{\tau}(\theta) = \frac{\int \tau(\theta) f(x|\theta) h(\theta) d\theta}{\int f(x|\theta) h(\theta) d\theta}$$

Bayes' estimator of $\tau(\theta)$ given above is under square error loss function (SELF).

3.7.5.1 PROPERTIES OF ESTIMATES OBTAINED BY BAYES' ESTIMATIONS

1. Whenever we obtain a Bayes' estimator that must be a function of minimal sufficient statistics.
2. Bayes' estimators are usually asymptotically consistent.
3. Bayes' estimators always tend to MLEs for large *n*.

3.8 INTERVAL ESTIMATION

We already stated that a point estimator is a single value of sample statistic used to estimate a population parameter. But point estimator may not always be expected to provide the exact value of the population parameter which makes the estimate unsatisfactory. Thus the technique of interval estimation provides the range of the true value of the parameter with a certain probability.

If 't' is the statistic used to estimate the parameter θ with a level of significance α, then $(1-\alpha)$ confidence limits for the parameter $\theta = t \pm S.E(t) \, . \, t_{\alpha/2}$, where $t_{\alpha/2}$ is the critical value and S.E. denotes the standard error of the statistic 't.'

The standard approach to obtain the confidence interval of population parameter involves the following steps:

Step 1: Select the test statistic' t.'

Step 2: Choose the level of significance.

Step 3: Obtain the standard error of the statistic 't,' i.e., S.E. (t).

Step 4: Obtain the $(1-\alpha)$ confidence limits for the parameter.

3.9 TESTING OF HYPOTHESIS

3.9.1 *HYPOTHESIS*

A hypothesis is a statement about the population developed for the purpose of testing is called Statistical hypothesis, for example, the mean monthly income for systems analysts is $3,625. Whereas the form of the probability distribution is known then any statement which refers to the values of the unknown parameters of that distribution is called a parametric hypothesis. Further, a hypothesis is defined in two ways: (i) simple hypothesis, and (ii) composite hypothesis. A parametric hypothesis which specifies completely all the parameters in a probability distribution is called a simple hypothesis and on the other hand which fails to completely specify the probability distribution is called a composite hypothesis. For example, $x_1, x_2, \ldots x_n$ are random samples from a normal distribution with mean μ and variance σ^2 which is known, then hypotheses $\mu = \mu_0$ is a simple hypothesis, whereas the hypothesis $\mu > \mu_0$ or $\mu < \mu_0$ or $\mu \neq \mu_0$ are the composite hypotheses. The Hypothesis testing is a stratagem which evolves on sample evidence and probability theory, used to determine whether the hypothesis is a rational statement and should not be rejected, or is irrational and should be rejected.

3.9.2 *TYPES OF HYPOTHESIS*

In testing of hypothesis, there are two types of hypothesis: (i) null hypothesis and (ii) Alternative hypothesis. A hypothesis is a tentative assumption about a population parameter which is tested for possible rejection under the consideration that it is true is known as a null hypothesis, and it is denoted by H_0. Whereas the opposite of what is stated in the null hypothesis or which differ from a given null hypothesis and is accepted when H_0 is rejected is called an Alternative hypothesis and is denoted by H_1. There are two types of the Alternative hypothesis: (i) one-sided Alternative hypothesis and (ii) two-sided Alternative hypothesis. Let H_0: $\mu = \mu_0$ then the Alternative hypothesis H_1: $\mu < \mu_0$ or $\mu > \mu_0$ are known as one-sided alternatives and the Alternative hypothesis H_1: $\mu \neq \mu_0$ is known as two-sided alternative.

Left tail	Right tail	two tail
$H_0 : \mu = \mu_0$	$H_0 : \mu = \mu_0$	$H_0 : \mu = \mu_0$
$H_1 : \mu < \mu_0$	$H_1 : \mu > \mu_0$	$H_1 : \mu \neq \mu_0$
(one-sided)	(one-sided)	(two-sided)

3.9.3 CRITICAL REGION AND ACCEPTANCE REGION

The sample space is divided into two mutually disjoint regions one is critical region, and the other one is acceptance region. Let random samples $x_1, x_2, \ldots x_n$ of the same size from a given population and compute the statistic $t(x)$ for the sample values. If the value of statistic lies in the critical region or rejection region, the null hypothesis H_0 is rejected, and if it falls in acceptance region, we reject H_1 and accept H_0. The critical region is denoted as w and acceptance region is denoted by $\overline{w}$.

3.9.4 TYPES OF ERRORS

The final made decision to accept or reject the null hypothesis H_0 is formed by the information based on sample values $x_1, x_2, \ldots x_n$ which is not necessarily be true for the respective population that leads some error in the testing of hypothesis. There are two types of errors (i) Type I error and (ii) Type II error.

Type I error is the rejection of true null hypothesis or rejecting H_0 when it is true. Type II error is acceptance of false hypothesis or accepting H_0 when it is false. Further, the probability of rejecting a null hypothesis H_0 when it is true, is called a size of type I error, and it is denoted by α known as a level of significance of the test. Thus the size of the type I error or the level of significance of the test or the size of critical region and is:

α = p [committing type I error]
α = p [Rejecting H_0 when H_0 is true]
α = p [Rejecting H_0 / H_0 is true]
$\alpha = p[\underline{X} \in w / H_0] = p_{\theta_0}(w)$

where w is the critical region.

3.9.4.1 REMARK

In practice, 0.05 and 0.01 are the commonly accepted values of the level of significance.

Similarly, the probability of committing a type II error is denoted by β, i.e.,

β = p [committing type II error]

β = p [Accepting H_0 when H_1 is true]

β = p [Accepting H_0 / H_1 is true]

$\beta = p[\underline{X} \in \overline{w} / H_1] = p_{\theta_0}(\overline{w})$

where $\overline{w}$ is the acceptance region.

3.9.5. POWER FUNCTION AND POWER OF A TEST

Power of the test is explained as the probability of rejectingH_0 when H_0 is false. Considered as a function of population parameter θ for all $\theta \in \Theta$, where Θ is the parameter space and is called the power function of the test. The power function denoted by $p(\theta)$ is:

$P(\theta)$ = p [Rejecting H_0 when H_1 is true]

$P(\theta) = p[\underline{X} \in w / H_1]$ = 1–p [Accepting H_0/H_1 is true] = $1-\beta$

The value of the power function at a particular value of the parameter is called the power of the test.

3.9.5.1 REMARKS

1. The graph between the parameter θ and the power function $P(\theta)$ for different values of the parameter θ is known as a power curve.
2. If $P(\theta) \geq \alpha$ then test is called unbiased for all Alternative hypothesis.

3.9.6 MOST POWERFUL CRITICAL REGION AND MOST POWERFUL TEST

The critical region w of size α for testing the simple null hypothesis $H_0: \theta = \theta_0$ against simple Alternative hypothesis $H_1: \theta = \theta_1$ is said to be most powerful critical region or best critical region of size α if

(i) $p[\underline{X} \in w / H_0] = \alpha$

(ii) $p[\underline{X} \in w / H_1] > p[\underline{X} \in w_1 / H_1]$

Whatever the other critical region w_1 holds condition (i).

The corresponding test is known as most powerful test. That is a test defined by most power critical is called most powerful test.

3.9.7 UNIFORMLY MOST POWERFUL CRITICAL REGION AND UNIFORMLY MOST POWERFUL TEST

The critical region w of size α for testing the simple hypothesis $H_0: \theta = \theta_0$ against composite Alternative hypothesis $H_1: \theta > \theta_0$ or $H_1: \theta > \theta_0$ $H_1: \theta \neq \theta_0$ is said to be uniform most powerful critical region of size α if

(i) $p[\underline{X} \in w / H_0] = \alpha$

(ii) $p[\underline{X} \in w / H_1] > p[\underline{X} \in w_1 / H_1]$ $\forall \theta > \theta_0$ or $\theta > \theta_0$ or $\theta \neq \theta_0$

Whatever, the other critical region w_1 hold condition (i). The corresponding test is known as uniformly most powerful test.

3.9.8. NEYMANN-PEARSON LEMMA (NP LEMMA)

Let $f(x;\theta)$ be a random sample of size n from a probability distribution having probability density function $f(x;\theta)$. Let L = $L(\theta; x_1 ... x_n) = L(\theta)$ be the likelihood function. Let $H_0: \theta = \theta_0$ $\forall$ $H_1: \theta = \theta_1$ [simple $\forall$ simple]. Let $L(\theta_0)$ and $L(\theta_1)$ be the likelihood function under H_0 and H_1 respectively then the critical region with size α is

$$w = \left\{ x \in s; \frac{L(\theta_1)}{L(\theta_0)} > k \right\}$$

where s is the sample space and k is any positive integer number.

3.9.9 HYPOTHESIS SETUPS FOR TESTING A POPULATION MEAN

The standard approach to carrying out a statistical test involves the following steps:

Step 1: State the null hypothesis and the alternate hypothesis.

Step 2: Select the level of significance.

Step 3: Select the test statistic.

Step 4: Formulate the decision rule.

Step 5: Make a decision and interpret the result.

3.9.10 TEST OF SIGNIFICANCE FOR SINGLE MEAN

If $x_1, x_2 \ldots x_n$ is a random sample of size n from a normal population with mean μ and variance σ^2, then the sample mean is distributed normally with μ and variance σ^2 / n. Under the null hypothesis H_0, that the sample has been drawn from a population with mean μ and variance $\overline{x}$, i.e., there is no significance difference between the sample mean $\overline{x}$, and population mean (μ), the test statistics for large sample is

$$Z = \frac{\overline{x} - \mu}{\sigma / \sqrt{n}} \sim N(0,1)$$

Under hypothesis $H_0 : \mu = \mu_0 \; \forall \; H_1 : \mu \neq \mu_0$ or $\mu > \mu_0$ or $\mu < \mu_0$. If $|z| > z_\alpha$, we reject the null hypothesis.

3.9.10.1 SOME NUMERICAL EXERCISE BASED ON TESTING SINGLE MEANS (Z TEST)

Example: The average IQ of a sample of 40 college students was found to be 105. Carry out a statistical test to determine whether the average IQ of college students is greater than 100, assuming that IQs are normally distributed. It is known from previous studies that the standard deviation of IQs among student is approximately 20.

Solution:

Step 1: State the null hypothesis and the alternate hypothesis.

$$H_0 : \mu = 100 \text{ vs } H_1 : \mu > 100$$

Step 2: Select the level of significance.
$\alpha = 0.05$ as stated in the problem
Step 3: Select the test statistic.
Use Z-distribution since σ is known and population being sampled is normally distributed.

The test statistic $Z = \frac{105 - 100}{20 / \sqrt{40}} \sim N(0,1)$

$$Z = 1.5811$$

Step 4: Formulate the decision rule.

Reject H_0 if $|Z| > Z_\alpha$

Step 5: Make a decision and interpret the result,

At a 5% level of significance, the tabulated value of Z is 1.6449. So the test statistic value of Z is greater than tabulated value of Z, so we reject the null hypothesis. Therefore it is reasonable to conclude that the average IQ of university students is greater than 100.

3.9.11 TEST OF SIGNIFICANCE BASED ON T-DISTRIBUTION FOR SINGLE MEAN

With these underlying assumptions we can apply t distribution for testing single mean:

(i) The parent population from which the sample is drawn is normal.
(ii) The sample observations are independent, i.e., the sample is random.
(iii) The population standard deviation is unknown.

Suppose a random sample $x_1, x_2 \ldots x_n$ of size n has been drawn from a normal population whose variance is unknown. The test statistics for large sample is

The test statistic $t = \dfrac{\overline{x} - \mu}{s / \sqrt{n}}$

Under hypothesis

$$H_0 : \mu = \mu_0 \ \forall \ H_1 : \mu \neq \mu_0 \text{ or } \mu > \mu_0 \text{ or } \mu < \mu_0$$

If $|t| > t_{\alpha, n-1}$, we reject null hypothesis.

3.9.11.1 SOME NUMERICAL EXERCISE BASED ON TESTING SINGLE MEANS (t-TEST)

Example: The annual rainfall in centimeters at Meghalaya weather station over the last ten years has been as follows:

17.2 281 25.3 26.2 30.7 19.2 23.4 27.5 29.5 31.6

A scientist at the weather station wishes to test whether the average annual rainfall has increased from its former long-term value of 23 cm. Test this hypothesis at the 5% level.

Solution:

Step 1: State the null hypothesis and the alternate hypothesis.

$$H_0 : \mu = 23 \text{ vs } H_1 : \mu > 23$$

Step 2: Select the level of significance.

$\alpha = 0.05$ as stated in the problem

Step 3: Select the test statistic.
Use t-distribution since σ is unknown.

The test statistic $$t = \frac{\bar{X} - 23}{20 / \sqrt{n}} \sim t_{10-1}$$

$$t = \frac{25.87 - 23}{\sqrt{2257} / \sqrt{10}} \sim t_9$$

$$t = 0.1910$$

Step 4: Formulate the decision rule.

Reject H_0 if $|t| > t_{\alpha,n-1}$

Step 5: Make a decision and interpret the result

At a 5% level of significance, the tabulated value of t with 9 degree of freedom is 1.833. Here the value of test statistic is lesser than tabulated value of t, so we accept the null hypothesis. Therefore, it is reasonable to conclude that the long-term average annual rainfall has not increased from its former long-term value of 23 cm.

KEYWORDS

- **best linear unbiased estimators**
- **Cramer Rao lower bound**
- **maximum likelihood estimator**
- **minimum variance bound**
- **minimum variance unbiased estimator**
- **uniform minimum variance unbiased estimator**

REFERENCES

Goon, A. M., Gupta, M. K., & Dasgupta, B., (1994). *An Outline of Statistical Theory* (Vols.1 & 2), World Press Private, Kolkata. ISBN 81-87567-34-1, pp. 1–771.

Hogg, R. V., & Craig, A. T., (1978). *Introduction to Mathematical Statistics, Sixth Edition,* Macmillan Publishing Co. Inc., New York, ISBN 0-02-978990-7, pp. 1–435.

Rao, C. R., (1974). *Linear Statistical Inference and Its Applications, Second Edition,* John Wiley & Sons, Inc., Printed in the United States of America, ISBN 0-471-21875-8, pp. 1–618.

Rohatgi, V. K., (1984). *An Introduction to Probability Theory and Statistics, Second Edition,* A Wiley-Interscience Publication, John Wiley & Sons, Inc., ISBN 978-0-471-34846-7, pp. 1–707.

CHAPTER 4

Assessment of Sampling Techniques: Theory and Application

FOZIA HOMA[1], MUKTI KHETAN[2], and SHWETA DIXIT[3]

[1]*Assistant Professor-Cum-Scientist, Department of Statistics, Mathematics and Computer Application, Bihar Agricultural University, Sabour, Bhagalpur, Bihar – 813210, India*

[2]*Assistant Professor, Department of Statistics, Sambalpur University, Sambalpur, Odisha – 768019, India*

[3]*Research Assistant Professor, School of Management, SRM Institute of Science and Technology, Chennai, Tamil Nadu – 603203, India*

4.1 INTRODUCTION

The need of statistical information seems endless in the modern era. Data are regularly collected to satisfy the necessity of information about the sets of elements. In the modern world, things are changing fast with time and essential for frequent collection of up to date data. The data are collected regularly to satisfy the requirement for information about the particular sets of elements. In this modern society, things are going fast with time, necessitating frequent collection of up-to-date data. Now the question arises how to get such data? The answer is to collect the data one needs on the basis of their objective. There are two methods are given in sampling theory for data collection: (i) complete enumeration (census) survey, and (ii) sample survey.

4.1.1 COMPLETE ENUMERATION (CENSUS) SURVEY

This is the oldest form of data collection. The first ever efforts for population enumerations were made in the city of Babylonia before 3800 B.C., followed by China before 3000 B.C. than Egypt near about 2500 B.C.

Modern census taking began with the enumeration of population Quebec in 1666. The periodic census as a regular function of government data from 1790 in the United States, followed by England and Wales in 1801 than Canada and 1881 in India. These are the national level census collected by the government for enumeration representation of each and every individual of a nation. Similarly, when we talk about enumeration survey, it will be the enumeration of each individual of a population to be studied according to one's objective. In other words, if the amount of information is assembled from every single unit in the population under consideration, then this procedure of collecting information is called census. As for instance, in India, the census is done in every 10th year since 1881 in which information is collected from each and every unit residing in a household. In other words, we can say a 100% sample, i.e., the total count of the population is known as census study.

4.2 SAMPLE SURVEY

We have an alternative to census survey that is sample survey. Sample survey gives a representative portrait of the population, without taking into consideration the entire population. If the information is collected by some chosen units from the population instead studying the whole population, then this method of data collection is called sample survey. In this survey method, only a part of the population is observed. It reduces the input assets such as cost, time, manpower, etc., of the survey. It is a common source for obtaining information on various social and economic aspects of society. Such techniques are extensively adopted by government bodies, nonprofit organizations or individuals for assessing different characteristics of national economy entailed for taking decisions regarding the imposition of taxes, price fixation, minimum wages, generation, and implication of several government policies, research institutes, universities, colleges, private market research companies, etc.

In other words, a sampling survey may be defined as an exercise that involves the collection of standardized data from a sample of units (e.g., persons, households, businesses, etc.) designed to represent a bigger picture of the population, in order to craft quantitative inferences about the parent population. Our knowledge, actions, and attributes are mainly survey based. This is almost true in both daily lives and in scientific research. Surveys may vary in requisites of the type of data collected, the techniques used for the collection of data, the design of the sampling units, and whether data is collected on repetition basis, whether it is the same sample or different samples. Sample surveys are now days widely acknowledged method for providing data on a widespread range of subjects for both research and administrative purposes. Large fraction of the quantitative information that is obtained today comes from sample surveys. Central, regional, and local government agencies are responsible for data collection. Such as estimates based on regular reports for unemployment, Income, and expenses, health indexes, poverty rates, agriculture, marketing, sales, etc. Generally, these estimates are based on sample data. National polls on political subjects are frequently reported in news reports. These reports are very general that only a small number reflect on the fact that almost everyone believes that something interesting and/or something practical can be said about a nation like India of half a past billion people on the basis of a sample of a few thousand. Sampling theory has a strong place in every branch of science due to its wider applicability (Figure 4.1).

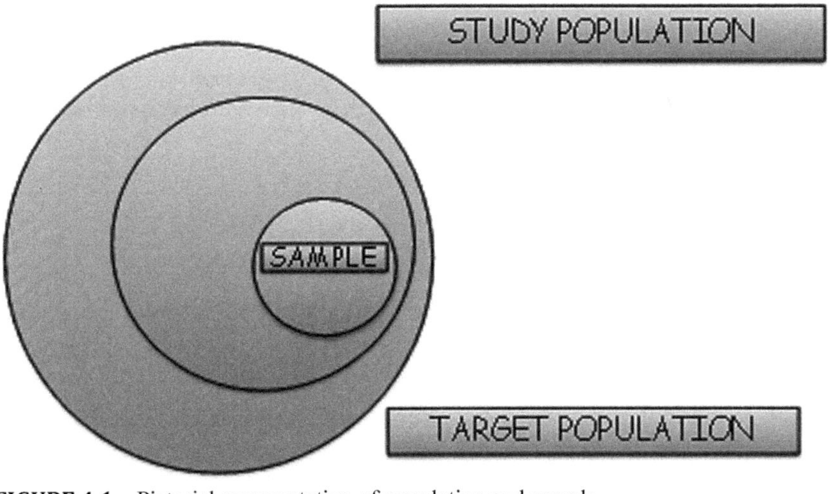

FIGURE 4.1 Pictorial representation of population and sample.

4.3 KEY ELEMENTS OF SAMPLE SURVEY

4.3.1 POPULATION

In statistics, the union of every sample units in a specific given region at a specific point of time is called the population. We can choose some sampling examples in the situation as, one wants to survey a government school for the gender composition, the sum of all the children registered in that school for the survey time will constitute the population of their interest, one wants to survey the medical facilities for cancer patients, the sum of all the patients registered in that hospital for the survey time will constitute the population of their interest. Similarly, if one wants to survey the working-age distributions in a government organization among employees, the sum of total number of working employees in the government organization for the survey period will constitute their population. Further, the population can be split into two subcategory, Finite, and Infinite population. If the units of the population are countable, then the population is called finite population otherwise infinite population. The total number of sampling units in the population is the population size, denoted generally by capital alphabet letter *N*.

4.3.2 SAMPLE

A subset of the population is called sample. For the above-mentioned examples; a collection of students in the government college, a collection of patients in the hospital, a group of employees in government organization which are selected according to some scientific procedure is called sample for the respective populations.

4.3.3 REPRESENTATIVE SAMPLE

If the selected sample contains all the relevant information of the parent population than that sample is called representative sample for that parent population. A representative sample should contain all the basic properties of the parent population. For above mentioned examples; if in government school (population) has 20% girls and 80% boys then the representative sample will be a sample having approximately 20% girls and 80% boys, if in hospital (population) has 40% cancer patients then the representative sample will be a sample having approximately 40% cancer patients and if in government office 65% young employee and 35% old employee then the representative sample will be a sample having approximately 65% young employee and 35% old employee etc.

4.3.4 SAMPLING FRAME

In sampling theory, a list containing each and every sampling units of the population is called sampling frame which provides the means of recognizing and contacting those sampling units. It is essential for any sort of survey and should be clearly specified for any population. For above-mentioned examples; all the students in the government school listed along with their registration number, all the patients in the hospital listed along with their ID and all employees in government organization listed along with employee ID are comprises the sampling frame for the respective examples.

4.3.5 SAMPLING UNIT

Simple elementary units or groups of such elementary units which are identifiable and observable besides being clearly defined are convenient for purposes of sampling, are sampling units. Sampling units for above motioned examples are; the student in the government school, the patient in the hospital and employee in government organization, are sampling units.

4.4. IMPORTANCE OF SAMPLING

4.4.1 REDUCED COST

Census enumeration is the method of collection which is theoretically convenient but practically complicated and quite expensive as cost is associated with each of the unit on the other hand in sample survey, we observed only a part of the population instead of enumeration of all population. Hence, cost of conducting a sample survey is lesser than the cost of a census survey. For the above-motioned examples, we observed only some students of government school, some patients in the hospital and some employee in government organization out of whole population; therefore cost will be reduce.

4.4.2 BETTER ASSOCIATION OF ACTIVITIES

Managing less is always better. In sample survey, the numbers of unit are less than census survey. Hence, it is easier to manage and deal with small samples, i.e., in earlier examples some students from government school, some patients from hospital, and some employees from government organization than to deal with all students from government school, all patients from hospital and all employees from government organization at a time.

4.4.3 MORE FEASIBLE

Sample surveys are necessary when the population units are destructive in basic nature. For examples, quality assessment of the crackers in the crackers company, assessment the life of the electric bulbs or in a medical experiment, feasibility is to use counts of blood (some ml only) for the diagnosis.

4.4.4 LESSER TIME

Sample surveys takes less time as compare to census surveys due to lesser number of units utilized in the surveys as each unit increase will increase associated time of the surveys. Therefore, in terms of the time consumed for the surveys, sample surveys are more beneficial in respect to census surveys.

4.5 STEPS INVOLVED IN SAMPLE SURVEYS

The major steps for conducting any sampling survey are given as below:

1. **What is the need of survey:** Need of the sample survey is the most important question of any researcher for any survey, and it is also the first step of any survey known as objective of the survey. The objective should be clear and understandable by all involved in the development and implementation of the survey.
2. **Population to be sampled:** Population is second most important step of the survey. The population is being decided according to the objective of the survey. In our example, population of students is to be sampled for a school survey whereas the population of patients has to be sampled for analyzing the medical facilities in a government hospital.
3. **Data to be collected:** Selection of data is a very important factor in sample surveys. It is essential that data should be relevant to the objective of the sample surveys. It should be remember that no essential data should be lost. Questions should be related to the

objective of the surveys, should be precise, concise, and least number of questions should be required in a good questionnaire.

4. **Degree of precision required:** Every sample surveys associated with time and cost. It is essential part to have the information of the degree of precision in the working data. One may take more samples if degree of precision is high. Numbers of sample surveys depends on the degree of precision and cost of the surveys.
5. **Measurement method:** The selection of measuring instrument and measuring method for data from the population should be specified. For example, data may be collected all the way through physical observations and measurements, personal interview, web-based enquiry, registration, transcription from records combination of any of these approaches, etc.
6. **Sample frame:** In sampling theory, the list of all sampling units is the sampling frame. The frame must cover the whole population without missing or without replicating a single unit.
7. **Sample selection:** The sample size is the core of the sampling survey. Sample size ideally should be as adequate and clear as possible. If the sample size is less, then it may not get satisfactory results. If the sample size is high, then it may increase cost of the surveys.
8. **The pre-test:** The questionnaire should be firstly test on a small scale, so that we can check whether the questionnaire is understood by sampling units or not. This is an informal study which is sufficient to examine the feasibility of the survey, also in terms of fieldwork.
9. **Organization of the field work:** There are several questions related to fieldwork organization such as, how to perform the sample surveys, handle administrative issues, deal with the problem of non-response error, response error, coverage error, measurement error, and missing observations, providing appropriate training to interviewer and surveyors, etc.
10. **Summary and analysis of data:** After the data collection, one should summarize and analyze the data according to the objective of the surveys.
11. **Information amplification for future surveys:** Ideally the results of any sample surveys should be beneficial for future work. That means any concluded sample survey will turned out a future guide for the surveys to be conducted in the outlook.

4.6 METHODS OF DATA COLLECTION

1. **Observations and measurements:** A surveyor interact the respondent by a face to face interaction. Surveyor observes the sampling unit as well as accounts the data for future analysis. He may use his previous experience for collection of data in a modified way. For instance, a rich women reporting that she is poor but having a one million car in her home may simply be observed and modified by the observer.
2. **Personal interview:** In this method of data collection the surveyor interacts with the respondent directly and asks questions which are the part of questionnaire. The data in the questionnaire is then filled by answers given by the respondents.
3. **Mail enquiry:** It is a simple method of the data collection. In this method, the questionnaire can be sent to the respondents by postal mail, e-mail, etc., with a request to complete the questionnaires and send it back. The internet facility is necessary for this method of data collection.
4. **Web-based enquiry:** In this method, we collect the data over internet based web pages. Several websites are there providing such services for online surveys nowadays. In this method, one has to convert their questionnaires in the websites format, and then the link can be sent to the selected respondents through email, WhatsApp or other social media handles. This is the most prevalent methods these days. Respondent can submit their response by clicking on this link, this click can direct the respondent to the website, and their responses can be registered online. The internet facility is a must have for this method of data collection.
5. **Registration systems:** This system compels respondent to register the data at specific place in certain time intervals. The best example for this method is recorded data on the number of births and deaths provided by the family members at their nearest city municipal offices. There are some government registration systems as well. One can give their details in the respective zonal offices.

4.7 PRINCIPLES OF SAMPLING THEORY

The main aim of sample survey is to create sampling more effective and to provide more accurate estimates of the population parameters keeping in mind the cost, time, and other existing resources for the survey. The sample survey theory is based mainly on three basic principles mentioned below:

(a) **Statistical Regularity:** According to King "*The law of statistical regularity lays down that a moderately large number of items chosen at random from a large group are almost sure on the average to possess the characteristics of the large group.*" This principle of regularity emphasize on the desirability and importance of choosing the sample randomly so that all the units got the same amount of chance of being appeared in the sample.
(b) **Validity:** In sampling design, principal of validity is meant that the sampling method should be enable one to obtain valid estimates and test about the population parameter. Probability sampling fulfills this requirement.
(c) **Optimization:** This principle emphasis on the efficiency and other resources of the sampling design. It consists of attaining a given efficiency level at the least possible price or attaining the most possible efficiency at fixed price.

4.8 METHOD OF SIMPLE RANDOM SAMPLE SELECTION

Random sample refers to the method of sample selection in which all item has a same chance of selection. Random sample not only depends on the selection method but also on the population size and nature. Proper care should be taken place to ensure the randomness of the selected. We can obtain random samples by the following two methods:

1. Lottery method; and
2. Random number table method.

4.8.1 LOTTERY METHOD

One of the simplest methods of random sample selection is a lottery method. In which one can select any number of sample units. Say, we wish to select 'n' sampling units from a population. We allocate n numbers from 1 to n; a single number for each sampling unit as an identification mark. Mark those numbers on paper chits. The chits should be as homogenous as possible in their shapes and sizes for keeping their physical identification same. The chits now can be kept in a container and thoroughly shuffled. Then 'n' chits are drawn at random one by one. The chits are thoroughly reshuffled before every next draw. After all the *n* selection, the corresponding 'n' candidates to those numbers on the chits drawn will comprise a random sample.

This selection method is quite independent of the population properties. This procedure of giving number to units on slips and selecting one slip after complete reshuffling is quite cumbersome. This method is quite time taking and difficult if the population is sufficiently large. So we adopt another method for random numbers selection.

4.8.2 RANDOM NUMBER TABLE METHOD

The most classic, convenient, and economical method of a random selection for a sample is the utilize the 'Random Number Tables.' The random number table has been constructed in a way that each digits *0, 1, 2, ..., 9* appears with approximately the equal number of occurrence and with great independency. If one has to select the sample from a population of size N (≤ 99), then two columns of random number table may be selected to give pairs from *00* to *99*. Similarly if N (≤ 999) or N (≤ 9999) and so on, then three columns or four columns and so on may be selected to get numbers from *000* to *999* or *0000* to *9999* and so on.

The method of the random sample selection consists of the following steps:

1. Classify the population units of size N from 1 to N.
2. Randomly select any page of the random number table and pick the numbers in any column or row randomly.
3. The population units communicating to the selected numbers constitute the random sample.

4.9 CLASSIFICATION OF SAMPLING TECHNIQUES

A sample selection is broadly classified into two major categories:

(a) Subjective Sampling: This is the non-probabilistic sampling method in which one selects the samples with a fixed purpose and the alternative of that selection of sample units entirely depends on the investigator choice of belief. This sampling technique usually suffers from the drawbacks of biasness and favoritism. That is why not offers a representative samples of the parent population. For instance, the researcher desires to give the picture of performance of the school, he may take students in the sample having higher marks in the class and completely ignores the students having just pass marks or students who fail the examinations.

(b) Probability Sampling: This is another scientific method of sample selection from the population. In this sampling scheme, every population unit has some specific pre-assigned chance or probability of being in the samples. In drawing a card from a pack of playing cards, each card has some pre-assigned chance to be in the sample. In sampling theory, there are several different type of probability sampling exists. Simple random sampling (SRS), stratified sampling, cluster sampling are few of them.

4.10 SIMPLE RANDOM SAMPLING (SRS)

This sampling methodology is the most basic sampling method and easy to use for the common people. This sampling method is applied for the homogenous population, i.e., the units to be sampled are of the same kind. In this methodology, a subpopulation of size 'n' known as the sample is picked up from the whole population of size "N" in a manner that every unit of the whole population will get an equal probability to be comprised in the subpopulation (Figure 4.2).

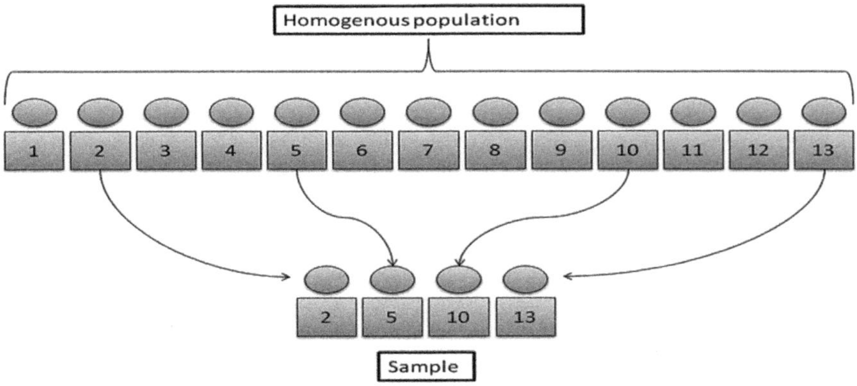

FIGURE 4.2 Diagrammatic representation of simple random sampling.

In SRS, the sampling units from the whole population may be included in the sample in two ways:

1. Simple Random Sampling Without Replacement (SRSWOR): In this scheme of sample selection, units are only once selected in the sample, and there is no chance of selection of a particular unit more than once. The sampling unit selected are not put back into the population. In this method, the possible distinct samples equals to ${}^{N}C_{n}$. For illustration, we consider a hypothetical population of size 5 having the values of sampling units as y: 12, 14, 15, 21, and 17. A random sub-population of size 2 is drawn from the population then the total of ${}^{5}C_{2}$ = 10 distinct samples will be drawn from the population. These samples will be (12, 14), (12,15), (12, 21), (12, 17), (14, 14), (14, 21), (14, 17), (15, 21), (15, 17) and (21, 17).

2. Simple Random Sampling with Replacement (SRSWR): In this scheme of sampling, there is chance that a specific sampling unit may be selected in the population more than once. The units once selected are placed back in the target population prior to the selection of the next unit. In this methodology, the number of samples will be N^n. In this methodology, the samples selected are not distinct.

For example, let us consider a population having a value of N = 4 and bearing values of sampling unit as y: 12, 16, 11, 17 and samples of dimension n = 2 has to be drawn from the population. The possible samples will be 4^2 = 16. The possible samples will be (12, 12), (12, 16), (12, 11), (12, 17), (16, 12), (16, 16), (16, 11), (16, 17), (11, 12), (11, 16), (11, 11), (11, 17), (17, 12), (17, 16), (17, 11) and (11, 17).

4.10.1 NOTATION AND TERMINOLOGY

Let us take into account a finite population of dimension N units, further suppose that 'y' is the variable under study. The capital letters are generally used for the description of the population, whereas the small letters refers the sample observations.

N: Refers the size of the Population.

n: The size of sub-population sampled from the population.

Y_i: It is the value of the study character for the ith (i = 1,2,...N) unit in the whole population.

y_i: It is the value of the study variable of the ith (i = 1,2,...n) unit sampled from the whole population.

$\overline{Y}_N = \frac{1}{N}\sum_{i=1}^{N} Y_i$: represents the population mean.

$\overline{y}_n = \frac{1}{n}\sum_{i=1}^{n} y_i$: corresponds to the sample mean.

$$S^2 = \frac{1}{N-1}\sum_{i=1}^{N}\left(Y_i - \overline{Y}^2\right)$$: Stands for population mean square.

$$s^2 = \frac{1}{n-1}\sum_{i=1}^{n}\left(y_i - \overline{y}^2\right)$$: symbolizes the sample mean square.

Theorem: In SRS, all the units have an equal chance of being included in the population, and the probability equals to 1/N.

Proof:

In SRSWR:

The figure of sampling units in the whole population have been always *N* as the sampling units are replaced back in the population after each draw, so the probability of selecting a sample at any draw is always 1/N.

In SRSWOR: From the population of size *N*, sampling units are drawn, and the probability of selection of unit at the first draw equals to $\frac{1}{N}$.

After the first draw, the number of sampling units in the whole population is (N–1), so the probability of selection of second unit from the available (N–1) units is $\frac{1}{N-1}$, and so on.

Here, we say E_1 be the event that specific unit is picked up in the 1st draw. i.e., $P(E_1) = \frac{1}{N}$.

At the next draw, i.e., 2nd draw, (N–1) units are available in the population for the sample selection. The possibility of selection of a unit in the second draw exists if and only if that unit is not picked up in the first draw and that unit is present in the leftover (N–1) units. The probability of non-selection of the unit at the first draw is $\left(1-\frac{1}{N}\right)=\left(\frac{N-1}{N}\right)$

Again, we suppose E_2 be the occasion that any specific unit is picked up at the 2nd draw.

$P(E_2)$ = probability that the specific unit is not picked up at the first draw and it is drawn from the remaining (N–1) units at the 2nd draw.

$P(E_2)$ = P(unit is not picked at 1st draw) *x* P(unit is selected at 2nd draw provided that it is not selected at the 1st draw)

(By compound probability theorem, since draws are independent)

$$P(E_2) = \left(\frac{N-1}{N}\right) \text{x} \left(\frac{1}{N-1}\right) = \frac{1}{N} = P(E_1)$$

Similarly, the probability of selecting a given unit at the 3rd draw.

$P(E_3)$ = probability that the specific sampling unit is not picked up at the first and second draws and it is drawn at the 3rd draw.

$P(E_3)$ = P(unit is not picked at 1st and 2nd draw) *x* P(it is drawn at 3rd draw provided it is not selected at the1st and 2nd draws)

$$= \left(\frac{N-1}{N}\right)\left(\frac{N-2}{N-1}\right) \text{x} \left(\frac{1}{N-2}\right) = \frac{1}{N} = P(E_1)$$

Proceeding in the same fashion,

We suppose E_r be the event that any specific sampling unit is picked up at the rth draw.

$P(E_r)$ = probability of selection of specific sampling unit at rth draw with condition that it is not chosen at any of the previous (r–1)

$P(E_r) = \prod_{i=1}^{r-1} P$ (it is not picked at ith draw) *x* P(it is drawn at rth draw provided it is not selected at the previous (r–1) draws)

$$P(E_r) = \prod_{i=1}^{r-1}\left[1-\frac{1}{N-(i-1)}\right] \text{x} \left(\frac{1}{N-(r-1)}\right) = \prod_{i=1}^{r-1}\left[\frac{N-i}{N-i+1}\right] \text{x} \left(\frac{1}{N-r+1}\right)$$

$$= \left[\left(\frac{N-1}{N}\right) \text{x} \left(\frac{N-2}{N-1}\right) \text{x} \left(\frac{N-3}{N-2}\right) \text{x} \ldots \text{x} \left(\frac{N-r+1}{N-r+2}\right)\right] \text{x} \left(\frac{1}{N-r+1}\right) = \frac{1}{N} = P(E_1).$$

This leads to very interesting and important properly of random sampling that the specific unit is drawn with same probability at any of the draw which is always equals to 1/N.

4.10.2. PROBABILITY OF SELECTION OF SPECIFIED UNIT IN THE SAMPLE

In SRSWOR:

Let us consider $s_1, s_2, \ldots, s_n$ be the units picked in the sample.

So, $P(s_1, s_2, \ldots s_n) = P(s_1).\ P(s_2) \ldots P(s_n)$; as the samples are selected independent of each other.

The probability of selection of a specified unit, i.e., ith unit say s_i in the sample can be done at any of the first draw, second draw, …, nth draw.

Thus the required probability is

$$P_1(i) + P_2(i) +. + P_n(i) = \frac{1}{N} + \frac{1}{N} + \ldots + \frac{1}{N} \text{ (n times)} = \frac{n}{N}$$

Again, let us consider $P_j(i)$ be the probability of selection of s_i unit at j^{th} draw

If $P(s_1) = \frac{n}{N}$, $P(s_2) = \frac{n-1}{N-1}$., $P(s_n) = \frac{1}{N-n+1}$,

So, the probability may be defined as

$$P(s_1, s_2, \ldots s_n) = \frac{n}{N}.\frac{n-1}{N-1}.\frac{n-2}{N-2}\cdots\frac{1}{N-n+1} = \frac{\lfloor n}{N(N-1)\ldots(N-n+1)} = \frac{1}{{}^{N}C_n}$$

In SRSWOR:

Let us suppose that $s_1, s_2, \ldots s_n$ be the units selected in the sample,

So, $P(s_1, s_2, \ldots s_n) = P(s_1).\ P(s_2)\ldots\ P(s_n)$; as the sample selection is independent of each other. So the required probability is

$$= \frac{1}{N}.\frac{1}{N}\cdots\frac{1}{N} = \frac{1}{N^n}.$$

Theorem: In SRSWOR, the sample mean is an unbiased estimate of population mean.

Proof: We have to prove that in SRSWOR, $E(\bar{y}_n) = \bar{Y}_N$.

$$E(\bar{y}_n) = E\left[\frac{1}{n}\sum_{i=1}^{n} y_i\right] = E\left[\frac{1}{n}\sum_{i=1}^{N} \alpha_i Y_i\right]$$

$$\text{where, } \alpha_i = \begin{cases} 1, \text{ if } i^{th} \text{unit is included in the sample} \\ 0, \text{ if } i^{th} \text{unit is not included in the sample} \end{cases}$$

$$E\left[\frac{1}{n}\sum_{i=1}^{n} \alpha_i Y_i\right] = \frac{1}{n}\sum_{i=1}^{n} E(\alpha_i) Y_i$$

Since α_i takes only two values, i.e., 1 and 0,

$E(\alpha_i) = 1.\ P(\alpha_i = 1) + 0.\ P(\alpha_i = 0) = 1.$ P(i^{th} unit is included in a sample of size n) + 0. P(^{ith} unit is not included in a sample of size n)

$$= E(\alpha_i) = 1.\ \frac{n}{N} + 0.\ P\left(1 - \frac{n}{N}\right) = \frac{n}{N}$$

Putting the value, we get

$E(\bar{y}_n) = \frac{1}{n}\sum_{i=1}^{N}\frac{n}{N}Y_i = \frac{1}{N}\sum_{i=1}^{N} Y_i = \bar{Y}_N$ as desired result.

Theorem: In SRSWOR, the sample mean square is an unbiased estimate of the population mean square.

Proof: Here, we have to proof E (s^2) = S^2

$$s^2 = \frac{1}{n-1}\sum_{i=1}^{n}\left(y_i - \bar{y}^2\right) = \frac{1}{n-1}\left[\sum_{i=1}^{n} y_i^2 - n\,\bar{y}^2\right]$$

$$= \frac{1}{n-1}\left[\sum_{i=1}^{n} y_i^2 - \frac{1}{n}\left(\sum_{i=1}^{n} y_i\right)^2\right]$$

$$= \frac{1}{n-1}\left[\sum_{i=1}^{n} y_i^2 - \frac{1}{n}\left(\sum_{i=1}^{n} y_i^2 + \sum_{i\neq j=1}^{n} y_i y_j\right)\right]$$

$$= \frac{1}{n-1}\left(1-\frac{1}{n}\right)\sum_{i=1}^{n} y_i^2 - \frac{1}{n(n-1)}\sum_{i\neq j=1}^{n} y_i y_j$$

$$= \frac{1}{n}\sum_{i=1}^{n} y_i^2 - \frac{1}{n(n-1)}\sum_{i\neq j=1}^{n} y_i y_j$$

$$E(s^2) = \frac{1}{n}E\left[\sum_{i=1}^{n} y_i^2\right] - \frac{1}{n(n-1)}E\left[\sum_{i\neq j=1}^{n} y_i y_j\right] \text{—*}$$

where,

$$E\left[\sum_{i=1}^{n} y_i^2\right] = E\left[\sum_{i=1}^{N} \alpha_i Y_i^2\right] \text{ and } E\left[\sum_{i=1}^{n} y_i y_j\right] = E\left[\sum_{i=1}^{N} \alpha_i \alpha_j Y_i Y_j\right]$$

$$\text{where, } \alpha_i = \begin{cases} 1, \text{ if i}^{\text{th}}\text{unit is included in the sample} \\ 0, \text{ if i}^{\text{th}}\text{unit is not included in the sample} \end{cases}$$

$$\text{and, } \alpha_i \alpha_j = \begin{cases} 1, \text{ if i}^{\text{th}}\text{unit is included in the sample} \\ 0, \text{ if i}^{\text{th}}\text{unit is not included in the sample} \end{cases}$$

$$E(\alpha_i) = \frac{\text{n}}{\text{N}} \text{ and } E(\alpha_i \alpha_j) = \frac{\text{n(n-1)}}{\text{N(N-1)}}$$

Substituting the values of E(α_i) and E(α_i α_j) in equation (*) we get,

$$E(s^2) = \frac{1}{n}\left[\frac{n}{N}\sum_{i=1}^{N} Y_i^2\right] - \frac{1}{n(n-1)}\left[\frac{n(n-1)}{N(N-1)}\sum_{i\neq j=1}^{N} Y_i Y_j\right]$$

$$E(s^2) = \frac{1}{N}\sum_{i=1}^{N} Y_i^2 - \frac{1}{N(N-1)}\sum_{i \neq j=1}^{N} Y_i Y_j$$

$$E(s^2) = \frac{1}{N}\left[\sum_{i=1}^{N} Y_i^2 - N\overline{Y}^2\right]$$

$E(s^2) = S^2$ as desired.

4.10.3 MERITS AND DEMERITS OF SIMPLE RANDOM SAMPLING

Merits:

1. Since, samples are selected randomly, i.e., having equal probability of inclusion in the sample for all the units, which completely eliminates personal bias.
2. A sample obtained through SRS is more representative as compare to that of sample obtained through purposive sampling.

Demerits:

1. The SRS method does not give efficient results for the heterogeneous population.
2. SRS may give chance to those sampling units which are widespread geographically, and the cost of collection of these data increase in terms of time, money, resources, etc.
3. SRS scheme requires updated sampling frame before selection of samples and sometimes it is too cumbersome to get the updated sampling frame.

4.11 STRATIFIED SAMPLING

The main goal in every estimation problem is to find an estimator of the population parameters (i.e., population mean, population variance, etc.), which can take care of all properties of the population under study. If the population under study is homogeneous with respect to the variable under study, then the method of SRS may provide better estimates of population parameters. Under homogeneous population with respect to characteristics under study, the drawn sample from SRS give representative sample. In case of heterogeneous population, SRS might not give a representative sample for the survey as well as may not give satisfactory results. In order to find a representative sample of the population in a heterogeneous population, we may apply SRS.

SRS reduces the heterogeneity of the whole population and may give reliable estimates of the population parameters. This sampling scheme is very useful and most commonly in sample surveys. For example; If we are interested to know about the average weight of the government school students of class 1 to class 12, the weight of changes a lot of class 1 to class 12. So, we divide the whole classes into four homogeneous blocks.

- Block 1 consist students of class 1 to 3,
- Block 2 consist students of class 4 to 6,
- Block 3 consist students of class 7 to 9,
- Block 4 consist students of class 10 to 12,

Further, draw the samples from each block by SRS and combine all the samples. Combination of these samples is the final sample for stratified random sampling for analysis.

4.11.1 DEFINITION

In this sampling scheme the whole heterogeneous population is partitioned into small homogenous groups or subpopulations in accordance to the characteristics under study. These small groups are called strata or block. Further, each subpopulation is treated as separate population and selection of samples is done by SRS independently from each stratum. Further, samples are drawn from each of the block or strata by any sampling method so that sample may be obtained which consists all the characteristics of the population under study. Such type of division of the whole heterogeneous population into a specific number of homogenous subpopulations is known as "stratification" and the method adopted is called as "stratified random sampling."

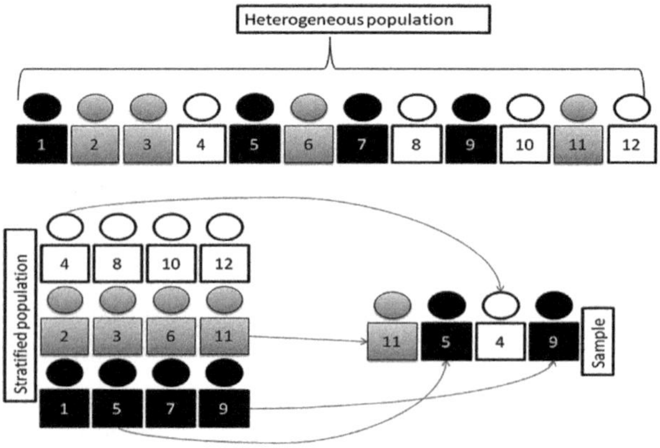

FIGURE 4.3 Diagrammatic representation of stratified sampling.

4.11.2 PRINCIPLES OF STRATIFICATION

The principles to be adopted in the stratification of heterogeneous population are stated as below:

(i) The block must be non-overlapping and should jointly include the entire population.

(ii) The stratification of the population must be arranged in such a way that block is homogenous in nature within themselves, but heterogeneous between themselves with respect to the study character.

Notations: Let us consider that the entire population under study be divided into k number of strata such that the size of ith stratum is N_i (i = 1, 2, …, k). From each block, independent samples of size n_i (i = 1, 2, …, k) are drawn using SRSWOR. Since the blocks are non-overlapping, we have

N_i: The total number of sampling units in the ith stratum.

n_i = The number of sampling units drawn with SRSWOR from i^{th} stratum.

$N = \sum_{i=1}^{k} N_i$: The total number of sampling units in the whole population.

$n = \sum_{i=1}^{k} n_i$: The total sample size drawn from the whole population.

k = The number of blocks or strata.

Y_{ij} = j^{th} sampling unit in i^{th} stratum in population. (j = 1, 2, …, N_i; i = 1, 2, …, k)

y_{ij} = j^{th} sampled unit in i^{th} stratum in sample. (j = 1, 2, …, n_i; i = 1, 2, …, k)

$$\overline{Y}_{N_i} = \frac{1}{N}\sum_{i=1}^{k} Y_{ij} = i^{th} \text{ stratum Population mean.}$$

$$\overline{Y}_N = \frac{1}{N}\sum_{i=1}^{k}\sum_{j=1}^{N_i} Y_{ij} = \frac{1}{N}\sum_{i=1}^{k} N_i\overline{Y}_{N_i} = \text{overall Population mean.}$$

$$\overline{Y}_{n_i} = \frac{1}{n}\sum_{i=1}^{k} y_{ij} = \text{The } i^{th} \text{ stratum sample mean.}$$

$$\overline{y}_n = \frac{1}{n_i}\sum_{i=1}^{k}\sum_{j=1}^{n_i} y_{ij} = \frac{1}{n}\sum_{i=1}^{k} n_i\overline{y}_{n_i} = \text{Overall sample mean.}$$

$$p_i = \frac{N_i}{N} : \text{The } i^{th} \text{ stratum weight.}$$

$$S_i^2 = \frac{1}{N_i - 1}\sum_{j=1}^{N_i}\left(Y_{ij} - \overline{Y}_{N_i}\right)^2 : \text{The } i^{th} \text{ stratum population mean square.}$$

$$s_i^2 = \frac{1}{n_i - 1} \sum_{j=1}^{n_i} \left(y_{ij} - \overline{y}_{n_i} \right)^2$$: The ith stratum sample mean square.

4.11.3 POPULATION MEAN ESTIMATION

Once the sample is selected, the next action is the population parameters estimation work. Among the different type of population parameters, population mean and the population variance is the most considered and common in use.

Sample mean should give an unbiased estimate of the population mean, i.e., $E(\overline{y}_n) = \overline{Y}_N$

$$E(\overline{y}_n) = E\left(\frac{1}{n} \sum_{i=1}^{k} n_i \overline{y}_{n_i} \right)$$

$$= \frac{1}{n} \sum_{i=1}^{k} n_i E\left(\overline{y}_{n_i} \right)$$

$$= \frac{1}{n} \sum_{i=1}^{k} n_i \overline{Y}_{N_i} \neq \overline{Y}_N$$

i.e., $\overline{y}_n$ gives the biased estimate of the population mean $\overline{Y}_N$. So we modify sample mean so as to acquire an unbiased estimator.

$\frac{n_i}{n} = \frac{N_i}{N}$ (weights of the blocks; is almost equal for the sample and population as the population with block is homogenous)

We modify the sample mean estimator as

$$\overline{y}_{st} = \frac{1}{N} \sum_{i=1}^{k} N_i \overline{y}_{n_i} = \sum_{i=1}^{k} p_i \overline{y}_{n_i} \ ; \ p_i = \frac{N_i}{N};$$

Weighted mean of strata sample mean or stratified sample mean.

Theorem: Stratified sample mean $\overline{y}_{st}$ gives an unbiased estimate of population mean $\overline{Y}_N$. i.e., $E(\overline{y}_{st}) = \overline{Y}_N$

Proof: Since the sample is selected from each block through SRS and we known that the sample mean is an unbiased estimate of the population mean in the case of SRS. Thus, we have,

$$E(\bar{y}_{st}) = E\left(\frac{1}{N}\sum_{i=1}^{k} N_i \bar{y}_{n_i}\right)$$

$$= \frac{1}{N}\sum_{i=1}^{k} N_i E\left(\bar{y}_{n_i}\right)$$

$$= \frac{1}{N}\sum_{i=1}^{k} N_i \bar{Y}_{N_i} = \bar{Y}_N$$

$$\therefore\ E(\bar{y}_{st}) = \bar{Y}_N$$

Which implies that sample mean of the stratified sampling provides an unbiased estimate of population mean.

4.11.4 POPULATION VARIANCE ESTIMATION

The sampling variance of the estimator $\bar{y}_{st}$ may be given as

$$V(\bar{y}_{st}) = V\left(\sum_{i=1}^{k} p_i \bar{y}_{n_i}\right)$$

$$V(\bar{y}_{st}) = \sum_{i=1}^{k} p_i^2\ V\left(\bar{y}_{n_i}\right) + 2\sum_{i=i'}^{k} p_i p_i^{'}\ \mathrm{cov}\left(\bar{y}_{n_i}, \bar{y}_{n_i^{'}}\right).$$

Since the sampling in each block is performed independently, hence the covariance term equals to zero, i.e., $\mathrm{cov}\left(\bar{y}_{n_i}, \bar{y}_{n_i^{'}}\right) = 0$ and the result reduces to the form

$$V(\bar{y}_{st}) = \sum_{i=1}^{k} p_i^2\ V\left(\bar{y}_{n_i}\right) \qquad \text{(a)}$$

As the sample in each block is selected using SRSWOR, we have

$$V(\bar{y}_{st}) = \sum_{i=1}^{k} p_i^2 \left(\frac{1}{n_i} - \frac{1}{N_i}\right) S_i^2 \qquad \text{(b)}$$

where S_i^2 is the ith stratum population mean square.

Deputing the result from equation (b) to equation (a), we have

$$V(\bar{y}_{st}) = \sum_{i=1}^{k} p_i^2 \left(\frac{1}{n_i} - \frac{1}{N_i}\right) S_i^2$$

Thus, we can see that $V(\bar{y}_{st})$ depends on S_i^2, the heterogeneity within the blocks or strata.

Number of strata: One should take as many strata as required to maintain the homogeneity of block or strata. Each block should have homogeneous nature within the block and heterogeneous nature between the block.

Sample size of each strata: Since the size of the sample persuades the estimates, it is necessary that the sample size of each block should be optimal. Decision of the sample size for each block or stratum is called allocation problem.

Allocation of sample size in different strata: In case of stratified sampling, the choice made in sample sizes n_1, n_2., n_k from different strata is done keeping in mind to avail the resources in effective way. The allocation is considered the important factors mentioned below:

1. The stratum size; N_i (i = 1,2,…,k)
2. The variability within the stratum; S_i^2 (i = 1,2,…,k)
3. The cost involved in collection of sampling unit in each stratum.

Good allocation is that, which achieves maximum precision with the minimum resources. The criteria for allocation may be done by minimizing the price of the survey for a given precision or maximizing the precision for a specified cost.

There are four different techniques of allocation of sample sizes in different strata in case of stratified sampling method mentioned below:

1. Equal Allocation: In this method of allocation, we select the sample size n_i to be the same for all strata. We draw samples of same size from all strata. This allocation is considered for the administrative or field work convenience. In this method, the sample size 'n' is partitioned equally among all strata. i.e.,

$$n_i = \frac{n}{k}; \text{ (i = 1, 2, …, k)}$$

Putting the value of n_i in the expression of variances of stratified sampling, i.e., $V(\bar{y}_{st})_{eq}$, we get $V(\bar{y}_{st})_{eq}$; variance of the stratified sampling in equal allocation.

$$V(\bar{y}_{st})_{eq} = \sum_{i=1}^{k} \left(\frac{k}{n} - \frac{1}{N_i}\right) p_i^2 S_i^2$$

2. Proportional Allocation: Bowley in 1926 recommended this method of allocation. This process of allocation is the most familiar in custom. This method of allocation is very easy, simple, and appropriate with practical viewpoint. When we are having information of only stratum size N_i ($i = 1, 2, \ldots, k$), the allocation of sample is done in proportional to their stratum size.

$n_i \propto N_i$ or

$n_i = CN_i$; where C is the constant for proportionality.

We take summation from the value 1 to k

$$\sum_{i=1}^{k} n_i = \sum_{i=1}^{k} CN_i$$

$$n = CN \Rightarrow C = \frac{n}{N}$$

putting the records of C, we get

$$n_i = \left(\frac{n}{N}\right) N_i \quad \text{or } n_i = \left(\frac{N_i}{N}\right) n = p_i n.$$

Putting the value of the term of $n_i = np_i$ in the expression of $V(\bar{y}_{st})$, we get $V(\bar{y}_{st})_{prop.}$; the variance of the stratified sampling bearing proportional allocation.

$$V(\bar{y}_{st})_{prop.} = \sum_{i=1}^{k} \left(\frac{1}{np_i} - \frac{1}{Np_i}\right) p_i^2 S_i^2 = \left(\frac{1}{n} - \frac{1}{N}\right) \sum_{i=1}^{k} p_i S_i^2.$$

3. Neymann Allocation: This method of allocation was given by Neymann in 1934. This allocation of sample is based on a joint consideration of stratum size and stratum variation. In this allocation procedure, it is supposed that the sampling cost per unit among different strata is the same and fixed. The sample size in different strata is allocated

$n_i \propto N_i Si$ or

$n_i = C' N_i S_i$ where C' is constant of proportionality.
We take summation from 1 to k

$$\sum_{i=1}^{k} n_i = \sum_{i=1}^{k} C' N_i S_i$$

$$n = C' \sum_{i=1}^{k} N_i S_i \text{ or } C' = \frac{n}{\sum_{i=1}^{k} N_i S_i}$$

substituting the value of the constant C' we get

$$n_i = \left(\frac{n}{\sum_{i=1}^{k} N_i S_i} \right) N_i S_i \ or \ n_i = n \frac{N_i S_i}{\sum_{i=1}^{k} N_i S_i} or \ n \frac{p_i S_i}{\sum_{i=1}^{k} p_i S_i}$$

Putting value of n_i in the equation of $V(\overline{y}_{st})$, we get $V(\overline{y}_{st})_{ney.}$; variance of the stratified sampling bearing Neymann allocation.

$$V(\overline{y}_{st})_{ney.} = \sum_{i=1}^{k} \left(\frac{\sum p_i S_i}{n p_i S_i} - \frac{1}{N p_i} \right) p_i^2 S_i^2 = \frac{1}{n} (p_i S_i)^2 - \frac{1}{N} \sum_{i=1}^{k} p_i S_i^2$$

$$V(\overline{y}_{st})_{ney.} = \frac{1}{n} (p_i S_i)^2 - \frac{1}{N} \sum_{i=1}^{k} p_i S_i^2$$

In applying this technique of allocation, some difficulty may arise because of the value of S_i being generally unknown. However, the stratum variances may be acquired from prior surveys or planned pilot survey.

4. Optimum Allocation: In this scheme of allocation, the size of the sample; n_i (i = 1,2,...k) from respective blocks or strata are decided with a vision to minimize variance, $V(\overline{y}_{st})$ for a pre-allotted cost for doing the survey work or to reduce cost of the survey for a particular value of $V(\overline{y}_{st})$. The simplest function of cost can be communicated as

$$C = C_0 + \sum_{i=1}^{k} n_i c_i$$

where the cost C_0 is acknowledged as the overhead cost which is constant and c_i is termed as the average cost of surveying per unit of the ith stratum. The cost c_i may depend upon nature and dimension of the units in the block or stratum.

To decide the optimal value of n_i, consider the function

$$\psi = v(\bar{y}_{st}) + \lambda C$$

$$\psi = \sum_{i=1}^{k} p_i^2 \left(\frac{1}{n_i} - \frac{1}{N_i} \right) S_i^2 + \lambda \left(C_0 + \sum_{i=1}^{k} n_i c_i \right)$$

For the optimum sample size, differentiate the function with respect to n_i and equate to zero. $\frac{d\psi}{dn_i} = -\frac{p_i^2 S_i^2}{n_i^2} + \lambda c_i = 0$

$$\lambda c_i = \frac{p_i^2 S_i^2}{n_i^2} \quad \text{or } n_i = \frac{p_i S_i}{\sqrt{\lambda}\sqrt{c_i}}$$

$$n_i \propto \frac{p_i S_i}{\sqrt{c_i}} \text{ or } n_i \propto \frac{N_i S_i}{\sqrt{c_i}} \text{ as } p_i = \frac{N_i}{N}$$

The expression for optimum sample size given as

$$n_i = n \frac{N_i S_i / \sqrt{c_i}}{\sum_{i=1}^{k} N_i S_i / \sqrt{c_i}}$$

Thus, the relation leads to important conclusions, large sample is required for a given stratum if:

a) The size of stratum is large, i.e., N_i is large.
b) The stratum has larger variability, i.e., S_i is large.
c) The cost involved in surveying per unit in stratum is cheaper, i.e., c_i is less.

Remarks:

- If the cost involved in survey per unit is same for all the strata, i.e., c_i is fixed, optimum allocation reduces to Neymann allocation.
- If c_i and S_i do not vary between strata, optimum allocation reduces to proportional allocation.

4.11.5 VARIANCE OF OPTIMUM ALLOCATION IS BASED ON PRINCIPLE OF OPTIMIZATION

(i) For given cost; and
(ii) For given precision

(i) For given cost: In the case, the assumption is that the cost is fixed, i.e., $C = C_0$ (say)

So, $C_0 = \sum_{i=1}^{k} c_i n_i$ and putting the value of n_i for optimum allocation

$$C_0 = \sum_{i=1}^{k} c_i \frac{p_i S_i}{\sqrt{\lambda}\sqrt{c_i}} = \frac{1}{\sqrt{\lambda}} \sum_{i=1}^{k} p_i S_i \sqrt{c_i}$$

$\sqrt{\lambda} = \frac{1}{C_0} \sum_{i=1}^{k} p_i S_i \sqrt{c_i}$; putting the value of $\sqrt{\lambda}$ in the expression of n_i, we get

$$n_i = \frac{p_i S_i C_0}{\sqrt{c_i} \sum_{i=1}^{k} p_i S_i \sqrt{c_i}}$$

This is the optimum allocation of sample size for the fixed cost.

Put the value of n_i in $V(\bar{y}_{st})$, we get $V(\bar{y}_{st})_{opt_c}$; variance of the stratified sampling for optimum allocation when the cost is fixed.

$$V(\bar{y}_{st})_{opt_c} = \sum_{i=1}^{k} \left(\frac{\sqrt{c_i} \sum_{i=1}^{k} p_i S_i \sqrt{c_i}}{p_i S_i C_0} - \frac{1}{Np_i} \right) p_i^2 S_i^2$$

(ii) For given variance: In this case, we assume that variance is predetermined, i.e., $V(\bar{y}_{st}) = V_0$ (say)

So, $V_0 = \sum_{i=1}^{k} \left(\frac{1}{n_i} - \frac{1}{N_i} \right) p_i^2 S_i^2$

Also, from optimum allocation we obtain, $n_i = \frac{p_i S_i}{\sqrt{\lambda}\sqrt{c_i}}$

Putting the value, we get

$$V_0 = \sum_{i=1}^{k} \left(\frac{\sqrt{\lambda}\sqrt{c_i}}{p_i S_i} - \frac{1}{Np_i} \right) p_i^2 S_i^2$$

Or, $\frac{1}{\sqrt{\lambda}} = \frac{\sum_{i=1}^{k} p_i S_i \sqrt{c_i}}{V_0 + \frac{1}{N}\sum_{i=1}^{k} p_i S_i^2}$; substitute this value in n_i, we get

$$n_i = \frac{p_i S_i}{\sqrt{c_i}} \frac{\sum_{i=1}^{k} p_i S_i \sqrt{c_i}}{V_0 + \frac{1}{N}\sum_{i=1}^{k} p_i S_i^2}$$

This is the optimum allocation of sample size for the fixed variance.

Putting the value of n_1 in the expression of $V(\bar{y}_{st})$, we get $V(\bar{y}_{st})_{opt_{var}}$; variance of the stratified sampling of optimum allocation for the fixed variance.

$$V(\bar{y}_{st})_{opt_{var}} = \sum_{i=1}^{k} \left(\frac{\sqrt{c_i} V_0 + \frac{1}{N}\sum_{i=1}^{k} p_i S_i^2}{p_i S_i \sum_{i=1}^{k} p_i S_i \sqrt{c_i}} - \frac{1}{Np_i} \right) p_i^2 S_i^2 .$$

4.11.6 MERITS AND DEMERITS OF THE STRATIFIED SAMPLING

Merits:

1. Samples collected through stratified random sampling may contain more characteristics of the population than the SRS.
2. It gives estimates with higher precision.
3. It is more concentrated geographically than SRS so there is reduced money, time, manpower, and work may be done with greater ease.
4. Stratified sampling reduces selection bias.

Demerits:

1. Stratified sampling may not be applied in every study.
2. The other drawback is sort of each unit of the entire population into single stratum accurate. In some practical situations, it is complicated to sort like in the case of sorting based on ethnicity, character, etc., and stratified sampling procedure renders into ineffective method.

4.12 CLUSTER SAMPLING

In many circumstances, the sampling frame may not available for the population; moreover, it is not easily prepared also (Figure 4.4). But the information may be obtainable for collections of elements so-called clusters. For example, the record of houses may be accessible, but not the persons inhabiting in them, inventory of individual farms may not be accessible, but the record of villages is generally available. Hence, in these conditions, houses or villages are known as clusters, and the selection is made of houses or villages in the sample. Such a sampling design is acknowledged as cluster sampling. In this sampling method, whole area enclosing the population is partitioned into undersized segments or clusters, and these clusters are occupied as the sampling units. In this sampling methodology, a sampling unit belongs to only one cluster, and no overlapping of units in different clusters are allowed.

In cluster design following procedure is pursued:

- Divide the entire population into clusters in accordance to some characterized guideline according to the objective of the work.
- Consider the clusters as the sampling unit.
- Draw a sample of clusters from the group of clusters of the population according to some method.
- Complete enumeration is done in the selected clusters of the sample.

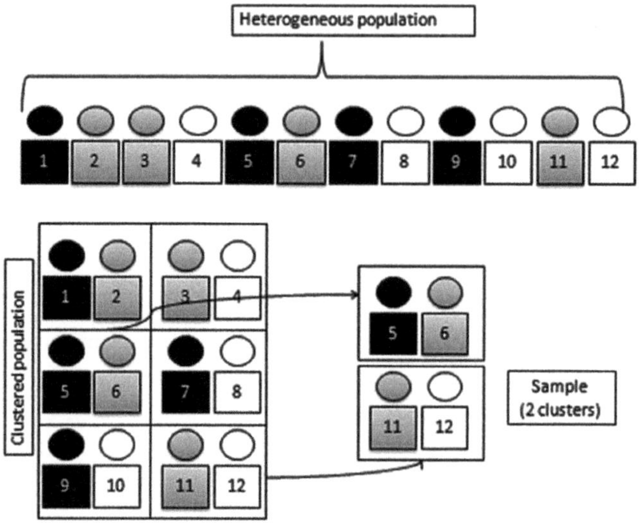

FIGURE 4.4 Diagrammatic representation of cluster sampling.

4.12.1 PREFERRED CLUSTER SAMPLING SITUATIONS

The situations in which the cluster sampling is preferred are:

i. Sampling frame is not readily accessible, and it is expensive and not practical to prepare it.
ii. Even if it is accessible, it is very complicated to locate the units of the whole population.
iii. Each of the sampling units must belong to one and only one cluster and all the clusters and should together constitute the whole population without any duplication or exclusion of the sampling units. This helps in the reduction of bias in the samples.

4.12.2 CONSTRUCTION AND SELECTION OF CLUSTERS

From the population, clusters are created in the approach that each cluster may represent the whole population, and the sampling units are so distributed in the clusters such that within clusters, sampling units are homogenous and between clusters, the units are homogenous. This technique is just the opposite as that in the construction of strata in stratified design.

There are two choices for the construction of clusters – Clusters may be of equal size, or they may vary in their sizes, i.e., unequal numbers of units in each cluster.

Various sampling procedures, i.e., SRS or stratified sampling or any other sampling may be applied by treating clusters themselves as the sampling units, i.e., complete enumeration is done in selected clusters.

Notations: For equal sizes clusters:

First, we consider the case of equal size clusters. Here, we consider that whole population is partitioned into *N* clusters and every cluster containing having M units. From the *N* clusters in the whole population, a sample of *n* clusters are selected by SRSWOR.

N represents the total clusters in the population.

n correspond to the clusters selected in the sample.

M is the figure of elements in each cluster.

y_{ij} symbolizes the value of the study variable of jth unit (j = 1,2,…, M) in the ith (i = 1,2,…N) cluster.

$\bar{y}_i = \frac{1}{M}\sum_{j=1}^{M} y_{ij}$ represents the ith cluster mean.

$\bar{\bar{y}}_n = \frac{1}{n}\sum_{i=1}^{n}\bar{y}_i$ symbolizes the mean of n cluster means.

$\bar{\bar{y}}_N = \frac{1}{N}\sum_{i=1}^{N}\bar{y}_i$ stands for the mean of N cluster means in the whole population.

$\bar{y}_{..} = \frac{1}{NM}\sum_{i=1}^{N}\sum_{j=1}^{M}y_{ij} = \bar{\bar{y}}_N$ represents the population mean.

$S_i^2 = \frac{1}{M-1}\sum_{j=1}^{M}\left(y_{ij} - \bar{y}_i.\right)^2$ is the mean square between elements in i[th].

$\bar{S}_w^2 = \frac{1}{M}\sum_{i=1}^{N}S_i^2$ is the mean square within the clusters.

$S_b^2 = \frac{1}{N-1}\sum_{i=1}^{N}\left(\bar{y}_{i.} - \bar{\bar{y}}_N\right)^2$: Mean square between clusters mean in the population.

$S^2 = \frac{1}{N-1}\sum_{i=1}^{N}\sum_{j=1}^{M}\left(y_{ij} - \bar{y}_{..}\right)^2$: Mean square between elements in the population.

Theorem: In cluster design of equal size clusters, sample mean is an unbiased estimate of the population mean.

Proof: we have to prove that $E(\bar{\bar{y}}_n) = \bar{\bar{y}}_N$

$$E(\bar{\bar{y}}_n) = E\left(\frac{1}{n}\sum_{i=1}^{n}\bar{y}_i\right)$$

$$= \frac{1}{n}\sum_{i=1}^{n}E(\bar{y}_i)$$

$$= E\left(\frac{1}{N}\sum_{i=1}^{N}\bar{y}_i\right)$$

$$= \frac{1}{n}\sum_{i=1}^{n}(\bar{\bar{y}}_N)$$

$E(\bar{\bar{y}}_n) = \bar{\bar{y}}_N$; hence proved.

Variance of Cluster Design: Since, n clusters are drawn from the population of N clusters using SRSWOR and we know that $v(\bar{y}_n)_{SRSWOR} = \frac{N-n}{nN} S_y^2$. Using this result of SRSWOR, variance of the sample mean in case of cluster design having fixed numbers of elements in each cluster is $v(\bar{\bar{y}}_N) = \frac{N-n}{nN} S_b^2$.

Unequal Clusters: In the previous segment, we have discussed the case when all the clusters are of same size, but practically, this may not always be feasible, and clusters may vary in their sizes.

In that case, we suppose there are N clusters in the population having unequal elements in each cluster.

Further, we consider M_i represents the elements in i^th^ cluster (i = 1,2,…,N).

Let $M_0 = \sum_{i=1}^{N} M_i$ is the sum of all the units in all the clusters in the whole population.

Per unit population mean is identified as $\bar{y}_{..}$

$$\bar{y}_{..} = \frac{1}{M_0} \sum_{i=1}^{N} \sum_{j=1}^{M} y_{ij} = \frac{1}{M_0} \sum_{i=1}^{N} M_i \bar{y}_{i.} \neq \bar{Y}_N$$

where $\bar{y}_{i.} = \frac{1}{M_i} \sum_{i=1}^{M_i} y_{ij}$ is the mean of ith cluster in case of unequal size clusters.

4.13 SYSTEMATIC SAMPLING

In the earlier sections, we have discussed the sampling design where samples are selected randomly. Here, in this section, we discuss a new sampling design in which only one unit is picked randomly, and the remaining samples are automatically chosen in accordance with some predefined pattern having systematic space between units in the samples. This type of sampling design having regular spacing units is known as systematic sampling design. It is functionally more handy than any other sampling design. In this sampling design, entire population of size N is serially nomenclatured from 1 to N in particular order, and a sample of size n is picked from the whole population in such a manner that N is expressible as the product of two integers k and n such that N = nk or k = N/n

A random sample 'i' is picked up such that 1< i < k and after that samples are selected after regular gap. The sample of size *n* withdrawn from the target population undertakes the units i, i+k, i+2k., i+(n–1)k. The number "i" selected randomly which decides the whole population is termed as Random Start. This type of sampling design is more suitable in the situations in which we have the complete knowledge of sampling frame. For instance, in the estimation of timer in forest survey, every k^{th} tree is picked in the sample from the whole population of the forest. Selection of cereals fields every k^{th} hectare distant for observation of incidence of pests (Figure 4.5).

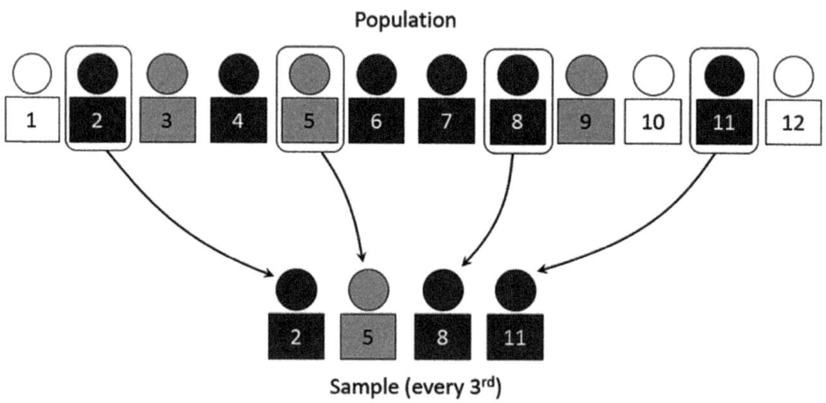

FIGURE 4.5 Diagrammatic representation of systematic sampling.

Notations: Let us consider y_{ij} is the jth element of the ith sample (i = 1, 2, …, k; j = 1, 2, …, n)

$\bar{y}_{sys.} = \bar{y}_{i.} = \frac{1}{n}\sum_{j=1}^{n} y_{ij}$ denotes the mean of ith systematic sample.

$\bar{Y}_N = \frac{1}{N}\sum_{i=1}^{k}\sum_{j=1}^{n} y_{ij} = \frac{1}{nk}\sum_{i=1}^{k}\sum_{j=1}^{n} y_{ij} = \frac{1}{k}\sum_{i=1}^{k}\bar{y}_{i.} = \bar{y}_{..}$ represent the Population mean.

$S^2 = \frac{1}{N-1}\sum_{i=1}^{k}\sum_{j=1}^{n}(y_{ij} - \bar{y}_{..})^2 = \frac{1}{nk-1}\sum_{i=1}^{k}\sum_{j=1}^{n}(y_{ij} - \bar{y}_{..})^2$ is the

Population mean square.

Systematic Random Samples

Random Start	Units in the sample	Probability	Mean
1	1 1+k. 1+(n–1)k	1/k	$\bar{y}_1$
2	2 2+k. 2+(n–1)k	1/k	$\bar{y}_2$
.	.	.	.
.	.	.	.
j	j j+k. j+(n–1)k	1/k	$\bar{y}_{j.}$
.	.	.	.
.	.	.	.
k	k 2k. nk	1/k	$\bar{y}_{k.}$

k rows of the table give the k systematic samples.

Theorem: Sample mean $\overline{y}_{sys.}$ gives the unbiased estimate of the population mean.

Proof: Since each one of the N sampling units in the population occurs once and the k samples and has an equal chance of inclusion in the sample and the probability of selection of ith unit (i = 1, 2, …, k) as systematic sample is 1/k.

$$\text{So, } E(\overline{y}_{sys.}) = E(\overline{y}_{i.}) = \frac{1}{k}\sum_{i=1}^{k}\overline{y}_{i.}$$

$$= \frac{1}{k}\sum_{i=1}^{k}\frac{1}{n}\sum_{j=1}^{n}y_{ij}$$

$$= \overline{y}_{..} = \overline{y}_N$$

$$E(\overline{y}_{sys.}) = \overline{y}_N \text{ Required result}$$

Example: Assume a hypothetical population of size 12 and a sub-population (sample) of size 3 is drawn from the target population.

Here, N = nk can be represented as 12 = 3x4, and the samples may be arranged as below:

1	5	9
2	6	10
3	7	11
4	8	12

Further, we draw a random number 'i' i.e., random start from 1 to 4 (say 3). Consequently the units of the third row bearing the serial numbers 3, 7, 11 will be selected in the sample.

But it not always possible that N is an integral multiple of 'k.' As if N = 22 and have to select 5 samples from that population. In this condition, $N \neq$ nk. For this type of situation, we opt circular systematic design.

4.13.1 CIRCULAR SYSTEMATIC SAMPLING

This systematic sampling design is applicable in the case at that time when $N \neq$ nk and the 'k' (sampling interval) is not uniquely defined. The usual systematic sampling is slightly modified, and samples are always obtained. In this case, we take 'k' as the integer nearest to N/n. After select of 'k,' draw a random number between 1 to k, and then systematic samples i, i+k, i+2k and so on in circular manner are picked up from the population until requisite number of samples are worn out from the population.

In this sampling methodology also, every sampling units has the equal probability of selection and the probability is always 1/N. This can be demonstrated from example given below:

Suppose N = 24 and n = 5 and k is the integer which is nearest to 24/5 = 4.8 and hence we take k = 5. The list of all the possible samples are given below:

S. No.	Samples	S. No.	Samples
1	(1,6,11,16,21)	13	(13,18,23,4,9)
2	(2,7,12,17,22)	14	(14,19,24,5,10)
3	(3,8,13, 18,23)	15	(15,20,1,6,11)
4	(4,9,14,19,24)	16	(16,21,2,7,12)
5	(5,10,15,20,1)	17	(17,22,3,8,13)
6	(6,11,16,21,2)	18	(18,23,4,9,14)
7	(7,12,17,22,3)	19	(19,24,5,10,15)
8	(8,13,18,23,4)	20	(20,1,6,11,16)
9	(9,14,19,24,5)	21	(21,2,7,12,17)
10	(10,15,20,1,6)	22	(22,3,8,13,18)
11	(11,16,21,2,7)	23	(23,4,9,14,19)
12	(12,17,22,3,8)	24	(24,5,10,15,20)

We can see from the table that probability of selection of any sample is always 1/24.

4.13.2 MERITS AND DEMERITS OF SYSTEMATIC SAMPLING

Merits:

1. This sampling scheme is functionally more handy and simple than any other sampling design.
2. Selection of samples is less costly as compared with other schemes.
3. Equally spaced samples are selected in the procedure hence it gives more precise estimates.

Demerits:

1. Systematic sampling will give highly biased results if the information with periodic features are associated.

4.14 PROBABILITY PROPORTIONAL TO SIZE SAMPLING

As the name itself is self-explanatory that in this sampling design sampling units are selected with the probability associated with the samples in proportional to their sizes. Which means giving higher weightage of selection to the larger units as compared to units bearing smaller clusters. In the previous sections, we have studies many sampling design in which every units have got equal probability of inclusion in the sample, but there are many circumstances, more efficient estimates are attained by allocating unequal probabilities of selection to the sampling units in the population. This type of methodology is termed as varying probability design of sampling.

Here, we consider y to be the study variable and x is some additional information known as auxiliary variable which is correlated to the study variable y, then the most common varying probability design is the design in which sampling units are drawn in accordance to their sizes. This category of sampling design is expressed as *probability proportional to size (pps) sampling*. This methodology give emphasis to the weightage of larger units in the population as its contribution in the population total is more as compared to that of smaller units. To illustrate this in our day to day life, let us consider an example of agricultural survey, in which yield depends on the area under farming and bigger farms add up more to the population total and consequently, area is considered as the size of the auxiliary information which is correlated to the yield of the production.

Two methods are available for the selection of sample in PPS sampling:

i. Cumulative total method; and
ii. Lahiri's method

4.14.1 CUMULATIVE TOTAL METHOD

Following steps are followed for the selection of sample:

- List all the units of the population by associating a number from 1 to N to every units of the population.
- Write the cumulative totals of the sizes as T_1, T_2. T_N.
- Selection of a random number R is done; such that $R \leq \sum_{i=1}^{N} X_i$
- If $X_1 + X_2 + X_3 + ..., +X_{i-1} \leq R \leq X_1 + X_2 + X_3 + ..., +X_{i-1} + X_i$ i.e., $T_{i-1} \leq R \leq T_i$ then i[th] unit will be selected.
- Continue the process until n samples are selected from the population.

This procedure can be demonstrated in the following table:

Unit	Size	Cumulative Totals		
1 2 i–1 i N	X_1 X_2 X_{i-1} X_i X_N	$T_1 = X_1$ $T_1 = X_1+X_2$ $T_{i-1} = \sum_{k=1}^{i-1} X_k$ $T_i = \sum_{k=1}^{i} X_k$ $T_N = \sum_{k=1}^{N} X_k$	Selection of a random number R is done such that R lies between 1 to T_N.	• If $T_{i-1} \leq R \leq T_i$ then i[th] unit is picked with its associated probability X_i/T_N; i = 1, 2. N. • Replicate the process until the desired number of samples (i.e., n units) are selected.

The ith unit is selected with the probability.

$$P_i = \frac{T_i - T_{i-1}}{T_N} = \frac{X_i}{T_N}$$

$P_i \,||\, X_i$; (T_N is the population constant which is a constant quantity.)

This process of sample selection requires note down of successive cumulative totals which is lingering process and tedious if we have to deal with the large population. To deal with this problem, Lahiri (1955) put forward another method which does not involve to pen down the cumulative totals.

4.14.2 LAHIRI'S METHOD:

Following steps are followed for the selection of sample in Lahiri's method:

- Selection of a pair of random number say (i, j) is done in such a way that $1 \leq j \leq M$ and $1 \leq j \leq M$ where M is the maximum of all sizes (X_1, X_2., X_N) i.e., M = Max X_i (i = 1,2,...N).
- If $j \leq X_i$, then the ith unit will be selected in the sample.
- If this condition is not satisfied, reject that pair of random number and choose another pair of random number and again check the above-mentioned condition.
- Repeat the process until *n* units are picked up in the sample.

Example: Consider the data set of 8 numbers of staff in the farm and its output.

Farm no.	No. of staff (in thousands), X	Production in metric tonnes), Y	Cumulative totals of size
1	3	25	$T_1 = 3$
2	4	30	$T_2 = 3+4 = 7$
3	7	17	$T_3 = 3+4+7 = 14$
4	10	25	$T_4 = 14+10 = 24$
5	12	20	$T_5 = 24+12 = 36$
6	6	10	$T_6 = 36+6 = 42$
7	2	8	$T_7 = 42+2 = 44$
8	3	7	$T_8 = 44+3 = 47$

4.14.3. SAMPLE SELECTION BY CUMULATIVE TOTAL METHOD

First draw: Select a random table between 1 to 47.

Suppose it is 25, $T_4 < 25 < T_5$

Y_5 is drawn and $Y_5 = 20$ enters the sample.

The process is repeated until required number of samples are obtained.

4.14.4 SELECTION OF SAMPLE BY LAHIRI'S METHOD

In this method; $M = \underset{i\,=\,1,2,\ldots 8}{Max} X_i = 12$

After that pair of random numbers (i, j) is selected such that $1 \leq i \leq 8$ and $1 \leq j \leq 12$

Table shows sample selection from Lahiri's method

Random no. $1 \leq i \leq 8$	Random no. $1 \leq j \leq 12$	Observation	Selection of sampling unit
2	5	$j = 5 > X_2 = 4$	Trial rejected
5	10	$j = 10 < X_5 = 12$	Trial selected(y_5)
7	8	$j = 8 > X_7 = 2$	Trial rejected
3	2	$j = 2 < X_3 = 7$	Trial selected(y_3)

Here, y_3 and y_5 are drawn in the sample.

4.15 MULTI-STAGE SAMPLING

As we have already discussed cluster sampling, in which all the units are observed of the selected clusters, if we increase the sample sizes, then the efficiency of cluster sampling decrease. In other words, we can say, if the number of elements within the clusters is low, then we get high efficiency, but if the number of elements within a cluster is high, then we get low efficiency. So, to cope up with the problem of low efficiency, we may go for subsampling technique. The procedure of selection of clusters is called the first stage sampling and clusters called first stage units and selection of elements of the selected clusters is called second stage sampling and elements within a cluster is called two stage units. The method is easily generalized to more than two stages, and in general, it is known as multi-stage sampling. For example, in a university survey, colleges may be considered as the first stage units, degree courses (BSc Agriculture or MSc Agriculture) within the college may be considered as the second stage units and students within programs i.e., degree courses may be considered as the third stage units.

4.15.1 STEPS INVOLVED IN THE SELECTION OF SAMPLE IN MULTI-STAGE SAMPLING

- The number of elements in the whole population is *NM.*
- The number of first stage units is *N* and size of all first units is M.
- The size of selected sample is *n.*
- The numbers of selected second stage units is 'm' from each first stage unit's *n.*
- Units from the first stage and second stage are picked with SRSWOR.

Note:

- Cluster sampling may be considered as the special case of two-stage sampling in the way that from the population of N clusters, a sample of n clusters chosen.
- If all clusters consist only one sample, we get SRSWOR.
- If $n = N$, we get stratified sampling.

Advantages:

The two-stage sampling is more handy than the one stage sampling. It will convert to one stage sampling when $m = M$. If we make clusters with small group of elements then it will be more efficient. This choice reduces the cost of the surveys.

4.15.2 SUCCESSIVE SAMPLING

Change is an inherent tendency of the environment. Some types of change straightway affect the living and surroundings of the human beings. Such variations depict the notice of human to recognize the patterns or rate of change at different points (occasions) of time or to know the quantity (real situation) at any given point of time (occasion) or simultaneously to know both the situations.

It is often seen that a population having a large number of elements remains unchanged on several occasions, but the values of the units change. An array of practical problems could fall in the arena of applied and environmental sciences where various characters opt to change over time with respect to different parameters. In such cases, successive (rotation) sampling is the most suitable statistical tool to provide the reliable estimates of population parameters on different occasions.

One time surveys may give the information only about the characteristic of the study variable and may not give information about the rate of change of the study variable characteristics from occasion to occasion and all estimates for all occasions. To get such information, we done sampling on consecutive occasions and a subset of the sample taken and at the same time as the rest of the sample is drawn afresh for using estimates for occasions. This procedure of sampling is known as successive (rotation) sampling, for example; yearly labor force surveys are done to estimate the numbers of people in employment and opinion surveys on political issue are conducted at intervals to know most preferable voter, etc.

The main feature of successive sampling (continuous survey) is the nomenclature of the sample on each phase or occasion. The configurations of the sample may be directed by the subsequent three types of arrangements:

(i) Selection of a new sample on every occasion (Repeated Sampling)
(ii) Selection of same sample on every occasion (Panel Sampling)
(iii) Partial selection of some units from previous occasion and some new units on the current occasion (Sampling on Successive Occasions or Rotation Sampling).

4.15.2 REPEATED SAMPLING

As the name itself is self-explanatory that in this scheme, sampling is repeated at every occasion and the main objective is to estimate the overall average at each occasion and a fresh sample is drawn from the same population at different occasions. Figure shows the sample structure of repeated sampling.

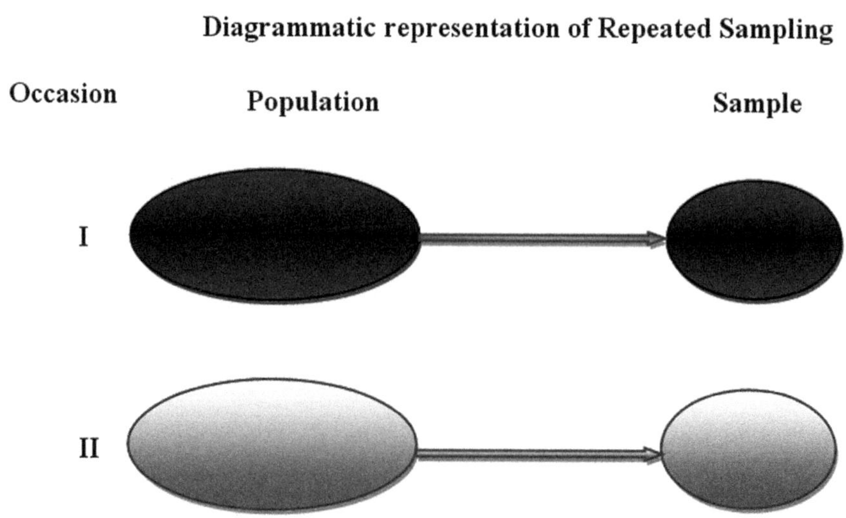

FIGURE 4.6 Diagrammatic representation of repeated sampling.

4.15.3 PANEL SAMPLING

In this sampling design, the main objective is to estimate the change included in the parameters due to the forces acting on the population. To achieve this, sample is retained at different occasions from the population to estimate the change.

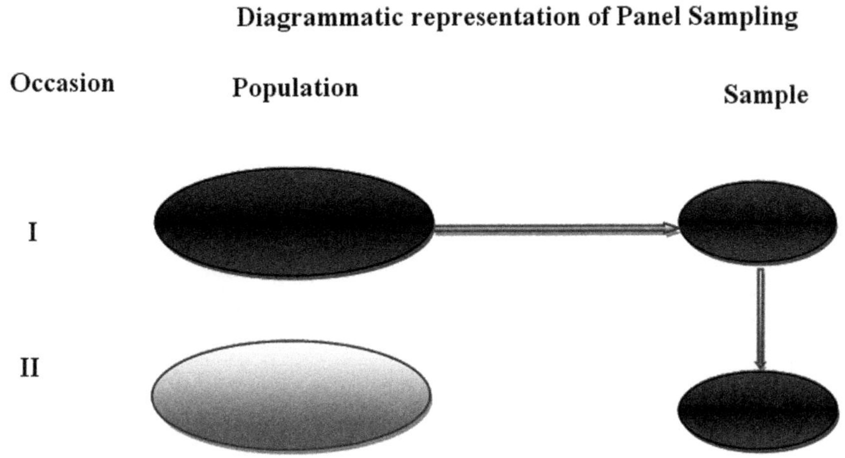

FIGURE 4.7 Diagrammatic representation of panel sampling.

4.15.4 ROTATION SAMPLING

In rotation sampling the objective is to estimate the average value for the most recent (current) occasion, the retention of a subset of the sample over occasions gives estimates as compared to other alternatives (Singh and Choudhary, 1986). This sampling is widely used when the population parameters are likely to change in accordance with time. There are different nomenclatures of this sample design in survey literature. Some researchers call this procedure as "Sampling on Successive Occasions with Partial Replacement of Units" while others call this as "Rotation Sampling" or "Rotation Designs for Sampling on Repeated Occasions" or "Sampling for Time Series."

Several interesting problems could fall in this arena, which we call as hunt of good rotation patterns. Some of the illustrations may be quoted in the following lines according to the different fields of study:

(i) Socio-Economic and Agricultural Field: (i) To know the employment status at time to time and the change in employment status over the period of time. (ii) To know the average purchasing power per household at time to time and to know the behavior of change in purchasing power over the period of time, and (iii) To know the agricultural production at time to time and to know the pattern of variation in agricultural production over the period of time, etc.

(ii) Demographic Field: To know the birth rate, death rate, infant mortality rate, migration patterns at time to time and to know the pattern of changes in these parameters over the period of time.

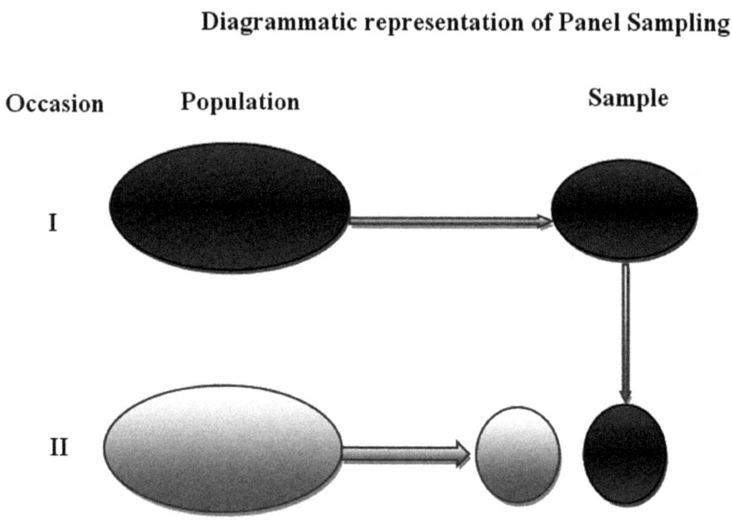

FIGURE 4.8 Diagrammatic representation of rotation sampling

4.16 ERRORS IN SAMPLE SURVEY

Difference between statistic (estimated value) and the parameter (true value) is known as error in sample surveys. Errors are very common problem in any type of surveys (census survey and sample survey). Conducting a complete survey without errors is not possible. Errors occur at every stage of the surveys. If we conduct any surveys in good manner, then the chances of errors will be reduced. Some errors can be reduced, but some are not. This error may be classified into two broad categories: (i) sampling error (ii) nonsampling error.

4.16.1 SAMPLING ERROR

This error arises when estimators based on sample observation are used to estimate the population parameters. As we know sample does not involve all the units of the target population, so there is a difference in the value of the estimator and population parameters. The difference between the sample and population values is considered as the sampling error. Sampling error occurs only in sample survey, not in census survey. If we increase the sample size, then the amount of sampling errors will reduce. In the case of census survey, the amount of sampling error is zero because in this case, we observed all the units in the target population.

4.16.2. NON-SAMPLING ERRORS

The non-sampling errors cannot be avoided in any type of survey. This error occurs at each step of the surveys. Non-sampling errors are increasing as the amount of sample increase. They're various types of non-sampling errors. Errors which arises due to reasons other than sampling are called *non-sampling errors*. Census survey does not include sampling error, but it is not free for non-sampling errors. The sample survey includes both, sampling errors and non-sampling errors. In some situations, the non-sampling errors may be higher and needs greater attention than the sampling errors. Broadly, non-sampling error may be categorized into five groups: specification errors, coverage errors, non-response, response errors, and processing errors.

(a) ***Specification errors*** arise at planning stage because of a variety of causes, e.g., poor quality of data with respect to the objectives of the surveys, omission or copying of the units, not proper care is taken in the method of schedules/interview, etc.

(b) ***Coverage errors:*** It represent a failure to include some units of the target population in sampling framework. There are two types of coverage errors (i) under coverage error (ii) over coverage error. Under coverage, error occur when an incomplete sampling frame is used to draw the sample. In other words, if we select the sample from the list, which is not including all the units of the population, then this error is called under coverage error. Whereas *over coverage errors* take place when the sampling frame consist of members that does not belong to target population. In other words, if we select the sample from the list, which consist extra units which is not a part of the population.

(c) ***Non-response errors***: Non-response means absence of response. The state which we are not getting any response from the respondents then this case is known as nonresponse. In such cases the results obtain from this data will be an inappropriate for any further conclusions. There are several techniques to deal with the problems of missing values such as

Hansen Hurwitz (1946) method and imputation methods, etc. There are two categories of non-response errors one is *Unit non-response,* and other one is the *Item non-response.* When we are not getting from the selected sample units, then this error is known as a *unit non-response error.* On the other hand, when we are getting partial responses from the respondents due to his unwillingness, ill understating or recall lapse for some particular questions will end up this item non-response error. In the situations where surveys based on sensitive issues respondents may refuse to give answers to some personal questions.

(d) ***Response Errors:*** Response error occurs when the selected sample unit giving wrong responses. In the case of sensitive issues such as their income level, questions related to their sexuality or some illegal acts the respondents tend to lie. To limit such errors, there is method in survey sampling is known as randomized response technique.

(e) ***Processing Errors:*** As the survey data may be edited for stability and for imputation because of the non-response in the data, errors may be introduced in the result. Observation and awareness of both computer and human checking may help in the minimization of this error. Proper handling of the data compilation also minimizes this processing error (Figure 4.9).

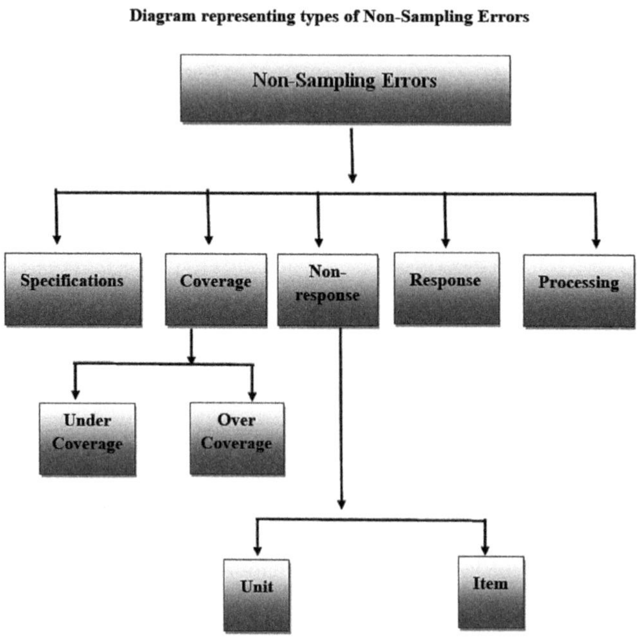

FIGURE 4.9 Diagram representing types of non-sampling errors.

KEYWORDS

- **cluster sampling**
- **multistage sampling**
- **simple random sampling**
- **stratified sampling**

REFERENCES

Gupta, S. C., & Kapoor, V. K., (2007). *Fundamentals of Applied Sciences*, page 1-708, Sultan Chand & Sons, Educational Publishers, New Delhi.

Kushwaha, K. S., & Kumar, R., (2009). *The Theory of Sample Surveys and Statistical Decisions*, page 321, New India Publishing Agency, New Delhi.

Singh, V. K., & Singh, L. B., (1999). Estimating the population mean using auxiliary information in the presence of non-response. In: Chattopadhyay, A., (ed.), *Mathematics and Statistics in Engineering and Technology*. Narosa Publishing House, New Delhi, India.

Sukhatme, P. V., & Sukhatme, B. V., (1970). *Sampling Theory of Surveys With Applications*, page 1-452, Asia Publishing House, India.

Thompson, S. K., (2012). *Sampling* (3rd edn.), page 1-446, Wiley Publication.

Yates, F., (1981). *Sampling Method for Censuses and Surveys* (4th edn.).page 1-458, Charles Griffin and Company Limited, London.

CHAPTER 5

Time Series

B. S. DHEKALE[1], K. P. VISHWAJITH[2], PRADEEP MISHRA[3], G. K. VANI[4], and D. RAMESH[5]

[1]*Assistant Professor, Sher-e-Kashmir University of Agriculture Sciences and Technology, Kashmir, India, E-mail: bhagyashreedhekale@yahoo.com*

[2]*Department of Agriculture Statistics, Bidhan Chandra Krishi Vishwavidyalaya, Nadia – 741252, India*

[3]*Assistant Professor (Statistics), College of Agriculture, JNKVV, Powarkheda, Hoshangabad (M.P.), 461110, India*

[4]*Assistant Professor (Agricultural Economics and F.M.), College of Agriculture, JNKVV, Jabalpur (M.P.), 482004, India*

[5]*Assistant Professor, Department of Statistics and Computer Applications, ANGRAU, Agricultural College, Bapatla, Andhra Pradesh, India*

5.1 DATA/INFORMATION

Statistics data is defined as the "facts or figures from which conclusions can be drawn." The data can also be defined as the study of characteristic varying over many individual objects, over the places, over the times, and so on. The data having its own characteristics, pattern, and variability which are needed to study before any statistical analysis. There are two types of data in econometrics: time series and cross-sectional.

5.1.1 TIME SERIES

In statistical analysis, most of the time we are dealing with data which is timely ordered, i.e., the same variable observed and measured at consecutive points of time. Time series data in which observations are taken or arranged chronologically, i.e., according to time points. Usually but not necessarily, the points of time are equally spaced. Thus, the production of rice in India over the years, accumulated rainfall of particular month over year, daily closing of specific stock recorded over last 10 weeks, birth, and mortality rate over the years in different states of India, the intensity of pests in cotton crop over the weeks, etc., are the examples of time series data.

5.1.2 CROSS-SECTIONAL

On the other hand, sometimes we deal with information about different individuals (or aggregates such as work teams, sales territories, stores, etc.) at the same point of time or during the same time period. A particular commodity, say pesticide price, at a given point of time over different retail shops, constitutes a cross-sectional data.

Data (Table 5.1A) on the productivity of rice in India from 2004–15 and monthly auction price of North Indian tea are examples of time series data where data is collected for a different months of year 2015 and arranged chronologically. On the contrast, data in Table 5.1B denotes the Production of wheat in different states of India in 2015–16, where data is collected at a particular point of time at various places which represent cross-sectional data.

TABLE 5.1 Time Series and Cross Section Data

A. Time series data				B. Cross section data	
Productivity of rice in India from 2004–15		Monthly auction price of North Indian tea		Production of wheat in different states of India in 2015–16	
Year	Productivity (Kg/hectare)	Month	Price (Rs./Kg)	Year	Production (000'tonnes)
2004	1658	Jan–15	125.00	Uttar Pradesh	30302
2005	1883	Feb–15	103.00	Punjab	16591
2006	1990	Mar–15	90.00	Madhya Pradesh	13133
2007	2024	Apr–15	137.00	Haryana	11117
2008	2095	May–15	147.00	Rajasthan	9275

TABLE 5.1 *(Continued)*

A. Time series data				B. Cross section data	
Productivity of rice in India from 2004–15		Monthly auction price of North Indian tea		Production of wheat in different states of India in 2015–16	
Year	Productivity (Kg/hectare)	Month	Price (Rs./Kg)	Year	Production (000'tonnes)
2009	2082	Jun–15	154.00	Bihar	5357
2010	2019	Jul–15	153.00	Gujarat	2944
2011	2121	Aug–15	153.50	Maharashtra	1181
2012	2312	Sep–15	145.00	West Bengal	896
2013	2374	Oct–15	142.00	Uttarakhand	858
2014	2319	Nov–15	146.00		
2015	2295	Dec–15	140.00		

A time series is a realization of a sequence of values of a variable indexed by time. It is a functional relationship between the time 't' and the realization of a characteristic 'v_t' at time period 't,' i.e., $V_t = f(t)$, where V_t denotes the value of variable or characteristic at time period t. Hence, t_1, $v_{t2}, \ldots, v_{tk}$ be the values for the variable at time period $t_1, t_2, t_3, \ldots, t_k$, respectively. The sequence of random variable $\{V_t: t = 0, \pm1, \pm2, \pm3, \ldots\}$ is called stochastic process and is a special type of stochastic process. Each element of the time series is treated as a random variable with a probability distribution. It is a bivariate distribution of two variables viz. the time (t) and the other one is value of the variable realized at time 't,' i.e., V_t.

Time series data can be continuous or discrete. Continuous time series data are recorded instantaneously, i.e., at every instant of time. For example, electrocardiograms, oscillograph which records harmonic oscillation of an audio amplifier. When data is recorded at regular or specific interval, known as discrete. Prices of a commodity for last six weeks or accumulated rainfall measured at regular interval, etc. Time series are best displayed in a scatter plot considering time as the independent variable known as historigram (Figures 5.1 and 5.2).

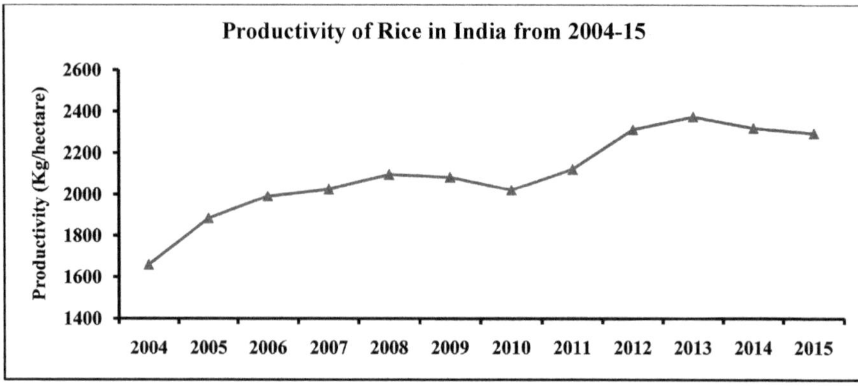

FIGURE 5.1 Graphical representation of productivity of rice in India from 2004–15.

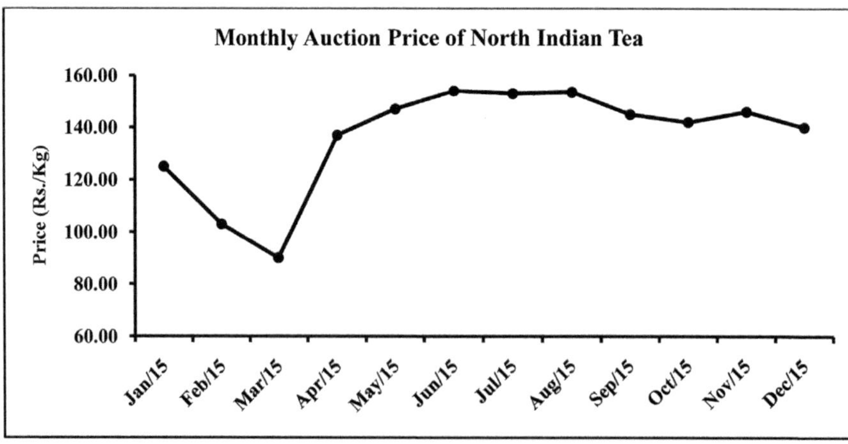

FIGURE 5.2 Graphical representation of monthly auction price of North Indian tea.

5.2 NEED TO STUDY THE TIME SERIES

Nowadays, time series modeling is a vibrant research area which grabbed attention of the researchers. The time series analysis also helps in isolating and measuring the various components of the time series. Time series also analysis helps in identifying the difference between the actual phenomenon and expectation of a realization. The main objective of time series analysis is to rigorously study the past phenomenon of a time series and to develop an appropriate model which describes the innate structure of the series which can be further used to generate future values for the series. Time series forecasting is another research arena based on time series modeling for future planning, particularly in business planning.

Time series can be categorized into stationary and non-stationary series. When observations of series fluctuate about a fixed mean level with constant variance for given period and the overall behavior of the series remains same over time, known as stationary time series. Whereas when time series does not fluctuate about fixed mean level and exhibits a floating or wandering behavior. A time series is weakly stationary if the mean function $\mu(V_t)$ is independent of t and the covariance functions $\gamma(V_{t+h,t})$ is independent of t for each h. On the other hand, a strictly stationary time series is defined if it satisfied the condition that $V_1, V_2, \ldots\ldots\ldots, V_n$ and $V_{1+h}, V_{2+h}, \ldots\ldots\ldots, V_{n+h}$ have the same joint distributions for all integer h and $n > 0$.

5.3 COMPONENTS OF TIME SERIES

A time series, in general, is supposed to be affected by four main components, which are categorized as:

(1) trend (secular trend);
(2) seasonal component;
(3) cyclical component; and
(4) irregular component.

The first three components are known as the systematic part and can be attributed to several factors. On the contrast, irregular component is a part, which can't be ascribed to any factors, any assigned cause. Usually but not necessarily, all-time series data have all the above components. A time series data may be comprised of two or more of the above components. The total food-grain production in India during the period of 1955–2015, population of India for the period 1950–2015, fertilizer use in India during the post-green revolution period, etc., are some of the examples of time series having secular trend, maybe with cyclical component and irregular components. But certainly, from these yearly time series, one cannot work out the seasonal components. While time series on accumulated rainfall for June to September from 1985–2015 exhibit secular trend, seasonal, and irregular component.

5.3.1 TREND COMPONENT

Trend is a long term overall movement in a time series which can be upward or downward but should be static. Let us consider the production of Rice in India from 1960–2015. The objective is to help the policymaker to formulate the policy according to production so that new policies can be made for benefit of farmer. From the figure 5.3, it can be seen that in spite of minor fluctuation, there is an overall tendency of the production of Rice to increase over time. This upward movement of the time series data is known as the secular trend. From Figure 5.3, it is clearly visible that clear idea that production is increasing. The example given below is of an upward trend. For example, suppose sale of soap A may decrease over a period of time because of better soap B coming to the market. This is an example of declining trend or downward trend. It is also be noted that secular trend need not to be necessarily linear, it can be non-linear, exponential, etc.

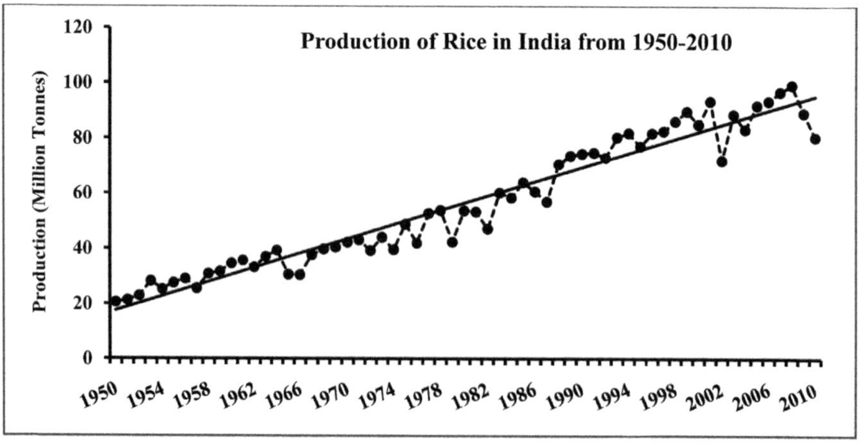

FIGURE 5.3 Production of rice in India.

5.3.2 CYCLICAL COMPONENT

The oscillatory movement of a time series with recurrent upward and downward movement continuing more than a year is known as cyclical movement of the time series. Cyclical component is a non-seasonal component describes the regular fluctuation. In cyclic variation length and size of cycle doesn't remain constant. The ups and downs in business activities are the effects of cyclical variation. A business cycle showing these oscillatory movements has to pass through four phases-prosperity, recession, depression, and recovery. In a business, these four phases are completed by passing one to another in this order. For example, the outbreak of locusts in different parts of India, outbreak of anthrax disease, fluctuation in prices of onion, etc., are some of the examples of cyclical movement of time series.

5.3.3 SEASONAL COMPONENT

Seasonal variations are the short-term fluctuation in a time series, which occur periodically in a year. Fluctuations are generally continues to repeat year after year. Seasonal fluctuations are repetitive and occur mainly due to seasonal variations, i.e., weather conditions and customs of people. Demand for raincoats are more in rainy season than winter and summer. Despite the consequences of trend sale of ice creams are more in summer and very little in winter season. Occurrence of particular diseases and pests because of congenial climatic condition and hosts, Because of changeover of season, viral fevers are more common. Occurrence of rainfall is more in June to September than any other months, etc., are the examples of seasonal component.

5.3.4 IRREGULAR COMPONENT

The irregular component is the leftover part when other components of the series (trend, seasonal, and cyclical) have been accounted for. Above mentioned three components viz. the secular trend, the seasonal component, and the cyclical component are due to assignable causes and can be isolated or separated using proper technique. As name indicates, these components follow no regularity in the occurrence pattern and fluctuations in time series that are short in duration, erratic in nature. Irregular component is generally unforeseen and beyond the control of human beings. Sometimes these irregular components may be considerably, high or may be negligible. It should be kept in mind that for forecasting purpose this irregular component of the time series is required to be minimized as far as possible.

5.4 TIME SERIES MODELS

The problem of combining the time series components is studied under time series modeling; the components may be added, multiplied or algebraically treated to get the values of the phenomenon at different point of time.

5.4.1 ADDITIVE MODEL

A time series can be thought of as a manifestation of additive effects of its components. As such a time series can be decomposed into its components:

$$U_t = T_t + S_t + C_t + E_t$$

where

T_t = the secular trend component at time period t;
S_t = the seasonal trend component at time period *t* and $\Sigma St = 0$;
C_t = the cyclical trend component at time period *t* and $\Sigma Ct = 0$; and
E_t = the error term at time period *t* and $\sum Et = 0$.

5.4.1.1 ASSUMPTION

The additive model assumes that the four components in a time series are independently operating in a time series. Thus, none of the components, i.e., trend, seasonal component or the cyclical component has any impact on the remaining components. In real life situation, it is hard to think of independence of these four components. In most of the business and economic time series, the components of time series are not independent to each other. However, for the simplicity in use, this model it is being extensively used.

5.4.2 MULTIPLICATIVE MODEL

The four components of time series, in classical approach, are assumed to be multiplicative in nature. That means U_t the value realized at time period *t* can be expressed as

$$U_t = T_t.S_t.C_t.E_t$$

where
T_t = the secular trend component at time period t;
S_t = the seasonal component at time period t;
C_t = the cyclical component at time period t; and
E_t = the error/irregular component corresponding to time period t.
Taking logarithm for both sides the series can be written as

$$\log U_t = \log T_t + \log S_t + \log C_t + \log E_t$$

or, $Y_t = T + S + C + E$, where $T = \log(T_t)$, $S = \log(S_t)$, $C = \log(C_t)$, $E = \log(E_t)$

Thus, a multiplicative model reduces to an additive model. In most of the business and economic time series the multiplicative model is found to be operative in nature.

In multiplicative or additive models one or more components of time series can be worked (by dividing or subtracting) out if one or more components of the series are known.

5.4.3 MIXED MODEL

It is not necessary that the different components of time series can be combined/decomposed only in the form of additive or multiplicative models. These four components can very well be combined/decomposed in different combinations of additive and multiplicative forms under different assumptions. Different forms of mixed models can be

$U_t = T_t.S_t + C_t + E_t$
$U_t = T_t.\ C_t + S_t + E_t$
$U_t = T_t + S_t.C_t + E_t$ and so on.

Existence of different components of time series can be accomplished by examining the nature of data provided. Nature of data gives us idea about the different components of time series. If the time series data is yearly then series can be comprised of three components, viz. the secular trend component, the cyclical component and the irregular component not seasonal while if data is given daily/weekly/fortnightly/monthly basis then seasonal components can be worked out. The existence of different components, by, and large, can visually be examined by drawing scattered diagram

A time series may comprise of different components, and these individual components can be isolated from the series if time series follows an additive model. In doing so different components in a time series are assumed to be independent of each other and effect of one components can be isolated from the time series by subtracting or dividing one or more components. In time series analysis, by examining the past performance of series and isolating the different components forecasting can be done.

5.5 INSPECTION OF TIME SERIES DATA

It is always keep in your mind to examine the time series data at first instance before going for any analysis. Sometimes time series data may contain one or more missing values. These values can be filled by using mean or median of a series depending on the data itself. It may contain outliers or extreme observations which may deflect analysis from original. Hence, it is always advisable to remove outlier using proper techniques.

Homogeneity of time series is prime importance. Suppose you are studying the production of rice in Bihar from 1990–2010. But in year 2000 Bihar state is divided and northern part is named as Bihar while southern part is known as Jharkhand. Due to this, the time series on production of rice may not be homogenous. There may be drastic differences in the values of production before and after the year 2000. If a long-term time series is homogeneous, then all variability and change in series can be worked out easily. Even sometimes according to various situations time series can be adjusted. Data pertaining to production would better be converted to production per units of area. Similarly, instead of taking monthly total sales of goods/salary/wages, it is better to divided each figure by the number of working days to make comparable on per day basis.

5.5.1 GRUBB'S TEST FOR DETECTING OUTLIERS

Time series data are often prone to the presence of outlier. An outlier in this context is an observation somewhat different from the rest of the observations, may be extreme in the series which can deviate the results of analysis. These outliers may be due to human (typological) error or due to particular cause (sudden increase in area under lentil due to increase in MSP in previous year). There are several ways to detect outliers.

Grubbs' test, sometimes referred to as the maximum normal residual test ESD method (extreme studentized deviate). It is used to detect outliers in univariate data distributions. It is based on the assumption that the data being analyzed were drawn from a normal distribution. Grubbs' test is a sequential iterative algorithm in that it detects only one outlier at a time and this outlier is deleted from the dataset, and the test is iterated again until no outliers are detected. The test is based on the difference of the mean of the sample and the most extreme data considering the standard deviation. Two-sided Grubbs test (Grubb, 1950) is often used to evaluate measurements, coming from a normal distribution of size *n,* which are suspiciously far from the main body of the data. It is the largest absolute deviation from the sample mean and is expressed in units of the sample standard deviation.

Grubbs' test is defined for the following hypothesis:

H_0: There are no outliers in the data set.

H_A: There is at least one outlier in the data set.

For a two-sided Grubb's test, the test statistic is defined as:

$$G = \frac{\max\limits_{i=1,\ldots n} \left| y_i - \overline{y} \right|}{s}$$

With $\overline{y}$ and s denoting the sample mean and standard deviation, respectively, calculated including the suspected outlier. The critical value of the Grubb's test is calculated as

$$C = \frac{(n-1)}{\sqrt{n}} \sqrt{\frac{t^2_{(\alpha/2n,n-2)}}{n-2+t^2_{(\alpha/2n,n-2)}}}$$

where $t_{(\alpha/2n,n-2)}$ denotes the critical value of the t-distribution with (n–2) degrees of freedom and a significance level of $\alpha/2n$. If $G > C$, then the suspected measurement is confirmed as an outlier.

Once outlier is detected, one may choose to exclude/replace the value from the analysis or one can go for transformation of data or may choose to keep the outlier. In our study, if only one outlier was detected, it was replaced by the median, which is often referred to as robust (i.e., small variability) in the presence of a small number of outliers and of course, it is the preferred measure of central tendency for skewed distributions. If the more number of outlier was detected due to particular cause, we used suitable transformation of data before further analysis.

5.5.2 RANDOMNESS TEST

Time series data consist of a systematic pattern (trend, cycle or seasonal) and random noise (error), which usually makes the pattern difficult to identify. Patterns in data series indicates the variations which are due to causes that come from outside the system. But inherent variations in a series or variations due to common cause will exhibit random behavior. Randomness in data series makes it difficult to identify the real pattern in a series and hence it is more important to check for randomness so that necessary efforts can be taken to reduce or cancel the effect due to random variation using smoothing techniques.

The present test for randomness is a non-parametric test based on the number of turning points used when sample size is large. The process is to count peaks and troughs in the series. A "peak" is a value greater than the two neighboring values, and a "trough" is a value, which is lower than of its two neighbors. Both the peak and trough are treated as turning points of the series. Thus, to get a turning point, one needs at least there data points. The number of turning points is clearly one less than the number of runs up and down in the series. The interval between two turning points is called a "phase." Three consecutive observations are required to define a turning point, μ_1, μ_2, μ_3. If the series is random, these three values could have occurred in any order of six possibilities. In only four of these ways would there be a turning point (when the greatest or least value is in the middle). Hence, the probability of a turning point in a set of three values is 2/3.

Let us consider now a set of values $\mu_1, \mu_2, ..., \mu_n$, and let us define a "marker" variable X_i by

$$X_i = \begin{cases} 1, & \begin{cases} \mu_i < \mu_{i+1} > \mu_{i+2} \\ \mu_i > \mu_{i+1} < \mu_{i+2} \end{cases} ; i = 1, 2, \ . \ . \ . \ , n - 2 \\ 0, & otherwise \end{cases}$$

The number of turning points p is then simply

$$p = \sum_{i=1}^{n-2} X_t$$

on simplification, one can workout

$$E(p) = \sum E(X_i) = \frac{2}{3}(n-2)$$

$$E\left(p^2\right) = E\left(\sum_{i=1}^{n-2} X_i\right)^2, \text{ on simplification}$$

$$E\left(p^2\right) = \frac{40n^2 - 144n + 131}{90} \text{ and}$$

$$V(p) = \frac{16n - 29}{90}$$

It can easily be verified that as the number of observation increases (n), the distribution of 'p' tends to normal. Thus, for testing the null hypothesis, i.e., series is random

We have the test statistic, $\tau = \frac{p - E(p)}{s_p} \sim N(0,1)$

where s_p is the standard deviation of 'p.'

Thus, if the calculated value of τ is greater than 1.96, we reject H_0 that the series is random otherwise accept it.

Example 1: Following is the data of area under tea production in Assam from 1971–2010. Confirm whether any outlier is present in data and is following any pattern?

Year	Area (ha)	Year	Area (ha)	Year	Area (ha)	Year	Area (ha)
1971	182325	1981	203038	1991	233284	2001	269154
1972	184244	1982	211320	1992	233658	2002	270683
1973	185113	1983	213007	1993	231942	2003	271589
1974	187408	1984	214741	1994	227120	2004	270475
1975	188794	1985	215117	1995	226281	2005	300502
1976	189338	1986	222618	1996	228205	2006	311822
1977	190621	1987	225783	1997	229843	2007	321319
1978	192427	1988	227517	1998	251625	2008	322214
1979	195459	1989	229428	1999	258455	2009	322214
1980	200569	1990	230363	2000	266512	2010	322214

Solution:

GRUBB'S TEST:

First, we will test the outliers using Grubbs' test. The null and alternate hypothesis are given below:

H_0: There are no outliers in the data set.

H_A: There is at least one outlier in the data set.

For a two-sided Grubb's test, the test statistic is defined as:

$$G = \frac{\max\limits_{i=1,\ldots n} |y_i - \bar{y}|}{s}$$

With $\bar{y}$ and s denoting the sample mean and standard deviation

$$\bar{y} = 236458.5 \text{ and } s = 42827.15$$

$$G = \frac{\max\limits_{i=1,\ldots n} |y_i - \bar{y}|}{s} = \frac{\max\limits_{i=1,\ldots n} |y_i - 236458.5|}{42827.15} = 2.002$$

Critical value of G is calculated by

$$C = \frac{(n-1)}{\sqrt{n}} \sqrt{\frac{t^2_{(\alpha/2n,n-2)}}{n-2+t^2_{(\alpha/2n,n-2)}}}$$

where $t_{(\alpha/2n,n-2)}$ denotes the critical value of the t-distribution with (n–2) degrees of freedom and a significance level of $\alpha/2n$.

$$\text{Value of } \begin{matrix} t_{(\alpha/2n,n-2)} = t_{(0.05,38)} = 2.02 \\ t^2_{(0.05,38)} = 4.09 \end{matrix}$$

$$C = \frac{(n-1)}{\sqrt{n}} \sqrt{\frac{t^2_{(\alpha/2n,n-2)}}{n-2+t^2_{(\alpha/2n,n-2)}}} = \frac{(40-1)}{\sqrt{40}} \sqrt{\frac{4.09}{40-2+4.09}} = 1.92$$

Hence as $G < C$, no outlier is present in data.

Randomness of data will be tested using turning point test. Null and alternate hypothesis is given as follows:

H_0: The series is random.

H_a: The series is not random.

The test statistic is given below:

$$\tau = \frac{p - E(p)}{s_p} \sim N(0,1)$$

$$\tau = \frac{p - E(p)}{s_p} = \frac{8 - 40}{10.7} = 4.26$$

As calculated value of τ is greater than 1.96, we reject H_0 that the series is random.

5.6 MEASUREMENT/ISOLATION OF TREND

In time series analysis trend isolation is one of the major tasks. Trend is considered as an additive component which contains information about the inherent variability of a time series. The trend is the long-term movement of a time series. Before proceeding for any analysis, first-time series data is needed to brought in to yearly figures (if not provided) by using proper technique. Trend of a series can be isolated by following methods.

(1) free hand curve fitting;
(2) method of semi-average;
(3) method of moving average; and
(4) method of mathematical curve fitting.

5.6.1 FREE HAND CURVE METHOD

The freehand curve is easiest and simplest method to study the trend of series. This method is devoid of complex mathematical manipulations and subjective in nature. In this method, by hand such curve is drawn by plotting the values of the series over a time. Linear or non-linear types of trends can also be explained using this method. The procedure for drawing freehand curve is as follows:

(i) At first plot, the original data are on a graph paper and carefully observed the direction of the plotted data.
(ii) Draw a smooth line the plotted points with free hand.

Let us take the data of yearly productivity of rice in India given in Figure 5.4 . If trends on these data drawn by two different persons, they may get trends as given below. Hence, while drawing the curve using this method, some precautions are needed to be taken.

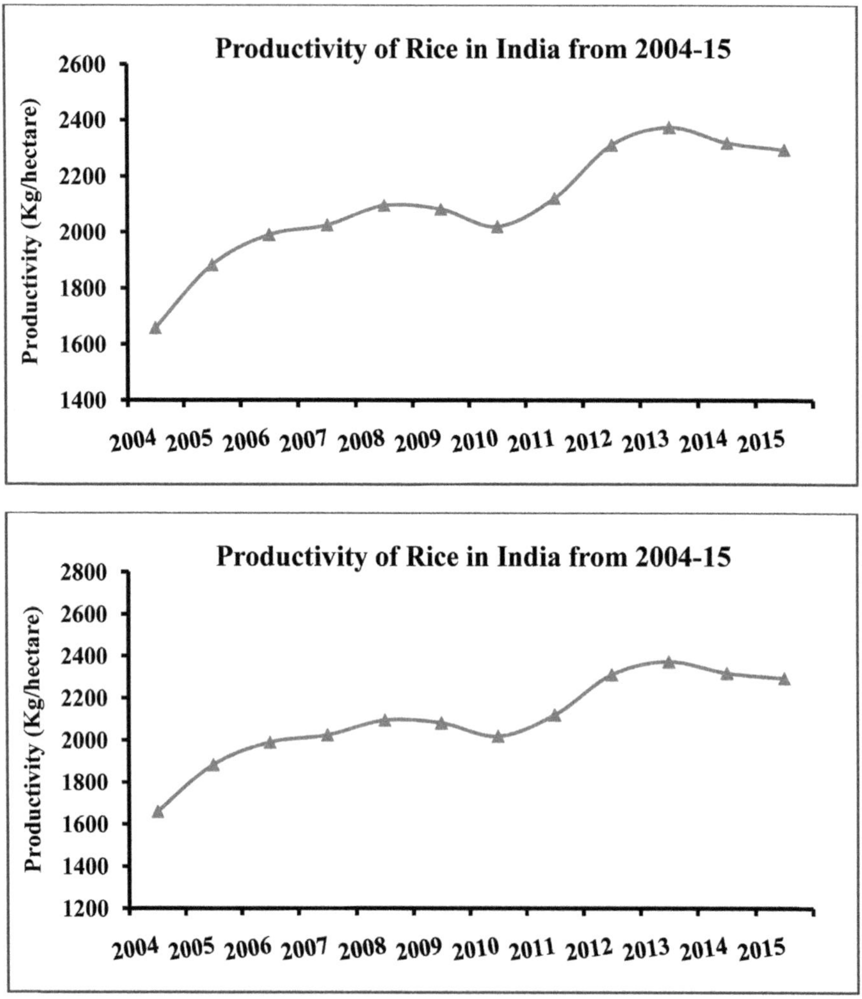

FIGURE 5.4 Productivity of rice in India, 2004–2015 (Yearly sales).

(i) Draw a curve as smooth as possible in such way that trend line divides the plot area into equal half, i.e., the areas below and above the trend line are equal to each other.

(ii) The vertical deviations of the data above the trend line must equal to the deviations below the line.

(iii) Sum of the squares of the vertical deviations of the observations from the trend should be minimum.

5.6.2 TREND BY THE METHOD OF SEMI-AVERAGES

This method is more objective than above method and can be used if a straight-line trend is to be obtained. The procedure to obtain trend is as follows:

1. Divide the data series into two equal halves.
2. Calculate average for each half. Each of these averages is shown against the mid-point of the half period; such that two points are obtained.
3. By joining these points, a straight-line trend is obtained.

The method is to be praised for its simplicity and flexibility. It is used to some extent in practical work. But if the time series consist of odd number of observations be then there arises the problem in dividing the whole series into equal halves. In such situations, general rule is to keep aside the middle most observation while partitioning whole time series into two equal halves.

Take an example of a time series data for the period 1984 to 2015, then the total number of observations will be 32, an even number, so we will divide the whole time series (1984–2015) in to two equal halves of periods between 1984–1999 and 2000–2015. On the other hand, if we are dealing with the time series data, 1984–2014, then the total number of observations is 31, an odd number. In this case, we shall keep aside the middle most observation, i.e., the observation corresponding to the year 1999 and frame two halves consisting of the observations corresponding to the period 1984–1998 and 2000–2015.

Example 1: Consider the example of productivity of rice in India during 2004–2015 (see Figure 5.5). Here the number of year is 12, an even number. So we shall have two periods 2004–2009 and 2010–2015 having corresponding mid-period of 2006–2007 and 2012–2013, respectively. The semi-average values are 65.88 (A) and 89.68 (B), respectively, for the two halves.

Productivity of Rice in India from 2004–15		
Year	Productivity (Kg/hectare)	HA
2004	1658	
2005	1883	
2006	1990	
2007	2024	1955.33
2008	2095	
2009	2082	
2010	2019	
2011	2121	
2012	2312	
2013	2374	2240
2014	2319	
2015	2295	

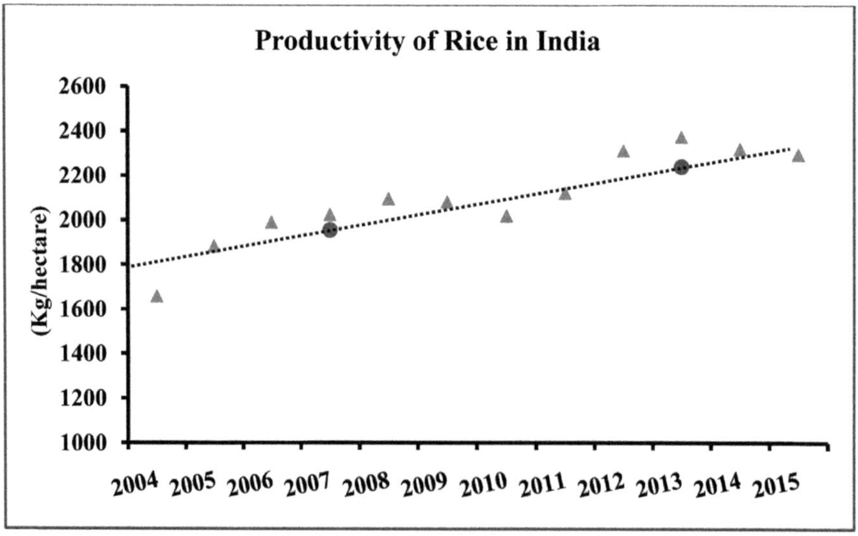

FIGURE 5.5 Rice productivity in India during 2004–2015.

5.6.3 METHOD OF MOVING AVERAGE

The moving average is a simple and flexible process of trend measurement which is quite accurate under certain conditions. This method establishes a trend by means of a series of successive averages covering overlapping periods of the data. The process of successively averaging, say, three years data, and establishing each average as the moving-average value of the central year in the group, should be carried throughout the entire series. For a five-item, seven-item or other moving averages, the same procedure is followed: the average obtained each time being considered as representative of the middle period of the group.

The choice of a 5-year, 7-year, 9-year, or other moving average is determined by the length of period necessary to eliminate the effects of the business cycle and erratic fluctuations. A good trend must be free from such movements, and if there is any definite periodicity to the cycle, it is well to have the moving average to cover one cycle period. Ordinarily, the necessary periods will range between three and ten years for general business series but even longer periods are required for certain industries.

If time series data consists of even number of observations then the moving average covers an even number of years, each average will still be representative of the midpoint of the period covered, but this mid-point will fall halfway between the two middle years. In the case of a four-year moving average, for instance, each average represents a point halfway between the second and third years. In such a case, a second moving average may be used to 'recenter' the averages. That is, if the first moving averages give averages centering half-way between the years, a further two-point moving average will recenter the data exactly on the years. This method, however, is valuable in approximating trends in a period of transition when the mathematical lines or curves may be inadequate. This method provides a basis for testing other types of trends, even though the data are not such as to justify its use otherwise.

Example 2: Using the data of sales of productivity of gram in India during 2000–2015 work out the 2 years, 3 years and 4 years moving average trend values.

Year	Productivity (Kg/ha)	Two-years moving average		Three-years moving average	Four-years moving average		Five-years moving average
		Two yearly moving average	Two period centered moving average		Four yearly moving average	Two period centered moving average	
2000	1691						
		1734.50					
2001	1778		1711.50	1689.33			

Year	Productivity (Kg/ha)	Two-years moving average		Three-years moving average	Four-years moving average		Five-years moving average
		Two yearly moving average	Two period centered moving average		Four yearly moving average	Two period centered moving average	
		1688.50			1669.00		
2002	1599		1646.00	1661.67		1649.13	1669.00
		1603.50			1629.25		
2003	1608		1586.75	1579.67		1600.38	1629.25
		1570.00			1571.50		
2004	1532		1554.75	1562.33		1566.00	1571.50
		1539.50			1560.50		
2005	1547		1545.25	1544.67		1583.13	1560.50
		1551.00			1605.75		
2006	1555		1611.50	1630.33		1637.75	1605.75
		1672.00			1669.75		
2007	1789		1730.25	1710.67		1701.00	1669.75
		1788.50			1732.25		
2008	1788		1790.50	1791.33		1753.75	1732.25
		1792.50			1775.25		
2009	1797		1777.25	1770.67		1739.88	1775.25
		1762.00			1704.50		
2010	1727		1689.25	1676.67		1679.50	1704.50
		1616.50			1654.50		
2011	1506		1581.75	1607.00		1628.25	1654.50
		1547.00			1602.00		
2012	1588		1567.25	1560.33		1606.50	1602.00
		1587.50			1611.00		
2013	1587		1631.25	1646.00		1632.63	1611.00
		1675.00			1654.25		
2014	1763		1698.00	1676.33			
		1721.00					
2015	1679						

5.6.4 METHOD OF LEAST SQUARE

As has been mentioned earlier, polynomial trend equations along with several other forms of trend equations can be fitted using ordinary least square technique. The credit of least-squares method is goes to Gauss in 1975 (Bretscher, 1995) but it was first published by Adrien-Marie Legendre in 1805 (Stigler, 1981). In the following sections, we shall find the method of fitting linear trend by taking suitable examples.

5.6.4.1 FITTING OF A LINEAR TREND

A straight-line trend is defined as

$$y_t = a_0 + a_1 t + e_t$$

where y_t is the realized value of the time series at time t, a_0 and a_1 are the parameters of the trend equation and e_t is the error component associated with y_t and e_t's are i.i.d $N(0, \sigma^2)$. According to the principle of least square,

$$S = \sum_t e_t^2 = \sum_t (y_t - \hat{y}_t)^2 = \sum_t (y_t - \hat{a}_0 - \hat{a}_1 t)^2$$

The sum of square (S) of residue is minimum when

$$\frac{\delta S}{\delta a_0} = 0 \text{ and } \frac{\delta S}{\delta a_1} = 0.$$

Which means

$$-2\sum_t (y_t - a_0 - a_1 t) = 0$$

and $$-2\sum_t (y_t - a_0 - a_1 t)t = 0. \quad (1)$$

The equations in (1) are called the normal equations. Thus from (1), we have

$$\sum_t y_t = \sum_t a_0 + \sum_t a_1 t = na_0 + a_1 \sum_t t$$

and

$$\sum_t ty_t = a_0 \sum_t t + a_1 \sum_t t^2$$

$\sum t$, $\sum t^2$, $\sum y_t$ and $\sum ty_t$ are the known quantities

From 1st equation we have

$$\sum_t y_t = na_0 + a_1 \sum_t t$$

$$a_0 = \frac{\sum_t y_t}{n} - \frac{a_1 \sum_t t}{n} = \bar{y}_t - a_1 \bar{t} \text{..........(2)}$$

and from the 2nd equation, we have

$$\sum_t ty_t = a_0 \sum_t t + a_1 \sum_t t^2$$

$$a_0 = \frac{\sum_t ty_t - a_1 \sum_t t^2}{\sum_t t} \text{...........(3)}$$

Equating (2) and (3) for a_0 we have

$$a_0 = \frac{\sum_t y_t}{n} - \frac{a_1 \sum_t t}{n} = \frac{\sum_t ty_t - a_1 \sum_t t^2}{\sum_t t}$$

$$or, \sum_t y_t \sum_t t - a_1 \sum_t t \sum_t t = n \sum_t ty_t - a_1 n \sum_t t^2$$

$$Or, a_1 n \sum_t t^2 - a_1 \sum_t t \sum_t t = n \sum_t ty_t - \sum_t y_t \sum_t t$$

$$Or, a_1 = \frac{n \sum_t ty_t - \sum_t y_t \sum_t t}{n \sum_t t^2 - \left(\sum_t t\right)^2} = \frac{\sum_t ty_t - \bar{y}_t \sum_t t}{\sum_t t^2 - \bar{t} \sum_t t} \qquad \text{..........(4)}$$

Putting the value of a_1 we have

$$a_0 = \frac{\sum_t ty_t - \dfrac{n\sum_t ty_t - \sum_t y_t \sum_t t}{n\sum_t t^2 - \left(\sum_t t\right)^2}\sum_t t^2}{\sum_t t}$$

$$= \frac{\sum_t ty_t\left[n\sum_t t^2 - \left(\sum_t t\right)^2\right] - \left[n\sum_t ty_t - \sum_t y_t \sum_t t\right]\sum_t t^2}{\left[n\sum_t t^2 - \left(\sum_t t\right)^2\right]\sum_t t}$$

$$= = \frac{\sum_t ty_t n\sum_t t^2 - \left(\sum_t t\right)^2 \sum_t ty_t - \sum_t ty_t n\sum_t t^2 + \sum_t y_t \sum_t t \sum_t t^2}{\left[n\sum_t t^2 - \left(\sum_t t\right)^2\right]\sum_t t} = \frac{\sum_t y_t \sum_t t^2 - \sum_t t \sum_t ty_t}{n\sum_t t^2 - \left(\sum_t t\right)^2}$$

By suitably changing the origin at the middle of the periods one can make $\sum_t t = 0$. Thereby $a_0 = \frac{1}{n}\sum_t y_t = \bar{y}_t$ .... (5) and from (4) we have

$$a_1 = \frac{\sum_t ty_t}{\sum_t t^2} \quad \text{............. (6)}$$

Thus the linear trend equation is given by

$$y_t = \frac{\sum_t y_t \sum_t t^2 - \sum_t t \sum_t ty_t}{n\sum_t t^2 - \left(\sum_t t\right)^2} + \frac{\sum_t ty_t - \bar{y}_t \sum_t t}{\sum_t t^2 - \bar{t}\sum_t t}t$$

And by making $\sum_t t = 0$ with suitable change of origin, we have the linear trend equation

$$y_t = \bar{y}_t + \frac{\sum_t ty_t}{\sum_t t^2}t\,.$$

Example 3: Fit linear trend line equation with the following time series data on production of total pulses ('000 tonnes). Also, find out the production in the year 2011.

Year	Production ('000 tonnes)
1996	14147.70
1997	12970.80
1998	14907.10
1999	13418.10
2000	11075.40
2001	13368.10
2002	11125.00
2003	14905.20
2004	13129.50
2005	13384.40
2006	14197.52
2007	14761.54
2008	14566.38
2009	14661.79
2010	18240.95

Solution:

The number of observation $n = 15$. Now by shifting the origin to 2003–2004, the time period reduces to –7, –6,–5, –4, –3, –2, –1, 0, 1, 2, 3, 4, 5, 6, 7. So we make the following table from the given data (see, Figure 5.6):

Year (n)	$t = n - 2003$	t^2	Production (Million tonnes) Y_t	ty_t	Trend values
1996	–7	49	14147.70	–99033.90	12577.16
1997	–6	36	12970.80	–77824.80	12769.56
1998	–5	25	14907.10	–74535.50	12961.96
1999	–4	16	13418.10	–53672.40	13154.36
2000	–3	9	11075.40	–33226.20	13346.76
2001	–2	4	13368.10	–26736.20	13539.16
2002	–1	1	11125.00	–11125.00	13731.56
2003	0	0	14905.20	0.00	13923.96

Year (n)	t = n –2003	t^2	Production (Million tonnes) Y_t	ty_t	Trend values
2004	1	1	13129.50	13129.50	14116.37
2005	2	4	13384.40	26768.80	14308.77
2006	3	9	14197.52	42592.56	14501.17
2007	4	16	14761.54	59046.16	14693.57
2008	5	25	14566.38	72831.89	14885.97
2009	6	36	14661.79	87970.73	15078.37
2010	7	49	18240.95	127686.62	15270.77
Total	0	280	208859.47	53872.26	

Let the linear trend equation be $Y_t = a_0 + a_1 t$, where t = Year 2003. Thus,

$$a_1 = \frac{\sum_t ty_t}{\sum_t t^2} = \frac{53872.26}{280} = 192.4009$$

$$\text{and } a_0 = \frac{1}{n}\left[\sum_t y_t\right] = \frac{1}{15}[208859.47] = 13923.96$$

So the trend line for the production of rape and mustard is

$$y_t = 13923.96 + 192.40t$$

$$or, y_{year} = 13923.96 + 192.40(y_{year} - 2003)$$

$$= -371455.09 + 192.40 y_{year}$$

So the production in the year 2011 is

$$y_{2011} = -371455.09 + 192.40(2011)$$

$$= 15463.17$$

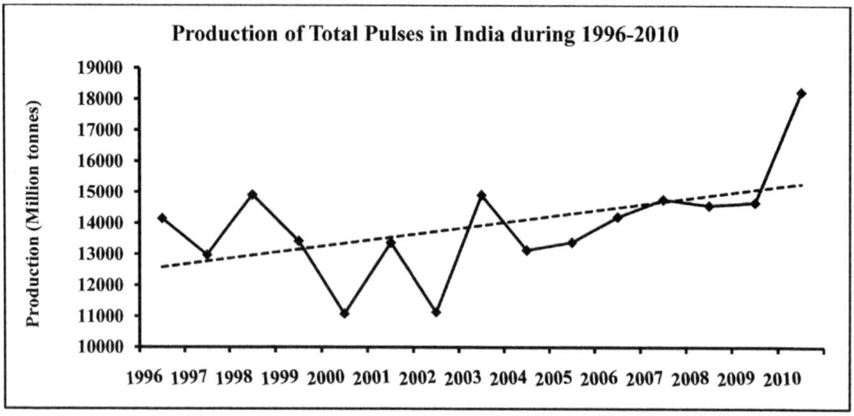

FIGURE 5.6 Production of total pulses in India during 1996–2010.

5.7 SEASONALITY

As in earlier section, we have seen the various techniques used in removal of trend, i.e., long term fluctuations from a time series. What about the short-term fluctuations? How these fluctuations are isolated from time series? Various techniques are available in literature for removal of seasonality. Some of the important techniques are given below:

1) method of simple averages;
2) ratio to trend method;
3) ratio-to-moving-average method; and
4) link relative method.

Here we will only discuss first method, i.e., Method of simple averages with example and other methods are only described.

5.7.1 METHOD OF SIMPLE AVERAGES

This is very simple method to isolate seasonal index. In this method first, arrange the data by years and then by months or quarters. Compute the averages for each month/quarter for all the years. Using all these averages find average again which is also known as grand average.

At last, the seasonal index for i^{th} month or quarter is computed by dividing each month/quarter average by grand average.

Example 4: The following table shows the monthly export of tea from north (M Kg.) India for 4 years. Compute the seasonal indices by simple averages method.

	2010	2011	2012	2013
January	9.24	8.10	7.68	8.42
February	8.24	7.10	7.72	7.46
March	10.29	9.15	7.95	9.74
April	6.04	4.90	6.03	3.57
May	6.66	5.52	7.67	5.83
June	6.65	5.51	7.96	6.11
July	10.40	9.26	10.19	12.09
August	13.80	12.66	13.69	13.18
September	14.83	13.69	14.63	14.58
October	12.37	11.23	13.21	12.45
November	15.62	14.48	15.03	15.19
December	17.75	16.61	16.93	18.04

Solution:

	2010	2011	2012	2013	Total	Monthly Average	Seasonal Index
January	9.24	8.10	7.68	8.42	33.44	8.36	79.39
February	8.24	7.10	7.72	7.46	30.52	7.63	72.46
March	10.29	9.15	7.95	9.74	37.13	9.28	88.15
April	6.04	4.90	6.03	3.57	20.54	5.14	48.76
May	6.66	5.52	7.67	5.83	25.68	6.42	60.97
June	6.65	5.51	7.96	6.11	26.23	6.56	62.27
July	10.40	9.26	10.19	12.09	41.94	10.49	99.57
August	13.80	12.66	13.69	13.18	53.33	13.33	126.62
September	14.83	13.69	14.63	14.58	57.73	14.43	137.05
October	12.37	11.23	13.21	12.45	49.26	12.31	116.94
November	15.62	14.48	15.03	15.19	60.32	15.08	143.21
December	17.75	16.61	16.93	18.04	69.33	17.33	164.60
Total					505.45	126.36	
Grand average					42.12	10.53	

5.7.2 RATIO TO TREND METHOD

The basic assumption of this method is that the seasonal variation for any given month/quarter is constant factor of the trend. First, compute the yearly averages for all the years. Then fit a mathematical model to the calculated yearly averages *viz.* quadratic or exponential, etc. and obtain the annual trend values by the method of least squares. By adjusting the trend equations, the monthly/quarterly trend values are obtained. Next, by assuming a multiplicative model, trend eliminated values are obtained by expressing the given time series values as percentage of the trend values.

5.7.2.1 MERITS

1. This method is easy to compute and understand.
2. Compared with the other method of monthly averages this method follows logical procedure for measuring seasonal variations.
3. It has an advantage over the ratio to moving average method that in this method we obtain ratio to trend values for each period for which data are available whereas it is not possible in ratio to moving average method.

5.7.2.2 DEMERITS

1. The main defect of the ratio to trend method is that if there are cyclical swings in the series, the trend whether a straight line or a curve can never follow the actual data as closely as a 12- monthly moving average does. So a seasonal index computed by the ratio to moving average method may be less biased than the one calculated by the ratio to trend method.

Example 4: Determine the seasonal variation by ratio-to-trend method for the following data

Year	1st Quarter	2nd Quarter	3rd Quarter	4th Quarter
2010	30	40	36	34
2011	34	52	50	44
2012	40	58	54	48
2013	54	76	68	62
2014	80	92	86	82

Solution: For determining the seasonal variation by ratio-to-trend method, first we need to determine the trend for the yearly data and then need to convert it to quarterly data.

Year	Yearly Totals	Yearly average (Y)	Deviation from mid-year (X)	XY	X^2	Trend values
2010	140	35	–2	–70	4	32
2011	180	45	–1	–45	1	44
2012	200	50	0	0	0	56
2013	260	65	1	65	1	68
2014	340	85	2	170	4	80
N = 5		$\sum Y = 280$		$\sum XY = 120.00$	$\sum X^2 = 10$	

The equation of the straight line trend is Y = a+bX

$$a = \frac{\sum Y}{N} = \frac{280.00}{5} = 56.00 \quad , \quad b = \frac{\sum XY}{\sum X^2} = \frac{120}{10} = 12.00$$

Quarterly increment = 12/4 = 3.

Estimation of quarterly trend values:

Now for 2010, trend values for the middle quarter, i.e., average of 2nd and 3rd quarter is 32. Quarterly increment is 3. So the trend value for the 2nd quarter is 32–3/2, i.e., 30.5 and for 3rd quarter is 32+3/2, i.e., 32.5. Trend values for the 1st quarter is 30.5–3, i.e., 27.5 and of 4th quarter is 33.5+3, i.e., 36.5. We thus get quarterly trend values as shown below:

	Trend values			
Year	1st Quarter	2nd Quarter	3rd Quarter	4th Quarter
2010	27.5	30.5	33.5	36.5
2011	39.5	42.5	45.5	48.5
2012	51.5	54.5	57.5	60.5
2013	63.5	66.5	69.5	72.5
2014	75.5	78.5	81.5	84.5

The given values are expressed as percentage of the corresponding trend values.

Hence for the 1stQtr of 2010, the percentage shall be (30/27.5)×100 = 109.9 and for 2nd Qtr. (40/30.5)×100 = 131.15, etc.

Trend values				
Year	**1st Quarter**	**2nd Quarter**	**3rd Quarter**	**4th Quarter**
2010	109.09	131.15	107.46	93.15
2011	86.08	122.35	109.89	90.72
2012	77.67	106.42	93.91	79.34
2013	85.04	114.29	97.84	85.52
2014	105.96	117.2	705.52	97.04
Total	463.84	591.41	1114.62	445.77
Average	92.76	118.28	222.92	89.15
SI Adjusted	92.05	117.36	102.12	88.46

Total of averages = 92.77+118.28+102.92+89.15 = 403.12.

As the total is more than 400 hence adjustment will be made by multiplying each average by 400/403.12, and final indices are obtained.

5.7.3 RATIO-TO-MOVING-AVERAGE METHOD

It eliminates the trend, cyclical, and irregular components from the original data (Y). In the following discussion, T refers to trend, C to cyclical, S to seasonal, and I to irregular variation. The numbers that result are called the typical seasonal index.

Example 5: Calculate seasonal indices by the ratio to moving average method, from the below-given data

Year	1st Quarter	2nd Quarter	3rd Quarter	4th Quarter
2010	70	72	60	65
2011	75	68	65	60
2012	78	62	63	68

Calculation of seasonal indices by ratio to moving average method						
Year	**Quarter**	**Given figure**	**4-nos. moving total**	**2 nos. moving total**	**4-nos. moving average**	**2-nos. moving average**
2010	**I**	70				
	II	72				

	III	60	267	539	67.38	99.07
	IV	65	272	540	67.50	100.74
2011	**I**	75	268	541	67.63	99.08
	II	68	273	541	67.63	100.92
	III	65	268	539	67.38	99.44
	IV	60	271	536	67.00	101.12
2012	**I**	78	265	528	66.00	100.38
	II	62	263	534	66.75	98.50
	III	63	271			
	IV	68				

Year	1st Quarter	2nd Quarter	3rd Quarter	4th Quarter
2010	-	-	99.07	100.74
2011	99.08	100.92	99.44	101.12
2012	100.38	98.5	-	-
Total	199.46	199.42	198.51	201.86
Average (a)	99.73	99.71	99.255	100.93
Seasonal Indices	99.83	99.81	99.35	101.03

Arithmetic average of average (a) = 399.625/4 = 99.90.

The seasonal indices will be obtain by expressing each quarterly average as percentage of 99.90.

Seasonal index of 1st quarter = 99.73/99.90*100 = 99.83.

Seasonal index of 2nd quarter = 99.71/99.90*100 = 99.81.

Seasonal index of third quarter = 99.26/99.90*100 = 99.35.

Seasonal index of fourth quarter = 100.93/99.90*100 = 101.03.

5.7.4 LINK RELATIVE METHOD

This method is slightly more complicated than other methods. This method is also known as Pearson's method. This method consists in the following steps.

1. The link relatives for each period are calculated by using the below formula:

$$Link\ relative\ for\ any\ period = \frac{Current\ periods\ figure}{Previous\ periods\ figure} \times 100$$

2. Calculate the average of the link relatives for each period for all the years using mean or median.
3. Convert the average link relatives into chain relatives on the basis of the first season. Chain relative for any period can be obtained by the chain relative for the first period is assumed to be 100.

$$\frac{Avg\ link\ relative\ for\ that\ period \times Chain\ relative\ of\ the\ previous\ period}{100}$$

4. Now the adjusted chain relatives are calculated by subtracting correction factor 'kd' from $(k+1)^{th}$ chain relative respectively.
   where k = 1, 2,.......11 for monthly and k = 1, 2, 3 for quarterly data.

   and $d = \frac{1}{N}\left[New\ chain\ relative\ for\ first\ period - 100\right]$

   where N denotes the number of periods
   i.e., $N = 12$ for monthly
   N = 4 for quarterly
5. Finally, calculate the average of the corrected chain relatives and convert the corrected chain relatives as the percentages of this average. These percentages are seasonal indices calculated by the relative link method.

5.7.4.1 MERITS

1. As compared to the method of moving average the relative link method uses data more.

5.7.4.2 DEMERITS

1. The relative link method needs extensive calculations compared to other methods and is not as simple as the method of moving average.
2. The average of link relatives contains both trend and cyclical components, and these components are eliminated by applying the correction.

Example 6: Following data gives the prices of pesticide during the different quarter of the year for the last 3 years. Using relative link method, calculate the quarterly seasonal indices for the prices of pesticide.

Year	Prices of Pesticide (Rs./kg.)			
	Q1	Q2	Q3	Q4
2003	4.75	4.50	5.00	4.85
2004	5.25	4.75	5.30	5.10
2005	5.50	5.00	5.50	5.30

5.7.4.3 *SOLUTION*

We know that link relative of any time period is given by the formula:

$$\frac{\text{Value of the current period}}{\text{Value of the previous period}} \times 100,$$

The value of the previous time period for the initial time period is not given. But for another time period, one can have the relative link values. The above problem can be solved in a stepwise manner.

Step 1: First calculate the link relatives are corresponding to each quarter of the given years from the given information.

Step 2: Calculate the average of the link relatives using the number of link relatives corresponding to each quarter

Step 3: In the 3rd step, calculate the chain relatives for each of the quarters with the help of the above average link relatives using the following formula:

$$\text{Chain relative for any quarter} = \frac{\text{LR for the quarter} \times \text{CR for the previous quarter}}{100}$$

Initially, the chain relative for the first period, here quarter is taken as 100.

Thus the CR for 2nd quarter =

$$\frac{\text{LR for 2nd quarter} \times \text{CR for 1st quarter}}{100} = \frac{\text{LR for 2nd quarter} \times 100}{100} = \text{LR for 2nd quarter}$$

Step 4: The new chain relative for the 1st quarter is obtained using the formula:

$$\frac{\text{LR for 1st quarter} \times \text{CR for 4th quarter}}{100} = \frac{108.05 \times 98.528}{100} = 106.455$$

Step 5: Correction factor d is calculated as follows:

$$d = \frac{106.455 - 100}{4} = \frac{6.455}{4} = 1.614$$

Step 6: Adjusted chain relatives are calculated for each quarter separately. Thus, adjusted CR for 1st quarter is CR(Q2) – 2d = 92.041–2x1.614 = 88.813, for quarter three is CR(Q3) – 3d = 97.229 and so on.

Step 7: Seasonal indices are calculated as $\frac{\text{adjusted CR for the quarter}}{\text{average of all the adjusted CR}} \times 100.$

Thus the seasonal index for third quarter is

$$\frac{\text{adjusted CR for 3rd quarter}}{\text{average of all the adjusted CR}} \times 100 = \frac{97.229}{94.53} \times 100 = 102.855.$$

We present the whole information in the following table:

Year	Link Relatives				Average
	Q1	Q2	Q3	Q4	
2003		94.74	111.11	97.00	
2004	108.25	90.48	111.58	96.23	
2005	107.84	90.91	110.00	96.36	
Total	216.09	276.12	332.69	289.59	
Average	108.05	92.04	110.90	96.53	
Chain relative (CR)	100	92.041	102.070	98.528	98.16
Adjusted CR	100	88.813	97.229	92.073	94.53
Seasonal Index	105.787	93.952	102.855	97.401	

So from the analysis, we can conclude that prices of pesticide is maximum during the 1st quarter followed by third quarter.

5.8 CYCLICAL FLUCTUATIONS

As discussed earlier, the oscillatory movement of a time series with recurrent upward and downward movement continuing more than a year is known as Cyclical Movement of the time series. The length and the intensity of oscillations or fluctuations may change from cycle to cycle. Thus, it is very difficult to have a cyclical movement with fixed period and intensity in real life. The measurement of cyclical fluctuation is generally done in two different approaches, viz, the residual approach and the periodogram analysis approach.

5.8.1 RESIDUAL APPROACH

The principle of residual approach lies in removing the other components viz, the trend, seasonal component and the irregular component from the given time series to get the cyclical component. Thus for a multiplicative model we have $U_t = T_t C_t S_t$ and by knowing the T_t and S_t component we can have $\frac{U_t}{T_t S_t} = C_t I_t$. Similarly, for additive model, the trend and the seasonal components are subtracted from the time series values to get the time series values with only the cyclical component and the irregular component. The I_t component is removed or smoothen by taking moving average of suitable periods. The T_t and S_t together are known as the normal value.

5.8.2 PERIODOGRAM ANALYSIS

Sahu and Das (2009) has explained this approach in his book, and the same is given here. Periodogram analysis starts with the residual time series values leaving the trend and the seasonal components from the observations. The essence of periodogram analysis is to known whether the residual time series values y_t contains a harmonic term with period μ or not; where μ is the period of cyclical variations. For a given trial period μ we can have $R_\mu^2 = A^2 + B^2$, known as intensity corresponding to the trial period μ, where $A = \frac{2}{n}\sum_{t=1}^{n} y_t Cos\frac{2\pi t}{\mu}$ and $B = \frac{2}{n}\sum_{t=1}^{n} y_t Sin\frac{2\pi t}{\mu}$, where n is the number of observations in the time series. Thus if λ be a trial period with amplitude α and the irregular component b_t, then the residual time series can be written as $y_t = aSin\frac{2\pi t}{\lambda} + b_t$, as usual, the irregular component b_t is assumed to be uncorrelated with $y_t = aSin\frac{2\pi t}{\lambda}$.

So we can have

$$A = \frac{2a}{n}\sum_t Sin\frac{2\pi t}{\lambda}Cos\frac{2\pi t}{\mu} + \frac{2a}{n}\sum_t b_t Cos\frac{2\pi t}{\mu}$$

$$= \frac{2a}{n}\sum_t Sin\alpha t\; Cos\beta t$$

$$= \frac{a}{n}\sum_t \left\{Sin(\alpha-\beta)t + Sin(\alpha+\beta)t\right\}$$

where $\alpha = \frac{2\pi}{\lambda}$ and $\beta = \frac{2\pi}{\mu}$

Thus

$$A = \frac{a}{n}\left\{ \frac{Sin\frac{n(\alpha-\beta)}{2} Sin\left\{\frac{(n+1)(\alpha-\beta)}{2}\right\}}{Sin\frac{(\alpha-\beta)}{2}} + \frac{Sin\frac{n(\alpha+\beta)}{2} Sin\left\{\frac{(n+1)(\alpha+\beta)}{2}\right\}}{Sin\frac{(\alpha+\beta)}{2}} \right\}$$

For large number of observation, the second term is always small. If β does not tends to μ i.e., if the trial period μ does not approach to the true period λ, then the first term will also be small. If β tends to α, i.e., μ approaches to true value λ, then

$$A = aSin\left(\frac{(n+1)(\alpha-\beta)}{2}\right) \times \left(\frac{Sin\frac{n(\alpha-\beta)}{2}}{\frac{n(\alpha-\beta)}{2}}\right) \Bigg/ \left(\frac{Sin\frac{(\alpha-\beta)}{2}}{\frac{(\alpha-\beta)}{2}}\right)$$

tends to $A = aSin\frac{(n+1)(\alpha-\beta)}{2}, \left(\because \frac{Sin\theta}{\theta} \to 1 \because \theta \to 0\right)$

Similarly,

$$B \to ACos(n+1)\left(\frac{\alpha-\beta}{2}\right)$$

$\beta \to \alpha$ is very small otherwise.

Thus we have $R_\mu^2 \to a^2$ & $\beta \to \alpha$

A number of trial periods around the true period R_μ^2 (guessed by plotting scattered diagram) are taken and corresponding R_μ^2 in each case are calculated. A graph of different R_μ^2 values corresponding to different trial values for $\mu(\lambda_1, \lambda_2, \lambda_3, ...)$ are drawn on a graph paper to get a periodogram. From the periodogram the true cycle period λ is obtained for a time series by equating the trial value corresponding to which the λ value is maximum. So the cyclical period is obtained by equating the period of cycle λ for which R_μ^2 is maximum.

By knowing the true period μ one may try to fit a sine-cosine curve with the values of the residual time series, i.e., y_t, and this is known as harmonic analysis.

Schuter and Walker test for periodogram analysis: The objective of this test is to test the significance of the periodicity worked out in periodogram analysis. Let us suppose at $\mu = \mu_0$ the $R_\mu^2 = A^2 + B^2$ is maximum. We want to test whether the periodicity worked out from the periodogram analysis is significant or not. That means we want to test $H_0 : \hat{\lambda} = \mu_0$.

We assume that y_t s are IIDN(0, σ^2).

This gives A ~ IIDN(0, $\frac{2\sigma^2}{n}$) and B~ IIDN(0, $\frac{2\sigma^2}{n}$). Thus the joint distribution of A and is given as $f_{AB}(A,B) = \frac{n}{4\pi\sigma^2} e^{\frac{n(A+B)}{4\sigma^2}}$.

Let us suppose A = Rcos θ and B = $R \sin \theta$; where 0<R<∞ and $0 < \theta < 2\pi$

Thus we have P(R^2> $R_{\mu_0}^2$) = $e^{\frac{nR_{\mu_0}^2}{4\sigma^2}}$

$$R_\mu^2 = \frac{4\sigma^2}{n} \text{ we have}$$

Writing

$$P(R^2 > KR_\mu^2) = e^{\frac{nR_{\mu_0}^2}{4\sigma^2}}$$

We have already defined that R^2 is the squared amplitude. Thus the probability of at least one squared amplitude is greater than K R_μ^2 is given by

$$p(\text{at least one squared amplitude} > KR_\mu^2)$$

$$= 1 - (1 - e^{-K})^m = f(K)(say)$$

where, m is the number of trial periods.

In practice generally we calculate $K^* = \frac{nR_\mu^2}{4\sigma^2}$ and find out f(K*) = $= 1 - (1 - e^{-K^*})^m$.

If f(K^*) <∞, then μ is significant that means the periodicity derived from periodogram analysis can be taken as an estimate of the true periodicity λ, and we cannot reject the null hypothesis.

5.9 IRREGULAR COMPONENT

As we know, the irregular component can occur at any time, and we can't define its amplitude. Hence, no mathematical model has been evolved to isolate the irregular component. One can isolate the irregular component using residual approach as used in cyclic fluctuations. Suppose data series is composed of trend (T_t), Seasonal (S_t) and Cyclic (C_t) component. Then for mathematical models, irregular variation can be isolated as $I_t = \frac{U_t}{T_t \times S_t \times C_t}$

Similarly for additive model, $I_t = U_t - T_t - S_t - C$

5.9.1 INDEX NUMBERS CONSTRUCTION

"Index number is a device that indicates the value of a variable at any given point of time as percentage of its value in the base year."

Spiegal explains "*an index number is a statistical measure designed to show changes in variable or a group of related variables with respect to time, geographical location or other characteristics.*"

There are usually two types of index, value index, and volume index. Price index is a value index while Area index and Quantity index are volume index.

Following is the classification of methods of index number construction.

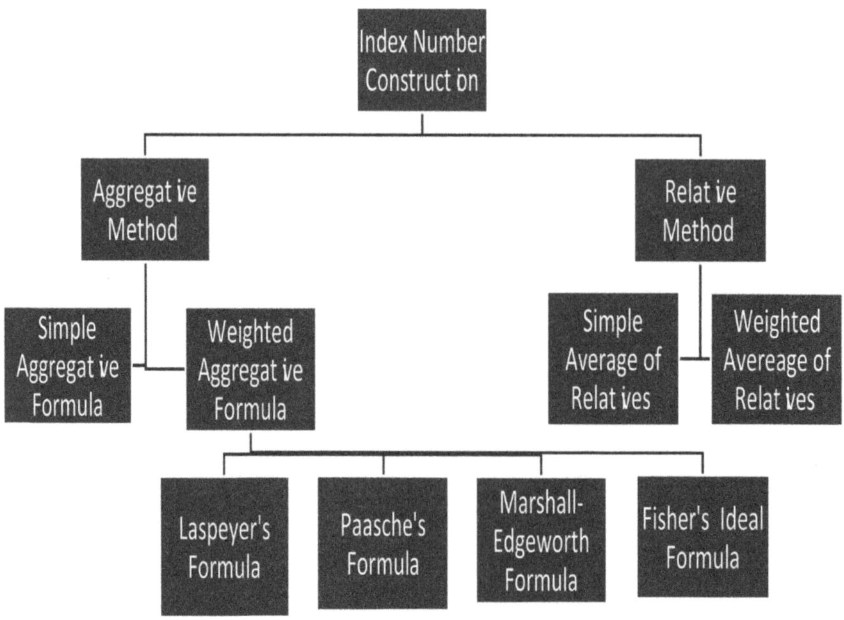

Let index number for tth year with '0' as base year be denoted by $I_{0,t}$, price by P, quantity by q and weight by w with 0 being subscript for base year, t being subscript for tth year (current year) and i being subscript for ith commodity.

Under aggregative method, simple aggregative method consists in aggregating the prices of all commodities of base and current year and diving the current year aggregate by base year aggregate followed by converting this ratio to percentage. Mathematically, this can be dented as following:

$$I_{0,t} = \frac{\sum_{i=1}^{n} P_{i,t}}{\sum_{i=1}^{n} P_{i,o}} \times 100$$

Here no weights are assigned to the prices being aggregated, and this remains a disadvantage for two reasons, first is that prices does not hold any meaning without quantities associated with it and second, all prices are denoted in same unit and if all units are different then mathematically aggregation is not possible.

Under relative method, simple average of relative method consists in averaging the price relatives unlike simply prices being averaged out for base and current year. To calculate price relatives, current year price of the commodity is divided by its base year price. Mathematically, it can be denoted as follows:

$$I_{0,t} = \sum_{i=1}^{n} I_{i,t} / n$$

where

$$I_{i,t} = \frac{P_t}{P_0} \times 100$$

This method too has same disadvantage of not considering weights but is better in terms of nullifying units being making relatives out of prices.

This disadvantage is being overcome under the weighted average of relatives method wherein weighted average is taken instead of a simple average. Mathematically it can be shown as below:

$$I_{0,t} = \sum_{i=1}^{n} w_i I_{i,t}$$

where

$$\sum_{i=1}^{n} w_i = 1 \text{ and } I_{i,t} = \frac{P_t}{P_0} \times 100$$

These two methods discussed under the head of relative methods, have one more disadvantage, computational burden is higher on account of calculating relatives for every year for all commodities.

Similar to the weighted average of relatives, under aggregative method, weighted aggregative formula, provides a computationally convenient way of preparing index number. Under this method, Laspeyer's and Paasche's formulae hold sway over all other formulae. This method consists of calculating the ratio of the weighted average of prices in current year to base year followed by expressing the same in percentage. A generic formula for this method is provided below:

$$I_{0,t} = \frac{\sum_{i=1}^{n} w_i P_{i,t}}{\sum_{i=1}^{n} w_i P_{i,o}} \times 100$$

The detailed explanation for each formulae under this method will be provided below.

1. ***Laspeyer's Formula:*** Under this formula, weights assigned are base year quantities produced/consumed/traded. Here, assumption is that the basket of commodities over the years since base year does not change significantly. The Laspeyer's formula is as provided below:

$$I_{0,t} = \frac{\sum_{i=1}^{n} q_{i,0} P_{i,t}}{\sum_{i=1}^{n} q_{i,0} P_{i,o}} \times 100$$

   By this formula, comparison of different year can be made because of same weight.

2. ***Paasche's Formula:*** Unlike the Laspayer's formula, this formula uses current year quantities as weights. This change makes it not useful for making inter-year comparisons. The Paasche's formula is as provided below:

$$I_{0,t} = \frac{\sum_{i=1}^{n} q_{i,t} P_{i,t}}{\sum_{i=1}^{n} q_{i,t} P_{i,o}} \times 100$$

3. ***Marshall-Edgeworth Formula*:** In this formula, instead of taking either current year or base year quantities, it takes average of current and base year quantities. This special weight assigned changes from year to year and therefore, inter-year comparison is not possible. The Marshall-Edgeworth formula is as provided below:

$$I_{0,t} = \frac{\sum_{i=1}^{n} \left(q_{i,t} + q_{i,0}\right) P_{i,t}}{\sum_{i=1}^{n} \left(q_{i,t} + q_{i,0}\right) P_{i,o}} \times 100$$

4. ***Fisher' Ideal Formula***: This formula is square root of two formulae namely, Laspeyer's formula and Paasche's formula. The Fisher' Ideal formula is as provided below:

$$I_{o,t} = \sqrt{\frac{\sum_{i=1}^{n} q_{i,0} P_{i,t}}{\sum_{i=1}^{n} q_{i,0} P_{i,o}} \frac{\sum_{i=1}^{n} q_{i,t} P_{i,t}}{\sum_{i=1}^{n} q_{i,t} P_{i,o}}} \times 100$$

This formula is called ideal because it passes time reversal and factor reversal test.

Caution in construction of index number

i. **Purpose of Index numbers under construction**: If purpose of construction of index number is not well defined then there will be problems in selection of items and weights. An index number serves at a time only some purpose but not all. If no purpose is specified, then it can be misleading at times.
ii. **Selection of items**: Based on purpose of index, items[1] are selected in such a way that these items are representative, tangible, having standard quality. Items not related to purpose should be excluded. Too large a list of items puts computation burden.

[1]In standard terminology items are called as 'Regimen.'

iii. **Choice of an appropriate average**: Index numbers are a type of average which in turn uses averages. There are many types of averages like arithmetic mean, geometric mean and harmonic mean. But due to computational ease, arithmetic mean is preferred over two others.
iv. **Assignment of weights (importance)**: Each item in the index is having its own importance and hence considering the purpose of index, each item must be provided appropriate weight to reflect the priority/preference/pattern existing in society/organization.
v. **Choice of base period**: The period with respect to which all comparisons and computation are made is called as base period. The period selected as base period must be economically a stable year.

Example:

	1971		1972	
Commodity	**Price**	**Quantity**	**Price**	**Quantity**
A	3	11	1	21
B	1	30	3	56
C	5	72	4	82
D	8	37	6	66
E	2	13	2	75
F	2	28	6	82
G	7	48	3	70
H	1	84	9	68
I	2	22	7	82

Summary of all calculations

Type of Index No.	Value	
Simple aggregative	132.26%	
Simple average of relative	242.35%	
Weighted average of relative	327.42%	255.83%
Laspeyer's Index	146.96%	
Paasche's Index	140.99%	
Marshall-Edgeworth Index	143.24%	
Fisher's Ideal Index	143.94%	

5.10 CLASSICAL LINEAR REGRESSION MODEL (CLRM)

The Classical Linear Regression Model (CLRM) was developed by Gauss in 1821 and since then has served as a norm or a standard against which may be compared the regression models that do not satisfy the Gaussian assumption. It is simply multiple linear regression model which having some more assumptions than linear regression. The assumptions of CLRM are given in the following subsections.

5.10.1 ASSUMPTIONS OF CLRM

1. The regression model is linear in parameter: The model should be correctly specified (has the correct variables and functional forms), has no measurement error, and has an additive error term.

   $Y_i = \beta_0 + \beta_1 X_{i1} + \cdots + \beta_k X_{ik} + \varepsilon_i$
2. Independent variables are non-stochastic, i.e., values taken by the X_s are fixed in repeated samples.
3. Error term have zero mean: For a given set of values of X_s, the expected value of random variable is zero.

   $E(\varepsilon_i) = 0$
4. All the independent variables are uncorrelated with the error term, i.e., Covariance between disturbances and independent variable must be equal to zero. That is, for all i = 1,..., *n*, j = 1,..., k,

   $$Cov(\varepsilon_i, X_{ij}) = E[(\varepsilon_i - E(\varepsilon_i))(X_{ij} - E(X_{ij}))]$$
   $$= E[\varepsilon_i (X_{ij} - E(X_{ij}))]$$
   $$= E(\varepsilon_i X_{ij}) - E(\varepsilon_i)E(X_{ij}) = E(\varepsilon_i X_{ij}) = 0$$
5. No autocorrelation between the errors.

   $Cov(\varepsilon_i, \varepsilon_j \mid X_i, X_j) = E(\varepsilon_i \varepsilon_j \mid X_i, X_j) = 0, \forall i, j$
6. Errors terms are homoskedastic in nature: Given the value of X_i, the variance of the disturbance is the same for all observations (Lai et al., 1978).

   $Var(\varepsilon_i \mid X_i) = E[\varepsilon_i - E(\varepsilon_i \mid X_i)]^2 = E(\varepsilon_i^2 \mid X_i) = \sigma^2$

   For heteroscedastic variance: variance of disturbance term is not constant for all observations

$$Var(\varepsilon_i \mid X_i) = \sigma_i^2$$

$$E[\varepsilon_i - E(\varepsilon_i \mid X_i)]^2 \neq E(\varepsilon_i^2 \mid X_i) = \sigma_i^2$$

7. The number of observation must be greater than number of parameters to be estimated, i.e., $n > j\ (= 1......k)$
8. The X values in a given sample must not be constant, i.e., $Var(X) \geq 0$.
9. The regression model should correctly be specified, i.e., the model should clearly spell out (1) the variables to be included (2) functional form of the model (3) probabilistic assumption about the variables.
10. Nonexistence of multicollinearity among the explanatory variables.

5.10.2 THE CO-EFFICIENT OF DETERMINATION (R^2)

The square of the multiple correlation coefficient is defined as the coefficient of determination and is denoted by R^2. The coefficient of determination R^2 defined as the proportion of total variation in the response variable (time) being explained by the fitted model is widely used.

$$R^2 = \frac{RgSS}{TSS} = (TSS - RSS)/TSS = 1 - \frac{\sum_{i=1}^{n} (Y_i - \hat{Y}_i)^2}{\sum_{i=1}^{n} (Y_i - \bar{Y})^2}$$

5.10.3 ADJUSTED R^2

As we introduce more and more variable in the model with the expectation of explaining the variation in the dependent variable in a better way, R^2 value is also to increase, if not, at least remains constant but never decreases. R^2 can be treated as non-decreasing function of number of explanatory variables.

At the same time, a dependent variable can very well be explained with varied number of explanatory variables in different regression models. Then how to judge which model has better explanatory power to explain variation in the dependent variable because the different R^2s have not come out from the models with same number of independent variables.

To overcome the problem of comparing different R^2s arising out of different regression models having same dependent variable but different numbers of independent variables adjusted R^2 ($\bar{R}^2$ or R^2_{adj}) has been worked out.

We have

$$R^2 = \frac{RgSS}{TSS} = (TSS - RSS)/TSS = 1 - \frac{\sum_{i=1}^{n} (Y_i - \hat{Y}_i)^2}{\sum_{i=1}^{n} (Y_i - \bar{Y})^2}$$

On the other hand

$$\bar{R}^2 = 1 - \text{RMS/TMS}$$

$$\bar{R}^2 = 1 - \frac{\left(\text{RSS}/_{n-k}\right)}{\left(TSS/_{n-1}\right)} = 1 - \frac{\left(\text{TSS-RgSS}/_{n-k}\right)}{\left(TSS/_{n-1}\right)}$$

$$= 1 - \left\{\frac{\left(\text{TSS}/_{n-k}\right)}{\left(TSS/_{n-1}\right)} - \frac{\left(\text{RgSS}/_{n-k}\right)}{\left(TSS/_{n-1}\right)}\right\}$$

$$= 1 - \left\{\frac{n-1}{n-k} - \frac{(n-1)R^2}{n-k}\right\}$$

$$= 1 - \frac{n-1}{n-k}(1 - R^2)$$

In regression model we have

(1) K > 2 thereby indicating that $\bar{R}^2 < R^2$ that means as the number of independent variables $\bar{R}^2$ increases less than R^2

(2) $\bar{R}^2 = 1 - \frac{n-1}{n-k}(1 - R^2)$ when $\bar{R}^2 = 1$, $\bar{R}^2 = 1$

When $R^2 = 0 \Rightarrow \bar{R}^2 = 1 - \frac{n-1}{n-k} = \frac{n-k-n+1}{n-k} = \frac{1-k}{n-k}$

if $k \geq 2$ the $\bar{R}^2$ is – ve.

Thus adjusted R^2 can be negative if k > 2 and $R^2 = 0$; when the value of the $\bar{R}^2$ become negative then it's value is taken as zero.

Thus is less than the R^2 and $\bar{R}^2$ can give negative value are the two main points against $\bar{R}^2$.

Now the question is which one is to use R^2 or $\bar{R}^2$? There are difference in opinion.

Scholars like Goldberger argued for modified $R^2 = (1-k/n)R^2$.

Comparison of two models can be made on the basis of the coefficient of determination (may or may not be with adjusted R^2) if and only if the sample size (n) and the dependent variable remain in the same form (i.e., one in Y form and other in ln(Y) can't be compared).

5.10.4 DEPENDABLE ESTIMATE OR MAXIMIZATION OF R^2?

In regression analysis, our objective is to obtain dependable estimate of the true population regression coefficient and infer statistically about these. But in many of the cases, the researchers become interested to try with various models and find out which of these models is giving maximum R^2 or $\overline{R}^2$ value and ultimately stick to that model. It is advisable to give more emphasis in finding out actual and theoretical/logical relevance of different variables to be considered in the model and their form of relationship. In the process, if one get better R^2, then it is well and good. But if one gets lower R^2 that does not necessarily prove that the model selected is bad one.

5.10.5 CONTRIBUTION OF EXPLANATORY VARIABLE

Starting with a set of variables, logical, and theoretical aspects will lead the researchers initially to include the core variables in the model. But how to judge the next variable to be included in the model? We have given a rough idea about this in Example 10.6 (Note). However, this can be done by judging the change in regression sum of squares due to inclusion of new variable into the model and can be measured in terms of significant improvement in R^2 value. Significance of inclusion of a new variable can be tested with the help of F test.

$$\mathrm{F} = \frac{(R^2_{new} - R^2_{old})/k'}{\left(1 - R^2_{new}\right)/n - k_{new}}$$

with

$$H_0 = \left(R^2_{new} - R^2_{old} = 0\right) \quad \text{at } (k', n-k)df$$

where k' is the number of new variables included in the model and k is the total no. of variables in the new model.

Generally when $\overline{R}^2$ is increased due to inclusion of a variable in the model then the variable is retained in the model. This happens (approximately) when the *t* value of the partial regression coefficient of the newly introduced variable is greater than unity in absolute value (against the test $H_0 : \beta = 0$).

Similarly, by introducing a group of variables if the F value increases by unity R^2 also increases (as thumb rule) and one can retain the new variables because these increase the explanatory power of the new model.

5.11 MULTICOLLINEARITY

One of the assumptions of CLRM model is that there are no exact linear relationships between the independent variables and the number of observations should be greater than the number of parameters to be estimated in the model. If these assumptions are violated, then multicollinearity exists. It means that when two or more of the predictors in a regression model are moderately or highly correlated, then we called multicollinearity is present. Multicollinearity is a kind of disturbance in the data, and if present in the data then reliable statistical inferences cannot be made (*Farrar and Glauber, 1967*). There could be two types of multicollinearity problems: Perfect and less than perfect/imperfect/near collinearity.

(i) Perfect multicollinearity: In this case the regression coefficients of the *X* variables are indeterminate and their standard errors infinite. One explanatory variable can be completely explained by a linear combination of other explanatory variables

(ii) Less than perfect/imperfect/near multicollinearity: If multicollinearity is less than perfect, then regression coefficients of the *X* variables can be determine but possesses large standard errors, which means the coefficients cannot be estimated with great precision. Multicollinearity among explanatory variables depends on the theoretical relationship between the variables. It is always remember that multicollinearity is sample phenomena.

5.11.1 WHAT ARE THE CONSEQUENCES OF MULTICOLLINEARITY

1. Estimates of parameter will remain unbiased, i.e., OLS is BLUE, and the Gauss-Markov theorem is still valid

2. The main consequence is that the variances (and, hence, standard errors) of some coefficient estimates can be large and precise estimation of variances is not possible.
3. The computed value of t-statistic can be small and may tend to be insignificant, and the overall coefficient of determination may be high.
4. The deletion of an explanatory variable may cause major changes in the estimates βs.
5. The overall fit of the equation will be largely unaffected.

5.11.2 DETECTION OF MULTICOLLINEARITY

There is no clear-cut criterion for estimating the multicollinearity of linear regression models. High correlation among variables does not necessarily imply multicollinearity. We can make a judgment by checking related statistics, such as tolerance value or variance inflation factor (VIF), Eigenvalue, and condition number, etc.

5.11.2.1 T RATIOS VS. R^2

In multiple regression, if we get high R^2 value but few regression coefficients are significant, i.e., 't' ratios are significant. In such cases, one may expect multicollinearity, but this is not an exact test. It is not precise test and also depends on other factors like sample size.

5.11.2.2 CORRELATION MATRIX OF REGRESSORS

Multicollinearity refers to a linear relationship among all or some the independent variables. Any pair of independent variables may not be highly correlated, but one variable may be a linear function of a number of others. In a three-variable regression, multicollinearity is the correlation between the two independent variables. Having high correlation is not a necessary condition for multicollinearity.

5.11.2.3 AUXILIARY REGRESSIONS OR VARIANCE INFLATION FACTOR (VIF)

Look at the extent to which a given explanatory variable can be explained by all the other explanatory variables in the equation. Run an OLS regression for each X as a function of all the other explanatory variables. For example, estimate the following:

$$X_{2i} = \alpha_1 + \alpha_3 X_{3i} + \ldots + \alpha_{K+1} X_{K+1i} + \varepsilon_i$$

where our hypothesis is that X_{2i} is a linear function of the other regressors (X_{3i}, etc.). We test the null hypothesis that the slope coefficients in this auxiliary regression are simultaneously equal to zero:

$$H_0 : \alpha_3 = \alpha_4 = \cdots = \alpha_{K+1} = 0$$

With the following F-test.

$$F = \frac{R_2^2 / K-1}{1 - R_2^2 / n-K}$$

where R_2^2 is the coefficient of determination with X_{2i} as the dependent variable, and K is the number of coefficients in the original regression. This is related to high VIFs discussed in the textbook, where

$$\text{VIFs} = \frac{1}{1 - R_2^2};$$

if VIF > 5, the multicollinearity is severe. But ours is a formal test.

If *VIF* > 10 is used to indicate the presence of multicollinearity between continuous variables. When the variables to be investigated are discrete in nature, Contingency Coefficient (CC) is used.

$$CC = \sqrt{\frac{\chi^2}{N + \chi^2}}$$

where, N is the total sample size.

If CC is greater than 0.75, the variables are said to be collinear.

5.11.2.4 FRISCH'S CONFLUENCE ANALYSIS (BUNCH-MAP ANALYSIS)

1. Regress the dependent variable on each of the explanatory variables separately. Examine the elementary regression equation on the basis of 'priori' and statistical criteria.
2. Select the most plausible elementary regression equation and insert gradually the additional variables.
3. Examine the effects of insertion of new-variable on individual coefficient, on their standard errors and on the overall R^2.
4. If the new variable improves R^2 without rendering the individual co-efficient unacceptable on the basis of priori consideration the variable is useful and retained.
5. If the new variable does not improve R^2 and does not affect the individual coefficient to a considerable extent, it is superfluous and not retained.
6. If the new variables affect considerably the sign or the values of the coefficient, it is detrimental.

5.11.3 REMEDIES OF MULTICOLLINEARITY

Several methodologies have been proposed to overcome the problem of multicollinearity.

1. *Do nothing:* Sometimes multicollinearity is not necessarily bad or unavoidable. If the R^2 of the regression exceeds the R^2 of the regression of an independent variable on other variables, there should not be much worry. Also, if the t-statistics are all greater than 2, there should not be much problem. If the estimation equation is used for prediction and the multicollinearity problem is expected to prevail in the situation to be predicted, we should not be concerned much about multicollinearity.
2. *Drop a variable(s) from the model:* This, however, could lead to *specification error.*
3. *Acquiring additional information:* Multicollinearity is a sample problem. In a sample involving another set of observations, multicollinearity might not be present. Also, increasing the sample size would help to reduce the severity of collinearity problem.
4. *Rethinking of the model:* Incorrect choice of functional form, specification errors, etc.

5. Prior information about some parameters of a model could also help to get rid of multicollinearity.
7. Use partial correlation and stepwise regression
8. Dropping variables/Additional or new data
9. Combining cross-sectional and time series data
10. Principal component method

5.12 LINEAR STATIONARY MODELS

Consider a time series $Y_1, Y_2, \ldots\ldots, Y_t$. Suppose the value for each Y is determined exactly by using some mathematical relationship then the series is said to be **deterministic. But if the** values of Y are determined through their probability distribution, then the series is said to be a **stochastic process.** Any stationary process $\{Z_t\}$ is comprised the sum of two components, i.e., (1) stochastic component, and (2) deterministic component. If the probability distribution remains same for all starting values of t throughout the process, then we can say that A **stationary stochastic process** is occurred. Such a process is completely defined by its mean, variance, and autocorrelation function (ACF). When a time series exhibits non-stationary nature, then first aim is to transform the non-stationary series into a stationary process.

If we consider only stationary processes it can be very much restrictive as most of economic variables are non-stationary in nature. But these models can be used as building blocks in more complicated nonlinear and/or non-stationary models like ARIMA.

5.12.1 STRICT STATIONARITY

1. A stochastic process $\{Y_t\}_{t=-\infty}^{\infty}$ is strict stationery if, for any given time finite integer r and for any set of subscripts $t_1, t_2, \ldots\ldots, t_r$ the joint distribution of $(Y_{t1}, Y_{t2}, \ldots\ldots, Y_{tr})$ depends only on $t_1 - t, t_2 - t, \ldots\ldots, t_r - t$ but not on t. The joint distribution of random variables in a strictly stationary stochastic process is time invariant. For example, the joint distribution of Y_1, Y_3, Y_7 is the same as the distribution of Y_{12}, Y_{16}, Y_{18}. This means they all have the same mean, variance, etc., assuming these quantities exist. The correlation between $Y_{t1}, Y_{t2}, \ldots\ldots Y_{tr}$ other than that the correlation between Y_t and Y_{tr} only depends on $t - t_r$ and not on t.

2. Let $\{Y_t\}_{t=-\infty}^{\infty}$ be strictly stationary and consider that $g(.)$ is also strictly stationary
3. If $\{Y_t\}_{t=-\infty}^{\infty}$ is and independently identically distributed (iid) sequence then it is strictly stationary.
4. Let $\{Y_t\}_{t=-\infty}^{\infty}$ be an iid sequence and let $X \sim N(0,1)$ independent of $\{Y_t\}_{t=-\infty}^{\infty}$. If suppose $Z_t = Y_t + X$, the sequence $\{Y_t\}_{t=-\infty}^{\infty}$ is not independent sequence but is an iid and is strictly stationary.
5. Covariance stationary: A stochastic process $\{Y_t\}_{t=-\infty}^{\infty}$ is covariance stationary if
   a. $E[Y_t] = \mu$ does not depend on *t*.
   b. $\text{var}(Y_t) = \sigma^2$ does not depend on *t*.
   c. $\text{cov}(Y, Y_{t-j}) = \gamma_j$ exists and is finite, depends only on *j* but not on *t* for $j = 0,1,2,...$

Suppose $Y_t \sim iid\ N(0,\sigma^2)$. Then $\{Y_t\}_{t=-\infty}^{\infty}$ is called a Gaussian white noise process and is denote $Y_t \sim GWN(0,\sigma^2)$

$$E[Y_t] = 0 \text{ independent of } t,$$

$$\text{var}(Y_t) = \sigma^2 \text{ independent of } t,$$

$$\text{cov}(Y, Y_{t-j}) = 0 \ (\text{for } j > 0) \quad \text{independent of } t \text{ for all j}$$

so that $\{Y_t\}_{t=-\infty}^{\infty}$ is covariance stationary.

5.12.2 WEAK STATIONARY

Since the definition of strict stationarity is generally too strict for everyday life a weaker definition of second order or weak stationarity is usually used. Weak stationarity means that mean, and the variance of a stochastic process do not depend on *t* (that is they are constant), and the autocovariance between Y_t and Y_{tr} only depends lag *t*.

a. $E[Y_t] = \mu$ does not depend on *t*.

b. $\text{var}(Y_t) < \infty$ for all *t*

c. $\text{cov}(Y, Y_{t-j}) = \gamma_j$ exists and is finite, only depends on $t - t_r$.

5.13 FITTING OF TIME-SERIES MODEL

In regression model, it is usually assumed that the error terms are assumed to be uncorrelated. This implies that the various observations within a series are statistically independent. However, this assumption is rarely met in practice. Usually, serial correlations in the observations often exist if the data are collected sequentially over time. That is, each observations of the observed data series $\{Y_t\}$, which being a family of random variables $\{ Y_t, t \in T\}$, where *T* is the index set, $T = \{ 0, \pm 1, \pm 2, \ldots,\}$ and apply standard time-series analysis technique to develop a model which will adequately represent the set of realizations and also their statistical relationship in a better way.

The statistical concept of correlation is to measure the relationships existing among the observations within the series. In these models, the values of correlations between the value of *Y* at time *t* (i.e., Y_t) and *Y* at earlier time periods (i.e., Y_{t-1}, Y_{t-2}, ...) are examined. The algebraic forms of Autoregressive and Moving average processes are

Autoregressive process

$$Z_t = C + \phi_1 Y_{t-1} + \varepsilon_t \tag{1}$$

Moving average process

$$Z_t = C - \theta_1 a_{t-1} + \varepsilon_t \tag{2}$$

A process involving past (time-lagged) *Y* terms is called an autoregressive (Abbreviated as AR) process. The longest time lag associated with a *Y* term on the right-hand side is called the AR order of the process. The equation (A) is thus an AR process of order one, abbreviated as AR(1). On the left-hand side, Y_t represents the set of possible observations on a time sequenced random variables Y_1. The coefficient ϕ_1 has a fixed numerical value which tells how Y_t is related to Y_{t-1}, C is a constant term related to the mean μ of the process. The constant term of an AR process is equal to the mean times the quantity one minus the sum of the AR coefficient, i.e., for an AR(1) process $C = \mu(1-\varphi_1)$.

The variable a_t stands for a random shock element at the time point, t. Although Y_t is related to Y_{t-1}, the relationship is not exact; it is probabilistic rather than deterministic. The random shock represents this probabilistic factor.

Now consider the process (B). The process with past (time-lagged) random shocks only are called moving average (abbreviated as MA) processes.

The longest time lag associated with an error term (i.e., a_t) is called MA order of the process. The equation (B) is an MA process of order one, abbreviated as MA(1). C is a constant term related to the mean (μ) of the process. For a pure MA model, C is equal to mean of the process, or, in symbol, $C = \mu$. The negative sign attached to θ_1 is merely a convention. It makes no difference whether we use negative or positive sign. The standard formula for calculating the auto-correlation coefficient is

$$r_k = \frac{\sum_{t=1}^{n-k}(Y_t - \overline{Y})(Y_{t-k} - \overline{Y})}{\sum_{t=1}^{n}(Y_t - \overline{Y})^2} \qquad k = 1, 2, 3, \ldots, \ldots, \ldots.$$

The above formula can be written more compactly since $\hat{Y}_t$ is defined as $Y_t - \overline{Y}$

$$r_k = \frac{\sum_{t=1}^{n-k}\hat{Y}_t\hat{Y}_{t-k}}{\sum_{t=1}^{n}\left(\hat{Y}_t\right)^2} \tag{3}$$

We use the symbol r_k for the estimated auto-correlation coefficient among the observations separated by k time periods within a time series. After calculating estimated autocorrelation coefficient, we plot them graphically for different lags in an estimated autocorrelation function (i.e., ACF).

The sample partial auto-correlation co-efficient can be estimated by using the following set of recursive equations.

$$\hat{\varphi}_{11} = r_1$$

$$\hat{\varphi}_{kk} = \frac{r_k - \sum \hat{\varphi}_{k-1,j} r_{k-j}}{1 - \sum_{j=1}^{k-1} \hat{\varphi}_{k-1,j} r_j} \tag{4}$$

k = 2, 3, …

where

$$\hat{\varphi}_{kj} = \hat{\varphi}_{k-1,j} - \hat{\varphi}_{kk}\hat{\varphi}_{k-1,k-j} \quad k = 3, 4, \ldots, \ldots; j = 1, 2, 3, \ldots, \ldots, k-1$$

$\hat{\varphi}_{kk}$ is the estimate of the true partial autocorrelation coefficient φ_{kk}, where r_k is the autocorrelation coefficient for k lags apart and $\hat{\varphi}_{kj}$ is the estimate of partial auto-correlation coefficient for *k* lags apart when the effect of *j* intervening lags has been removed.

5.13.1 BOX-JENKINS AUTO-REGRESSIVE INTEGRATED MOVING AVERAGE (ARIMA) MODELS

Box-Jenkins time-series models written as ARIMA (p,d,q) was first popularized by Box and Jenkins in 1976. This model take care of three types of processes, namely autoregressive of order p; differencing to make a series stationary of degree d and moving average of order q as this method applies only to a stationary time series data. Stationary time series follows strong stationarity condition implies that the mean, variance, and covariance are constant. Most economic data increase over time, which makes the mean of the economic time series change over time, producing a non-stationary time series. This situation necessitates the transformation of a non-stationary time series to a stationary one. When the data are non-stationary, to be brought into stationary by the methods like differencing, i.e., $W_t = Y_t - Y_{t-1}$. The series W_t is called the first differences of Y_t, and the second difference of the series is $V_t = W_t - W_{t-1}$. In many cases first differencing is sufficient to bring about a stationary mean and second differencing is done in few cases only.

5.13.2 TEST FOR STATIONARITY

The stationarity requirement ensures that one can obtain useful estimates of the mean, variance, and ACF from a sample. If a process has a mean that is changing in each time period, one could not obtain useful estimates since only one observation available per time period. This necessitates testing any observed series of data for stationarity. There are three ways to determine whether the above-mentioned stationarity requirement is met.

i. Examine the realization visually to see if either the mean or the variance appear to change over time.
ii. Examine the estimated AR coefficient to see if it satisfies the stationary condition. In case of AR(1) process the condition for stationary is that absolute value of ϕ_1 must be less than one, or symbolically, $|\phi_1| < 1$. In practice, one don't know ϕ_1; therefore, one apply the condition to $\hat{\varphi}_1$ (i.e., estimate of ϕ_1) rather than ϕ_1.

iii. For an MA(1) process the corresponding condition is that the absolute value of θ_1 must be less than one. Which is called the condition of invariability, or in symbols $|\theta_1| < 1$.

Examine the estimated ACF to see if the auto-correlations move rapidly towards zero. In practice, "rapidly" means that the absolute t-values of the estimated auto-correlations should fall below roughly 1.6 by about lags 4 or 5. These numbers are only guidelines and are not absolute rules. If the ACF does not fall rapidly to zero, we should suspect a non-stationary mean and consider differencing of the data.

5.13.3 METHODOLOGY RELATED TO ARIMA MODEL

ARIMA modeling consists of three operational steps:

i. Identification;
ii. Estimation; and
iii. Diagnostic checking.

(i) Identification: At the identification stage, compare the estimated ACF and PACF's to find a match. Choose, as a tentative model, the ARMA process whose theoretical ACF and PACF best match the estimated ACF and PACF. In choosing a tentative model, one should keep in mind the principle of parsimony. The most important general characteristics of theoretical ACF and PACF of AR and MA models are:

1. A stationary AR process has a theoretical ACF and decays or "damps out" toward zero. But it has theoretical PACF that cuts off sharply to zero after few spikes. The lag length of the last PACF spike equals the AR order (p) of the process.
2. A MA process has a theoretical ACF that cuts off to zero after a certain number of spikes. The lag length of the last ACF spike equals the MA order (q) of the process. The theoretical PACF decays or "dies out" toward zero.

The general characteristics of theoretical ACF and PACF of five common stationary process, viz. AR(1), AR(2), MA(1), MA(2) and ARMA(1,1) are summarized in the following Table 5.2 (Pankratz, 1983).

TABLE 5.2 Detailed Characteristic of Five Common Stationary Processes

S. No.	Process	ACF	PACF
1.	AR(1)	Exponential decay: (i) On the positive side if $\phi_1 > 0$; (ii) Alternating in sign starting on negative side if $\phi_1 < 0$	Spike at lag 1, then cuts off to zero (i) Spike is positive if $\phi_1 >$ 0; (ii) Spike is negative side if $\phi_1 < 0$
2.	AR(2)	A mixture of exponential decays or a damped sine wave. The exact pattern depends on the signs and sizes of ϕ_1 and ϕ_2	Spikes at lags 1 and 2, then cuts off to zero
3.	MA(1)	Spike at lag 1, then cuts off to zero. (i) spike is positive if $\theta_1 < 0$ and (ii) spike is negative is $\theta_1 > 0$	Damps out exponentially: (i) alternating in sign, starting on the positive side, if $\theta_1 < 0$; (ii) on the negative side if $\theta_1 > 0$
4.	MA(2)	Spikes at lags 1 and 2, then cuts off to zero	A mixture of exponential decays or a damped sine wave. The exact pattern depends on the signs and sizes of θ_1 and θ_2.
5.	ARMA (1,1)	Exponential decay from lag 1: (i) sign of p_1 = sign of $\phi_1 - \theta_1$ (ii) all one sign if $\phi_1 > 0$ (iii) alternating sign if $\phi_1 < 0$	Exponential decay from lag 1: (i) $\phi_{11} = p_1$ (ii) all one sign if $\theta_1 > 0$ (iii) alternating in sign if $\theta_1 < 0$.

(iii) Estimation: Estimating the parameters for Box-Jenkins models are a quite complicated and based on non-linear least squares and maximum likelihood estimation. ARIMA models which include only AR terms are special cases of linear regression models; hence they can be fitted by ordinary least squares. AR forecasts are a linear function of the coefficients as well as a linear function of past data. In principle, least-squares estimates of AR coefficients can be exactly calculated from autocorrelations in a single "iteration." ARIMA models which include MA terms are similar to regression models, but can't be fitted by ordinary least squares. Forecasts are a linear function of past data, but they are *nonlinear* functions of coefficients. One can't fit MA models using ordinary multiple regression because there's no way to specify ERRORS as an independent variable in model. MA models, therefore, require a nonlinear estimation algorithm to be used. For this reason, the parameter estimation should be left to a high-quality software program that fits Box and Jenkins models (Bleikh, 2014).

(iv) Diagnostic Checking: At the identification stage of the Box-Jenkins time series methodology, a tentative model based on the patterns of ACF and PACF can be selected. The parameters of such a model are estimated at the estimation stage. Now at the final stage of ARIMA model building, namely the diagnostic checking stage it is necessary to test the suitability of the selected model. For this purpose, we need to test independence of residuals. This can be done using various methods either visual inspection or using some tests like chi-square test or Ljung-Box Q test.

5.13.3 TEST FOR INDEPENDENCE OF ERRORS

In this step, one can see whether the chosen model fits the data reasonably well. One simple test of the chosen model is to see if the residuals estimated from this model white noise. A white noise process is one which exhibits mean zero and no correlation between its values at different times. The graphs for the ACF and PACF of the ARIMA residuals are included within the lines representing two standard errors to either side of zero. If the values extend beyond two standard errors are statistically significant at approximately a = 0.05, and show evidence of serial dependence. If the residuals are white noise or independent, one can accept the particular model; if not, one start the process afresh, thus the Box and Jenkins methodology is an iterative process.

To summarize the three stages of ARIMA model building, the parameters of the tentatively selected ARIMA model at the identification stage are estimated at the estimation stage, and the adequacy of the chosen model is tested at the diagnostic checking stage. If the model is found to be inadequate, the three stages are repeated until satisfactory ARIMA model is selected for representing the time-series observations under consideration.

5.13.3.1 LJUNG-BOX Q TEST

The Ljung-Box Q statistic use to test whether a series of observations over time are random and independent. If observations are not independent, one observation can be correlated with a different observation k time units later, a relationship called autocorrelation This is a test for autocorrelated errors. It is an improved version of the Box-Pierce test. The Ljung-Box Q statistic tests the null hypothesis that autocorrelations up to lag k equal zero (i.e., the data values are random and independent up to a certain number of lags – in this case, 12). If the LBQ is greater than a specified critical value, autocorrelations for one or more lags might be significantly different from zero, indicating the values are not random and independent over time.

5.13.4 MODEL SELECTION CRITERIA

Among the competitive Box-Jenkins model best model is selected on the basis of minimum of Akaike's Information Criterion (AIC) and Bayesian Information Criterion (BIC), maximum R^2, minimum root mean square error (RMSE), minimum mean absolute percentage error (MAPE), minimum of maximum average percentage error (MaxAPE), minimum of maximum absolute error (MaxAE). Any model which has fulfilled most of the above criteria is selected. This section provides definitions of the goodness-of-fit measures used in time series modeling.

5.13.4.1 AKAIKE'S INFORMATION CRITERION (AIC)

Denoting by v^*, the estimate of white noise variance σ^2, obtained by fitting the corresponding ARIMA model, the AIC consists in computing the statistic,

$$AIC(p,q) = Ln\ v^*\ (p,q) + (2\ /n)\ (p+q),$$

where p and q are the order of AR and MA processes respectively, and n is the number of observations in the time-series.

5.13.4.2 BAYESIAN INFORMATION CRITERION (BIC)

Bayesian Information Criterion (BIC) is computed as,

$$BIC(p,q) = Ln\ v^*\ (p,q) + (p+q)\ [\ Ln\ (n)\ /n\ ].$$

A modification to BIC is the Schwarz – BIC (Cromwell et.al., 1994), given by $SC(p,q) = n * Ln\ v^*(p+q)+(p+q)\ Ln\ n$. The lower the values of these statistics, the better is the selected model.

5.13.4.3 R-SQUARE

An estimate of the proportion of the total variation in the series that is explained by the model. This measure is most useful when the series is stationary. Positive values mean that the model under consideration is better than the baseline model.

$$R^2 = \frac{\sum_{i=1}^{n}\left(\hat{X}_i - \overline{X}\right)^2}{\sum_{i=1}^{n}\left(X_i - \overline{X}\right)^2}$$

5.13.4.4 ROOT MEAN SQUARE ERROR (RMSE)

The square root of mean square error. A measure of how much a dependent series varies from its model-predicted level, expressed in the same units as the dependent series.

$$RMSE = \sqrt{\frac{\sum_{t=1}^{n}\left(X_t - \hat{X}_t\right)^2}{n}}$$

5.13.4.5 MEAN ABSOLUTE PERCENTAGE ERROR (MAPE)

A measure of how much a dependent series varies from its model-predicted level. It is independent of the units used and can, therefore, be used to compare series with different units.

$$MAPE = \frac{\sum_{t=1}^{n}\left|\frac{X_t - \hat{X}_t}{X_t}\right|}{n} \text{x}100$$

5.13.4.6 MEAN ABSOLUTE ERROR (MAE)

Measures how much the series varies from its model-predicted level. MAE is reported in the original series units.

$$MAE = \frac{\sum_{t=1}^{n} \left| X_t - \hat{X}_t \right|}{n}$$

5.13.4.7 MAXIMUM ABSOLUTE PERCENTAGE ERROR (MAXAPE)

The largest forecasted error, expressed as a percentage. This measure is useful for imagining a worst-case scenario for your forecasts.

$$MaxAPE = 100 \max \left(\left| \frac{X_t - \hat{X}_t}{X_t} \right| \right)$$

5.13.4.8 MAXIMUM ABSOLUTE ERROR (MAXAE)

The largest forecasted error expressed in the same units as the dependent series. Like MaxAPE, it is useful for imagining the worst-case scenario for the forecasts. Absolute percentage error may occur at different series points–for example, when the absolute error for a large series value is slightly larger than the absolute error for a small series value. In that case, the maximum absolute error will occur at the larger series value, and the maximum absolute percentage error will occur at the smaller series value.

$$MaxAE = \max \left(\left| X_t - \hat{X}_t \right| \right)$$

5.13.5 FORECASTING OF TIME SERIES USING ARIMA

One of the reasons for the popularity of the ARIMA modeling is its success in forecasting. To forecast the values of a time series, the basic Box and Jenkins strategy is as follows:

a. First, examine the stationary. This step can be done by computing the ACF and partial autocorrelation (PACF) or by a normal root analysis.
b. If the time series is not stationary, make it stationary, maybe through differencing one or more times.
c. The ACF and PACF of the stationary time series are then computed to find out if the series is purely autoregressive or purely of the

moving average type or a mixture of the two and tentative model is then estimated.

e. The residuals from this model are examined to find out if they are white noise. If the residuals are white noise, the tentative model is probably a good approximation to the underlying stochastic process. If these are not, the process is repeated afresh. Therefore, the Box and Jenkins method is iterative. The model finally selected can be used for forecasting.

Example 7: Following is the data of area under tea production in Assam from 1971–2010. We will fit the ARIMA model for the data and forecast the production up to 2020.

Year	Area (ha)	Year	Area (ha)	Year	Area (ha)	Year	Area (ha)
1971	182325	1981	203038	1991	233284	2001	269154
1972	184244	1982	211320	1992	233658	2002	270683
1973	185113	1983	213007	1993	231942	2003	271589
1974	187408	1984	214741	1994	227120	2004	270475
1975	188794	1985	215117	1995	226281	2005	300502
1976	189338	1986	222618	1996	228205	2006	311822
1977	190621	1987	225783	1997	229843	2007	321319
1978	192427	1988	227517	1998	251625	2008	322214
1979	195459	1989	229428	1999	258455	2009	322214
1980	200569	1990	230363	2000	266512	2010	322214

Solution: Analysis of the above data is done using SPSSv.16 and *R* software. To fit ARIMA model, the first step is to check the stationarity of the data by plotting ACF. One can check the stationarity using Augmented Dickey-Fuller Test (ADF), which is a unit root test. Unit roots can cause unpredictable results in your time series analysis. Unit root is present in data is a null hypothesis for the test. One can do ADF test in *r* software. A significant result of the test suggests that rejection of the null hypothesis.

For given data, we fitted ADF test using *r* software, and results are as follows:

Augmented Dickey-Fuller Test in 'r'

Data: area.

Dickey-Fuller = –1.1182, Lag order = 3, p-value = 0.9076.

Alternative hypothesis: stationary.

Seeing the results, we can say that p-value is nonsignificant this means unit root is present in data and conclude that data is not stationary.

In SPSS we can plot ACF graph and check stationarity in data (Figure 5.7).

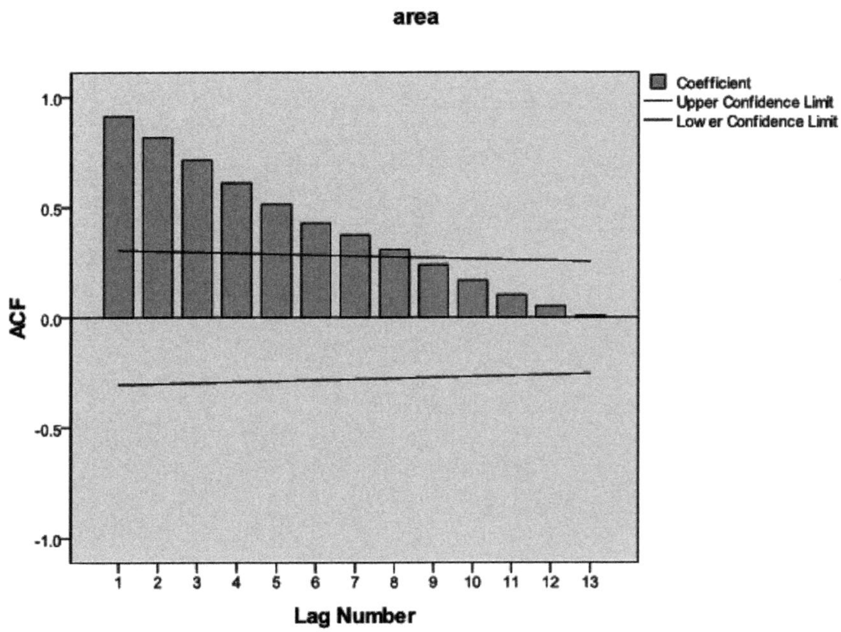

FIGURE 5.7 ACF of the original data.

From Figure 5.7, it is seen that ACF of the data having strong persistence across all the lags and ACFs are not rapidly falling to zero. Hence, data is not stationary. Both ADF test and ACF graphs conclude that differencing is needed to make data stationary. Results of ADF test after one differencing in *r* and SPS are given below,

Augmented Dickey-Fuller Test in 'r'

Data: diff(Area)

Dickey-Fuller = –4.0977, Lag order = 3, p-value = 0.01628

Alternative hypothesis: stationary

ADF test for differenced data is significant, and we can reject null hypothesis that unit root is present.

In SPSS, we have plotted ACF graphs of differenced data. After differencing the ACF of the data are within the limit (Figure 5.8) and hence we can say that data is stationary. This data is further used for ARIMA fitting with difference 1.

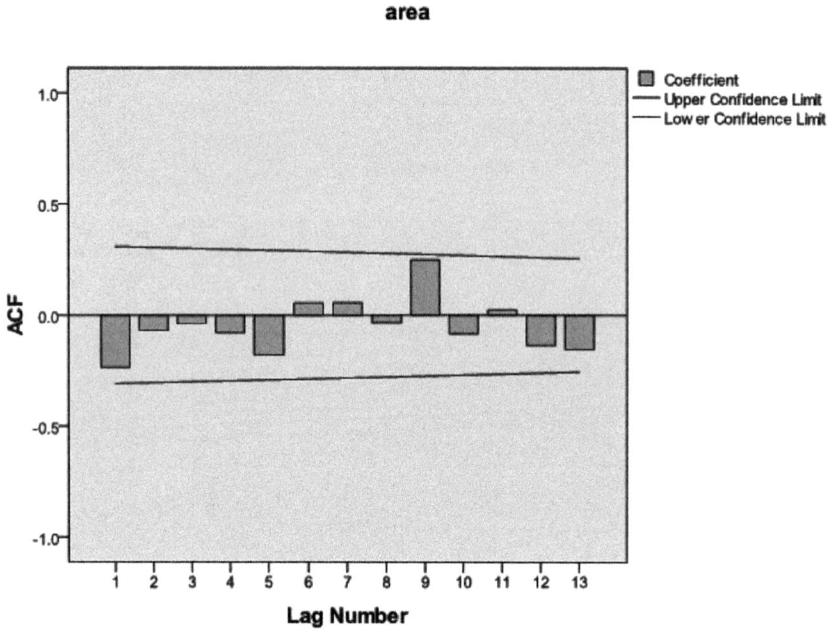

FIGURE 5.8 ACF of the difference data.

Various ARIMA models were tried starting from p = 1 and q = 1, we found that ARIMA (5,1,3) was best fit. As all the AR and MA coefficients are significant (Tables 5.3 and 5.4). This model is selected for further analysis.

TABLE 5.3 ARIMA Analysis for Tea Data

ARIMA Model Parameters						
			Estimate	**SE**	**t**	**Sig.**
ARIMA (5,1,3)	**AR**	Lag 1	0.815	0.309	2.634	0.013
		Lag 2	0.651	0.315	2.067	0.047
		Lag 3	–1.060	0.317	–3.346	0.002
		Lag 4	–1.024	0.257	–3.983	0.001
		Lag 5	0.453	0.211	2.142	0.040
	Difference		1			
	MA	Lag 1	0.915	0.420	2.186	0.049
		Lag 2	0.606	0.323	1.874	0.070
		Lag 3	–0.845	0.394	–2.147	0.040

TABLE 5.4 ARIMA (5,1,3) Model Statistics

Model Statistics							
Model	**Model Fit statistics**						
	R^2	**RMSE**	**MAPE**	**MAE**	**MaxAPE**	**MaxAE**	**Normalized BIC**
ARIMA(5,1,3)	0.98	6016.87	1.22	3033.50	8.48	21334.33	18.16

Residual of this model is plotted for the same model, and it can be seen that the ACF and PACF are within the limits (Figures 5.9 and 5.10). From Table 5.5, it is also concluded that Ljung Box test is nonsignificant. Hence, this model is finally selected for the forecasting purpose. Is observed that area under tea in Assam is going to increase and by 2020, the forecasted area under tea cultivation is 3.54 lakh ha.

TABLE 5.5 ARIMA (5,1,3) Residual Statistics

Residual Analysis			
Model	**Ljung-Box Q (18)**		
	Statistics	**DF**	**Sig.**
ARIMA(5,1,3)	10.007	10	0.070

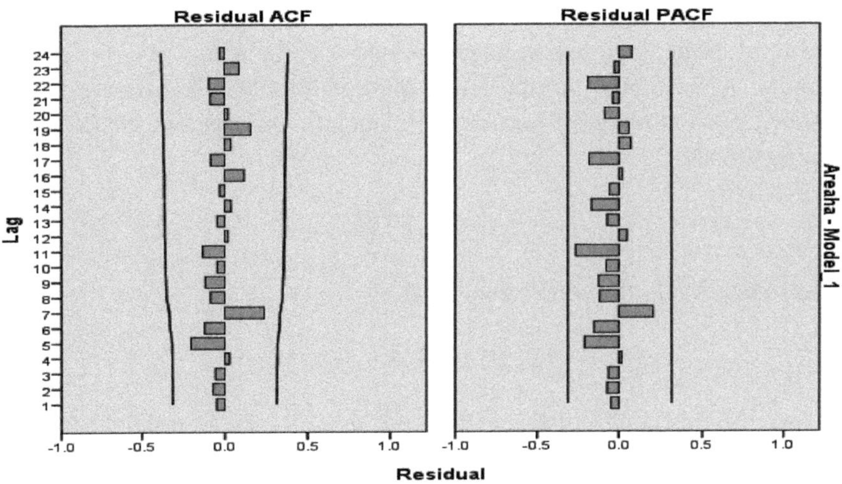

FIGURE 5.9 ACF and PACF of Residuals from ARIMA (5,1,3) model.

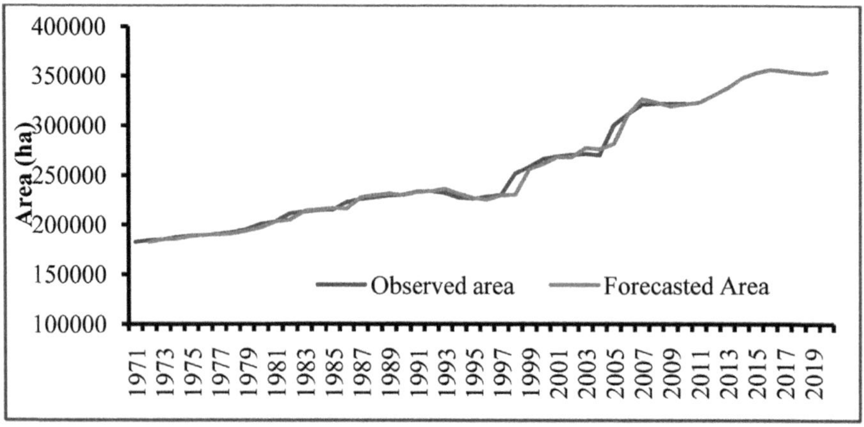

FIGURE 5.10 Forecasting of area under tea of Assam using ARIMA(5,1,3) model.

5.14 ARCH AND GARCH MODELS

If data is volatile over the time, i.e., variance is not stationary then linear Gaussian models cannot be applied. Box and Jenkins models can be applied in situation where data is not stationary in terms of mean. In case of nonstationary variance, generally, ARCH (Autoregressive Conditional Variance) model is applied.

Engle (1982) showed that it is possible to simultaneously model the mean and variance of a series. As preliminary step to understanding Engle's methodology, it should be noted that the conditional forecasts are vastly superior to unconditional forecasts. To elaborate, suppose we estimate the stationary model

$$y_t = a_0 + a_1 y_{t-1} + \mu_t \quad (5)$$

To forecast y_{t+1}, the conditional mean to y_{t+1} is

$$E(y_{t+1}/y_t) = a_0 + a_1 y_t \quad (6)$$

If we use this conditional mean to forecast y_{t+1}, the forecast error variance is

$$E\left[(y_{t+1} - a_o - a_1 y_t)^2\right] = E_t \varepsilon_{t+1}^2 = \sigma^2 \quad (7)$$

Instead, if unconditional forecasts are used, the unconditional forecast is always the long-run mean of the $\{y_t\}$ sequence that is equal to $a_0/(1-a_1)$. The unconditional forecast error variance is

$$E\{[y_{t+1}-a_0/(1-a_1)]^2\}=E\left[(\varepsilon_{t+1}+a_1\varepsilon_t+a_1^2\varepsilon_{t-1}+a_1^3\varepsilon_{t-2}+....)^2\right]=\sigma^2/(1-a_1^2) \quad (8)$$

Since $1/(1-a_1^2)>1$, the unconditional forecast has a greater variance than the conditional forecasts (since they take into account the known current and past realization of series) are preferable.

Similarly, if the variance of (u_t) is not constant, we can estimate any tendency for sustained movements in the variance using ARMA model. For example, let $(\hat{u}_t)$ denote the estimated residuals from the model $y_t = a_0 + a_1 y_{t-1} + \mu_t$ so that the conditional variance of y_{t+1} is

$$Var(y_{t+1}/y_t)=E\left[(y_{t+1}-a_0-a_1y_t)^2\right]=E_t(\hat{u}_{t+1}^2)=\sigma^2 \quad (9)$$

Thus far, we have set $E_t(\hat{u}_{t+1}^2)$ equal to σ^2. Now suppose that the conditional variance is not constant. One simple strategy is to model the conditional variance of y_t using square of estimated residuals

$$\sigma^2{}_t = a_0 + a_1\hat{u}_{t-1}^2 \quad (10)$$

If the values of a_1, a_2,....., a_n all equal to zero. The estimated variance is simply the constant a_0. Otherwise, the conditional variance of y_t evolves according to the autoregressive process given by above equation can come from an autoregression, an ARMA model, or a standard regression model.

This linear specification of Eq. (10) is known as the ARCH (1) model. The full model would be

$$y_t = \beta_0 + \beta_1 y_{t-1} + ... + \beta_k y_{t-k} + \mu_t ......14.7\,;\ \mu_t \sim N(0,\sigma^2)$$

where

$$\sigma_t^2 = \alpha_0 + \alpha_1 \hat{u}_{t-1}^2$$

This can be extended to the general case where the error variance depends on q lags of squared error terms.

$$\sigma_t^2 = \alpha_0 + \alpha_1\hat{\mu}_{t-1}^2 + \alpha_2\hat{\mu}_{t-2}^2 + ... + \alpha_q\hat{\mu}_{t-q}^2 \quad (11)$$

The simplest example from the class of multiplicative conditionally heteroskedasticity models proposed by Engle (1982) is

$$\varepsilon_t = v_t \sqrt{\alpha_0 + \alpha_1 \varepsilon_{t-1}^2} \tag{12}$$

where, v_t = white noise process such that $\sigma_v^2 = 1$; v_t and u_{t-1} are independent of each other, and α_0, α_1 are constant such that α_0>0 and 0<α_1<1. This is similar to earlier process but in multiplicative forms.

5.14.1 THE GARCH MODEL

After the introduction of ARCH model by Engel, various extensions of ARCH model have been suggested. However, ARCH model has some drawbacks. Firstly, when the order of ARCH model is very large, estimation of a large number of parameters is required which is cumbersome. Secondly, the conditional variance of ARCH(q) model has the property that unconditional ACF of squared residuals; if it exists, decays very rapidly compared to what is typically observed, unless maximum lag q is large.

To overcome these difficulties, Bollerslev (1986) proposed the Generalized ARCH (GARCH) model in which conditional variance is also a linear function of its own lags. This model is also a weighted average of past squared residuals, but it has declining weights that never go completely to zero. It gives parsimonious models that are easy to estimate and, even in its simplest form, has proven surprisingly successful in predicting conditional variances Bollerslevs (1986) extended the Engle's original work by developing a technique that allows the conditional variance to be dependent on previous own lags like an ARMA process. Now let the error process be such that

$$\sigma_t^2 = \alpha_0 + \alpha_1 \mu_{t-1}^2 + \beta \sigma_{t-1}^2 \tag{13}$$

This is a GARCH (1,1) model, which is like an ARMA (1,1) model for the variance equation. We can also write

$$\sigma_{t-1}^2 = \alpha_0 + \alpha_1 \mu_{t-2}^2 + \beta \sigma_{t-2}^2 \tag{14}$$

$$\sigma_{t-2}^2 = \alpha_0 + \alpha_1 \mu_{t-3}^2 + \beta \sigma_{t-3}^2 \tag{15}$$

Substituting Eq. (14), in Eq. (13) for σ_{t-1}^2 and σ_{t-2}^2 in the above equations we get

$$\sigma_t^2 = \alpha_0 + \alpha_1\mu_{t-1}^2 + \beta\alpha_0 + \alpha_1\beta\mu_{t-2}^2 + \beta^2\left(\alpha_0 + \alpha_1\mu_{t-3}^2 + \beta\sigma_{t-3}^2\right) \tag{16}$$

$$\sigma_t^2 = \alpha_0\left(1+\beta+\beta^2\right) + \alpha_1\beta\mu_{t-1}^2\left(1+\beta L+\beta^2L^2\right) + \beta^3\sigma_{t-3}^2 \tag{17}$$

Substituting Eq. (15) in Eq. (16) for σ_{t-2}^2 in the above equations, we get

$$\sigma_t^2 = \alpha_0 + \alpha_1\mu_{t-1}^2 + \beta\alpha_0 + \alpha_1\beta\mu_{t-2}^2 + \beta^2\left(\alpha_0 + \alpha_1\mu_{t-3}^2 + \beta\sigma_{t-3}^2\right) \tag{18}$$

$$\sigma_t^2 = \alpha_0\left(1+\beta+\beta^2\right) + \alpha_1\beta\mu_{t-1}^2\left(1+\beta L+\beta^2L^2\right) + \beta^3\sigma_{t-3}^2 \tag{19}$$

An infinite number of successive substitutions would yield

$$\sigma_t^2 = \alpha_0 + \alpha_1\mu_{t-1}^2 + \beta\alpha_0 + \alpha_1\beta\mu_{t-2}^2 + \alpha_0\beta^2 + \alpha_1\beta^2\mu_{t-3}^2 + \beta^3\sigma_{t-3}^2 \tag{20}$$

$$\sigma_t^2 = \alpha_0\left(1+\beta+\beta^2+........\right) + \alpha_1\beta\mu_{t-1}^2\left(1+\beta L+\beta^2L^2+......\right) + \beta^\infty\sigma_0^2 \tag{21}$$

So GARCH (1,1) model can be written as an infinite order of ARCH model.

$$\sigma_t^2 = \alpha_0 + \alpha_1\mu_{t-1}^2 + \alpha_2\mu_{t-2}^2 + ... + \alpha_q\mu_{t-q}^2 + \beta_1\sigma_{t-1}^2 + \beta_2\sigma_{t-2}^2 + ... + \beta_p\sigma_{t-p}^2 \tag{22}$$

The GARCH (1,1) model can be extended to GARCH (p, q)

$$\sigma_t^2 = \alpha_0 + \sum_{i=1}^{q}\alpha_i\mu_{t-i}^2 + \sum_{j=1}^{p}\beta_j\sigma_{t-j}^2 \tag{23}$$

This GARCH (p, q) model – called GARCH (p, q) – allows for both autoregressive and moving average components in the heteroscedastic variance. If we set p = 0 and q = 1, it is clear that the first order ARCH (1) model is simply a GARCH (0, 1) model. If all β_1 equal zero, the GARCH (p, q) model is equivalent to an ARCH (q) model. The benefits of the GARCH model should be clear; a high order ARCH model may have more parsimonious GARCH representation that is much easier to identify and estimate. This is particularly true since all coefficients in Eq. (23) must be positive. Moreover, to ensure that the conditional variance is finite, all characteristic roots of Eq. (23) must lie inside the unit circle. Clearly, the more parsimonious model will entail fewer coefficient restrictions.

The key feature of GARCH models is that the conditional variance of the disturbance of the (y_t) sequence constitutes an ARMA process. Hence, it is to be expected that the residuals from a fitted ARMA model should display this characteristic pattern. To explain, suppose we estimate (y_t) as an ARMA process. If our model of (y_t) is adequate, the ACF and PACF of the residuals should be indicative of white noise process. However, the ACF of squared residuals can help in identifying the order of the GARCH process.

Equation (13) looks very much like an ARMA (q, p) process in the $(u_t)^2$ sequence. If there is conditional heteroskedasticity, the correlogram should be suggestive of such a process. The technique to construct the correlogram of the squared residuals is as follow:

5.14.2 TESTING FOR ARCH/GARCH EFFECT

Step 1: Estimate the (y_t) sequence using the best-fitting ARMA model (or regression model)

$$y_t = \beta_1 + \beta_2 x_{2t} + ... + \beta_k x_{kt} + \mu_t$$

Step 2: Then square the residuals, and regress them on q own lags to test for ARCH of order q, i.e., run the regression

$\hat{\mu}_t^2 = \gamma_0 + \gamma_1 \hat{\mu}_{t-1}^2 + \gamma_2 \hat{\mu}_{t-2}^2 + ... + \gamma_q \hat{\mu}_{t-q}^2 + v_t$, obtain R^2 from this regression

Step 3: The test statistic is defined as TxR^2 (the number of observations multiplied by the coefficient of multiple correlations) from the last regression, and is distributed as a $\chi^2(q)$.

Step 4: The null and alternative hypotheses are

$$H_0: \gamma_1 = 0 \text{ and } \gamma_2 = 0 \text{ and } \gamma_3 = 0 \text{ and ...... and } \gamma_q = 0$$

$$H_1: \gamma_1 \neq 0 \text{ or } \gamma_2 \neq 0 \text{ or } \gamma_3 \neq 0 \text{ or ...... or } \gamma_q \neq 0$$

Step 5: If the value of the test statistic is greater than the critical value from the χ^2 distribution, then reject the null hypothesis.

Step 6: Note that the ARCH test is also sometimes applied directly to returns instead of the residuals from Stage 1 above.

5.14.3 FORECASTING VARIANCES USING GARCH MODELS

Producing conditional variance forecasts from GARCH models uses a very similar approach to producing forecasts from ARMA models.

Consider the following GARCH (1,1) model:

$$\sigma_t^2 = \alpha_0 + \alpha_1 \mu_{t-1}^2 + \beta \sigma_{t-1}^2$$

It is needed is to generate forecasts of $\sigma_{T+1}^2 \mid \Omega_T, \sigma_{T+2}^2 \mid \Omega_T, \ldots\ldots, \sigma_{T+s}^2 \mid \Omega_T$ where Ω_T denotes all information available up to and including observation T.

Adding one to each of the time subscripts of the above conditional variance equation, and then two, and then three would yield the following equations

$\sigma_{T+1}^2 = \alpha_0 + \alpha_1 + \beta \sigma_T^2; \sigma_{T+2}^2 = \alpha_0 + \alpha_1 + \beta \sigma_{T+1}^2; \sigma_{T+3}^2 = \alpha_0 + \alpha_1 + \beta \sigma_{T+2}^2$ and so on.

Diagnostic checking: Diagnostic checking of GARCH model is also done in the same way as in the case of a univariate ARIMA model using visual inspection of ACF and PACF graphs of residuals and Ljung-Box Q test.

Model Selection Criteria's: Among the competitive GARCH model best model is selected on the basis of a minimum of AIC, BIC, Mean Error (ME), RMSE, Mean absolute error (MAE), Mean Percentage Error (MPE), MAPE and the maximum value of R^2. Any model which has fulfilled most of the above criteria is selected as the best GARCH model.

ARCH Example (Franses, 1998): Fitting of a time-series model for describing the average monthly European spot price of black pepper 1970.10–1996.04 in US dollars per million tons.

From Figure 5.13, it can be observed that there are periods where the price fluctuates heavily. Furthermore, there seem to be larger price increases than there are large negative price changes. Hence, the pepper price seems to have positive skewness. An estimated autocorrelation function is given in Table 5.6.

TABLE 5.6 An Estimated Autocorrelation Function

Series	$\hat{\rho}_1$	$\hat{\rho}_2$	$\hat{\rho}_3$
EACF of x_t	0.338	0.024	0.012
EACF of $\hat{y}^2_t$	0.136	0.194	0.01.

Example (Paul et al., 2009; IASRI): All India data of monthly export of spices during the period April 2000 to December 2006 are obtained from www.indiastat.com and given below. Fit the data using the suitable model and validate the fitted model.

SN	Date	Spice Export	SN	Date	Spice Export	SN	Date	Spice Export
1	Apr–00	137.95	28	Jul–02	118.62	55	Oct–04	135.19
2	May–00	129.95	29	Aug–02	132.59	56	Nov–04	148.11
3	Jun–00	123.51	30	Sep–02	121.42	57	Dec–04	143.66
4	Jul–00	123.28	31	Oct–02	142.20	58	Jan–05	143.00
5	Aug–00	123.97	32	Nov–02	126.62	59	Feb–05	162.24
6	Sep–00	124.68	33	Dec–02	107.91	60	Mar–05	189.11
7	Oct–00	110.52	34	Jan–03	135.11	61	Apr–05	166.77
8	Nov–00	134.18	35	Feb–03	110.45	62	May–05	162.09
9	Dec–00	126.61	36	Mar–03	128.65	63	Jun–05	162.19
10	Jan–01	141.00	37	Apr–03	117.94	64	Jul–05	160.79
11	Feb–01	163.51	38	May–03	110.75	65	Aug–05	175.06
12	Mar–01	163.32	39	Jun–03	106.18	66	Sep–05	178.36
13	Apr–01	136.18	40	Jul–03	104.54	67	Oct–05	187.73
14	May–01	96.05	41	Aug–03	93.80	68	Nov–05	160.29
15	Jun–01	123.24	42	Sep–03	122.46	69	Dec–05	192.91
16	Jul–01	119.42	43	Oct–03	131.32	70	Jan–06	169.11
17	Aug–01	109.23	44	Nov–03	129.90	71	Feb–06	176.98
18	Sep–01	99.94	45	Dec–03	163.96	72	Mar–06	221.39
19	Oct–01	142.34	46	Jan–04	119.06	73	Apr–06	187.32
20	Nov–01	95.94	47	Feb–04	104.83	74	May–06	222.72
21	Dec–01	90.44	48	Mar–04	223.73	75	Jun–06	232.33
22	Jan–02	105.77	49	Apr–04	152.74	76	Jul–06	235.40
23	Feb–02	110.72	50	May–04	156.13	77	Aug–06	267.57

SN	Date	Spice Export	SN	Date	Spice Export	SN	Date	Spice Export
24	Mar–02	123.59	51	Jun–04	167.17	78	Sep–06	272.59
25	Apr–02	148.44	52	Jul–04	166.81	79	Oct–06	245.04
26	May–02	177.88	53	Aug–04	149.59	80	Nov–06	292.45
27	Jun–02	129.99	54	Sep–04	158.54	81	Dec–06	256.74

(www.iasri.res.in/sscnars/socialsci/16-AR_1%20GARCH_1,1__SAS.pdf).

Solution: From the total 81 data points, first 77 data points corresponding to the period April 2000 to August 2006 are used for building the model and remaining are used for validation purpose. A perusal of the data shows that, during the period from April 2004 to February 2006, these varied between Rs.135.19 crores and Rs.192.91 crores. Then the spices export suddenly jumped to the level of Rs.221.39 crores in March 2006, which was followed by a sudden dip to as low as Rs. 187.32 crores in the very next month. All this clearly shows that volatility was present during March 2006. A similar type of presence of volatility was noticed at several other time-epochs, like May 2001, September 2001.

Fitting of GARCH Model Using SAS Version 9.2:

Title 'Monthly Spices Data April 2000 to August 2006';

data spices;

input time y;

cards;

1	137.95	40	104.54
2	129.95	41	93.8
3	123.51	42	122.46
4	123.28	43	131.32
5	123.97	44	129.9
6	124.68	45	163.96
7	110.52	46	119.06
8	134.18	47	104.83
9	126.61	48	223.73
10	141	49	152.74
11	163.51	50	156.13
12	163.32	51	167.17
13	136.18	52	166.81

14	96.05	53	149.59
15	123.24	54	158.54
16	119.42	55	135.19
17	109.23	56	148.11
18	99.94	57	143.66
19	142.34	58	143
20	95.94	59	162.24
21	90.44	60	189.11
22	105.77	61	166.77
23	110.72	62	162.09
24	123.59	63	162.19
25	148.44	64	160.79
26	177.88	65	175.06
27	129.99	66	178.36
28	118.62	67	187.73
29	132.59	68	160.29
30	121.42	69	192.91
31	142.2	70	169.11
32	126.62	71	176.98
33	107.91	72	221.39
34	135.11	73	187.32
35	110.45	74	222.72
36	128.65	75	232.33
37	117.94	76	235.4
38	110.75	77	267.57
39	106.18		

```
run;
proc print data = spices; run;
/*Plot of the series*/
proc sgplot data = spices (obs = 77); series x = time y = y;
xaxis values = (0 to 84 by 12); run;
quit
/*seasonal trend is present in the following graph*/
```

Output of GRAPH of Original Data

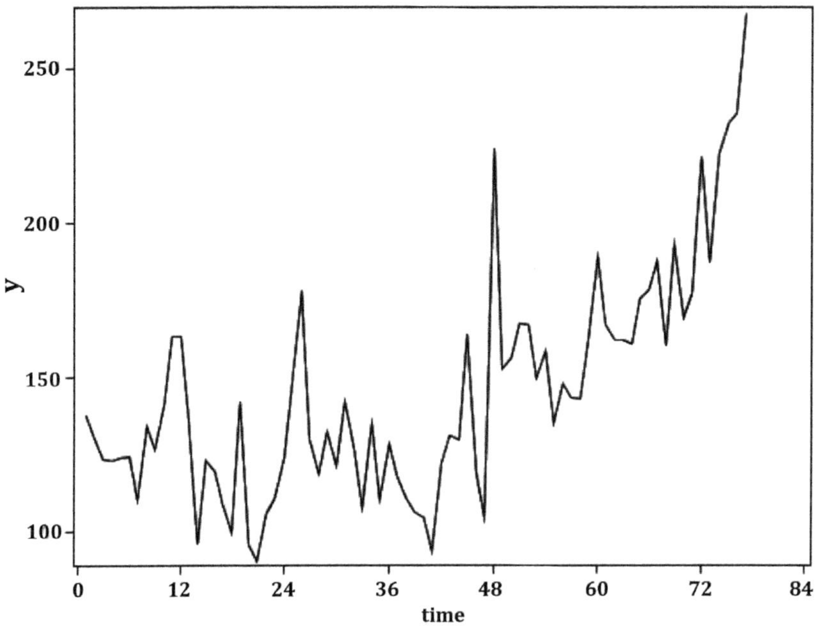

/* Test for Heteroscedasticty*/ ods html;
ods graphics on;
proc autoreg data = spices;
model y = /nlag = 1 archtestdwprob
method = ml; output out = r r = yresid;
run; quit; ods graphics off; ods html;

Output in table formats:

TABLE 5.7 Q and LM Tests for ARCH Disturbances

Order	Q	Pr> Q	LM	Pr> LM	Order	Q	Pr>Q	LM	Pr> LM
1	19.0861	<.0001	35.1593	<.0001	**7**	42.3082	<.0001	48.2483	<.0001
2	31.9060	<.0001	42.9105	<.0001	**8**	42.4390	<.0001	48.4019	<.0001
3	38.7479	<.0001	46.2432	<.0001	**9**	42.4813	<.0001	48.7728	<.0001
4	39.6084	<.0001	46.4692	<.0001	**10**	42.4981	<.0001	48.9755	<.0001
5	42.1106	<.0001	47.5614	<.0001	**11**	42.7323	<.0001	48.9770	<.0001
6	42.1426	<.0001	47.6035	<.0001	**12**	42.7925	<.0001	49.7264	<.0001

TABLE 5.8 Maximum Likelihood Estimate

SSE	47263.9749	**DFE**	75
MSE	630.18633	**Root MSE**	25.10351
SBC	722.444292	**AIC**	717.756681
MAE	18.6043037	**AICC**	717.918843
MAPE	12.9539443	**Regress R-Square**	0.0000
Durbin-Watson	2.3850	**Total R-Square**	0.5311

TABLE 5.9 Parameter estimates of AR(1) model for spice data

Variable	DF	Estimate	Standard Error	t Value	Approx. Pr > |t|
Intercept	1	150.1369	12.4876	12.02	<.0001
AR1	1	–0.7764	0.0858	–9.05	<.0001

The Q statistics test for changes in variance across time by using lag windows ranging from 1 through 12 is given in the Table 5.7. These tests strongly indicate heteroscedasticity, with $p < 0.0001$ for all lag windows.

The Lagrange multiplier (LM) tests also indicate that a long memory period *generalized autoregressive conditional heteroscedasticity* GARCH model is one approach to modeling time series with heteroscedastic errors. From Table 5.9 we can observe AR(1) is significant, so we can fit AR(1) mean equation and GARCH(1,1) variance equation (Figure 5.11).

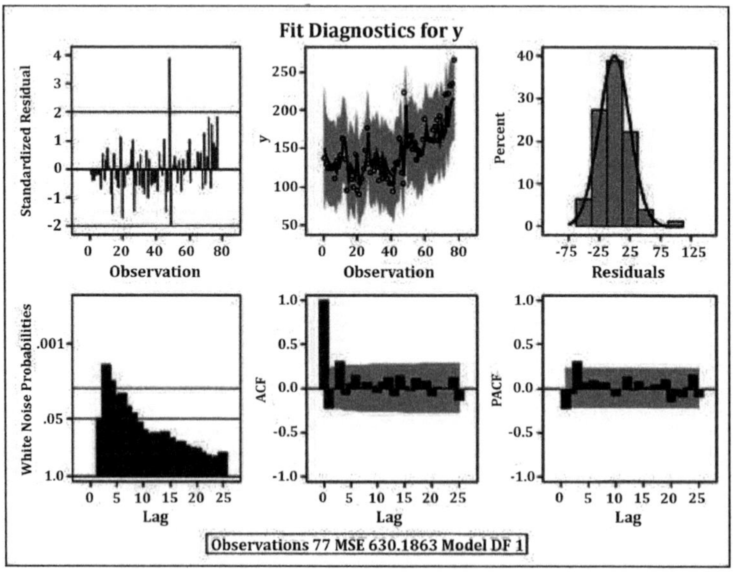

FIGURE 5.11 Diagnostic plots for spice data after fitting AR(1) model for spice data.

From the Figure 5.11 , we observe that ACF and PACF are insignificant at 5% level, thereby confirming that mean equation is correctly specified.

ods html;

ods graphics on;

/*-- AR(1)-GARCH(1,1) model for the *Y* series --*/ proc autoreg data = spices;

model *y* = /nlag = 1 garch = (q = 1,p = 1) maxit = 50;

output out = out_spicescev = vhat p = predlcl = lclucl = ucl *r* = r_garch-cpev = cpev; run;

quit;

odsgraphis off; ods html;

ods graphics on;

/*-- AR(1)-GARCH(1,1) model for the *Y* series --*/ proc autoreg data = spices;

model *y* = /nlag = 1 garch = (q = 1,p = 1) maxit = 50;

output out = out_spicescev = vhat p = predlcl = lclucl = ucl *r* = r_garch-cpev = cpev; run;

quit;

odsgraphis off; ods html;

Output of the AR(1)-GARCH(1,1)

TABLE 5.10 Ordinary Least Square Estimates for AR(1)-GARCH(1,1) model

SSE	100794.43	DFE	76
MSE	1326	Root MSE	36.41761
SBC	775.491877	AIC	773.148072

TABLE 5.11 AR(1)-GARCH(1,1) model estimation for spice data

SSE	47333.4508	Observations	77
MSE	614.72014	Uncond Var	0.00
Log Likelihood	–350.93743	Total R-Square	0.5304
SBC	723.593878	AIC	711.874851
MAE	18.5615167	AICC	712.719921
MAPE	12.9027544	Normality Test	18.9517
		Pr>ChiSq	<.0001

TABLE 5.12 Parameter estimated of AR(1)-GARCH(1,1,) model for spice data

Variable	DF	Estimate	Standard Error	t Value	Approx. Pr > |t|
Intercept	1	149.4223	11.2830	13.24	<.0001
AR1	1	–0.7888	0.0883	–8.93	<.0001
ARCH0	1	144.0878	96.9414	1.49	0.1372
Variable	DF	Estimate	Standard Error	t Value	Approx. Pr> |t|
ARCH1	1	0.8183	0.4530	1.81	0.0708
GARCH1	1	0.2112	0.2251	0.94	0.3482

Though estimate of GARCH(1) is slightly significant (from Table 5.12) but we can say AR(1)-GARCH(1,1) is a good fit because AIC (from Table 5.11) is smaller than the AIC of AR(1) (from Table 5.8 and Figure 5.12).

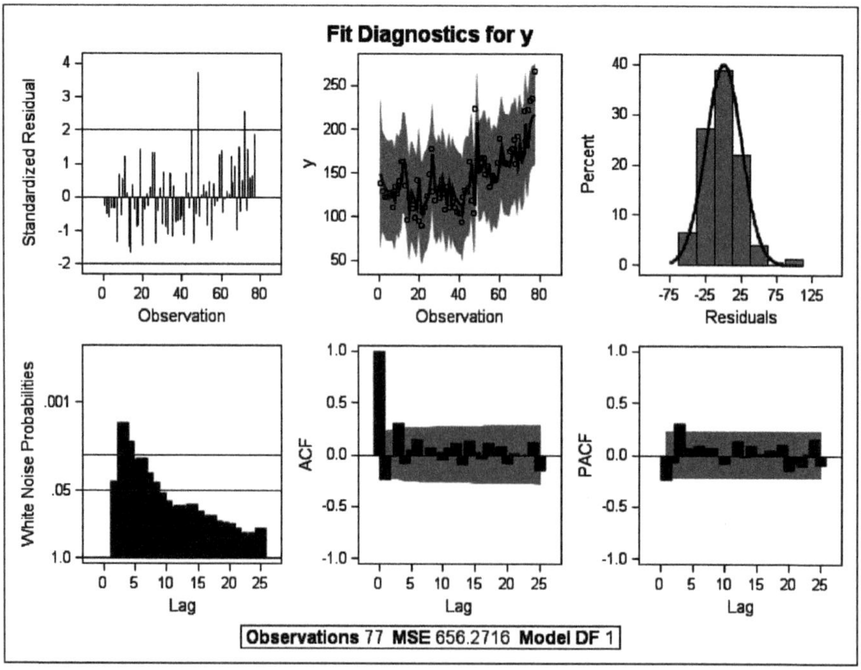

FIGURE 5.12 Diagnostic plots for spice data after fitting AR(1)-GARCH(1,1) model for spice data

Figure 5.12 shows that autocorrelation functions ACF and PACF are insignificant at 5% level, thereby confirming that the mean and variance equations are correctly specified.

```
/*GRAPH of Fitted series along with data*/
Title 'Monthly Export of Spices from India'; proc sgplot data = out_spices;
scatter x = time y = y; series x = time y = pred;
xaxis label = 'Time (April, 2000 To August, 2006)'; yaxis label = 'Actual
And Predicted (in Rs. Crore)'; run;
quit;
```

The following graph (Figure 5.13) shows that the fitted GARCH model is able to capture the volatility present in the data set.

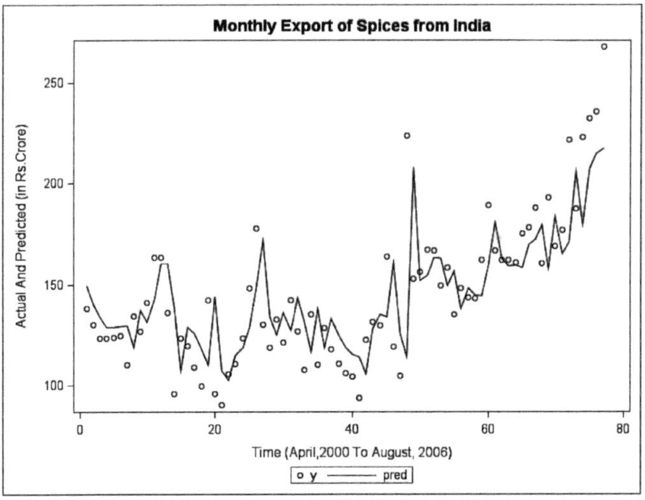

FIGURE 5.13 Actual and Predicted Monthly Export of Spices from India through fitting AR(1)-GARCH(1,1) model.

```
/* Conditional standard deviation of fitted AR(1)-GARCH(1,1) */ data seout;
set out_spices; shat = sqrt(vhat); run;
proc print data = seout; run;
title 'conditional Standard Deviations';
proc sgplot data = seoutnoautolegend;
series x = time y = shat/ lineattrs = (color = black);
xaxis values = (0 to 84 by 12);
run; quit;
```

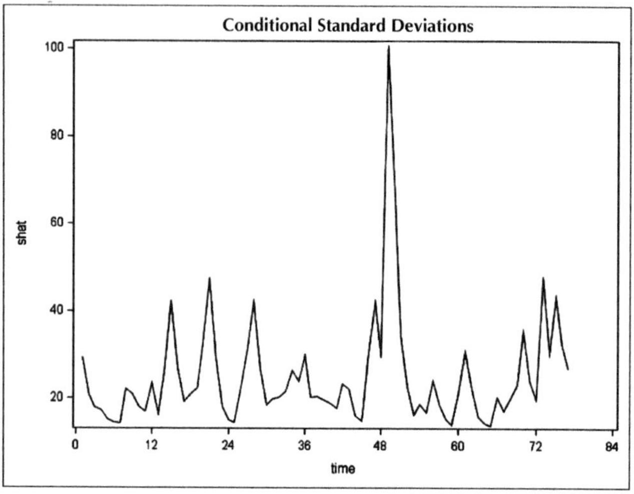

FIGURE 5.14 Conditional standard deviation of fitted AR(1)-GARCH(1,1)

```
/*Forecast data for four months*/ data b;
y = .;
do time = 78 to 81; output; end; run;
data b;
merge spices b; by time;
run;
proc print data = b; run;
proc autoreg data = b;
model y = /nlag = 1 garch = (q = 1,p = 1) maxit = 50;
output out = out_spices1 cev = vhat p = predlcl = lclucl = ucl r = r_garch-
cpev = cpev; run;
proc print data = out_spices1; run;
```

TABLE 5.13 Forecasts of Export of Spices (in Rs. Crore) for Fitted AR(1)-GARCH(1,1) Model

Months	Actual Value	One-step-ahead forecast (Static)	Multi-step ahead forecast (Dynamic)
September 2006	272.59	242.62(48.66)	242.62(48.66)
October 2006	245.04	224.57(25.44)	222.94(50.81)
November 2006	292.45	210.16(16.69)	207.41(52.93)
December 2006	256.74	198.17(13.85)	195.17(55.51)

Figures in "( )" are corresponding standard errors. For multi-step ahead, the forecast error variance usually increases with respect to the increase in a number of steps ahead for forecasting.

Mean and variance equations are in Table 5.13:

$$y_t = 149.42 - 0.789 y_{t-1} + \sigma_t^2$$
$$(11.283)\ (0.088)$$

where $\varepsilon_t = \sigma_t \eta_t$ and σ^2 satisfies the variance equation

$$\sigma_t^2 = 144.088 + 0.818 \sigma_{t-j}^2 + 0.211 \mu_{t-i}^2$$
$$(96.941)\ (0.453)\quad (0.225)$$

5.15 MARKOV CHAIN ANALYSIS

A Markov chain is a stochastic model describing a sequence of possible events in which the probability of each event depends only on the state attained in the previous event. Markov process is named after the Russian mathematician Andrey Markov, is a stochastic process that satisfies the Markov property (Richard, 2009). It is a process with either a discrete state space or discrete index set (Soren, 2003).

In case of forecasting area under different crops grown in a state/country, it is clear that different series are related by the fact that total area available for cultivation (including fallow land) remains more or less same every from year to year. It is also true that many areas specialize in the production of the particular crop and are unlikely to shift to cultivation of other crops due to soil and climatic conditions prevailing in the region. In this case, the Markov chain technique can be successfully used and which considers area under different crops as part of a unified system and then forecasts the area under different crops. Markov chain analysis is one such technique which takes care of the above considerations. The Markov chain assumes that area under a given crop in this year is the sum of area retained by that crop itself and the area gained by this crop from other crops due to the process of substitution. If we assume area under a given crop in t[th] year be $A_{j,t}$ and probability of ith crop substituting j[th] crop be $P_{i,j}$, then $A_{j,t}$ can be shown as a function of the area under different crops in t–1th year as following:

$$\sum_{j=1}^{m} A_{j,t-1} P_{ij} + \in_{j,t} = A_{j,t} \qquad (24)$$

This is true for all crops (j = 1, 2,…,i,…,m) and $\in_j$ is the random error term for t[th] period forecast of jth crop area.

In matrix form, this can be represented as follows

$$(A_{1,t-1} \quad \cdots \quad A_{i,t-1} \cdots \quad A_{m,t-1}) \begin{pmatrix} P_{11} & \cdots & P_{m1} \\ \vdots & \ddots & \vdots \\ P_{1m} & \cdots & P_{mm} \end{pmatrix} = \begin{pmatrix} A_{1,t} \\ \vdots \\ A_{m,t} \end{pmatrix} \quad (25)$$

$$A_{t-1}P = A_t \quad (26)$$

Here, *P* matrix is called First order transitional probability matrix. This matrix is a critical component for generating the forecast. Markov chain has the property that given the *P* matrix and initial state values (A_{t-1}), the forecast will be generated as follows:

$$A_t = A_0 P^t \quad (27)$$

To generate the transition probability matrix (TPM), we use the properties of Markov chain. The first property is taken from Eq. (24) and second property is that row total of *P* matrix must add to one. Now to estimate the TPM we utilize linear goal programming framework. In this framework, we minimize the sum of squared errors. Since the framework is a linear one and therefore a linear approximation is to split the error term into positive and negative deviations and as a result minimize the sum of positive and negative deviations. The positive deviations are denoted by $v_{i,t}$ and the negative ones by $u_{i,t}$, where i denotes i[th] crop.

$$Min \sum_{t=1}^{s} \sum_{i=1}^{n} (u_{i,t} + v_{i,t}) + 0 * \sum_{i=1}^{n} \sum_{j=1}^{m} P_{ij}$$

s.t.

$$\sum_{j=1}^{m} A_{j,t-1}P_{ij} + u_{i,t} - v_{j,t} = A_{j,t} \;\; \forall\, t-1 \; and \; i$$

$$\sum_{j=1}^{m} P_{ij} = 1 \;\; \forall \; i$$

$$P_{ij} \geq 0 \;\; \forall \, i, j$$

This goal programming problem is solved through Excel Solver and an illustration is provided below.

Table 5.14 provides the acreage data for Rabi crops grown in India for 1966–76.

TABLE 5.14 Rabi Crops Area in India for 1966–76

Year	Oilseed	Cereals	Pulses	Coarse Cereals	Total
1966–67	4979	14156	13802	9594	42531
1967–68	5533	16523	14120	10109	46285
1968–69	5145	17618	12920	10112	45795
1969–70	5556	18478	13553	9869	47456
1970–71	5808	19876	13859	9004	48547
1971–72	6276	20810	13859	9360	50305
1972–73	5468	21076	12729	7759	47032
1973–74	6109	20381	14271	8697	49458
1974–75	6399	19942	13269	9190	48799
1975–76	6132	22488	15051	8686	52356

Source: Department of Economic and Statistics (DES), Government of India.

Since the total area under Rabi crops is changing from year to year and therefore to make the total equal from year to year, every row is divided by its respective row total to convert whole values to proportion. After converting data to proportions, we get Table 5.15.

TABLE 5.15 Cropping Pattern in Rabi Season in India for 1966–76

India	Oilseed	Cereals	Pulses	Coarse Cereals	Total
1966–67	0.117063	0.33284	0.324511	0.225586308	1
1967–68	0.119541	0.356983	0.305076	0.218399248	1
1968–69	0.112343	0.384719	0.282132	0.220806502	1
1969–70	0.117069	0.389372	0.285591	0.207967026	1
1970–71	0.119632	0.409425	0.285473	0.18546956	1
1971–72	0.12476	0.413673	0.275496	0.186071892	1
1972–73	0.116253	0.448124	0.270653	0.164969234	1
1973–74	0.123515	0.412094	0.288541	0.175850394	1
1974–75	0.131127	0.408644	0.271901	0.188328714	1
1975–76	0.117115	0.429509	0.28748	0.165895986	1

Source: Author's own calculations.

Using the above table of proportions, further calculations will be made. Table 5.16 provides all formulae required in generating Markov first order TPM. To begin with, first prepare list of positive and negative deviations in cell range A26: A105 followed by a list of all probabilities of TPM from cell range A106: A121. The respective cell ranges B26: B105 and B106: B121 will contain respective variable values. So, finally B26: B121 will house final values of decision variables. This programme will have to be solved using Excel solver available in the Data tab of the excel sheet. Table 5.17 provides an output of goal programming. Figure 5.15 provides a screenshot of Excel solver. Table 5.18 provides the final TPM we wanted. This table is a transpose of Table 5.17 multiplied by 100. All rows of Table 5.18 sums to 100% (Table 5.19).

TABLE 5.16 Formulae Used in Generating TPM

Crops	Oilseed	Cereals	Pulses	Coarse Cereals	
Oilseed		= B106	= B107	= B108	= B109
Cereals		= B110	= B111	= B112	= B113
Pulses		= B114	= B115	= B116	= B117
Coarse Cereals		= B118	= B119	= B120	= B121
Total		= SUM(J14: J17)	= SUM(K14: K17)	= SUM(L14: L17)	= SUM(M14: M17)

Objective function = SUM(B26: B105)

Constraints	Left hand side values	Sign	Right hand side values
1	= SUMPRODUCT($J$14: $M$14,B14: E14)+B26-B62	=	= B15
2	= SUMPRODUCT($J$14: $M$14,B15: E15)+B27-B63	=	= B16
3	= SUMPRODUCT($J$14: $M$14,B16: E16)+B28-B64	=	= B17
4	= SUMPRODUCT($J$14: $M$14,B17: E17)+B29-B65	=	= B18
5	= SUMPRODUCT($J$14: $M$14,B18: E18)+B30-B66	=	= B19
6	= SUMPRODUCT($J$14: $M$14,B19: E19)+B31-B67	=	= B20
7	= SUMPRODUCT($J$14: $M$14,B20: E20)+B32-B68	=	= B21

TABLE 5.16 *(Continued)*

8	= SUMPRODUCT($J$14: $M$14,B21: E21)+B33-B69	=	= B22
9	= SUMPRODUCT($J$14: $M$14,B22: E22)+B34-B70	=	= B23
10	= SUMPRODUCT($J$15: $M$15,B14: E14)+B35-B71	=	= C15
11	= SUMPRODUCT($J$15: $M$15,B15: E15)+B36-B72	=	= C16
12	= SUMPRODUCT($J$15: $M$15,B16: E16)+B37-B73	=	= C17
13	= SUMPRODUCT($J$15: $M$15,B17: E17)+B38-B74	=	= C18
14	= SUMPRODUCT($J$15: $M$15,B18: E18)+B39-B75	=	= C19
15	= SUMPRODUCT($J$15: $M$15,B19: E19)+B40-B76	=	= C20
16	= SUMPRODUCT($J$15: $M$15,B20: E20)+B41-B77	=	= C21
17	= SUMPRODUCT($J$15: $M$15,B21: E21)+B42-B78	=	= C22
18	= SUMPRODUCT($J$15: $M$15,B22: E22)+B43-B79	=	= C23
19	= SUMPRODUCT($J$16: $M$16,B14: E14)+B44-B80	=	= D15
20	= SUMPRODUCT($J$16: $M$16,B15: E15)+B45-B81	=	= D16
21	= SUMPRODUCT($J$16: $M$16,B16: E16)+B46-B82	=	= D17
22	= SUMPRODUCT($J$16: $M$16,B17: E17)+B47-B83	=	= D18
23	= SUMPRODUCT($J$16: $M$16,B18: E18)+B48-B84	=	= D19
24	= SUMPRODUCT($J$16: $M$16,B19: E19)+B49-B85	=	= D20
25	= SUMPRODUCT($J$16: $M$16,B20: E20)+B50-B86	=	= D21
26	= SUMPRODUCT($J$16: $M$16,B21: E21)+B51-B87	=	= D22
27	= SUMPRODUCT($J$16: $M$16,B22: E22)+B52-B88	=	= D23

TABLE 5.16 *(Continued)*

37	= SUMPRODUCT($J$17: $M$17,B14: E14)+B53-B89	=	= E15
38	= SUMPRODUCT($J$17: $M$17,B15: E15)+B54-B90	=	= E16
39	= SUMPRODUCT($J$17: $M$17,B16: E16)+B55-B91	=	= E17
40	= SUMPRODUCT($J$17: $M$17,B17: E17)+B56-B92	=	= E18
41	= SUMPRODUCT($J$17: $M$17,B18: E18)+B57-B93	=	= E19
42	= SUMPRODUCT($J$17: $M$17,B19: E19)+B58-B94	=	= E20
43	= SUMPRODUCT($J$17: $M$17,B20: E20)+B59-B95	=	= E21
44	= SUMPRODUCT($J$17: $M$17,B21: E21)+B60-B96	=	= E22
45	= SUMPRODUCT($J$17: $M$17,B22: E22)+B61-B97	=	= E23
46	= J18	=	1
47	= K18	=	1
48	= L18	=	1
49	= M18	=	1

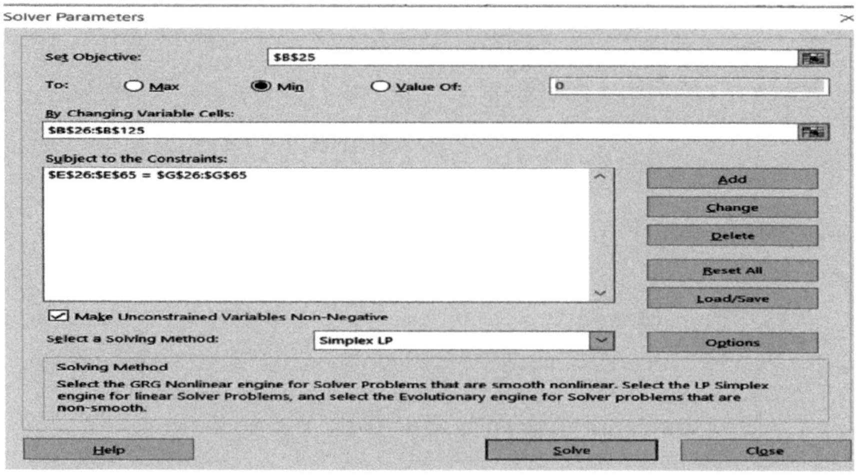

FIGURE 5.15 Screenshot of Excel solver in Windows 10 Microsoft Excel 2013.

TABLE 5.17 Final Output of Goal Programming

Crops	Oilseed	Cereals	Pulses	Coarse Cereals
Oilseed	0.00000	0.13965	0.22514	0.00000
Cereals	1.00000	0.64785	0.00000	0.15524
Pulses	0.00000	0.19548	0.39537	0.44764
Coarse Cereals	0.00000	0.01702	0.37950	0.39712
Total	1.00000	1.00000	1.00000	1.00000

TABLE 5.18 Markov First Order TPM for Rabi Crops in India (1966–76)

Crops	Oilseed	Cereals	Pulses	Coarse Cereals	Total
Oilseed	0%	100%	0%	0%	100%
Cereals	14%	65%	20%	2%	100%
Pulses	22.51%	0%	40%	38%	100%
Coarse Cereals	0%	16%	45%	40%	100%

TABLE 5.19 Forecast Vs. Actual, a Case of Rabi Pulses in India (1966–86)

	Forecast	Actual						
Year	Oilseed	Cereals	Pulses	Coarse Cereals	Oilseed	Cereals	Pulses	Coarse Cereals
1967–68	5002	16658	12165	8706	5533	16523	14120	10109
1968–69	5065	17146	11963	8357	5145	17618	12920	10112
1969–70	5088	17470	11822	8150	5556	18478	13553	9869
1970–71	5101	17671	11738	8021	5808	19876	13859	9004
1971–72	5110	17795	11685	7940	6276	20810	13859	9360
1972–73	5116	17872	11653	7891	5468	21076	12729	7759
1973–74	5119	17919	11633	7860	6109	20381	14271	8697
1974–75	5121	17948	11620	7841	6399	19942	13269	9190
1975–76	5123	17967	11613	7829	6132	22488	15051	8686
1976–77	5123	17978	11608	7822	6081	22326	12914	7928
1977–78	5124	17985	11605	7817	6895	23356	13388	7926
1978–79	5124	17989	11603	7815	7035	24736	13487	7961
1979–80	5124	17992	11602	7813	6581	24029	12089	8330
1980–81	5125	17993	11601	7812	7400	24009	12035	7435
1981–82	5125	17994	11601	7811	8137	23980	13507	7850

TABLE 5.19 *(Continued)*

	Forecast	Actual						
Year	Oilseed	Cereals	Pulses	Coarse Cereals	Oilseed	Cereals	Pulses	Coarse Cereals
1982–83	5125	17995	11601	7811	7213	25418	12681	8026
1983–84	5125	17995	11600	7810	7686	26962	12533	7640
1984–85	5125	17996	11600	7810	7784	25555	12332	7735
1985–86	5125	17996	11600	7810	7496	24901	13526	7916

Now, we are in possession of TPM and this we will use to generate a forecast for the next 10 years. Table 5.19 provides the forecasted versus actual values so that side by side comparison can be easily made. It can be observed that forecast remains flat (without much variation) from 1980–81 onwards for many series. This is due to the Markov chain attaining what is called stationary state. Once Markov chain attains stationary state then the TPM coefficients in all the rows will be same, and all further forecast will be same. This fixed probability row vector is denoted by W. Mathematically this is represented as follows:

$$\lim_{n \to \infty} P^n = W$$

And this fixed probability vector has the property that upon pre-multiplying with TPM it yields the same fixed probability vector. This fixed probability vector shows the proportion of area under different crops when Markov chain reaches the stationary state or equilibrium in the long run.

KEYWORDS

- **autocorrelation function**
- **auxiliary regressions or variance inflation factor**
- **chain relative**
- **classical linear regression model**
- **contingency coefficient**
- **variance inflation factor**

REFERENCES

Bleikh, H. *Y.*, & Youngm, W. L., (2014). *Time Series Analysis and Adjustment: Measuring, Modeling, and Forecasting for Business and Economics*. Ashgate Publishing.

Bollerslev, T., (1986). Generalized autoregressive conditional heteroscedasticity, *J. Econometrics*, 31–307.

Box, G. E. P., Jenkins, G. M., & Reinsel, G. C., (1976). *Time Series Analysis: Forecasting and Control* (2nd edn.). Holden-Day, San Francisco.

Bretscher, O., (1995). Linear Algebra With Applications (3rd edn.). Upper Saddle River, NJ: Prentice Hall.

Engle, R. F., (1982). *Autoregressive conditional heteroscedasticity with estimates of the variance of United Kingdom inflation. Econometrica, 50–987.*

Farrar, D. E., & Glauber, R. R., (1967). Multicollinearity in regression analysis: The problem revisited. *Review of Economics and Statistics, 49(1), 92–107.*

Lai, T. L., Robbins, H., & Wei, C. *Z.*, (1978). Strong consistency of least squares estimates in multiple regression. *PNAS, 75(7), 3034–3036.*

Pankratz, A., (1983). Forecasting With Univariate Box-Jenkins Models- Concepts and Cases. John Wiley & Sons, New York.

Paul, R. P., & Ghosh, H., (2009). GARCH nonlinear time series analysis for modeling and forecasting of India's volatile spices export data. *Journal of the Indian Society of Agricultural Statistics, 63*(2), 123–131.

Richard, S., (2009). Basics of Applied Stochastic Processes. *Springer Science & Business Media,* DOI: 10.1007/978-3-540-89332-5.

Sahu, P. K., & Das, A. K., (2009). *Agriculture and Applied Statistics-II.* Kalyani Publisher, Ludhiana, India.

Soren, A., (2003). Applied Probability and Queues. *Springer Science & Business Media.*

Stigler, S. M., (1981). Gauss and the invention of least squares, *Ann. Stat., 9(3), 465–474.* doi: 10.1214/aos/1176345451.

CHAPTER 6

Overview of MS Excel, SPSS, Minitab, and R

PRADEEP MISHRA[1], R. B. SINGH[2], G. K. VANI[3], G. F. AHMED[4], and SUPRIYA[5]

[1]*Assistant Professor (Statistics), College of Agriculture, JNKVV, Powarkheda, (M.P.) – 461110, India, E-mail: pradeepjnkvv@gmail.com*

[2]*Professor, Department of Mathematics and Statistics, College of Agriculture, JNKVV, Jabalpur (M.P.) – 482004, India*

[3]*Assistant Professor (Agricultural Economics & F.M.), College of Agriculture, JNKVV, Jabalpur (M.P.) – 482004, India*

[4]*Assistant Professor, College of Agriculture, JNKVV, Powarkheda, (M.P.) – 461110, India*

[5]*Assistant Professor, College of Agriculture Campus, Kotwa, Azamgarh, NDUAT, Faizabad, India*

6.1 MS EXCEL STEPS FOR ANALYSIS

6.1.1 Example 1: Following data give the yield (Y) corresponding to 10 yield components (X_1, X_2, X_3, X_4, X_5, X_6, X_7, X_8, X_9, X_{10}) for certain agricultural crop. Work out the ***Descriptive Statistics*** like mean, standard deviation, standard error, ranger minimum, and maximum value.

X1	X2	X3	X4	X5	X6	X7	X8	X9	X10	Y
43	5	1.6	3	1,00,000	3	24	2	9	24	14.49459
47	5	1.2	2	75,000	4	30	2	8	22	27.4552
42	5	2.4	2	1,50,000	6	27	2	8	25	39.68635
58	3	1.6	2	1,20,000	7	35	0	4	27	21.8179
45	5	2.8	2	2,00,000	8	32	2	7	26	28.17015
47	6	3.2	2	2,50,000	5	30	2	3	26	14.98783

X1	X2	X3	X4	X5	X6	X7	X8	X9	X10	Y
42	5	2.4	2	1,60,000	7	22	0	4	27	17.0145
66	3	2.4	2	1,40,000	4	45	2	8	24	41.18381
52	5	1.2	2	1,10,000	4	34	0	5	24	16.10352
44	6	4	2	3,00,000	4	29	0	4	25	15.90416
52	4	0.8	4	1,10,000	8	28	2	3	22	23.26946
62	2	2	2	1,80,000	6	46	2	4	27	19.49062
64	3	12	3	3,50,000	7	41	2	4	28	31.15193
62	3	8	3	2,50,000	6	44	2	7	28	23.33775
35	6	6	5	3,00,000	5	16	2	4	29	26.64379
35	6	5.2	2	3,70,000	4	12	2	3	25	30.11387
34	6	4.4	2	2,90,000	3	10	2	4	25	21.87816
76	5	6	2	3,00,000	6	52	2	4	24	31.03547
52	5	0.8	3	2,00,000	6	31	0	4	27	13.48353
58	4	0.4	3	1,90,000	5	37	2	4	26	10.84107
56	4	0.8	3	1,00,000	4	35	0	4	24	16.11716
42	5	4	2	2,00,000	5	27	2	6	26	31.02111
65	4	3.2	2	2,60,000	9	51	2	4	22	23.17923
27	6	1.2	2	1,80,000	2	10	0	4	25	23.54457
35	6	3.2	2	3,10,000	8	17	0	2	27	16.45729

Solution: For getting descriptive statistics

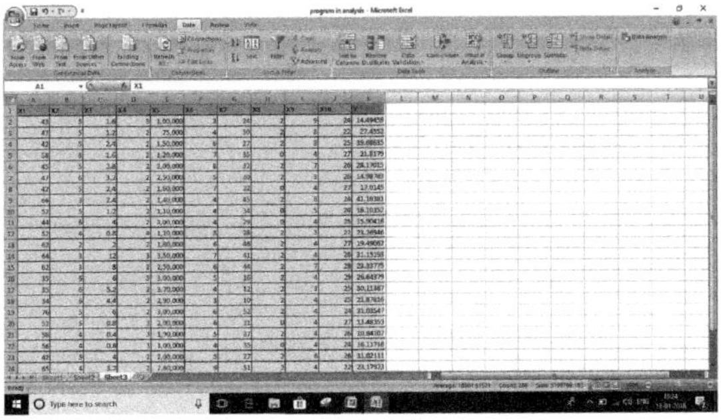

Then go to data → Click on Data analysis

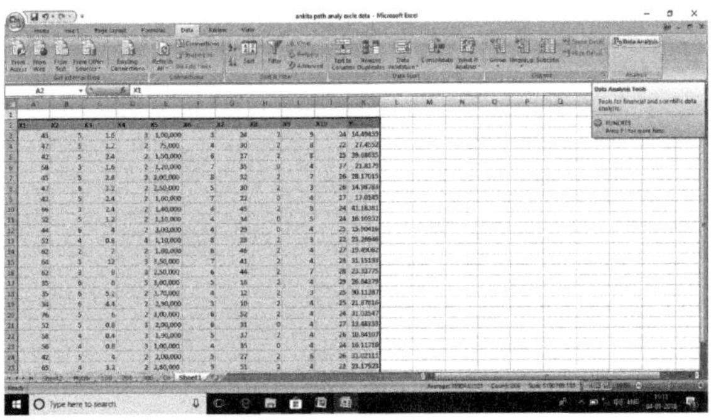

After clicking on data analysis following dialog box appear on the screen

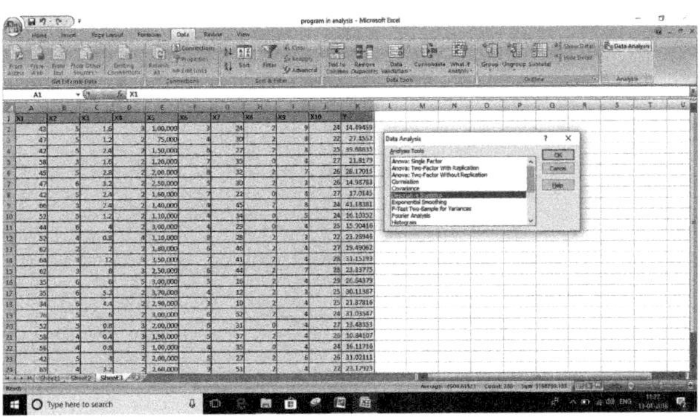

Select data in input range and select output range, where result need and tick (√) summary statistics and click OK.

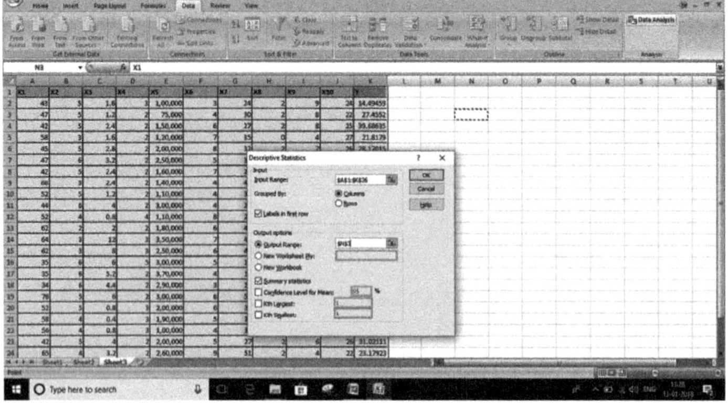

Deceptive statistics of all variable appear on the screen

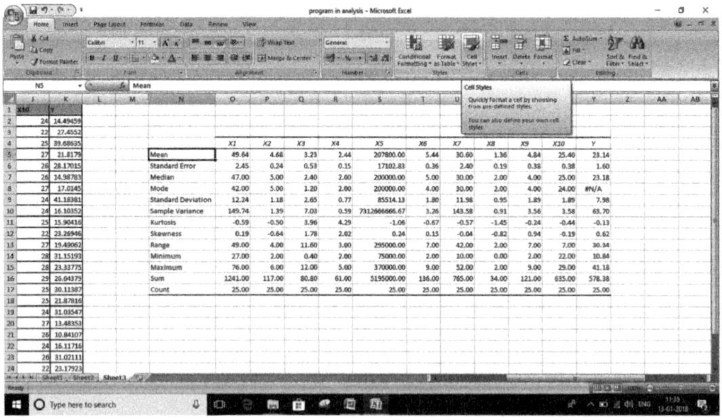

6.1.2 ***Example 2:*** Following data give the yield (Y) corresponding to 10 yield components (X_1, X_2, X_3, X_4, X_5, X_6, X_7, X_8, X_9, X_{10}) for certain agricultural crop. Work out the ***Correlation Matrix***.

X1	X2	X3	X4	X5	X6	X7	X8	X9	X10	Y
43	5	1.6	3	1,00,000	3	24	2	9	24	14.49459
47	5	1.2	2	75,000	4	30	2	8	22	27.4552
42	5	2.4	2	1,50,000	6	27	2	8	25	39.68635
58	3	1.6	2	1,20,000	7	35	0	4	27	21.8179
45	5	2.8	2	2,00,000	8	32	2	7	26	28.17015
47	6	3.2	2	2,50,000	5	30	2	3	26	14.98783
42	5	2.4	2	1,60,000	7	22	0	4	27	17.0145
66	3	2.4	2	1,40,000	4	45	2	8	24	41.18381

X1	X2	X3	X4	X5	X6	X7	X8	X9	X10	Y
52	5	1.2	2	1,10,000	4	34	0	5	24	16.10352
44	6	4	2	3,00,000	4	29	0	4	25	15.90416
52	4	0.8	4	1,10,000	8	28	2	3	22	23.26946
62	2	2	2	1,80,000	6	46	2	4	27	19.49062
64	3	12	3	3,50,000	7	41	2	4	28	31.15193
62	3	8	3	2,50,000	6	44	2	7	28	23.33775
35	6	6	5	3,00,000	5	16	2	4	29	26.64379
35	6	5.2	2	3,70,000	4	12	2	3	25	30.11387
34	6	4.4	2	2,90,000	3	10	2	4	25	21.87816
76	5	6	2	3,00,000	6	52	2	4	24	31.03547
52	5	0.8	3	2,00,000	6	31	0	4	27	13.48353
58	4	0.4	3	1,90,000	5	37	2	4	26	10.84107
56	4	0.8	3	1,00,000	4	35	0	4	24	16.11716
42	5	4	2	2,00,000	5	27	2	6	26	31.02111
65	4	3.2	2	2,60,000	9	51	2	4	22	23.17923
27	6	1.2	2	1,80,000	2	10	0	4	25	23.54457
35	6	3.2	2	3,10,000	8	17	0	2	27	16.45729

Solution: From the given data we calculate the following correlation coefficient and prepare the correlation table.

X1	X2	X3	X4	X5	X6	X7	X8	X9	X10	Y
43	5	1.6	3	1,00,000	3	24	2	9	24	14.49459
47	5	1.2	2	75,000	4	30	2	8	22	27.4552
42	5	2.4	2	1,50,000	6	27	2	8	25	39.68635
58	3	1.6	2	1,20,000	7	35	0	4	27	21.8179
45	5	2.8	2	2,00,000	8	32	2	7	26	28.17015
47	6	3.2	2	2,50,000	5	30	2	3	26	14.98783
42	5	2.4	2	1,60,000	7	22	0	4	27	17.0145
66	3	2.4	2	1,40,000	4	45	2	8	24	41.18381
52	5	1.2	2	1,10,000	4	34	0	5	24	16.10352
44	6	4	2	3,00,000	4	29	0	4	25	15.90416
52	4	0.8	4	1,10,000	8	28	2	3	22	23.26946
62	2	2	2	1,80,000	6	46	2	4	27	19.49062
64	3	12	3	3,50,000	7	41	2	4	28	31.15193
62	3	8	3	2,50,000	6	44	2	7	28	23.33775
35	6	6	5	3,00,000	5	16	2	4	29	26.64379
35	6	5.2	2	3,70,000	4	12	2	3	25	30.11387
34	6	4.4	2	2,90,000	3	10	2	4	25	21.87816

X1	X2	X3	X4	X5	X6	X7	X8	X9	X10	Y
76	5	6	2	3,00,000	6	52	2	4	24	31.03547
52	5	0.8	3	2,00,000	6	31	0	4	27	13.48353
58	4	0.4	3	1,90,000	5	37	2	4	26	10.84107
56	4	0.8	3	1,00,000	4	35	0	4	24	16.11716
42	5	4	2	2,00,000	5	27	2	6	26	31.02111
65	4	3.2	2	2,60,000	9	51	2	4	22	23.17923
27	6	1.2	2	1,80,000	2	10	0	4	25	23.54457
35	6	3.2	2	3,10,000	8	17	0	2	27	16.45729

For getting correlation matrix

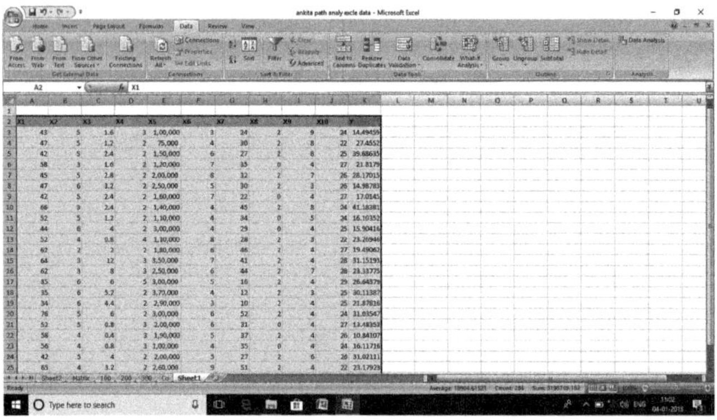

Then go to data → Click on Data analysis

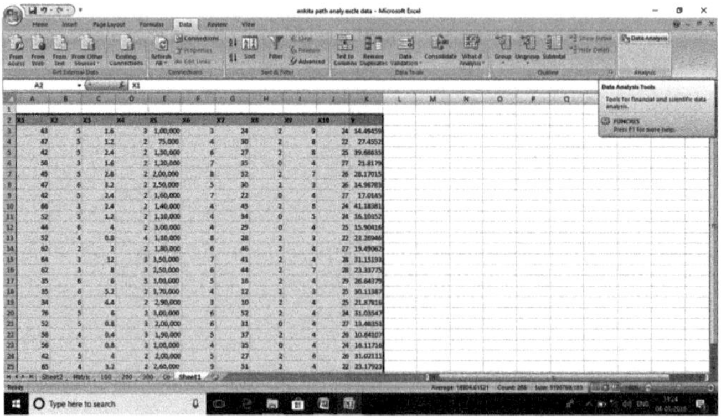

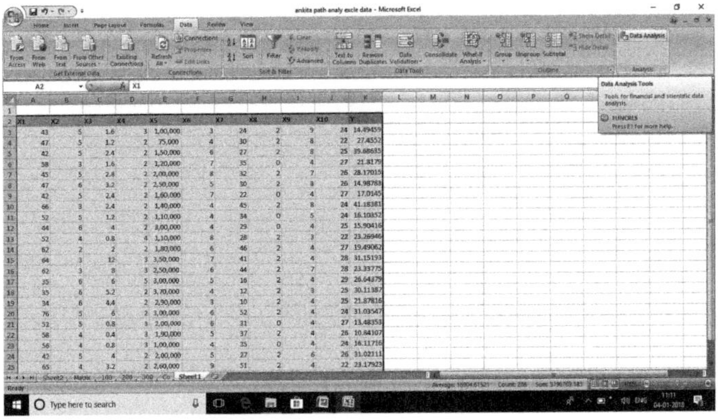

After clicking on data analysis following dialog box appear on the screen.

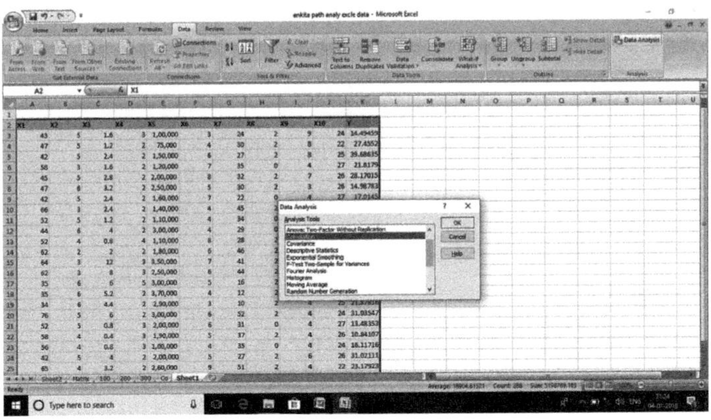

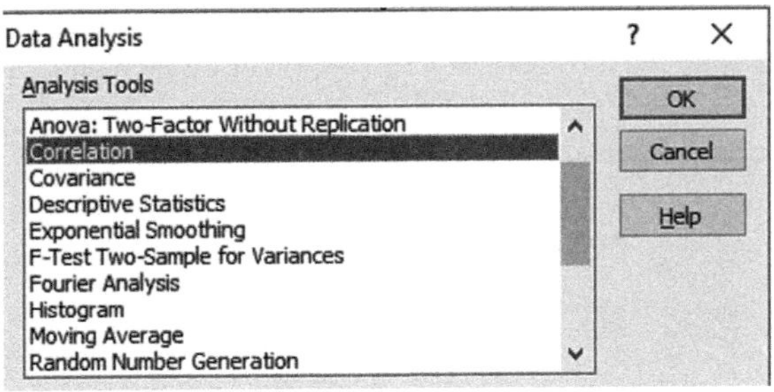

After clicking on correlation following dialog box appear on the screen and select the data in input range and put output range, where correlation matrix is needed.

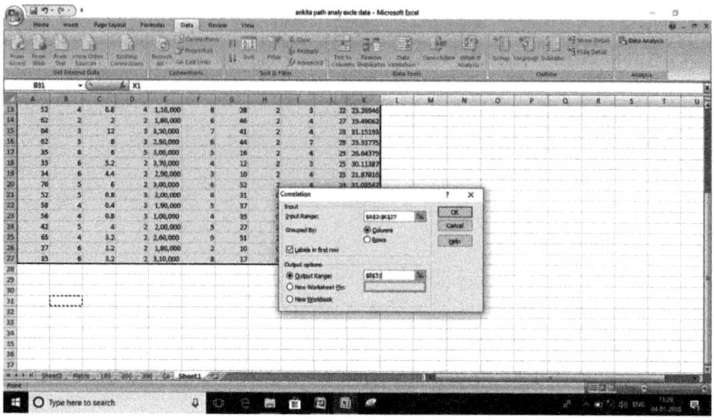

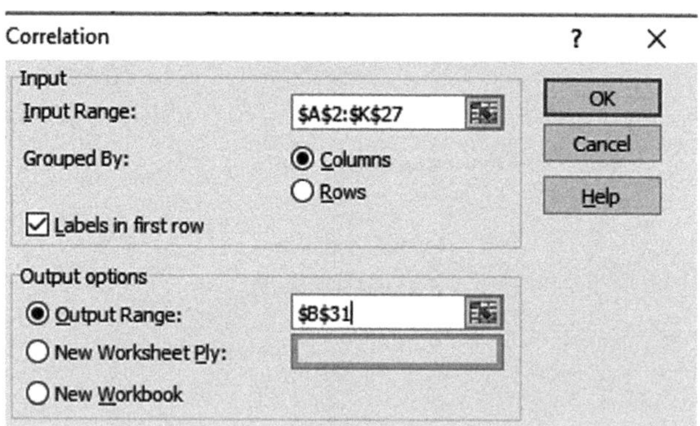

After clicking on OK correlation matrix will appear on the screen.

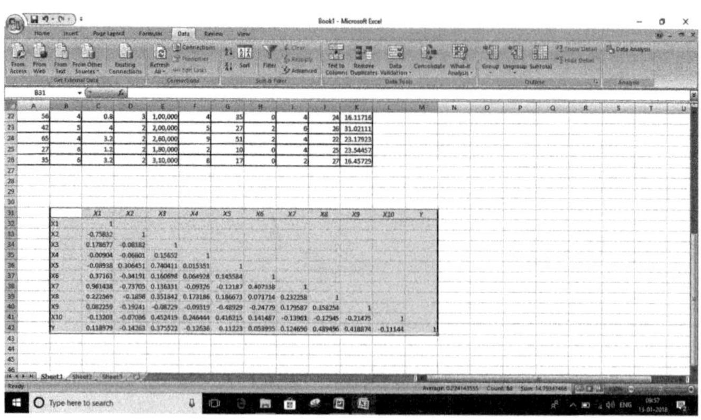

6.1.3 Example 3: Following data give the yield (Y) corresponding to 10 yield components (X_1, X_2, X_3, X_4, X_5, X_6, X_7, X_8, X_9, X_{10}) for certain agricultural crop. Work out the ***Regression Analysis*** for below data.

Solution: For getting regression analysis.

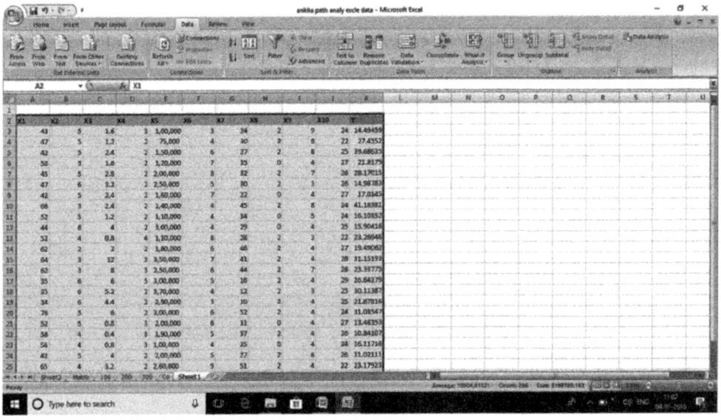

Then go to data → Click on Data analysis → Regression

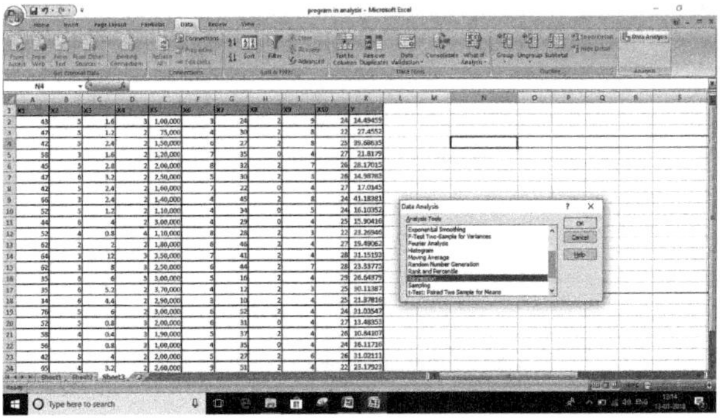

After clicking on regression below dialog box appear on the screen.

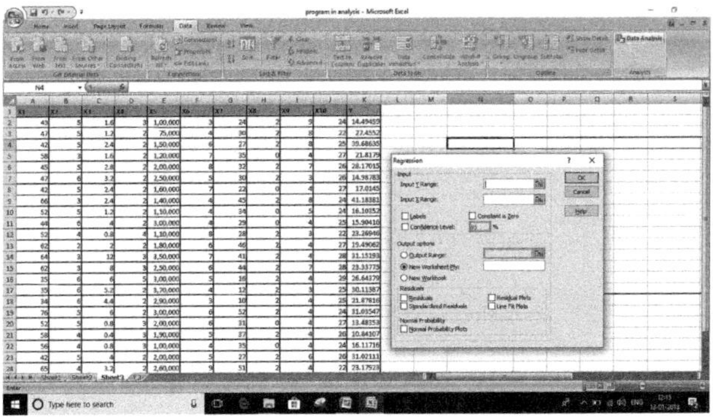

In Input *Y* range select dependent variable data and Input *X* range select independents variable data and tick (√) on label, residual, and normal probability plot.

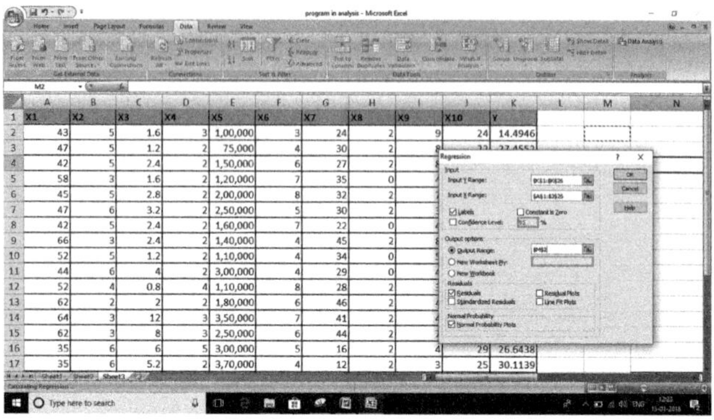

After clicking on OK, results will appear on the screen.

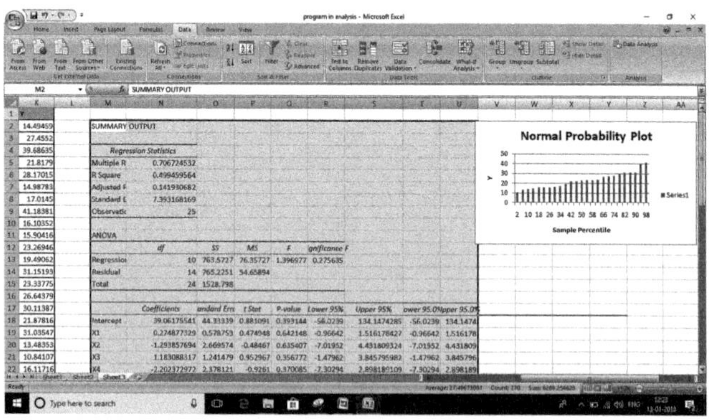

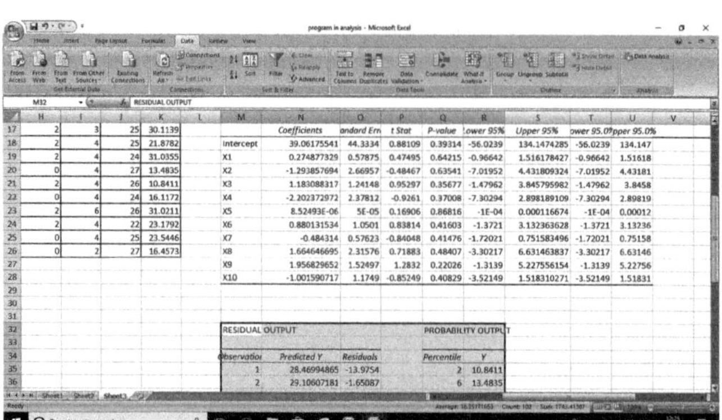

6.1.4 Example 4: Use t-test for test the significance of Two Series Assuming Equal Variance.

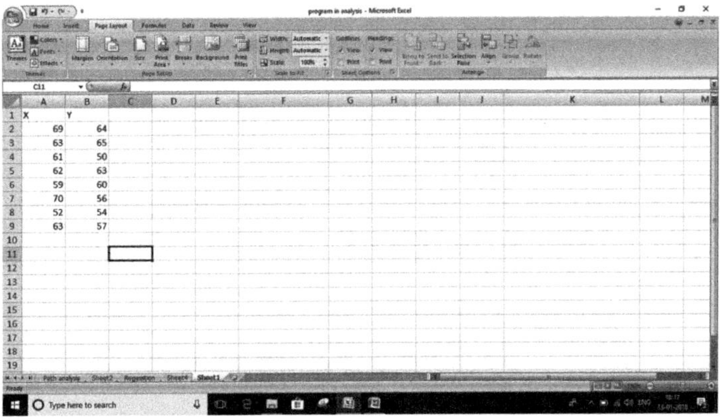

Then go to data → Click on Data analysis → t-Test: Two-Sample Assuming Equal Variances

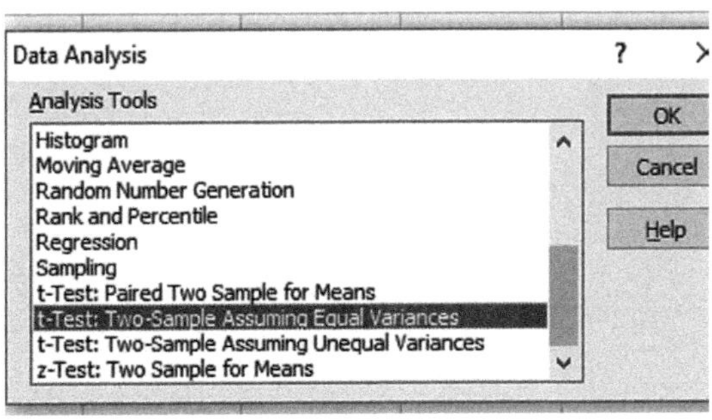

After clicking on OK, following dialog box appear on the screen.

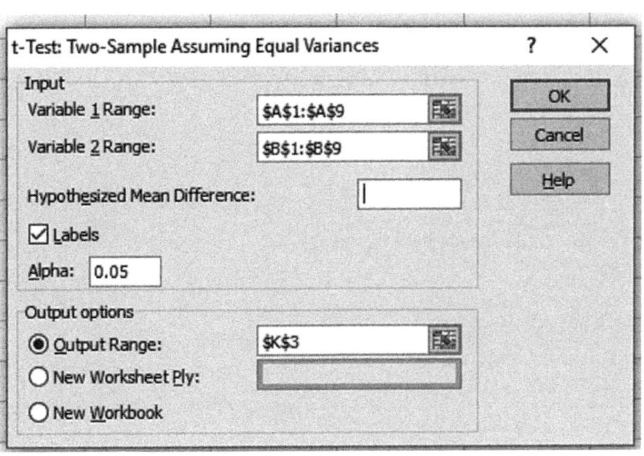

After clicking on OK, results will appear on the screen.

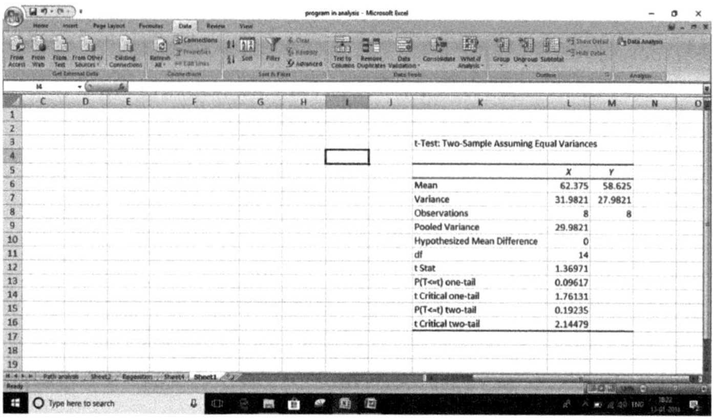

6.1.5 *Example 5:* F-test to Compare Two Variances for two variables

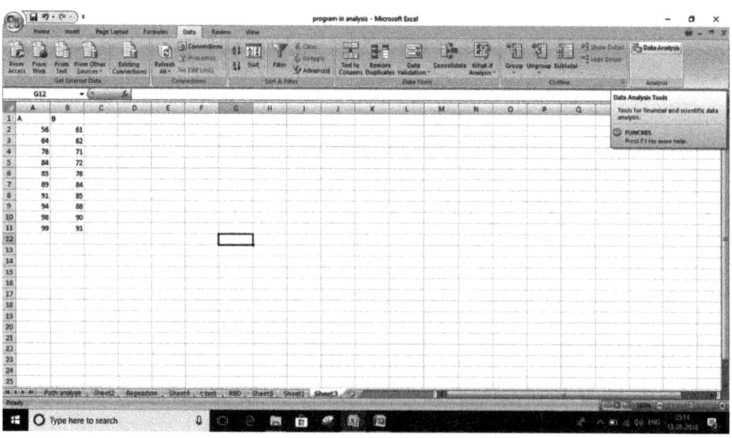

Then go to data → Click on Data analysis → F-Test to Compare Two Variances

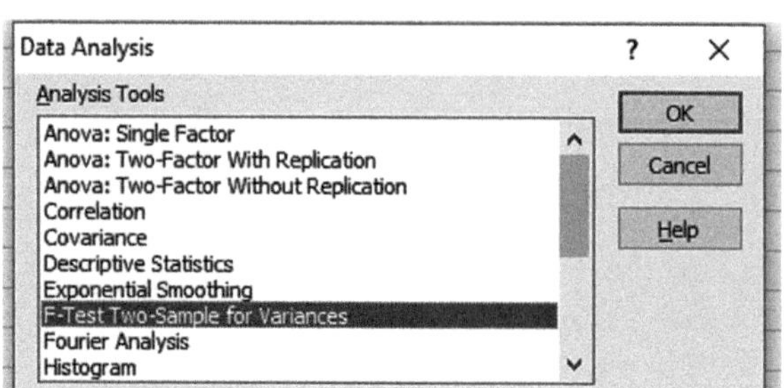

After this click on OK, following dialog box appear on screen.

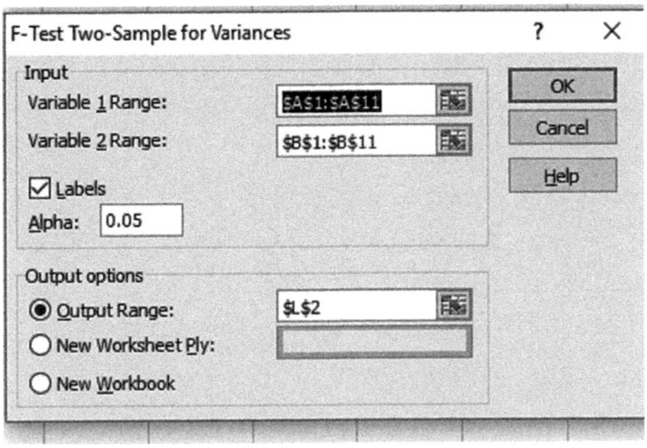

Click on OK and result will appear on the screen.

F-Test Two-Sample for Variances		
	A	*B*
Mean	83.8	78.2
Variance	201.7333	125.2889
Observations	10	10
df	9	9
F	1.610145	
P(F<=f) one-tail	0.244531	
F Critical one-tail	3.178893	

6.1.6 Example 6: Following data give the yield (Y) corresponding to 10 yield components (X_1, X_2, X_3, X_4, X_5, X_6, X_7, X_8, X_9, X_{10}) for certain agricultural crop. Work out the path of the correlations of yield with the yield components.

X1	X2	X3	X4	X5	X6	X7	X8	X9	X10	Y
43	5	1.6	3	1,00,000	3	24	2	9	24	14.49459
47	5	1.2	2	75,000	4	30	2	8	22	27.4552
42	5	2.4	2	1,50,000	6	27	2	8	25	39.68635
58	3	1.6	2	1,20,000	7	35	0	4	27	21.8179
45	5	2.8	2	2,00,000	8	32	2	7	26	28.17015
47	6	3.2	2	2,50,000	5	30	2	3	26	14.98783
42	5	2.4	2	1,60,000	7	22	0	4	27	17.0145
66	3	2.4	2	1,40,000	4	45	2	8	24	41.18381

X1	X2	X3	X4	X5	X6	X7	X8	X9	X10	Y
52	5	1.2	2	1,10,000	4	34	0	5	24	16.10352
44	6	4	2	3,00,000	4	29	0	4	25	15.90416
52	4	0.8	4	1,10,000	8	28	2	3	22	23.26946
62	2	2	2	1,80,000	6	46	2	4	27	19.49062
64	3	12	3	3,50,000	7	41	2	4	28	31.15193
62	3	8	3	2,50,000	6	44	2	7	28	23.33775
35	6	6	5	3,00,000	5	16	2	4	29	26.64379
35	6	5.2	2	3,70,000	4	12	2	3	25	30.11387
34	6	4.4	2	2,90,000	3	10	2	4	25	21.87816
76	5	6	2	3,00,000	6	52	2	4	24	31.03547
52	5	0.8	3	2,00,000	6	31	0	4	27	13.48353
58	4	0.4	3	1,90,000	5	37	2	4	26	10.84107
56	4	0.8	3	1,00,000	4	35	0	4	24	16.11716
42	5	4	2	2,00,000	5	27	2	6	26	31.02111
65	4	3.2	2	2,60,000	9	51	2	4	22	23.17923
27	6	1.2	2	1,80,000	2	10	0	4	25	23.54457
35	6	3.2	2	3,10,000	8	17	0	2	27	16.45729

Solution: From the given data we calculate the following correlation coefficient and prepare the correlation table.

X1	X2	X3	X4	X5	X6	X7	X8	X9	X10	Y
43	5	1.6	3	1,00,000	3	24	2	9	24	14.49459
47	5	1.2	2	75,000	4	30	2	8	22	27.4552
42	5	2.4	2	1,50,000	6	27	2	8	25	39.68635
58	3	1.6	2	1,20,000	7	35	0	4	27	21.8179
45	5	2.8	2	2,00,000	8	32	2	7	26	28.17015
47	6	3.2	2	2,50,000	5	30	2	3	26	14.98783
42	5	2.4	2	1,60,000	7	22	0	4	27	17.0145
66	3	2.4	2	1,40,000	4	45	2	8	24	41.18381
52	5	1.2	2	1,10,000	4	34	0	5	24	16.10352
44	6	4	2	3,00,000	4	29	0	4	25	15.90416
52	4	0.8	4	1,10,000	8	28	2	3	22	23.26946
62	2	2	2	1,80,000	6	46	2	4	27	19.49062
64	3	12	3	3,50,000	7	41	2	4	28	31.15193
62	3	8	3	2,50,000	6	44	2	7	28	23.33775

X1	X2	X3	X4	X5	X6	X7	X8	X9	X10	Y
35	6	6	5	3,00,000	5	16	2	4	29	26.64379
35	6	5.2	2	3,70,000	4	12	2	3	25	30.11387
34	6	4.4	2	2,90,000	3	10	2	4	25	21.87816
76	5	6	2	3,00,000	6	52	2	4	24	31.03547
52	5	0.8	3	2,00,000	6	31	0	4	27	13.48353
58	4	0.4	3	1,90,000	5	37	2	4	26	10.84107
56	4	0.8	3	1,00,000	4	35	0	4	24	16.11716
42	5	4	2	2,00,000	5	27	2	6	26	31.02111
65	4	3.2	2	2,60,000	9	51	2	4	22	23.17923
27	6	1.2	2	1,80,000	2	10	0	4	25	23.54457
35	6	3.2	2	3,10,000	8	17	0	2	27	16.45729

For getting correlation matrix

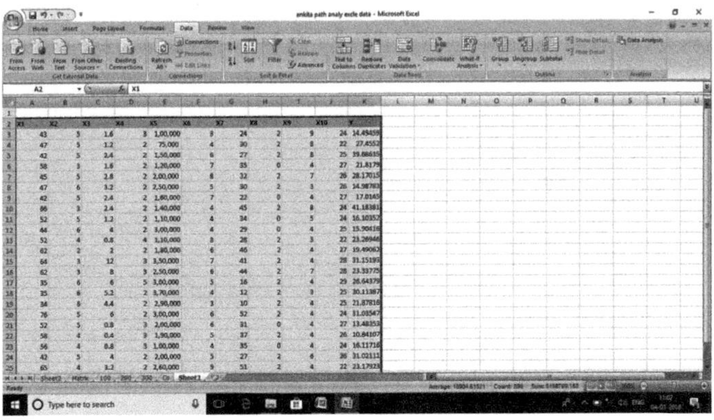

Then go to data → Click on Data analysis.

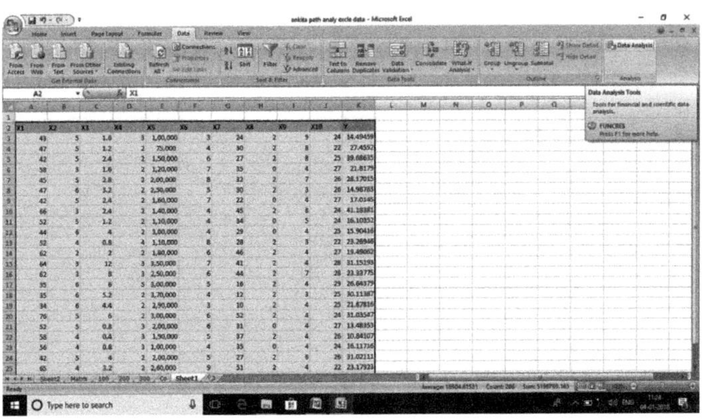

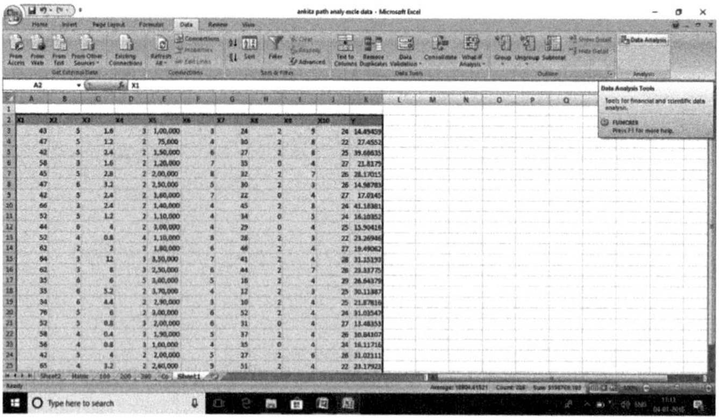

After clicking on data analysis following dialog box appear on the screen.

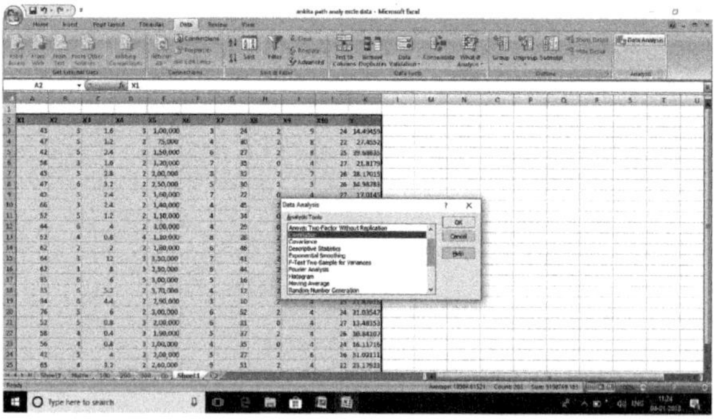

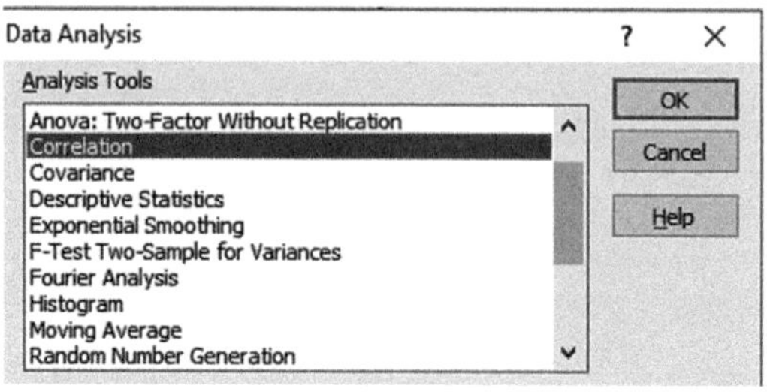

After clicking on correlation following dialog box appear on the screen and select the data in input range and put output range, where correlation matrix is needed.

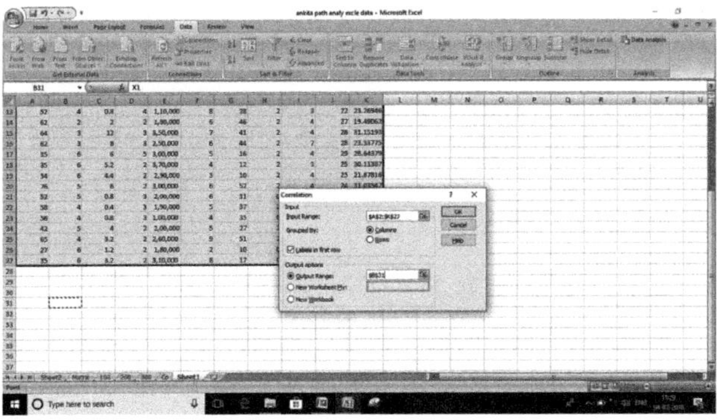

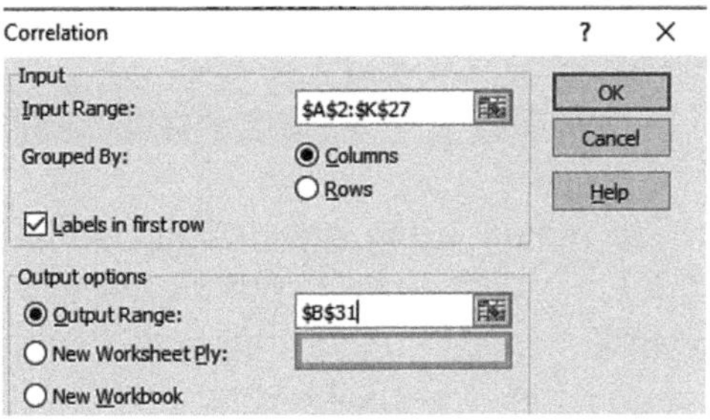

After clicking on OK correlation matrix will appear on the screen.

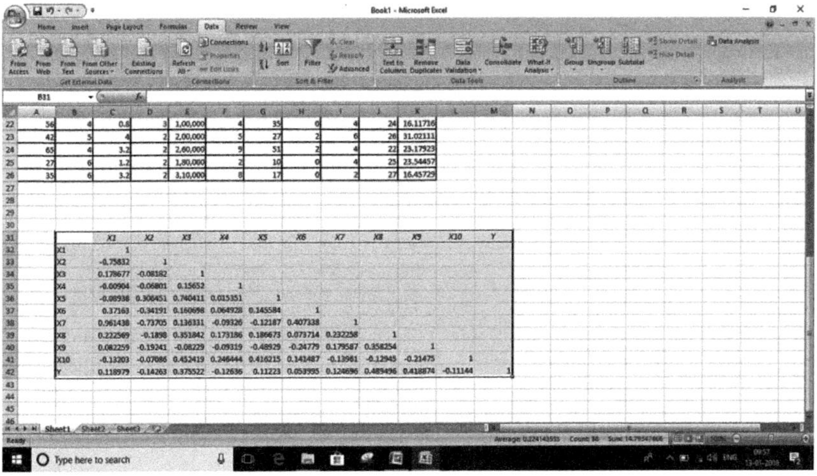

Then arrange the correlation matrix for further process.

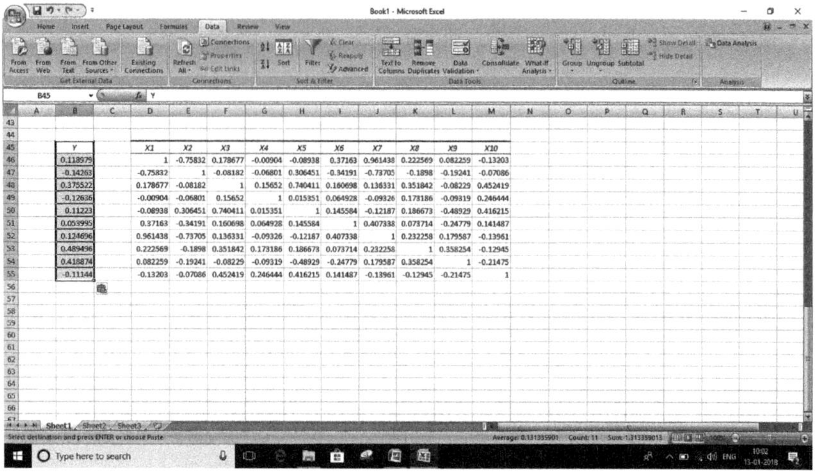

Then use function option using equal sign and select MINVERSE formula and select X_1 to X_{10} data before putting formula select the output range for inverse matrix.

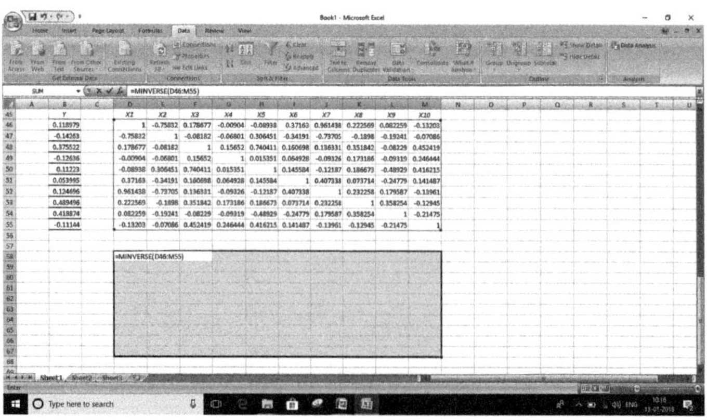

Use F2 followed by Ctrl+Shift+Enter button for getting inverted matrix as given below:

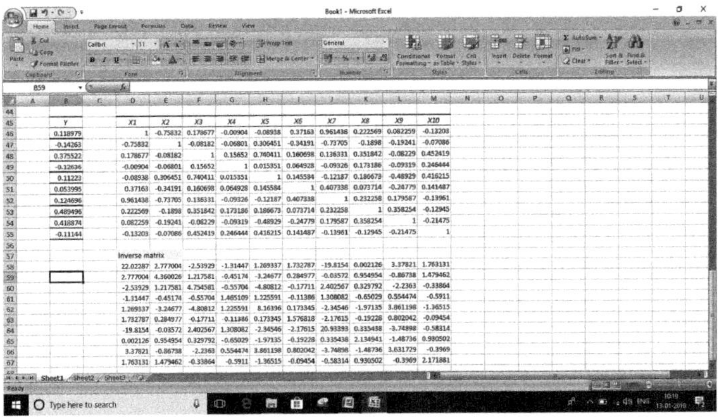

Next step is multiplication of inverse matrix and *Y* to get correlation matrix. Go to formula and click on insert function.

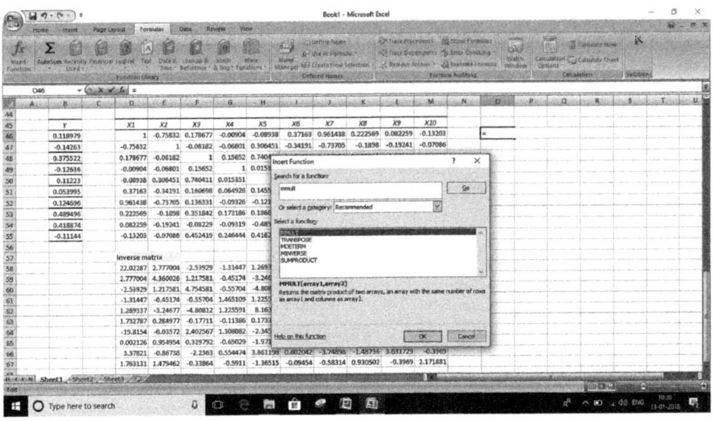

So for that use MMULT option and select the inverse matrix and *Y* data.

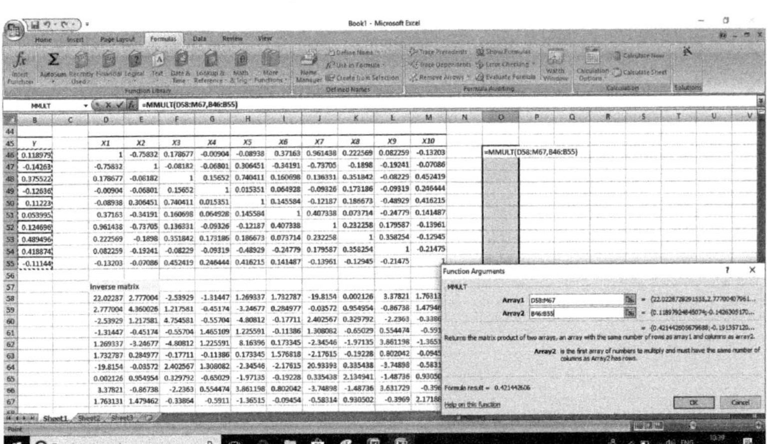

Use F2 followed by Ctrl+Shift+Enter button for getting direct effect as given below:

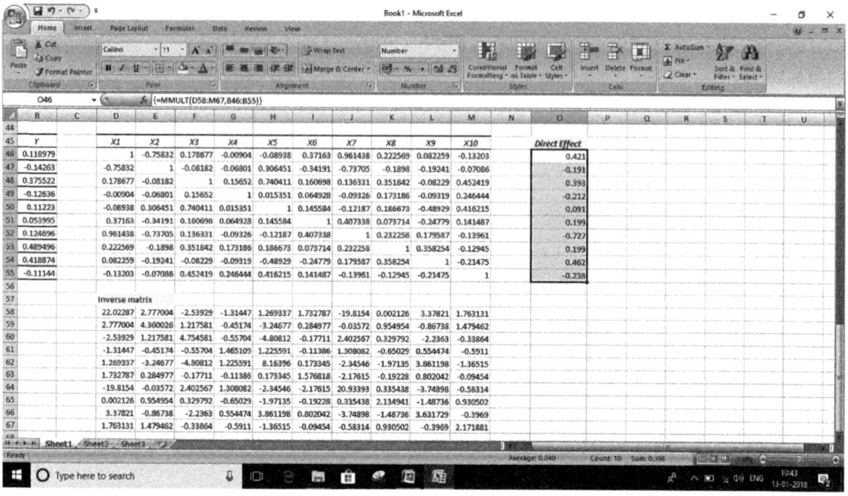

For getting the path coefficient for indirect effects multiply the inverse matrix and direct effect values.

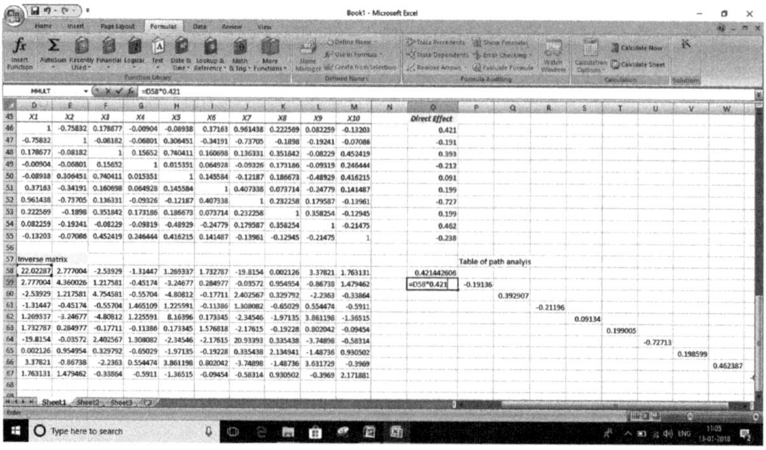

Using this for all, we will get final table for path analysis:

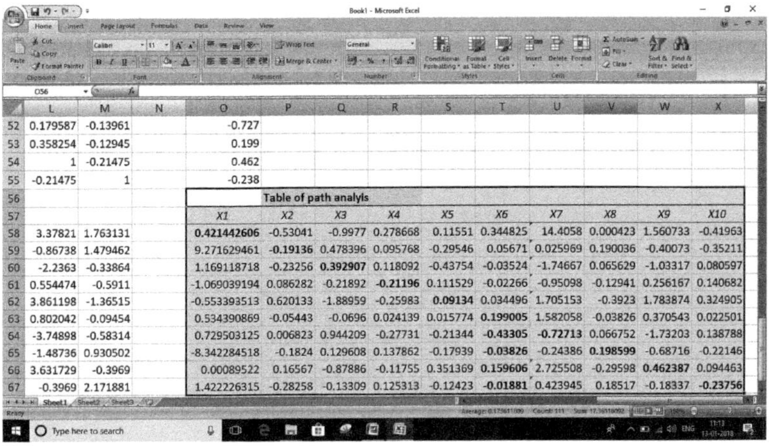

The diagonal elements (written in bold letters) of the table are the direct effects whereas the off-diagonal elements are the indirect effects.

6.2 SPSS STEPS FOR ANALYSIS

6.2.1 Example 1: Following data give the yield (Y) corresponding to 10 yield components (X_1, X_2, X_3, X_4, X_5, X_6, X_7, X_8, X_9, X_{10}) for certain agricultural crop. Work out the ***Descriptive Statistics*** like mean, standard deviation, standard error, ranger minimum, and maximum value.

X1	X2	X3	X4	X5	X6	X7	X8	X9	X10	Y
43	5	1.6	3	1,00,000	3	24	2	9	24	14.49459
47	5	1.2	2	75,000	4	30	2	8	22	27.4552
42	5	2.4	2	1,50,000	6	27	2	8	25	39.68635
58	3	1.6	2	1,20,000	7	35	0	4	27	21.8179
45	5	2.8	2	2,00,000	8	32	2	7	26	28.17015
47	6	3.2	2	2,50,000	5	30	2	3	26	14.98783
42	5	2.4	2	1,60,000	7	22	0	4	27	17.0145
66	3	2.4	2	1,40,000	4	45	2	8	24	41.18381
52	5	1.2	2	1,10,000	4	34	0	5	24	16.10352
44	6	4	2	3,00,000	4	29	0	4	25	15.90416

X1	X2	X3	X4	X5	X6	X7	X8	X9	X10	Y
52	4	0.8	4	1,10,000	8	28	2	3	22	23.26946
62	2	2	2	1,80,000	6	46	2	4	27	19.49062
64	3	12	3	3,50,000	7	41	2	4	28	31.15193
62	3	8	3	2,50,000	6	44	2	7	28	23.33775
35	6	6	5	3,00,000	5	16	2	4	29	26.64379
35	6	5.2	2	3,70,000	4	12	2	3	25	30.11387
34	6	4.4	2	2,90,000	3	10	2	4	25	21.87816
76	5	6	2	3,00,000	6	52	2	4	24	31.03547
52	5	0.8	3	2,00,000	6	31	0	4	27	13.48353
58	4	0.4	3	1,90,000	5	37	2	4	26	10.84107
56	4	0.8	3	1,00,000	4	35	0	4	24	16.11716
42	5	4	2	2,00,000	5	27	2	6	26	31.02111
65	4	3.2	2	2,60,000	9	51	2	4	22	23.17923
27	6	1.2	2	1,80,000	2	10	0	4	25	23.54457
35	6	3.2	2	3,10,000	8	17	0	2	27	16.45729

Solution: For getting descriptive statistics

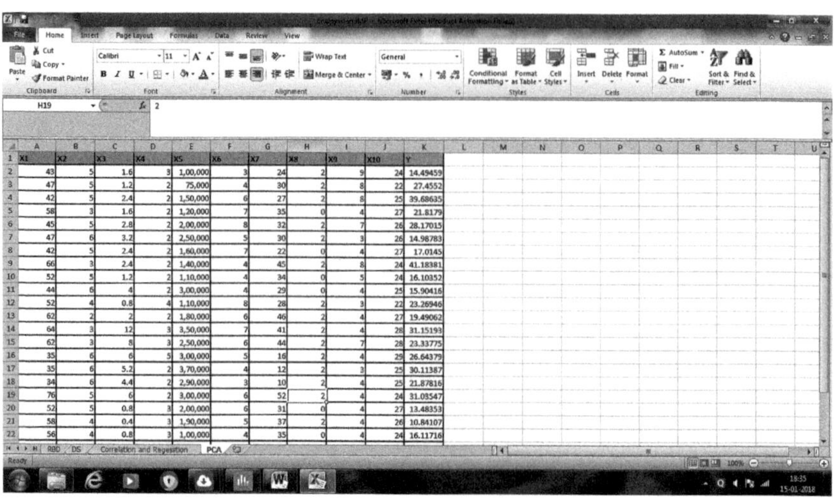

Then copy and paste the data in SPSS sheet or Import first.

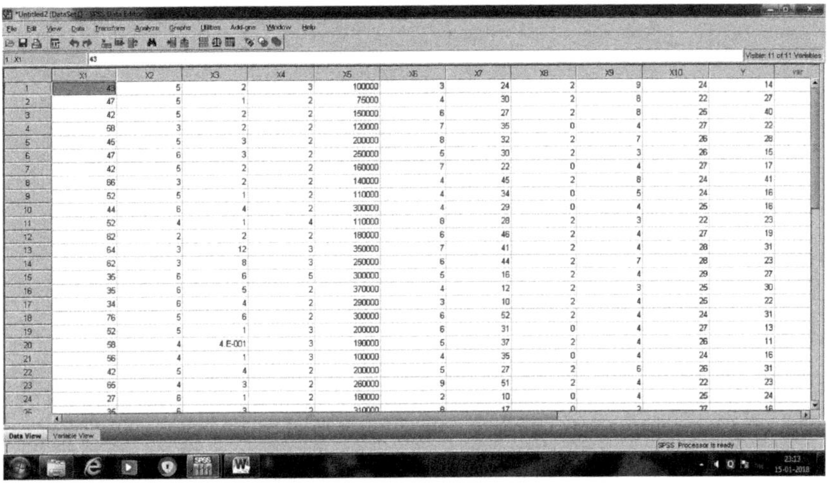

Then go to analysis → Descriptive statistics → Descriptive. After click on descriptive following dialog box appear on screen:

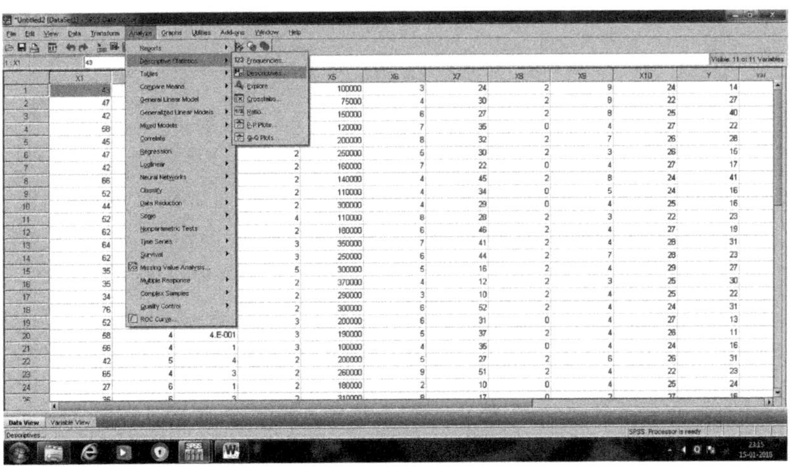

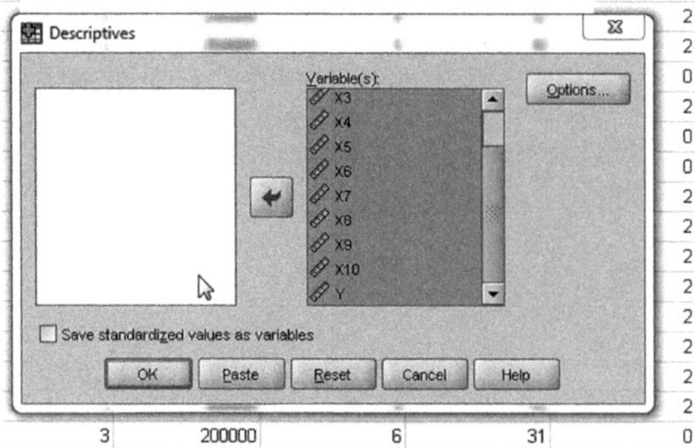

Then go to Options button for selection of descriptive measures:

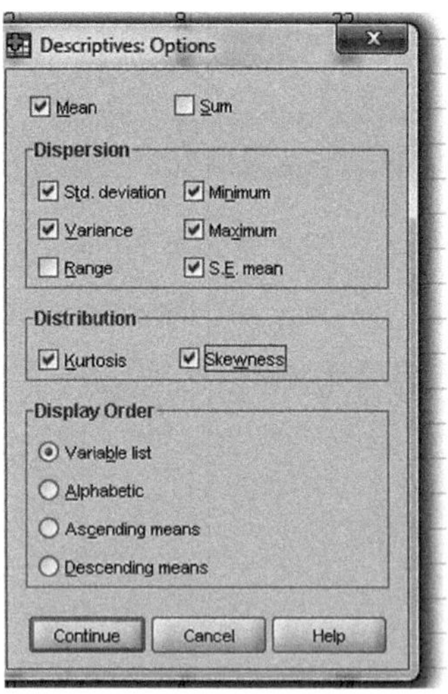

Click Continue and press OK button result will appear on screen.

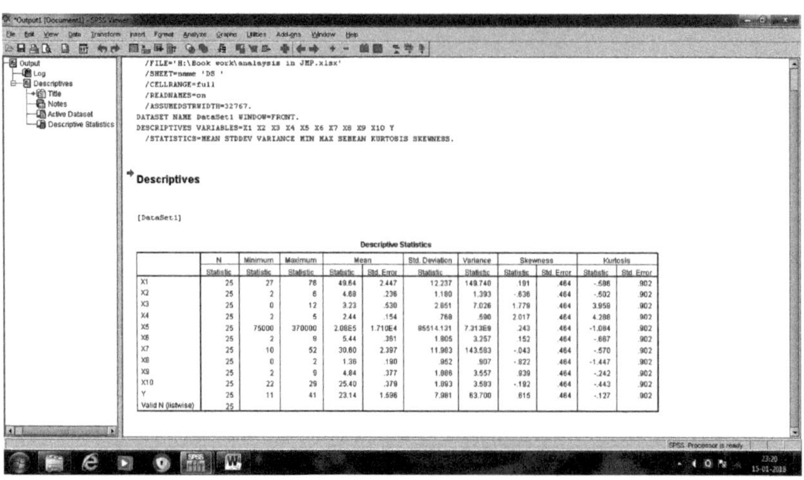

6.2.2 Example 2: Following data given (**Section *6.2.1 – Example 1***) the yield (Y) corresponding to 10 yield components (X_1, X_2, X_3, X_4, X_5, X_6, X_7, X_8, X_9, X_{10}) for certain agricultural crop. Work out the ***Correlation Analysis*** for below data.

Solution: For getting correlation analysis

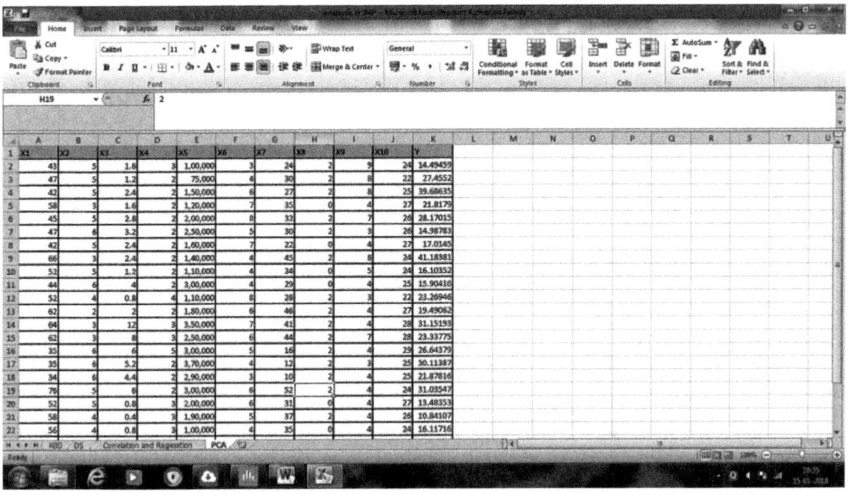

Then copy and paste the data in SPSS sheet or Import first.

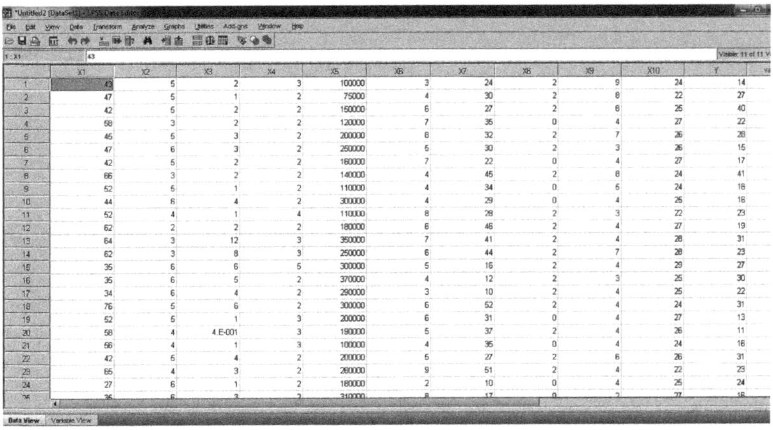

Then go to analysis → correlate → Bivariate. After click on Bivariate following dialog box appear on screen.

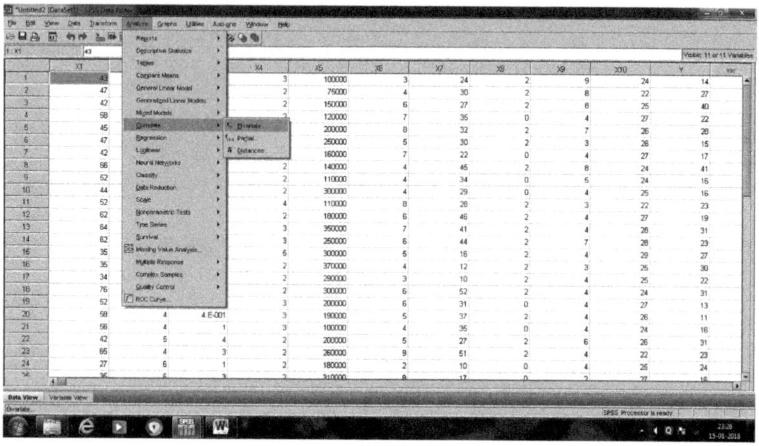

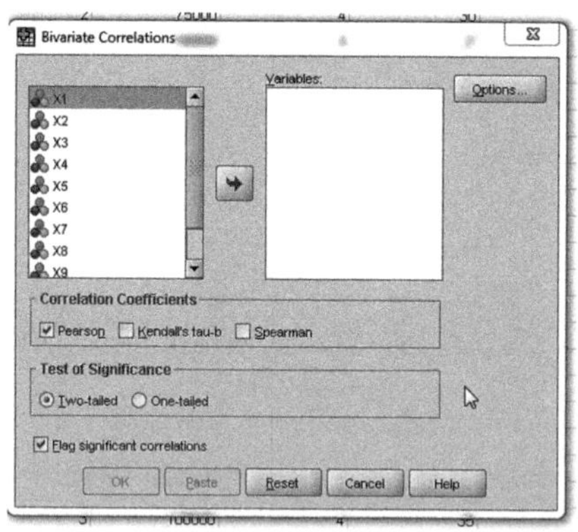

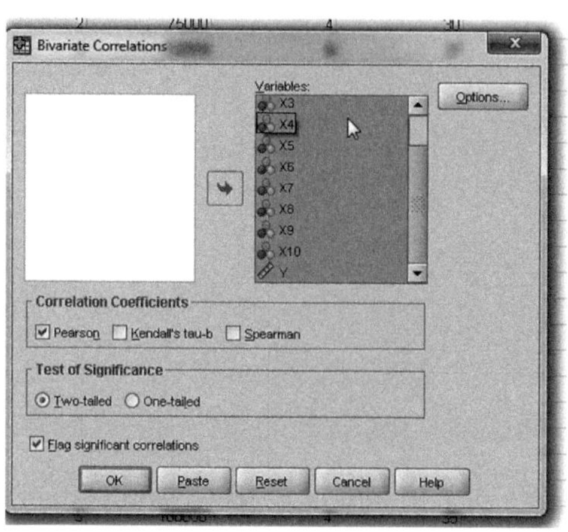

Then click on OK and result will appear on screen.

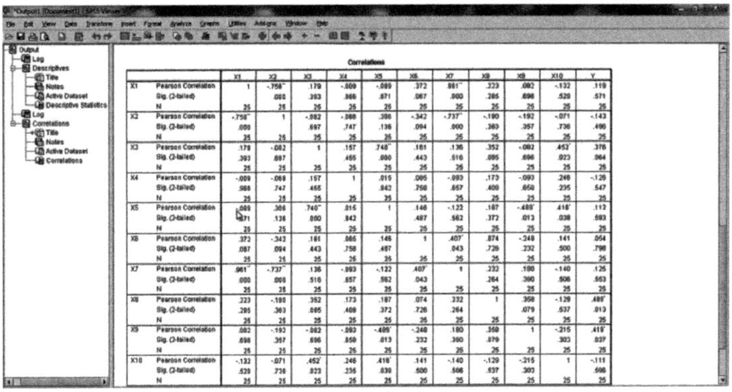

Correlations

		X1	X2	X3	X4	X5	X6	X7	X8	X9	X10	Y
X1	Pearson Correlation	1	-.758**	.179	-.009	-.089	.372	.961**	.223	.082	-.132	.119
	Sig. (2-tailed)		.000	.393	.966	.671	.067	.000	.285	.696	.529	.571
	N	25	25	25	25	25	25	25	25	25	25	25
X2	Pearson Correlation	-.758**	1	-.082	-.068	.306	-.342	-.737**	-.190	-.192	-.071	-.143
	Sig. (2-tailed)	.000		.697	.747	.136	.094	.000	.363	.357	.736	.496
	N	25	25	25	25	25	25	25	25	25	25	25
X3	Pearson Correlation	.179	-.082	1	.157	.740**	.161	.136	.352	-.082	.452*	.376
	Sig. (2-tailed)	.393	.697		.455	.000	.443	.516	.085	.696	.023	.064
	N	25	25	25	25	25	25	25	25	25	25	25
X4	Pearson Correlation	-.009	-.068	.157	1	.015	.065	-.093	.173	-.093	.246	-.126
	Sig. (2-tailed)	.966	.747	.455		.942	.758	.657	.408	.658	.235	.547
	N	25	25	25	25	25	25	25	25	25	25	25
X5	Pearson Correlation	-.089	.306	.740**	.015	1	.146	-.122	.187	-.489*	.416*	.112
	Sig. (2-tailed)	.671	.136	.000	.942		.487	.562	.372	.013	.038	.593
	N	25	25	25	25	25	25	25	25	25	25	25
X6	Pearson Correlation	.372	-.342	.161	.065	.146	1	.407*	.074	-.248	.141	.054
	Sig. (2-tailed)	.067	.094	.443	.758	.487		.043	.726	.232	.500	.798
	N	25	25	25	25	25	25	25	25	25	25	25
X7	Pearson Correlation	.961**	-.737**	.136	-.093	-.122	.407*	1	.232	.180	-.140	.125
	Sig. (2-tailed)	.000	.000	.516	.657	.562	.043		.264	.390	.506	.553
	N	25	25	25	25	25	25	25	25	25	25	25
X8	Pearson Correlation	.223	-.190	.352	.173	.187	.074	.232	1	.358	-.129	.489*
	Sig. (2-tailed)	.285	.363	.085	.408	.372	.726	.264		.079	.537	.013
	N	25	25	25	25	25	25	25	25	25	25	25
X9	Pearson Correlation	.082	-.192	-.082	-.093	-.489*	-.248	.180	.358	1	-.215	.419*
	Sig. (2-tailed)	.696	.357	.696	.658	.013	.232	.390	.079		.303	.037
	N	25	25	25	25	25	25	25	25	25	25	25
X10	Pearson Correlation	-.132	-.071	.452*	.246	.416*	.141	-.140	-.129	-.215	1	-.111
	Sig. (2-tailed)	.529	.736	.023	.235	.038	.500	.506	.537	.303		.596
	N	25	25	25	25	25	25	25	25	25	25	25

6.2.3. Example 3: Following data given (**Section *6.2.1 – Example 1***) the yield (Y) corresponding to 10 yield components (X_1, X_2, X_3, X_4, X_5, X_6, X_7, X_8, X_9, X_{10}) for certain agricultural crop. Work out the ***Regression Analysis*** for below data.

Solution: For getting Regression analysis

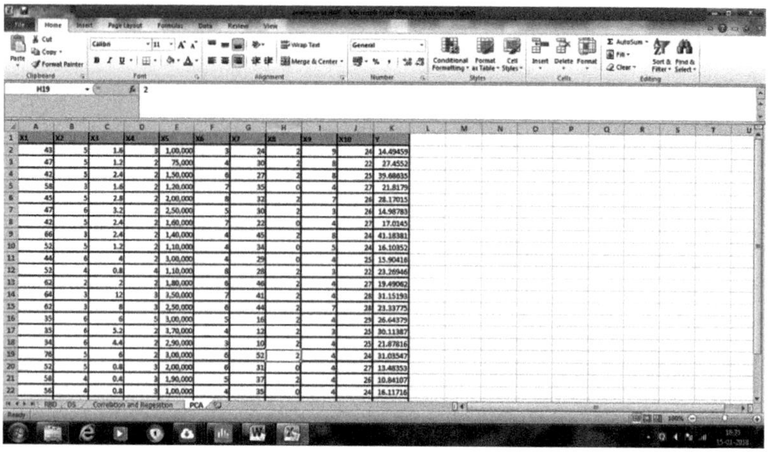

Then copy and paste the data in SPSS sheet or Import first.

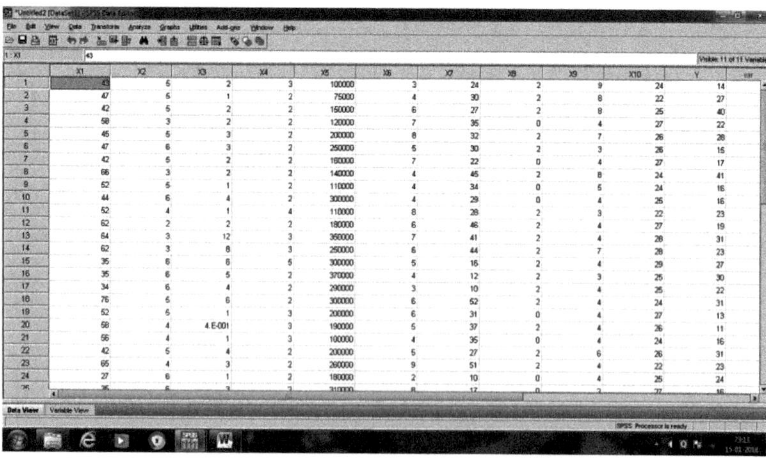

Then go to analysis →Regression→Linear Regression. After click on Linear Regression following dialog box appear on screen

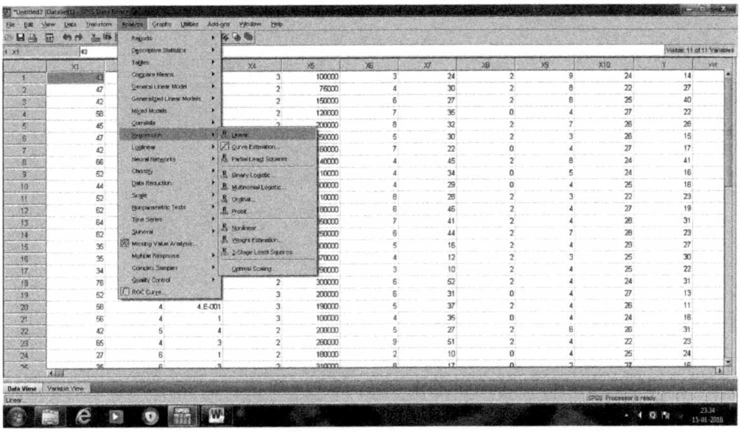

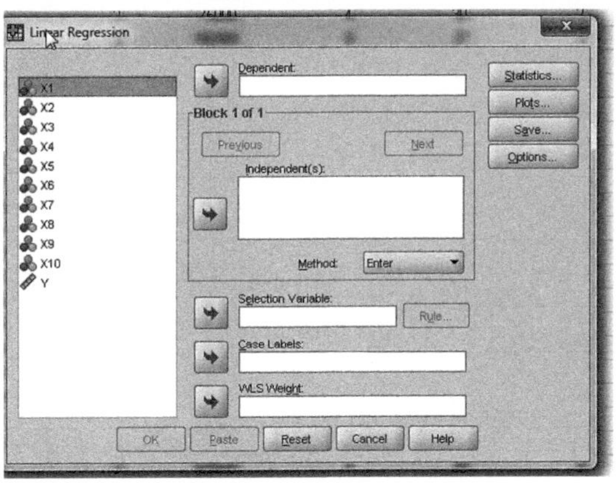

Select the Dependent and Independent variables.

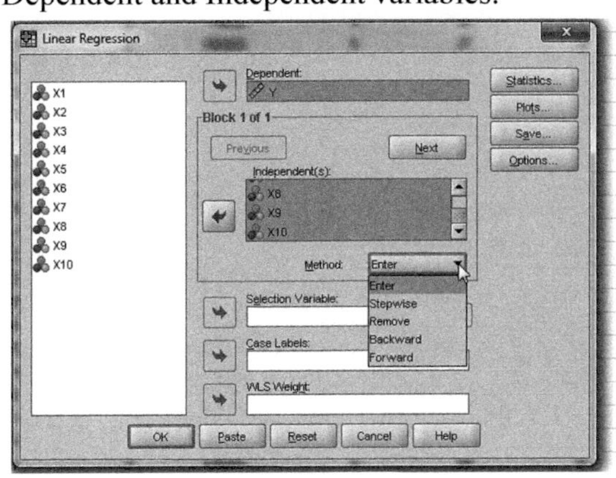

By selecting methods, we can run stepwise, backward, and forward regression as well.

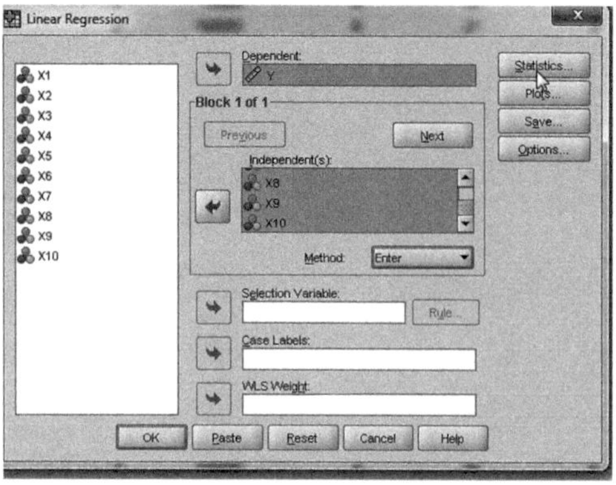

Click on OK and result will appear on screen.

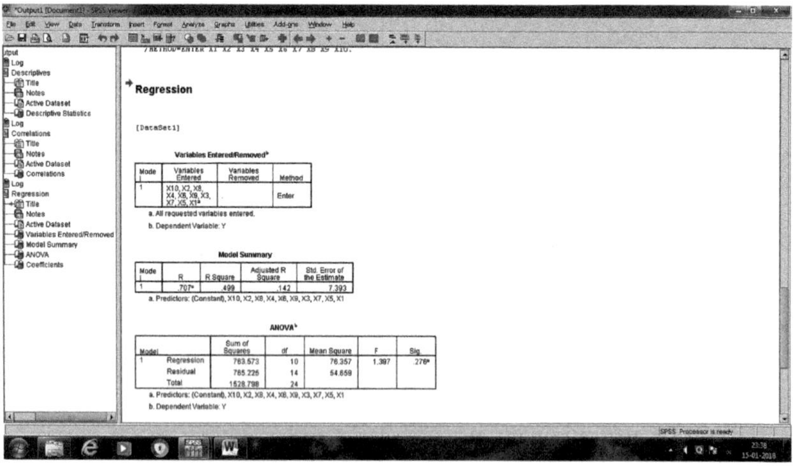

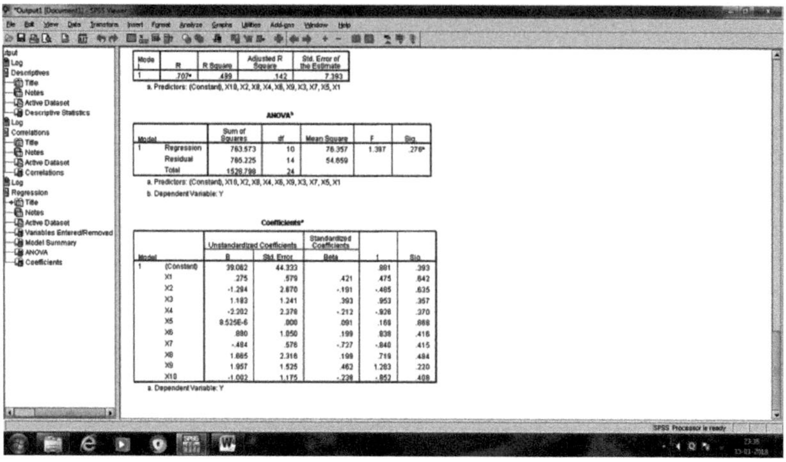

6.2.4 Example 4: RBD Design: A field trial conduct in soybean in seven different varieties and each variety is having four replications. The data are given below in table:

Treatment	R1	R2	R3	R4
T1	34.12	32.32	31.21	30.12
T2	36.44	28.00	29.32	29.55
T3	27.77	25.60	27.80	29.70
T4	30.90	30.01	34.00	32.07
T5	28.75	31.49	29.03	22.32
T6	20.39	22.99	25.78	27.40
T7	27.23	29.58	30.48	26.50

Solution:

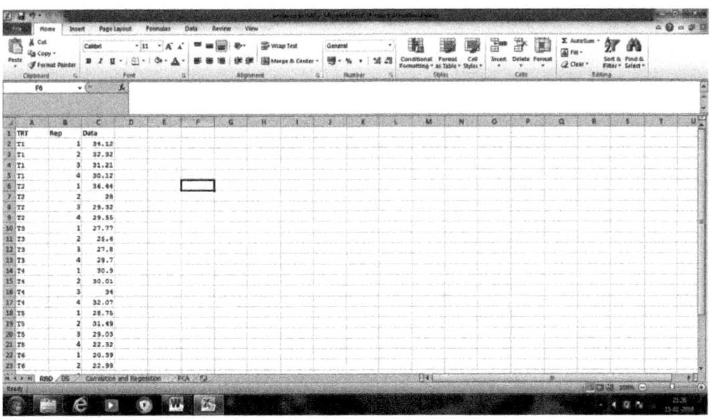

Copy and paste the data in SPSS sheet.

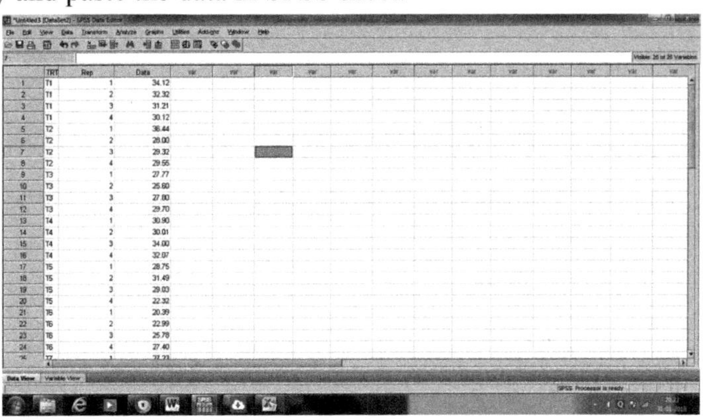

Go to analyze → General Linear Model → Univariate. After clicking on Univariate following dialog box appear on screen:

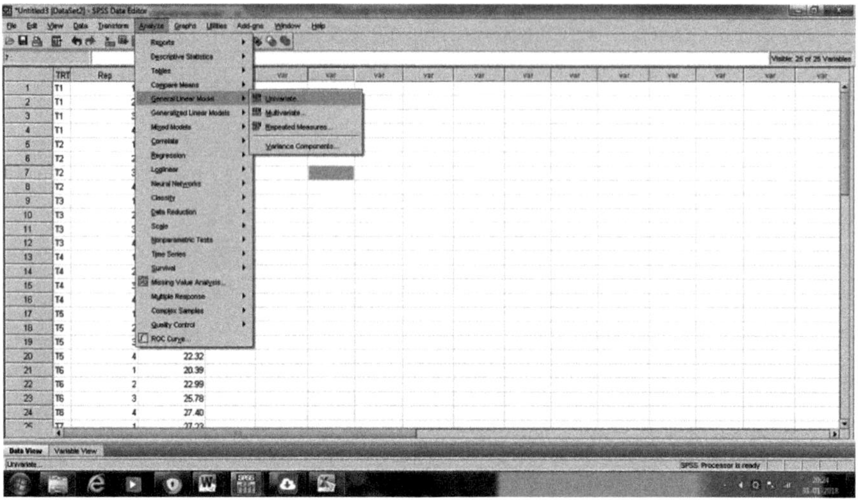

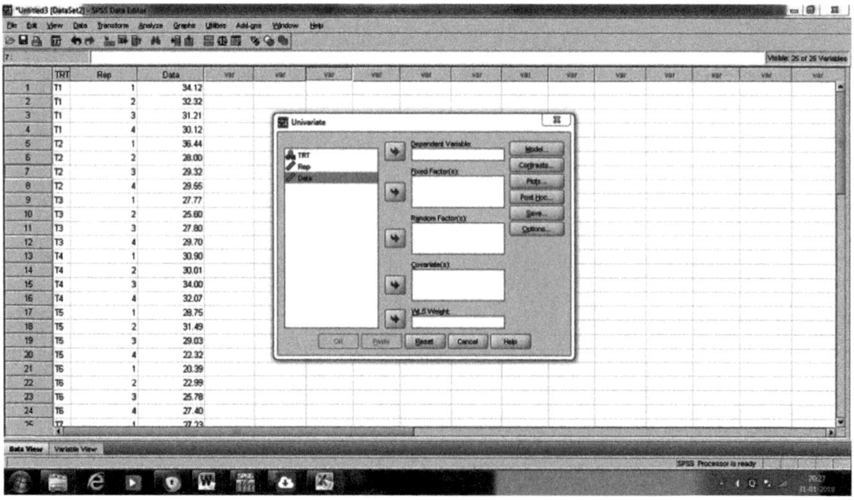

Put variable in Dependent variable in Data section and Fixed factor put treatment and replication.

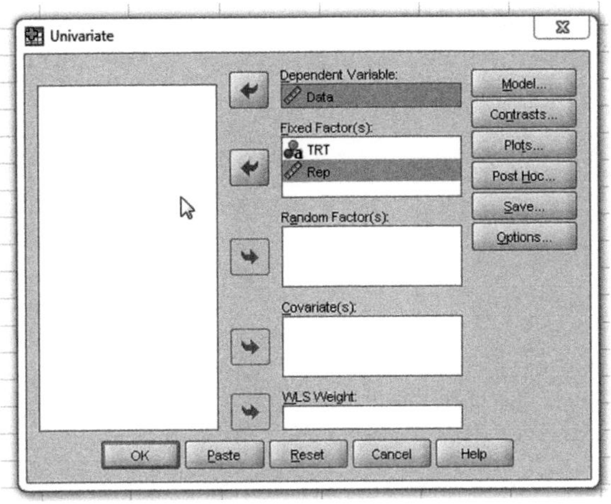

Then go to Model, after clicking on Model, dialog box appear on screen.

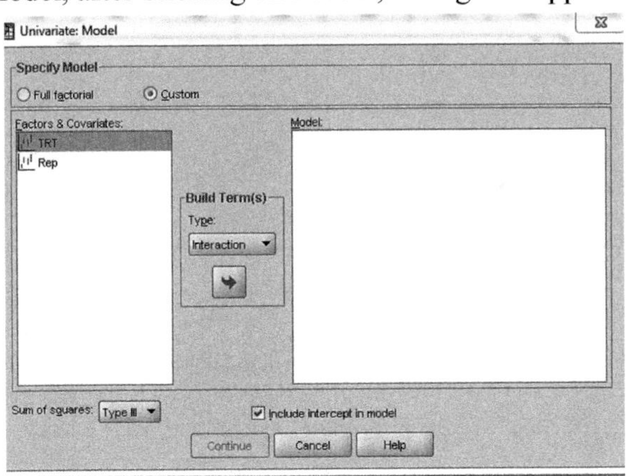

Put TRT and Rep in model section.

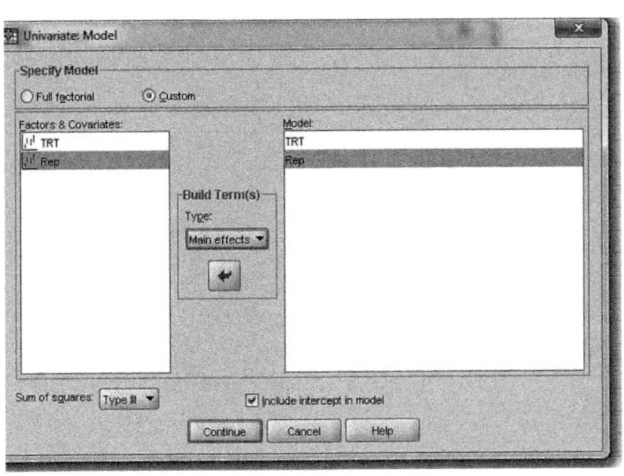

Then click on Continue and select any one of the Post Hoc option to select the desired multiple pairwise comparison procedure from the following screen:

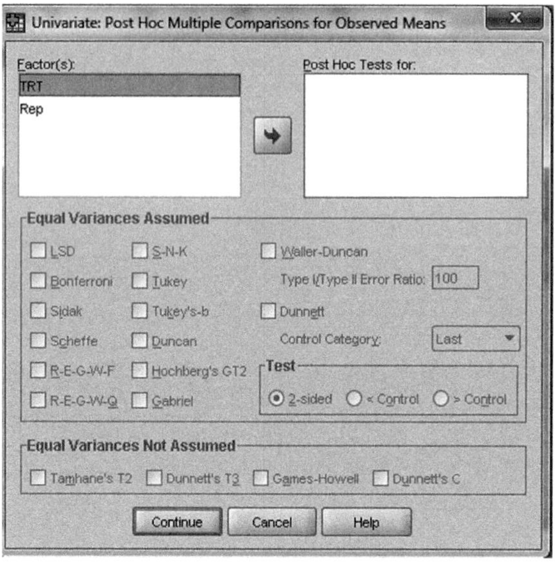

Put treatment in Post Hoc Tests for:

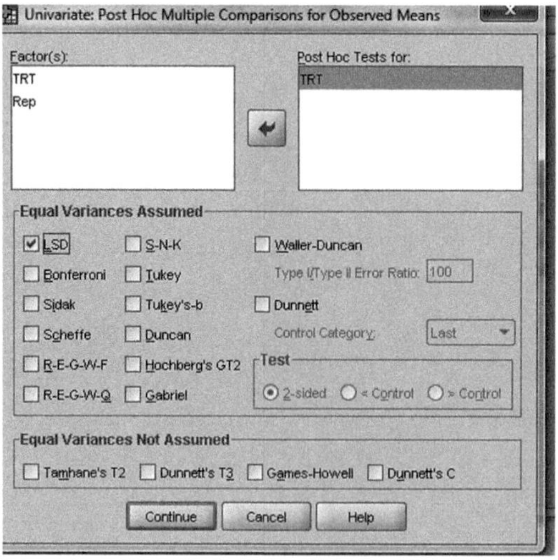

Click on Continue and OK button and result will appear on the screen.

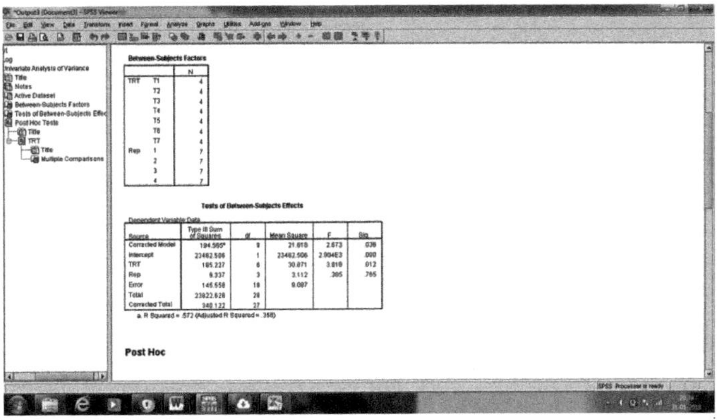

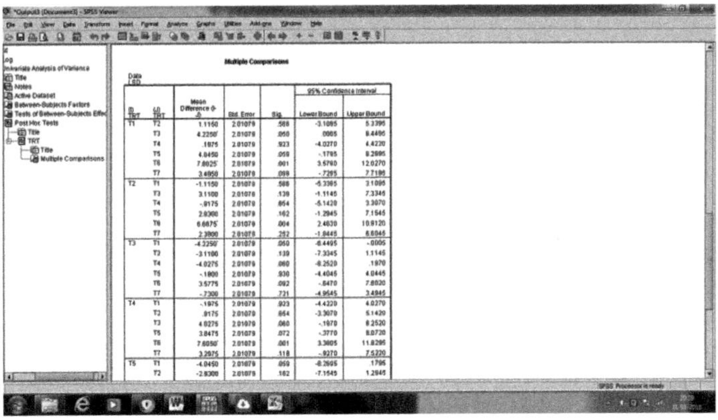

6.2.6 *Example 6:* ARIMA modeling for production of rapeseed and mustard in India.

Year	Area (In '000 Hectare)	Production (In '000 Tonne)
1950–1951	2071	762
1951–1952	2401	943
1952–1953	2105	858
1953–1954	2244	872
1954–1955	2439	1037
1955–1956	2556	860
1956–1957	2539	1043
1957–1958	2412	933
1958–1959	2447	1042
1959–1960	2910	1063

Year	Area	Production
	(In '000 Hectare)	(In '000 Tonne)
1960–1961	2883	1347
1961–1962	3168	1346
1962–1963	3127	1303
1963–1964	3046	915
1964–1965	2910	1474
1965–1966	2913	1298
1966–1967	3006	1228
1967–1968	3244	1568
1968–1969	2870	1347
1969–1970	3172	1564
1970–1971	3323	1976
1971–1972	3614	1433
1972–1973	3319	1808
1973–1974	3457	1704
1974–1975	3680	2252
1975–1976	3339	1936
1976–1977	3128	1550
1977–1978	3584	1650
1978–1979	3544	1860
1979–1980	3470	1428
1980–1981	4113	2304
1981–1982	4399	2382
1982–1983	3827	2207
1983–1984	3874	2608
1984–1985	3987	3073
1985–1986	3982	2680
1986–1987	3719	2605
1987–1988	4508	3370
1988–1989	4832	4877
1989–1990	4967	4125
1990–1991	5782	5229
1991–1992	6553	5863
1992–1993	6193	4803
1993–1994	6289	5328
1994–1995	6058	5758
1995–1996	6547	6000

Year	Area	Production
	(In '000 Hectare)	(In '000 Tonne)
1996–1997	6545	6658
1997–1998	7041	4703
1998–1999	6513	5664
1999–2000	6027	5790
2000–2001	4477	4190
2001–2002	5073	5083
2002–2003	4544	3880
2003–2004	5428	6291
2004–2005	7316	7593
2005–2006	7276	8131
2006–2007	6790	7438
2007–2008	5826	5834
2008–2009	6298	7201
2009–2010	5588	6608
2010–2011	6901	8179
2011–2012	5894	6604
2012–2013	6363	8029
2013–2014	6646	7877
2014–2015	5799	6282

Solution:

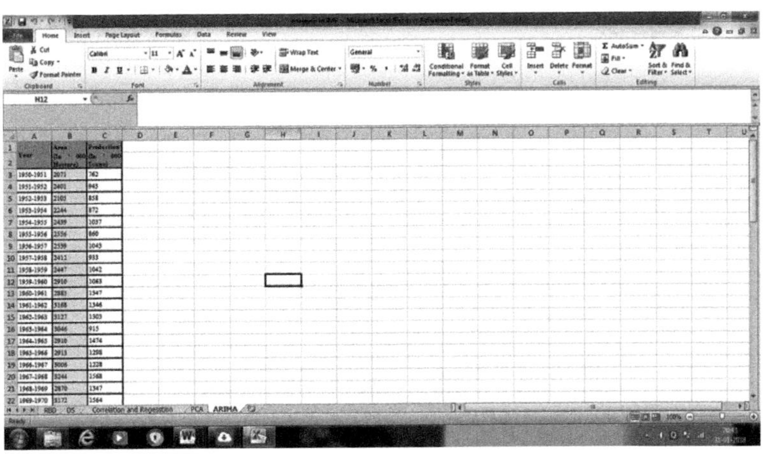

Transfer or import the data into SPSS sheet.

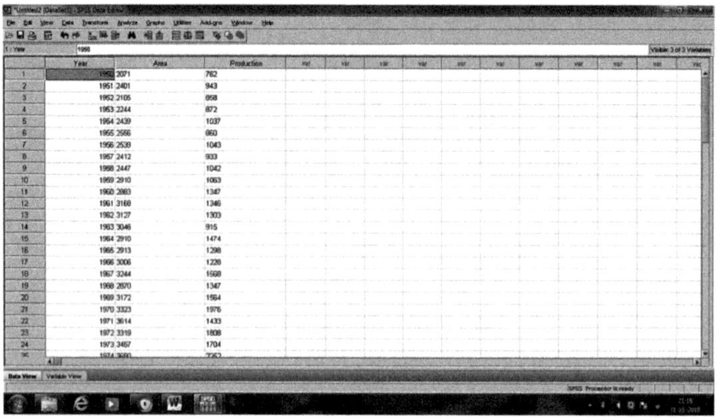

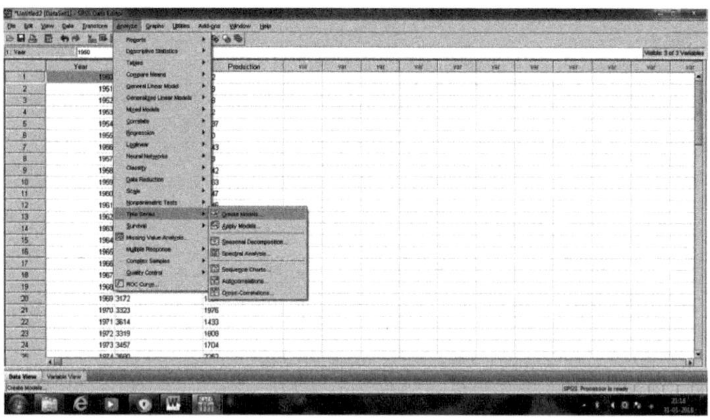

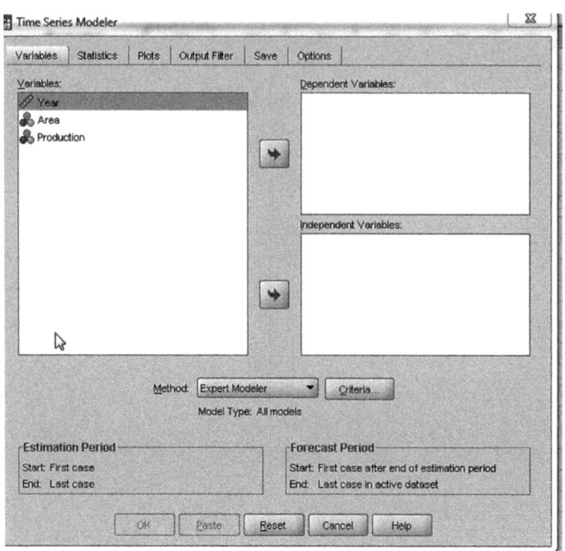
Time Series Modeler
Variables
Statistics
Plots
Output Filter
Save
Options
Variables:
Year
Area
Production
Dependent Variables:
Independent Variables:
Method: Expert Modeler
Criteria...
Model Type: All models
Estimation Period
Start: First case
End: Last case
Forecast Period
Start: First case after end of estimation period
End: Last case in active dataset
OK
Paste
Reset
Cancel
Help

Put variables in independent and dependent variable and Method select ARIMA. After that go to Criteria to select (p,d,q).

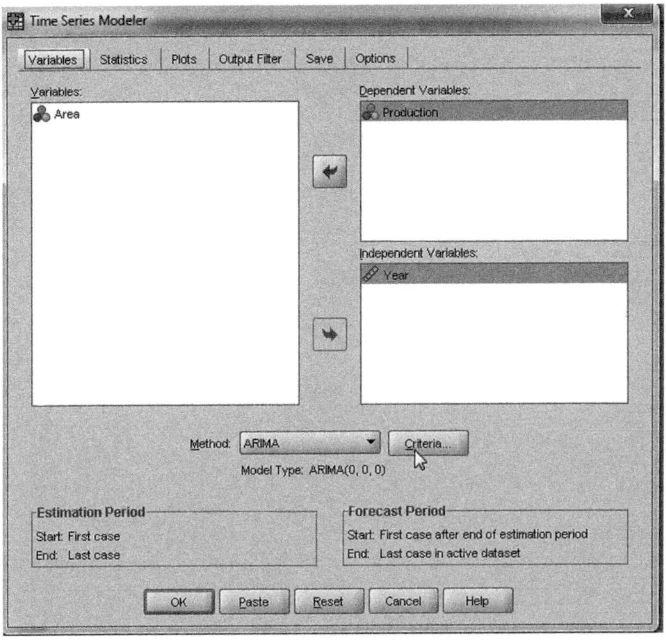

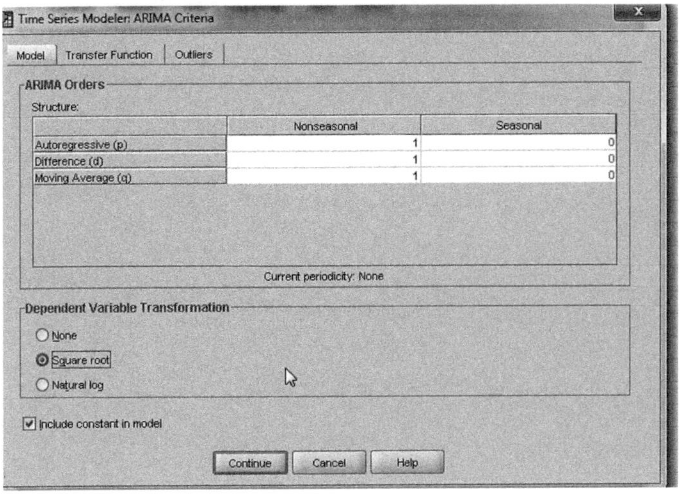

Click on Continue. After that go to the Statistics Section. Select the required model selection criteria.

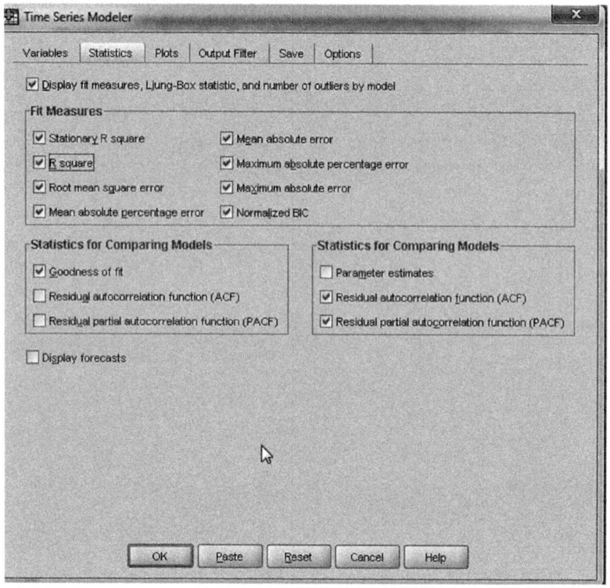

After that go to plot Section.

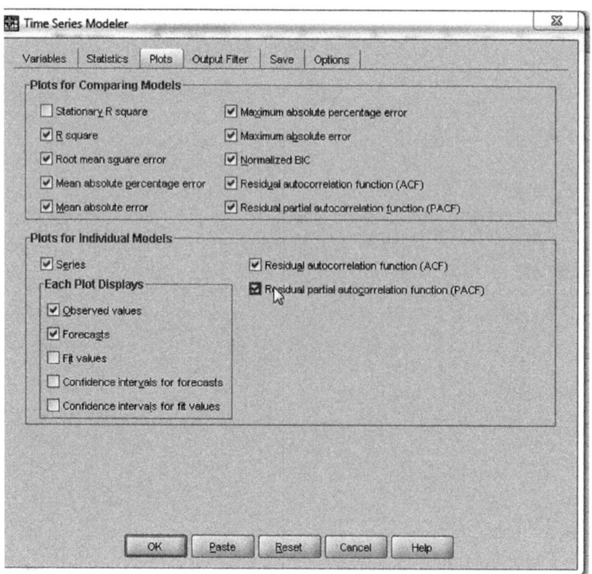

After that go to Output filter Section.

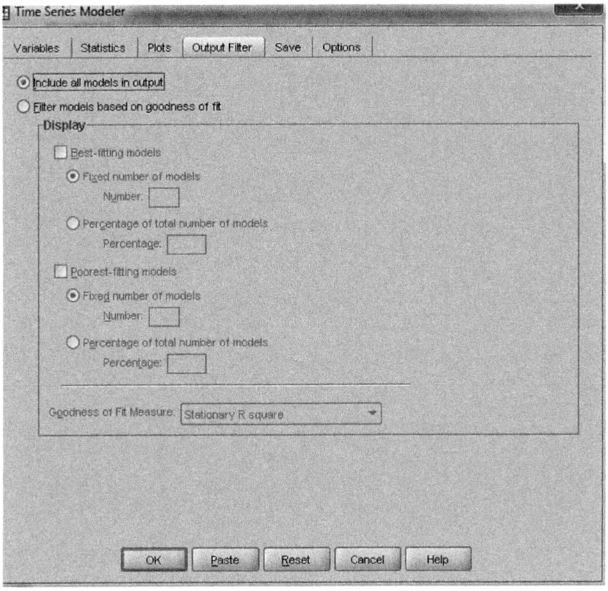

After that go to save Section.

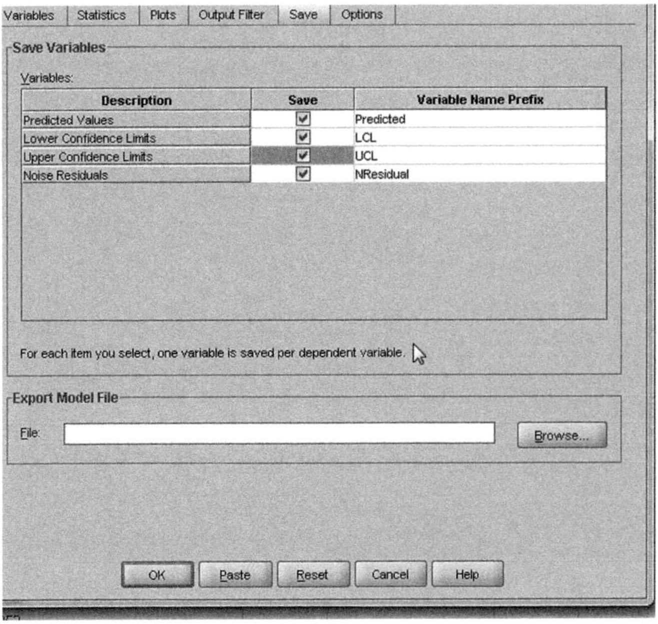

After that go to Option Section.

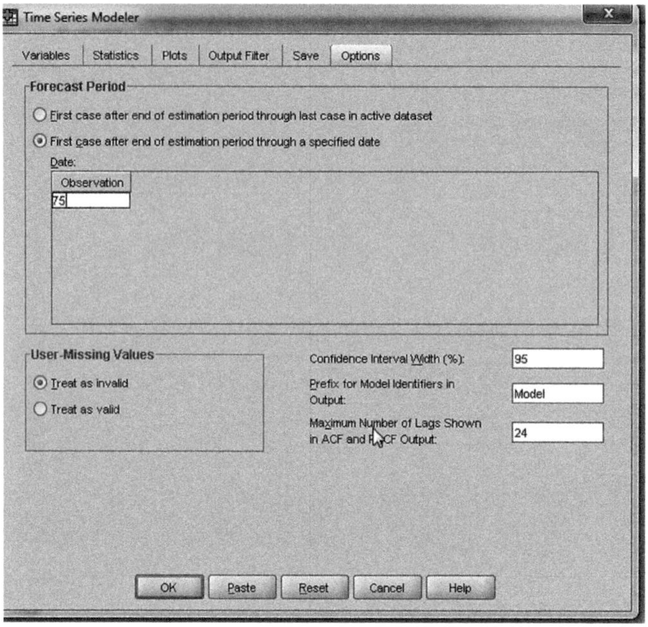

Click on OK button and result will appear for model (1,1,1). Like that we can run for other model also for comparison.

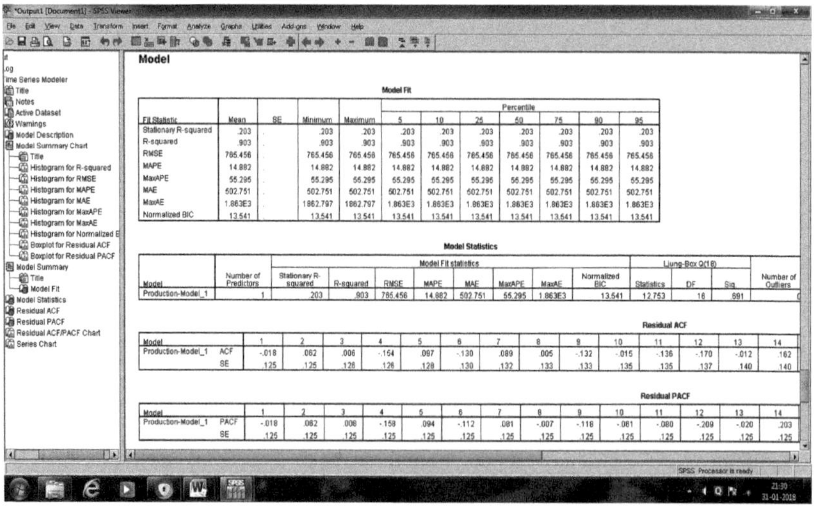

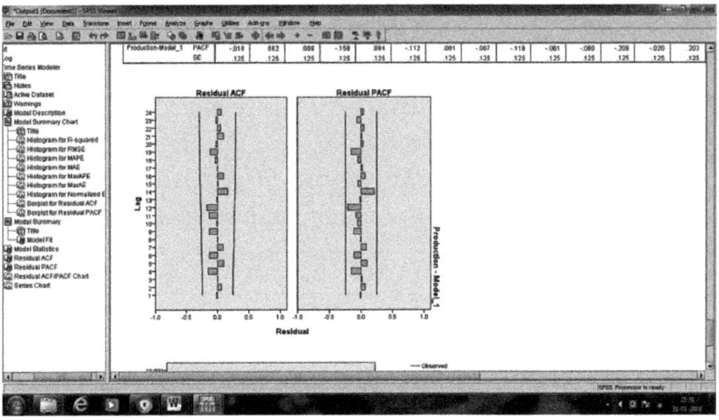
Residual ACF
Residual PACF
Lag
Residual
Production - Model_1

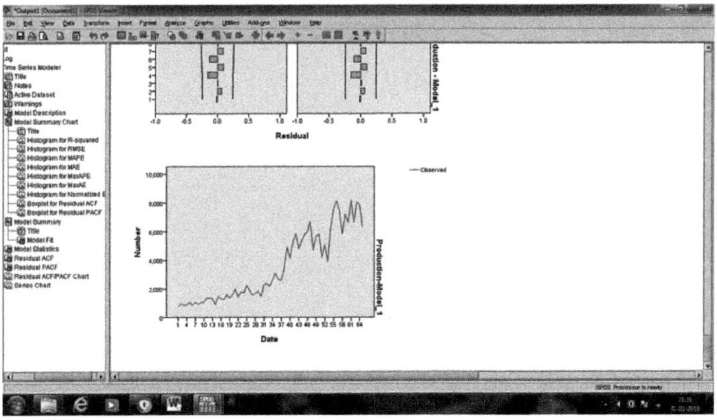
Residual
Number
Date
Observed
Production-Model_1

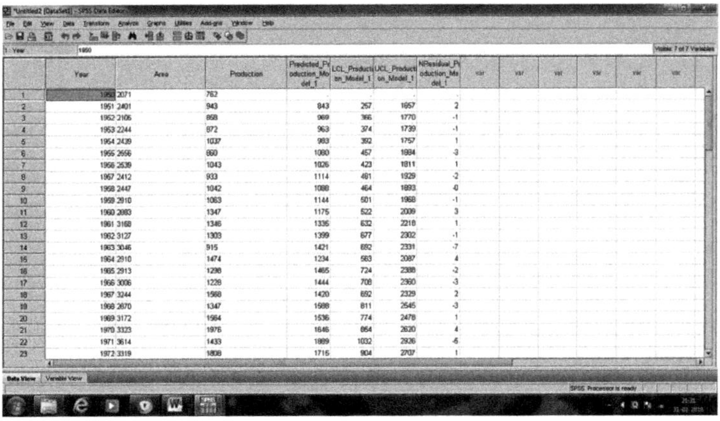
Year
Area
Production

6.3 MINITAB STEPS FOR ANALYSIS

6.3.1 Example 1: Following data give the yield (Y) corresponding to 10 yield components (X_1, X_2, X_3, X_4, X_5, X_6, X_7, X_8, X_9, X_{10}) for certain agricultural crop. Work out the ***Descriptive Statistics*** like (Mean, standard deviation, standard error, ranger minimum, and maximum value).

X1	X2	X3	X4	X5	X6	X7	X8	X9	X10	Y
43	5	1.6	3	1,00,000	3	24	2	9	24	14.49459
47	5	1.2	2	75,000	4	30	2	8	22	27.4552
42	5	2.4	2	1,50,000	6	27	2	8	25	39.68635
58	3	1.6	2	1,20,000	7	35	0	4	27	21.8179
45	5	2.8	2	2,00,000	8	32	2	7	26	28.17015
47	6	3.2	2	2,50,000	5	30	2	3	26	14.98783
42	5	2.4	2	1,60,000	7	22	0	4	27	17.0145
66	3	2.4	2	1,40,000	4	45	2	8	24	41.18381
52	5	1.2	2	1,10,000	4	34	0	5	24	16.10352
44	6	4	2	3,00,000	4	29	0	4	25	15.90416
52	4	0.8	4	1,10,000	8	28	2	3	22	23.26946
62	2	2	2	1,80,000	6	46	2	4	27	19.49062
64	3	12	3	3,50,000	7	41	2	4	28	31.15193
62	3	8	3	2,50,000	6	44	2	7	28	23.33775
35	6	6	5	3,00,000	5	16	2	4	29	26.64379
35	6	5.2	2	3,70,000	4	12	2	3	25	30.11387
34	6	4.4	2	2,90,000	3	10	2	4	25	21.87816
76	5	6	2	3,00,000	6	52	2	4	24	31.03547
52	5	0.8	3	2,00,000	6	31	0	4	27	13.48353
58	4	0.4	3	1,90,000	5	37	2	4	26	10.84107
56	4	0.8	3	1,00,000	4	35	0	4	24	16.11716
42	5	4	2	2,00,000	5	27	2	6	26	31.02111
65	4	3.2	2	2,60,000	9	51	2	4	22	23.17923
27	6	1.2	2	1,80,000	2	10	0	4	25	23.54457
35	6	3.2	2	3,10,000	8	17	0	2	27	16.45729

Solution: For getting descriptive statistics,

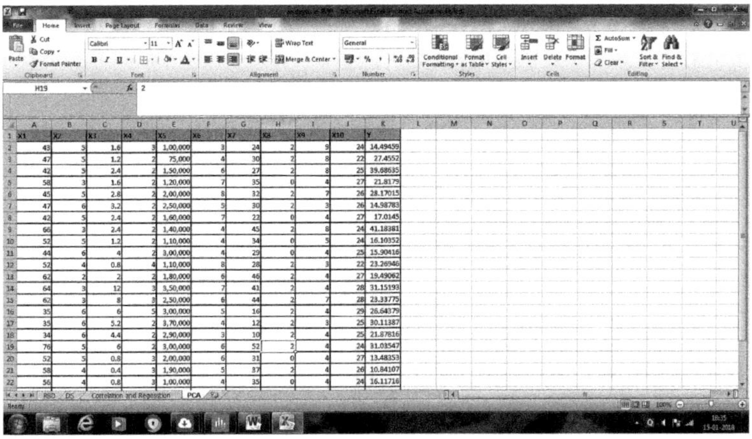

Then copy and paste the data in MINITAB worksheet first.

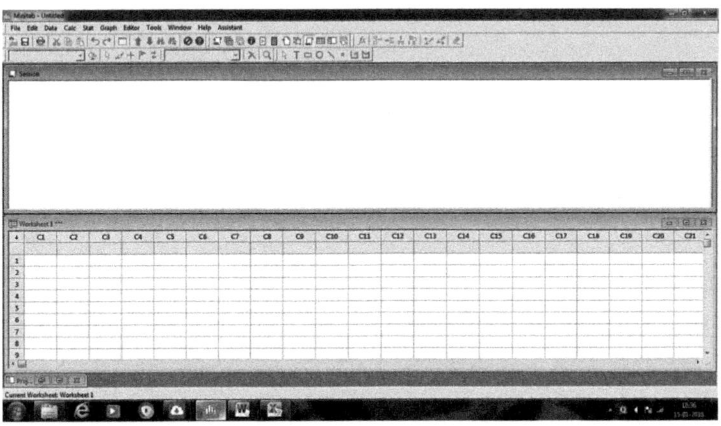

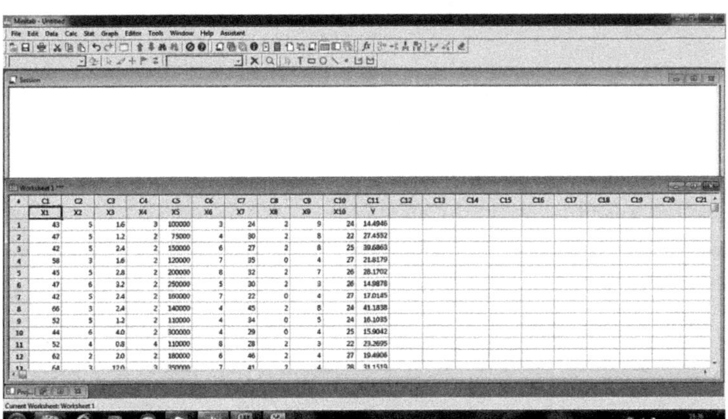

After pasting the data in MINITAB, go to Stat option and click on display descriptive statistics. After clicking on this, the following dialog box appears on the screen.

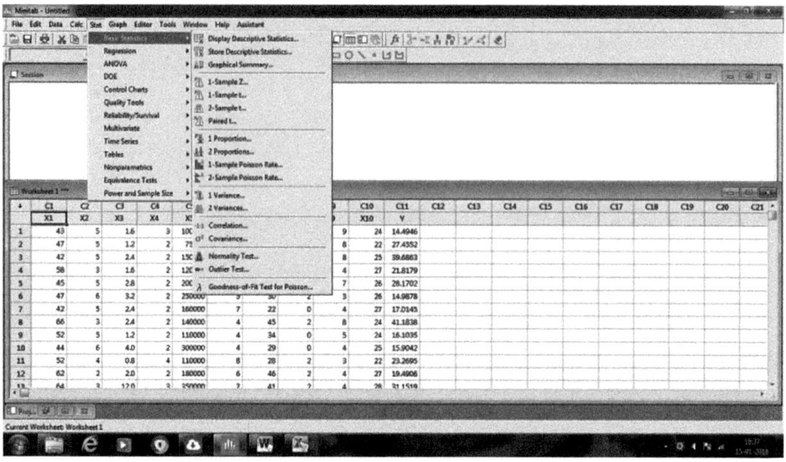

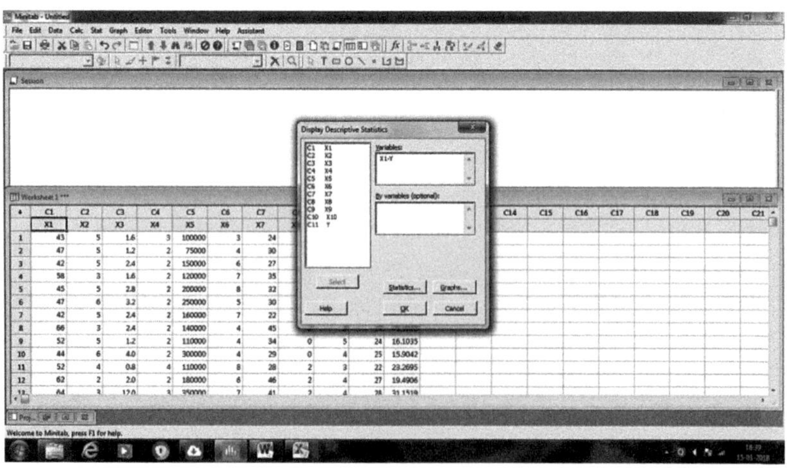

Select the variable, which need to calculate descriptive statistics and go to statistics option to choose descriptive measure. And click on OK for go to main window.

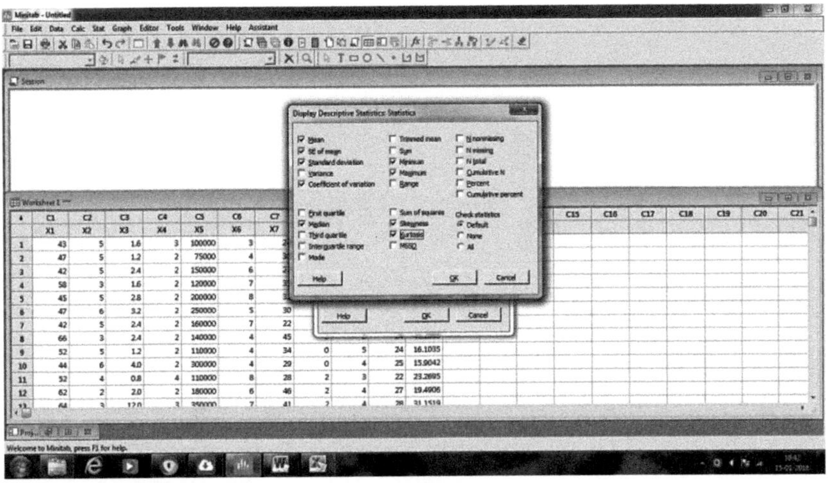

Descriptive Statistics: X_1, X_2, X_3, X_4, X_5, X_6, X_7, X_8, X_9, X_{10}, Y.

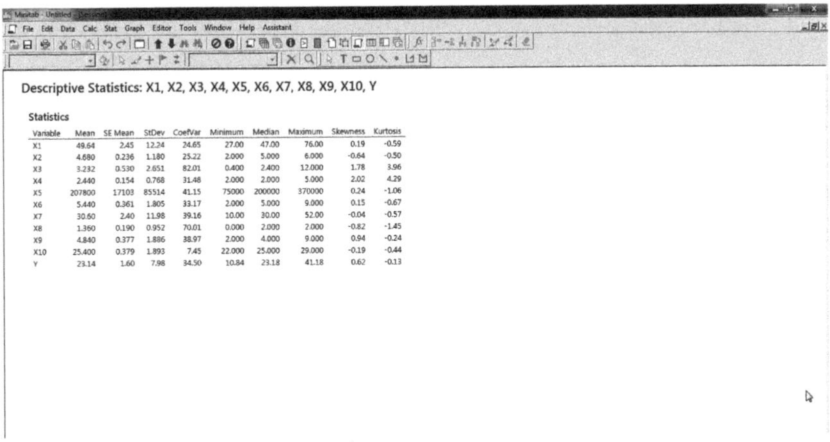

Descriptive Statistics: X1, X2, X3, X4, X5, X6, X7, X8, X9, X10, Y

Statistics

Variable	Mean	SE Mean	StDev	CoefVar	Minimum	Median	Maximum	Skewness	Kurtosis
X1	49.64	2.45	12.24	24.65	27.00	47.00	76.00	0.19	-0.59
X2	4.680	0.236	1.180	25.22	2.000	5.000	6.000	-0.64	-0.50
X3	3.232	0.530	2.651	82.01	0.400	2.400	12.000	1.78	3.96
X4	2.440	0.154	0.768	31.48	2.000	2.000	5.000	2.02	4.29
X5	207800	17103	85514	41.15	75000	200000	370000	0.24	-1.06
X6	5.440	0.361	1.805	33.17	2.000	5.000	9.000	0.15	-0.67
X7	30.60	2.40	11.98	39.16	10.00	30.00	52.00	-0.04	-0.57
X8	1.360	0.190	0.952	70.01	0.000	2.000	2.000	-0.82	-1.45
X9	4.840	0.377	1.886	38.97	2.000	4.000	9.000	0.94	-0.24
X10	25.400	0.379	1.893	7.45	22.000	25.000	29.000	-0.19	-0.44
Y	23.14	1.60	7.98	34.50	10.84	23.18	41.18	0.62	-0.13

6.3.2 ***Example 2:*** Following data given (**Section *6.3.1 – Example 1***) in the yield (Y) corresponding to 10 yield components (X_1, X_2, X_3, X_4, X_5, X_6, X_7, X_8, X_9, X_{10}) for certain agricultural crop. Work out the ***Correlation Analysis*** for below data.

Solution: For getting correlation analysis,

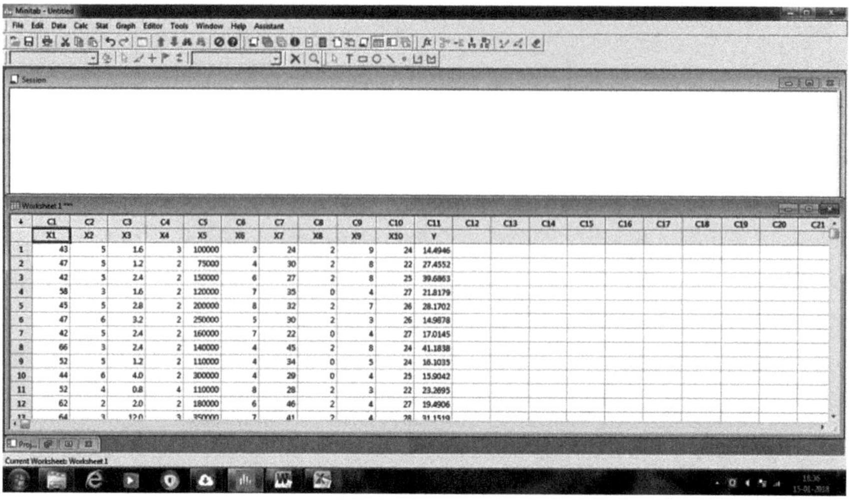

After pasting the data in MINITAB, go to Stat option and click on correlation. After clicking on this, the following dialog box appear on the screen.

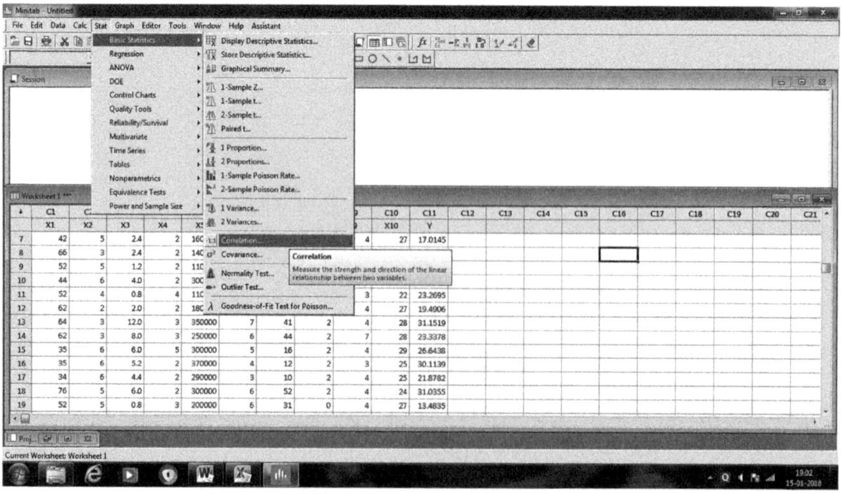

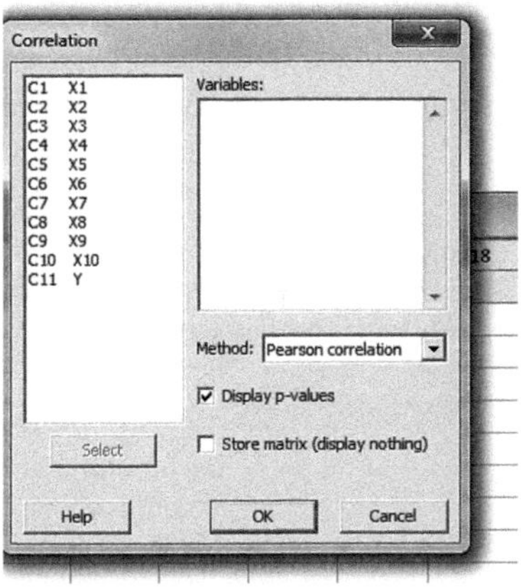

Select all variable and put into variables section.

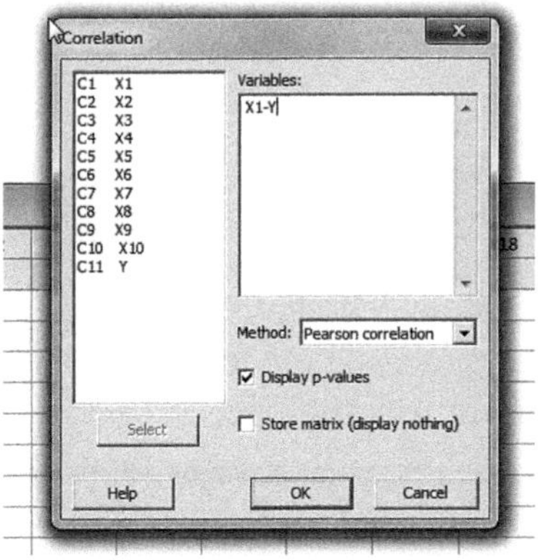

After selecting variable, click on OK button and results will appear on the screen.

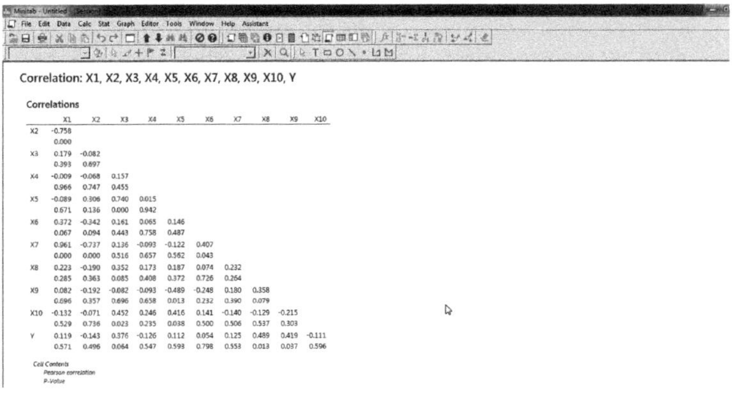

Correlation: X1, X2, X3, X4, X5, X6, X7, X8, X9, X10, Y

Correlations

	X1	X2	X3	X4	X5	X6	X7	X8	X9	X10
X2	-0.758									
	0.000									
X3	0.179	-0.082								
	0.393	0.697								
X4	-0.009	-0.068	0.157							
	0.966	0.747	0.455							
X5	-0.089	0.306	0.740	0.015						
	0.671	0.136	0.000	0.942						
X6	0.372	-0.342	0.161	0.065	0.146					
	0.067	0.094	0.443	0.758	0.487					
X7	0.961	-0.737	0.136	-0.093	-0.122	0.407				
	0.000	0.000	0.516	0.657	0.562	0.043				
X8	0.223	-0.190	0.352	0.173	0.187	0.074	0.232			
	0.285	0.363	0.085	0.408	0.372	0.726	0.264			
X9	0.082	-0.192	-0.082	-0.093	-0.489	-0.248	0.180	0.358		
	0.696	0.357	0.696	0.658	0.013	0.232	0.390	0.079		
X10	-0.132	-0.071	0.452	0.246	0.416	0.141	-0.140	-0.129	-0.215	
	0.529	0.736	0.023	0.235	0.038	0.500	0.506	0.537	0.303	
Y	0.119	-0.143	0.376	-0.126	0.112	0.054	0.125	0.489	0.419	-0.111
	0.571	0.496	0.064	0.547	0.593	0.798	0.553	0.013	0.037	0.596

Cell Contents
Pearson correlation
P-Value

6.3.3 Example 3: Following data given (**Section *6.3.1 – Example 1***) in the yield (Y) corresponding to 10 yield components (X_1, X_2, X_3, X_4, X_5, X_6, X_7, X_8, X_9, X_{10}) for certain agricultural crop. Work out the ***Regression Analysis*** for below data.

Solution: For getting regression analysis

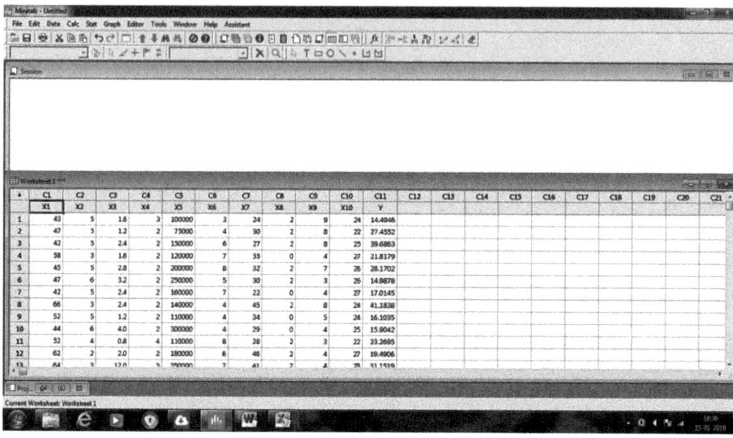

After pasting the data in MINITAB, go to Stat option and click on regression and go to Fit regression model option. After clicking on this, the following dialog box appears on the screen.

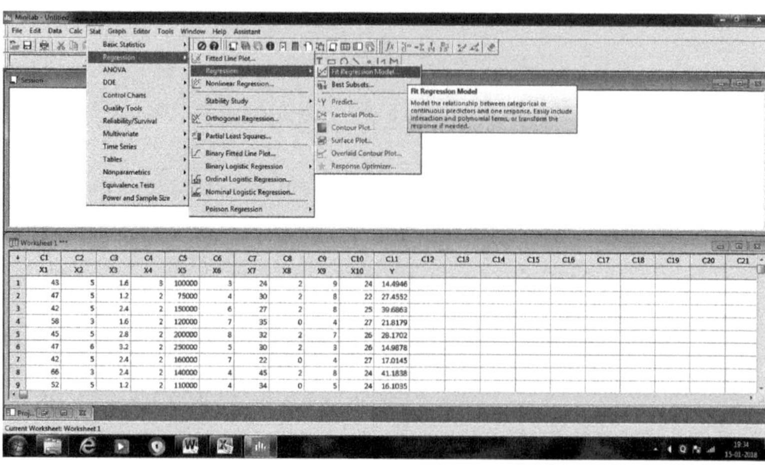

After clicking on Fit regression model, following dialog box appear on the screen.

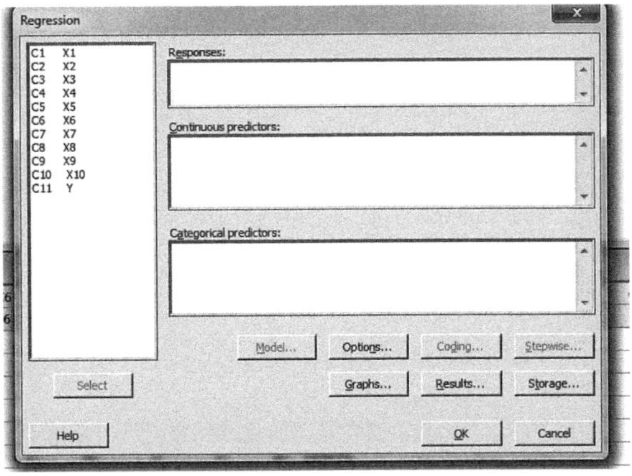

In responses put dependent variable and continuous predictors put independent variables. Click on OK button.

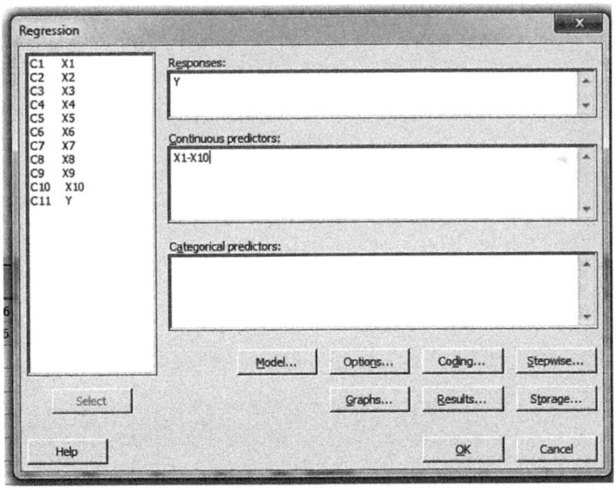

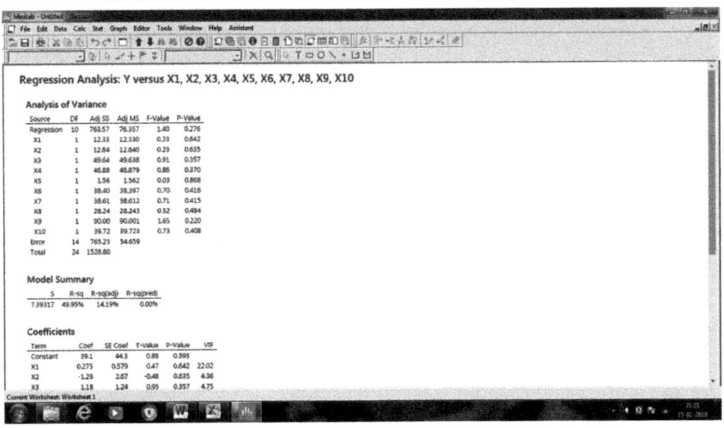

Regression Analysis: Y versus X1, X2, X3, X4, X5, X6, X7, X8, X9, X10

Analysis of Variance

Source	DF	Adj SS	Adj MS	F-Value	P-Value
Regression	10	763.57	76.357	1.40	0.276
X1	1	12.33	12.330	0.23	0.642
X2	1	12.84	12.840	0.23	0.635
X3	1	49.64	49.638	0.91	0.357
X4	1	46.88	46.879	0.86	0.370
X5	1	1.56	1.562	0.03	0.868
X6	1	38.40	38.397	0.70	0.416
X7	1	38.61	38.612	0.71	0.415
X8	1	28.24	28.243	0.52	0.484
X9	1	90.00	90.001	1.65	0.220
X10	1	39.72	39.723	0.73	0.408
Error	14	765.23	54.659		
Total	24	1528.80			

Model Summary

S	R-sq	R-sq(adj)	R-sq(pred)
7.39317	49.95%	14.19%	0.00%

Coefficients

Term	Coef	SE Coef	T-Value	P-Value	VIF
Constant	39.1	44.5	0.88	0.393	
X1	0.275	0.579	0.47	0.642	22.02
X2	-1.29	2.67	-0.48	0.635	4.36
X3	1.18	1.24	0.95	0.357	4.75

Current Worksheet: Worksheet 1

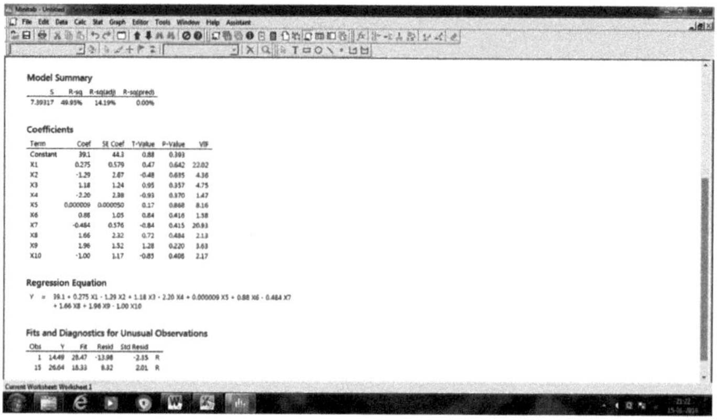

6.3.4 *Example 4:* Use t-test for test the significance of two series assuming equal variance.

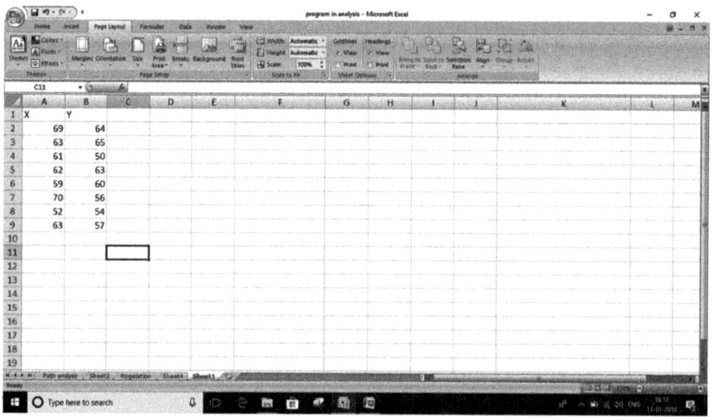

Transfer the data into Minitab worksheet.

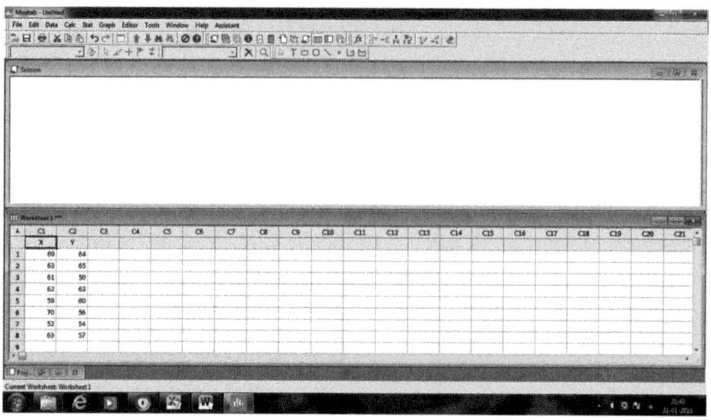

Then go to Stat→Basic Statistics→2 – Sample t-Test:

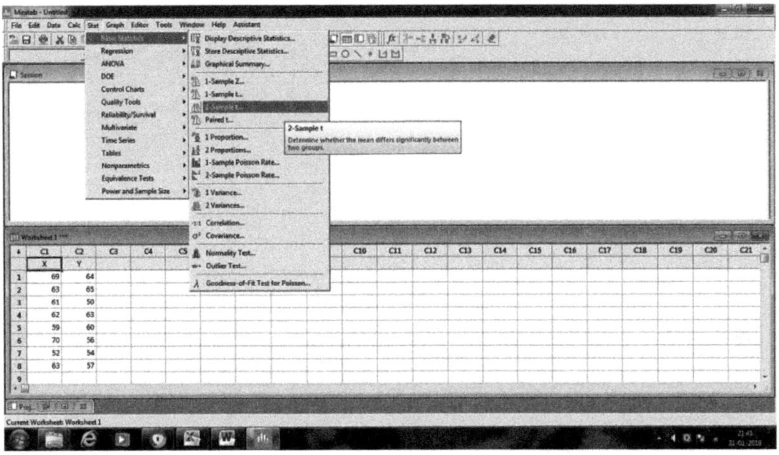

After clicking on 2 – Sample t-Test: following dialog box appear on screen.

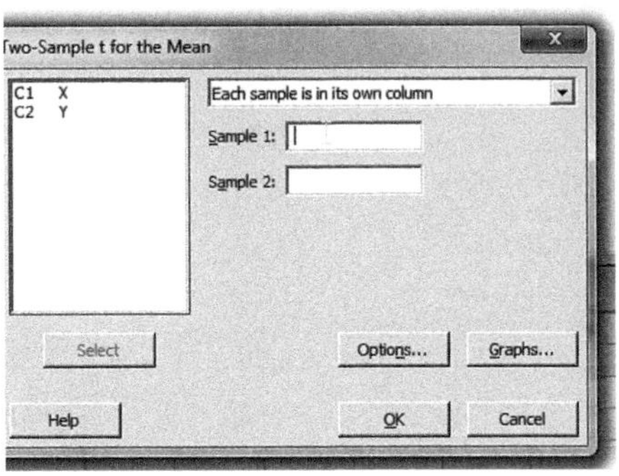

Put variable on Sample 1 and 2.

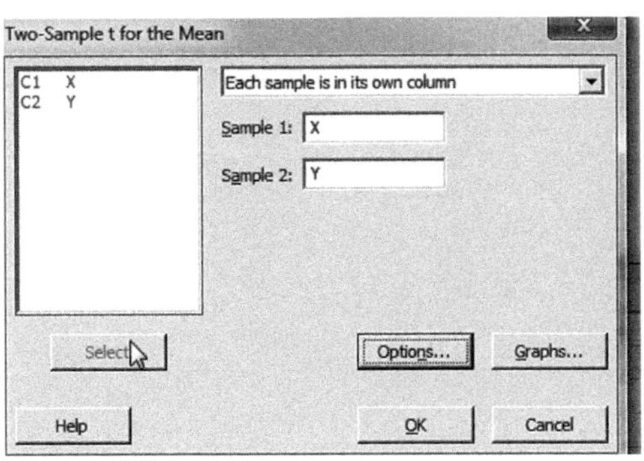

Click on OK and result will appear on screen.

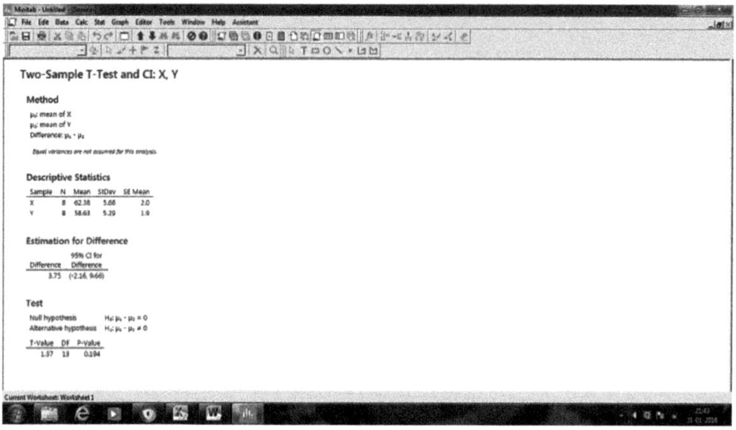

6.3.5 *Example 5: RBD Design:* A field trial conduct in soybean in seven different varieties and each variety is having four replications. The data are given below in table:

Treatment	R1	R2	R3	R4
T1	34.12	32.32	31.21	30.12
T2	36.44	28.00	29.32	29.55
T3	27.77	25.60	27.80	29.70
T4	30.90	30.01	34.00	32.07
T5	28.75	31.49	29.03	22.32
T6	20.39	22.99	25.78	27.40
T7	27.23	29.58	30.48	26.50

Solution:

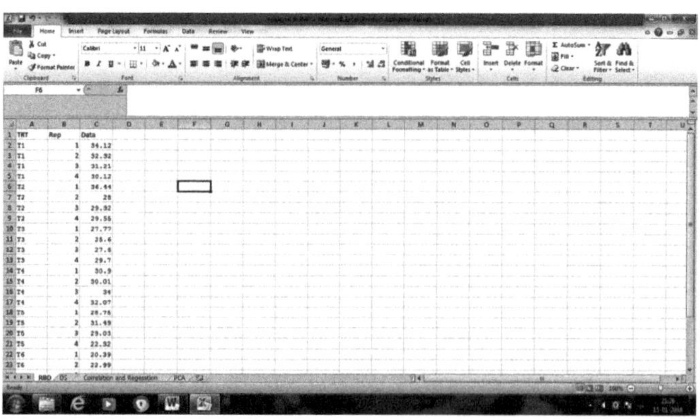

Copy and paste the data in MINITAB sheet.

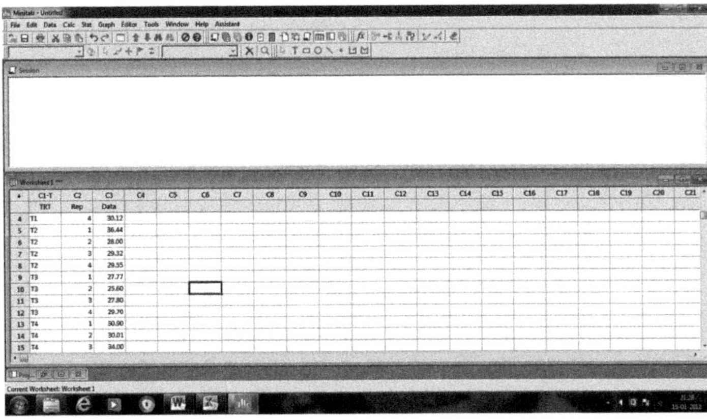

After transferring the data in MINITAB sheet, go to Stat → ANOVA → General Linear Model → Fit general Linear Model. After clicking on Fit General linear model following dialog box appear on the screen.

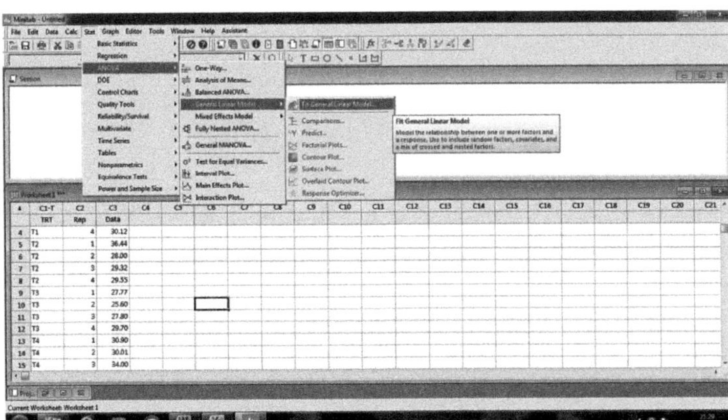

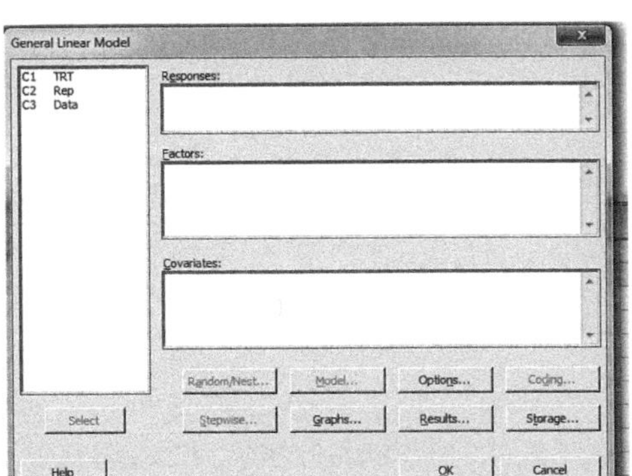

Put data in responses section and TRT and Rep put into factors section and click on OK button.

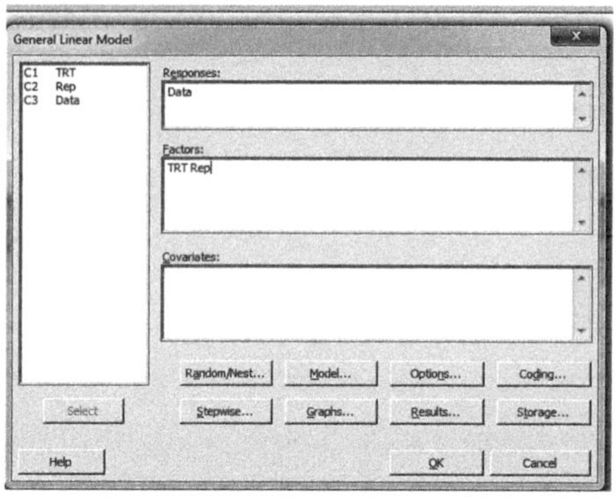

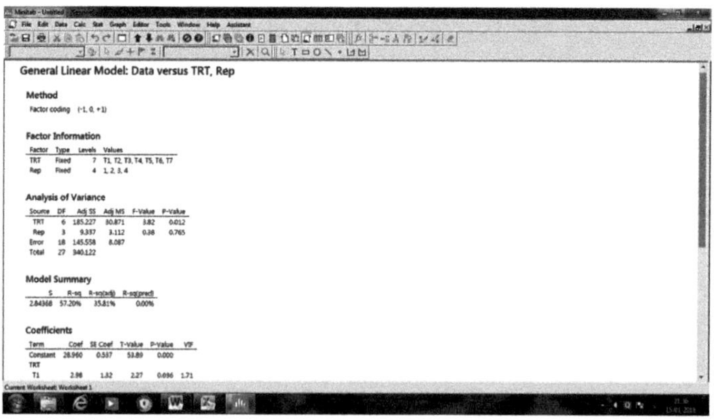

For seeing the treatment difference using Tukey's mean separation methods. We need to run further process.

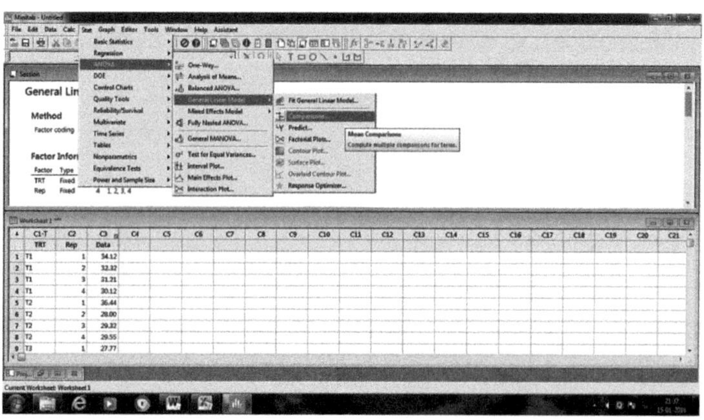

After clicking on Comparison, following dialog box appear on screen. Tick (√) on treatments (TRT). Click on OK.

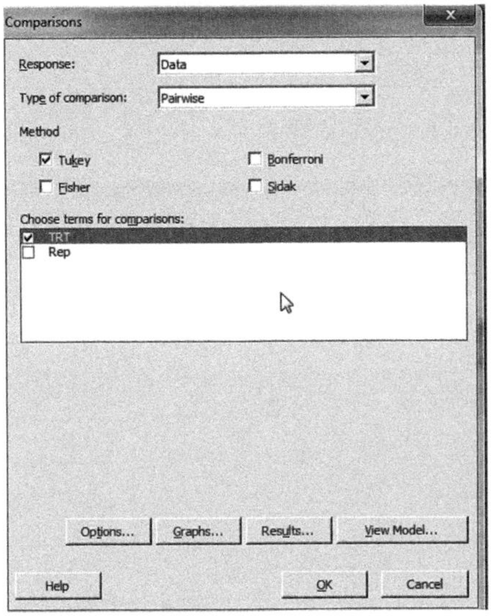

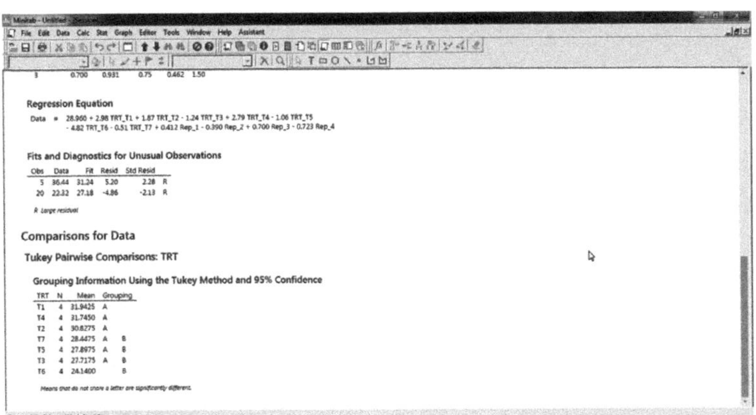

Comparisons for Data

Tukey Pairwise Comparisons: TRT

Grouping Information Using the Tukey Method and 95% Confidence

TRT	N	Mean	Grouping	
T1	4	31.9425	A	
T4	4	31.7450	A	
T2	4	30.8275	A	
T7	4	28.4475	A	B
T5	4	27.8975	A	B
T3	4	27.7175	A	B
T6	4	24.1400		B

Means that do not share a letter are significantly different.

6.3.6 Example 6: Following data given (**Section *6.3.1 – Example 1***) in the yield (Y) corresponding to 10 yield components (X_1, X_2, X_3, X_4, X_5, X_6, X_7, X_8, X_9, X_{10}) for certain agricultural crop. Work out the ***Principal Component (PCA) Analysis*** for below data.

Solution: For getting PCA analysis,

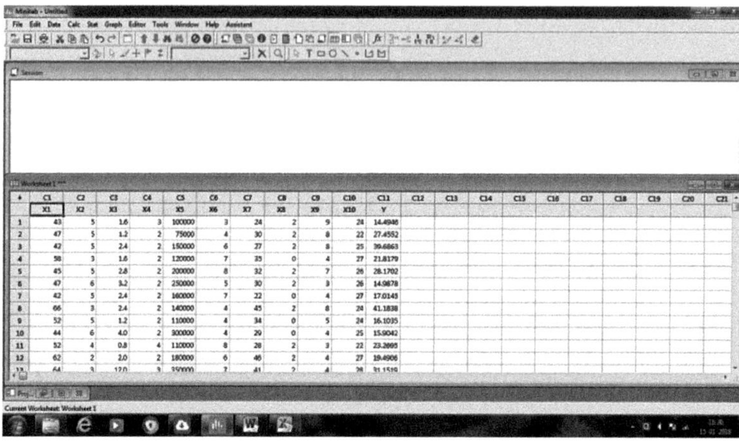

After pasting the data in MINITAB, go to Stat option and click on Multivariate and go to PCA option. After clicking on this, the following dialog box appear on the screen.

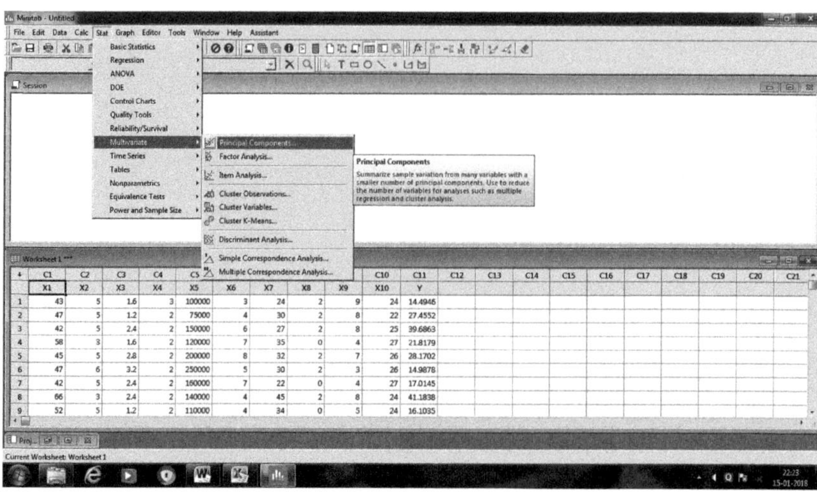

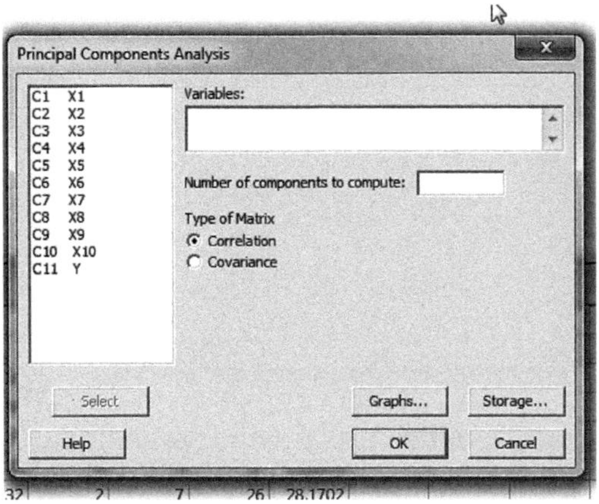

Select the X1–X10 on variables as well as give 'Number of components to compute' and click on Graphs

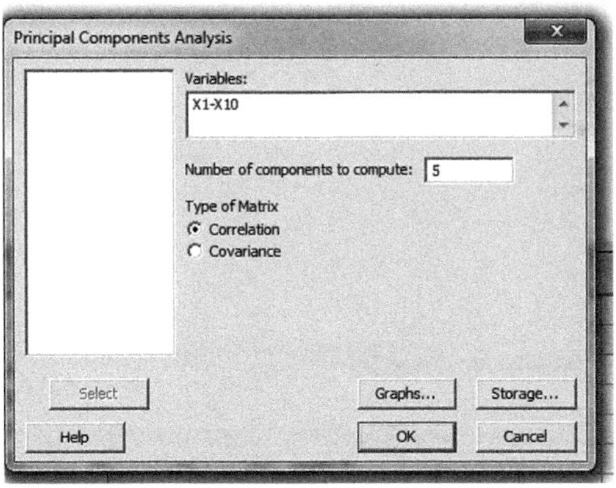

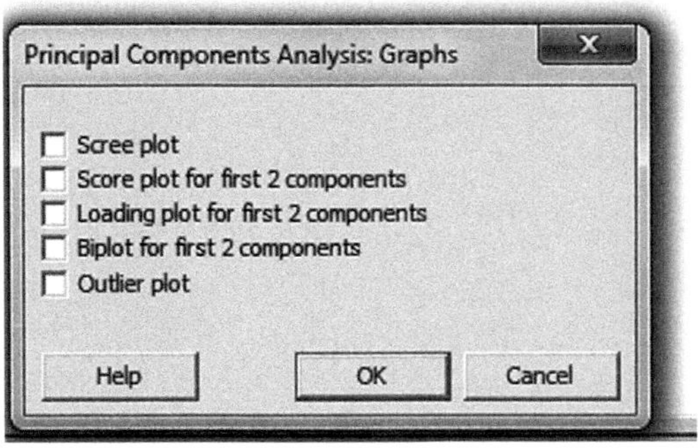

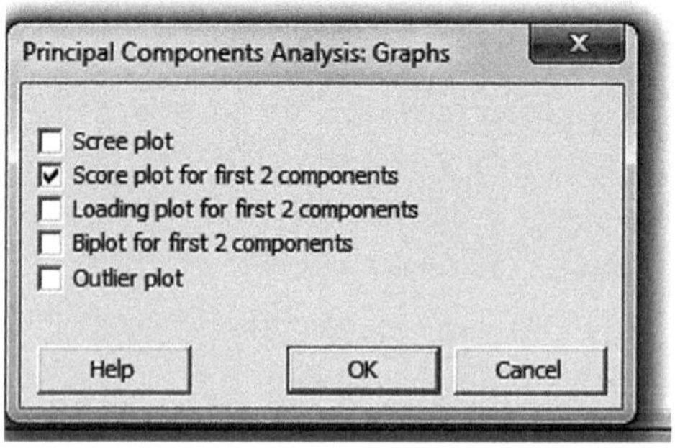

After clicking on OK button results will appear on the screen.

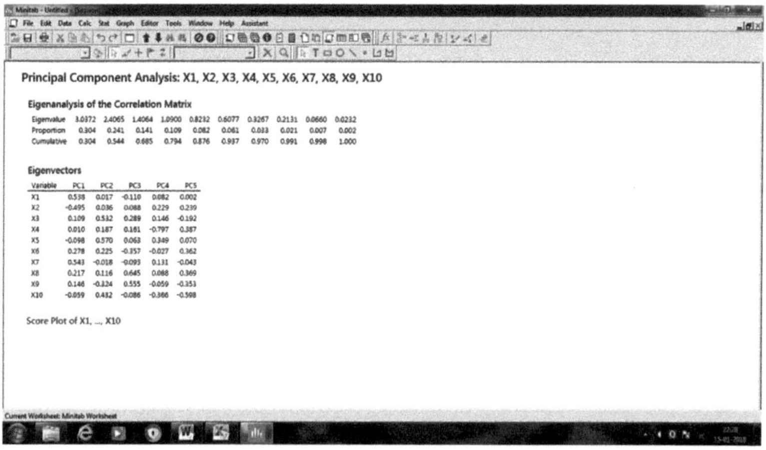

Principal Component Analysis: X1, X2, X3, X4, X5, X6, X7, X8, X9, X10

Eigenanalysis of the Correlation Matrix

Eigenvalue	3.0372	2.4065	1.4064	1.0900	0.8232	0.6077	0.3267	0.2131	0.0660	0.0232
Proportion	0.304	0.241	0.141	0.109	0.082	0.061	0.033	0.021	0.007	0.002
Cumulative	0.304	0.544	0.685	0.794	0.876	0.937	0.970	0.991	0.998	1.000

Eigenvectors

Variable	PC1	PC2	PC3	PC4	PC5
X1	0.538	0.017	-0.110	0.082	0.002
X2	-0.495	0.036	0.088	0.229	0.239
X3	0.109	0.532	0.289	0.146	-0.192
X4	0.010	0.187	0.161	-0.797	0.387
X5	-0.098	0.570	0.063	0.349	0.070
X6	0.278	0.225	-0.357	-0.027	0.362
X7	0.543	-0.018	-0.093	0.131	-0.043
X8	0.217	0.116	0.645	0.088	0.369
X9	0.146	-0.324	0.555	-0.059	-0.353
X10	-0.059	0.432	-0.086	-0.366	-0.598

Score Plot of X1, ..., X10

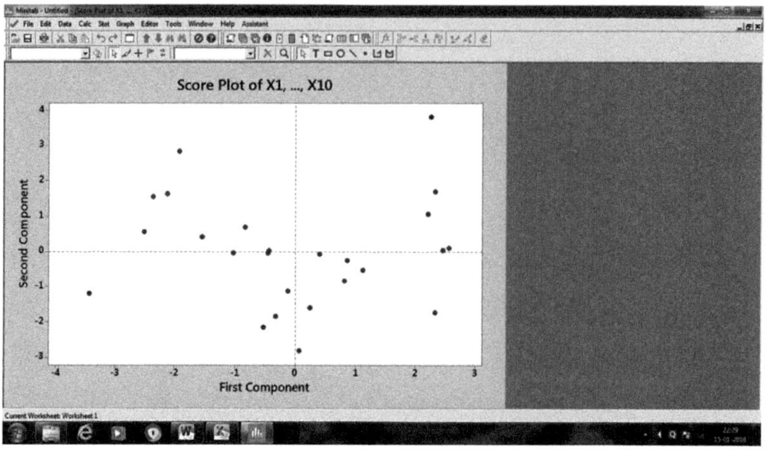

6.3.7 Example 7: Following data give the yield (Y) corresponding to 10 yield components (X_1, X_2, X_3, X_4, X_5, X_6, X_7, X_8, X_9, X_{10}) for certain agricultural crop. Work out the ***Factor Analysis***.

X1	X2	X3	X4	X5	X6	X7	X8	X9	X10	Y
43	5	1.6	3	1,00,000	3	24	2	9	24	14.49459
47	5	1.2	2	75,000	4	30	2	8	22	27.4552
42	5	2.4	2	1,50,000	6	27	2	8	25	39.68635
58	3	1.6	2	1,20,000	7	35	0	4	27	21.8179
45	5	2.8	2	2,00,000	8	32	2	7	26	28.17015
47	6	3.2	2	2,50,000	5	30	2	3	26	14.98783
42	5	2.4	2	1,60,000	7	22	0	4	27	17.0145
66	3	2.4	2	1,40,000	4	45	2	8	24	41.18381
52	5	1.2	2	1,10,000	4	34	0	5	24	16.10352
44	6	4	2	3,00,000	4	29	0	4	25	15.90416
52	4	0.8	4	1,10,000	8	28	2	3	22	23.26946
62	2	2	2	1,80,000	6	46	2	4	27	19.49062
64	3	12	3	3,50,000	7	41	2	4	28	31.15193
62	3	8	3	2,50,000	6	44	2	7	28	23.33775
35	6	6	5	3,00,000	5	16	2	4	29	26.64379
35	6	5.2	2	3,70,000	4	12	2	3	25	30.11387
34	6	4.4	2	2,90,000	3	10	2	4	25	21.87816
76	5	6	2	3,00,000	6	52	2	4	24	31.03547
52	5	0.8	3	2,00,000	6	31	0	4	27	13.48353
58	4	0.4	3	1,90,000	5	37	2	4	26	10.84107
56	4	0.8	3	1,00,000	4	35	0	4	24	16.11716
42	5	4	2	2,00,000	5	27	2	6	26	31.02111
65	4	3.2	2	2,60,000	9	51	2	4	22	23.17923
27	6	1.2	2	1,80,000	2	10	0	4	25	23.54457
35	6	3.2	2	3,10,000	8	17	0	2	27	16.45729

Solution: From the given data we calculate the following Factor Analysis.

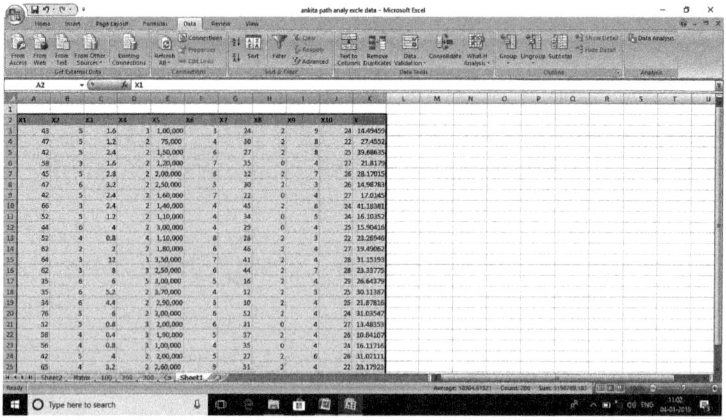

First transfer data into MINITAB worksheet

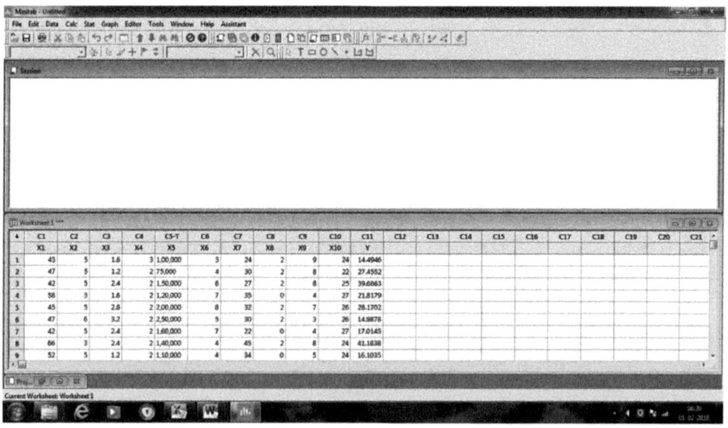

After transfer the data → Stat →Multivariate→ Factor Analysis.

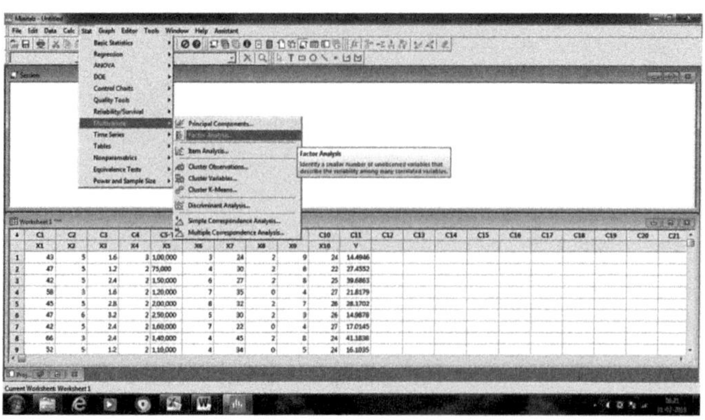

After clicking on Factor Analysis, dialog box appear on screen.

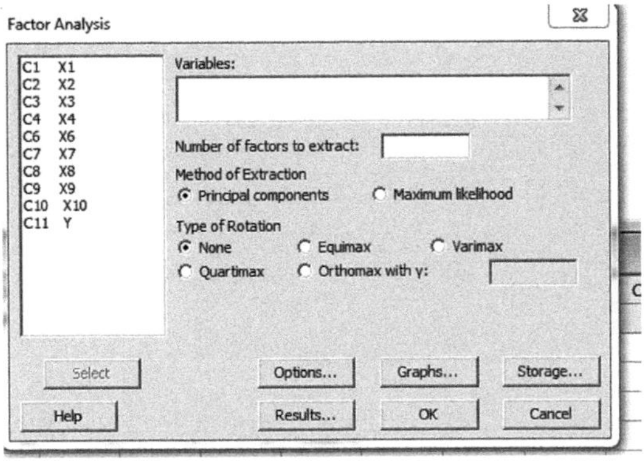

After put X1–X10 required variable section or choose required variable for analysis

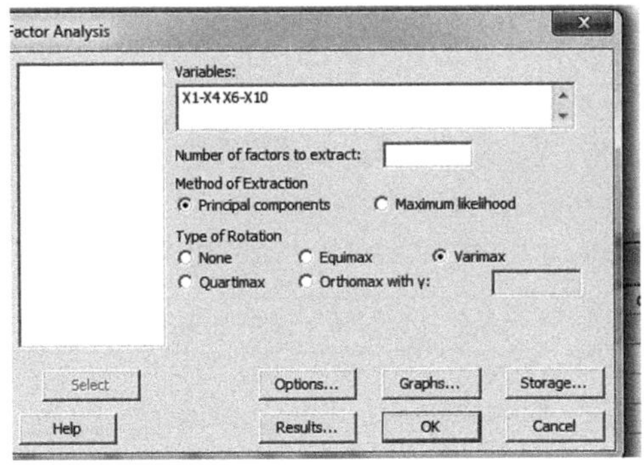

Then Click on OK button and result will appear on screen.

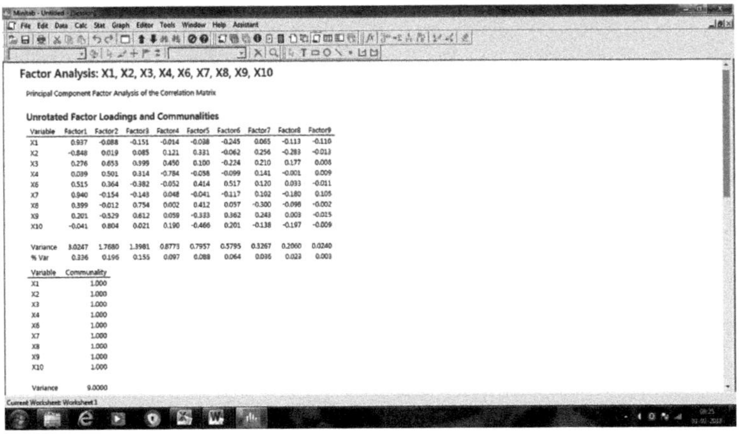

Factor Analysis: X1, X2, X3, X4, X6, X7, X8, X9, X10

Principal Component Factor Analysis of the Correlation Matrix

Unrotated Factor Loadings and Communalities

Variable	Factor1	Factor2	Factor3	Factor4	Factor5	Factor6	Factor7	Factor8	Factor9
X1	0.937	-0.088	-0.151	-0.014	-0.038	-0.245	0.065	-0.113	-0.110
X2	-0.848	0.019	0.085	0.121	0.331	-0.062	0.256	-0.283	-0.013
X3	0.276	0.653	0.399	0.450	0.100	-0.224	0.210	0.177	0.008
X4	0.039	0.501	0.314	-0.784	-0.058	-0.099	0.141	-0.001	0.009
X6	0.515	0.364	-0.382	-0.052	0.414	0.517	0.120	0.033	-0.011
X7	0.940	-0.154	-0.143	0.048	-0.041	-0.117	0.102	-0.180	0.105
X8	0.399	-0.012	0.754	0.002	0.412	0.057	-0.300	-0.098	-0.002
X9	0.201	-0.529	0.612	0.059	-0.333	0.362	0.243	0.003	-0.015
X10	-0.041	0.804	0.021	0.190	-0.466	0.201	-0.138	-0.197	-0.009
Variance	3.0247	1.7680	1.3981	0.8773	0.7957	0.5795	0.3267	0.2060	0.0240
% Var	0.336	0.196	0.155	0.097	0.088	0.064	0.036	0.023	0.003

Variable	Communality
X1	1.000
X2	1.000
X3	1.000
X4	1.000
X6	1.000
X7	1.000
X8	1.000
X9	1.000
X10	1.000
Variance	9.0000

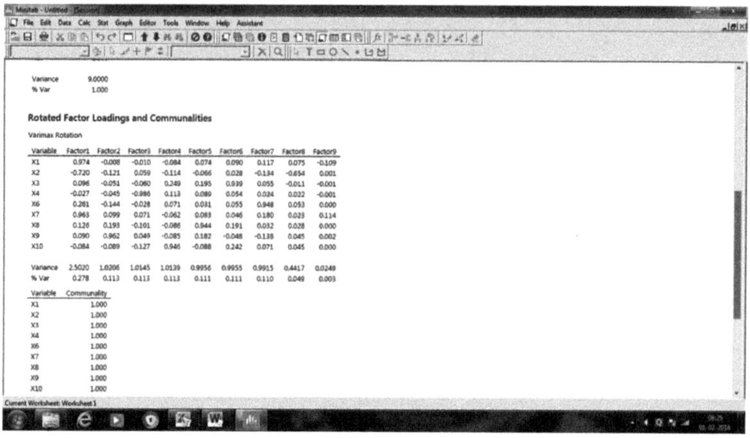

Variance 9.0000
% Var 1.000

Rotated Factor Loadings and Communalities

Varimax Rotation

Variable	Factor1	Factor2	Factor3	Factor4	Factor5	Factor6	Factor7	Factor8	Factor9
X1	0.974	-0.008	-0.010	-0.084	0.074	0.090	0.117	0.075	-0.109
X2	-0.720	-0.121	0.059	-0.114	-0.066	0.028	-0.134	-0.654	0.001
X3	0.096	-0.051	-0.060	0.249	0.195	0.939	0.055	-0.011	-0.001
X4	-0.027	-0.045	-0.986	0.113	0.089	0.054	0.024	0.022	-0.001
X6	0.261	-0.144	-0.028	0.071	0.031	0.055	0.948	0.053	0.000
X7	0.963	0.099	0.071	-0.062	0.083	0.046	0.180	0.023	0.114
X8	0.126	0.193	-0.101	-0.086	0.944	0.191	0.032	0.028	0.000
X9	0.090	0.962	0.049	-0.085	0.182	-0.048	-0.138	0.045	0.002
X10	-0.084	-0.089	-0.127	0.946	-0.088	0.242	0.071	0.045	0.000
Variance	2.5020	1.0206	1.0145	1.0139	0.9956	0.9955	0.9915	0.4417	0.0249
% Var	0.278	0.113	0.113	0.113	0.111	0.111	0.110	0.049	0.003

Variable	Communality
X1	1.000
X2	1.000
X3	1.000
X4	1.000
X6	1.000
X7	1.000
X8	1.000
X9	1.000
X10	1.000

Current Worksheet: Worksheet 1

6.4 R STEPS FOR ANALYSIS

6.4.1 ***Example 1:*** Following data give the yield (Y) corresponding to 10 yield components (X_1, X_2, X_3, X_4, X_5, X_6, X_7, X_8, X_9, X_{10}) for certain agricultural crop. Work out the ***Descriptive Statistics*** like mean, standard deviation, standard error, ranger minimum, and maximum value.

X1	X2	X3	X4	X5	X6	X7	X8	X9	X10	Y
43	5	1.6	3	1,00,000	3	24	2	9	24	14.49459
47	5	1.2	2	75,000	4	30	2	8	22	27.4552
42	5	2.4	2	1,50,000	6	27	2	8	25	39.68635
58	3	1.6	2	1,20,000	7	35	0	4	27	21.8179
45	5	2.8	2	2,00,000	8	32	2	7	26	28.17015
47	6	3.2	2	2,50,000	5	30	2	3	26	14.98783
42	5	2.4	2	1,60,000	7	22	0	4	27	17.0145
66	3	2.4	2	1,40,000	4	45	2	8	24	41.18381
52	5	1.2	2	1,10,000	4	34	0	5	24	16.10352
44	6	4	2	3,00,000	4	29	0	4	25	15.90416
52	4	0.8	4	1,10,000	8	28	2	3	22	23.26946
62	2	2	2	1,80,000	6	46	2	4	27	19.49062
64	3	12	3	3,50,000	7	41	2	4	28	31.15193
62	3	8	3	2,50,000	6	44	2	7	28	23.33775
35	6	6	5	3,00,000	5	16	2	4	29	26.64379

X1	X2	X3	X4	X5	X6	X7	X8	X9	X10	Y
35	6	5.2	2	3,70,000	4	12	2	3	25	30.11387
34	6	4.4	2	2,90,000	3	10	2	4	25	21.87816
76	5	6	2	3,00,000	6	52	2	4	24	31.03547
52	5	0.8	3	2,00,000	6	31	0	4	27	13.48353
58	4	0.4	3	1,90,000	5	37	2	4	26	10.84107
56	4	0.8	3	1,00,000	4	35	0	4	24	16.11716
42	5	4	2	2,00,000	5	27	2	6	26	31.02111
65	4	3.2	2	2,60,000	9	51	2	4	22	23.17923
27	6	1.2	2	1,80,000	2	10	0	4	25	23.54457
35	6	3.2	2	3,10,000	8	17	0	2	27	16.45729

Solution: For getting descriptive statistics

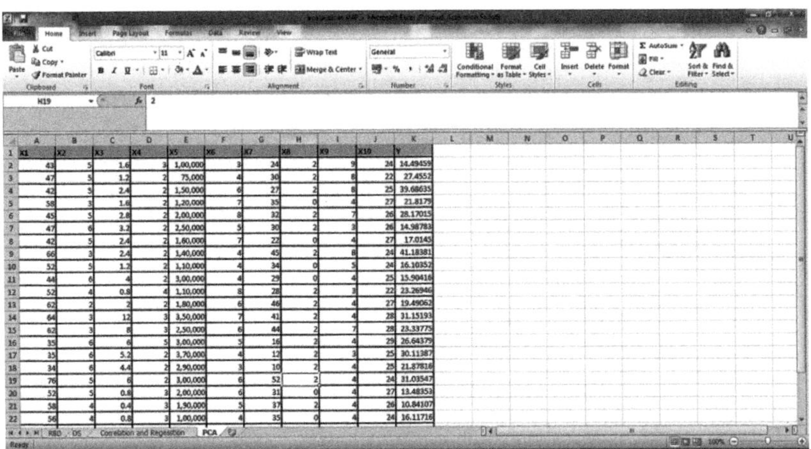

Solution:

Step 1: Showing the data entered for analysis in the excel. We need to import this into *R* for performing analysis. File should be in csv format to import it into *R*. Following code will import the file to *R* software.

data=read.csv("file:///C:/Users/Lolipop/Documents/DISCRIPITVE_DATA.csv")

Step 2: Use summary (data) to get the summary statistics like Minimum, Maximum, Mean, Median, 1st and 3rd quartile as shown below.

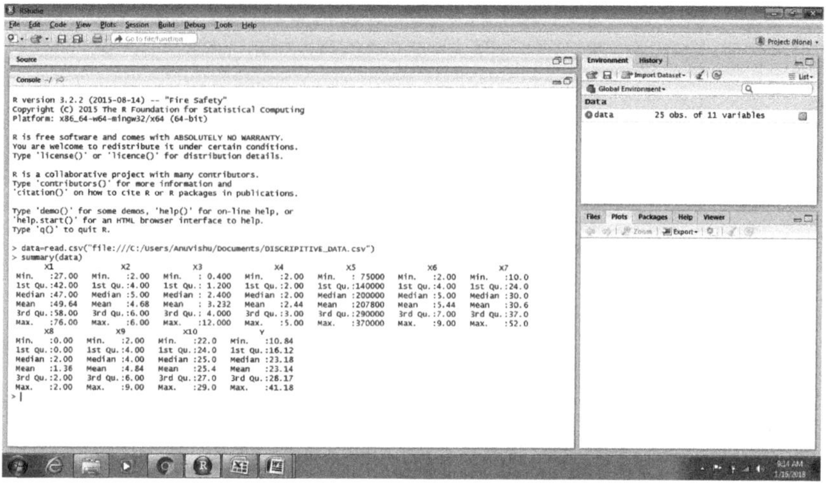

Note:

1) If you want summary statistics for particular variable in the data say x1, then use data summary(data$x1) or use attach(data) which facilitate to use the objects in the database by simply giving their names and use summary(x1).
2) We can also use individual code to get statistics like mean(x1), median(x1), min(x1), max(x1), var(x1), sd(x1), etc. as shown below.

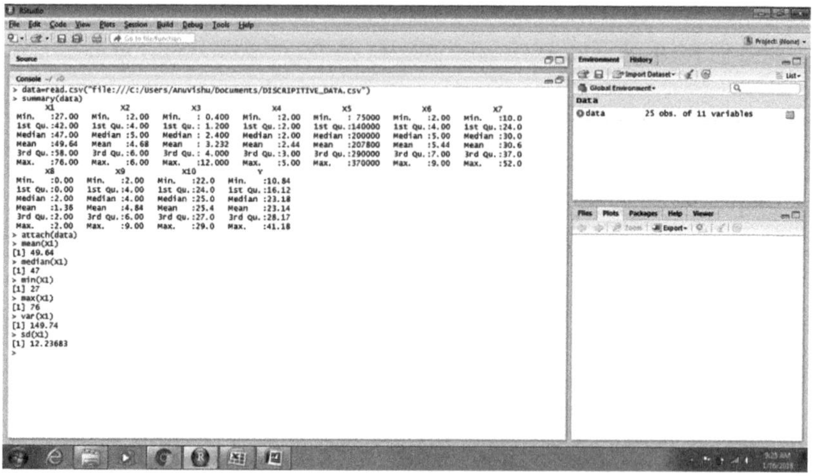

6.4.2 Example 2: *Correlation in R:* Find the correlation coefficient between area and production of rapeseed and mustard in India.

Year	Area	Production
	(In '000 Hectare)	(In '000 Tonne)
1950–1951	2071	762
1951–1952	2401	943
1952–1953	2105	858
1953–1954	2244	872
1954–1955	2439	1037
1955–1956	2556	860
1956–1957	2539	1043
1957–1958	2412	933
1958–1959	2447	1042
1959–1960	2910	1063
1960–1961	2883	1347
1961–1962	3168	1346
1962–1963	3127	1303
1963–1964	3046	915
1964–1965	2910	1474
1965–1966	2913	1298
1966–1967	3006	1228
1967–1968	3244	1568
1968–1969	2870	1347
1969–1970	3172	1564
1970–1971	3323	1976
1971–1972	3614	1433
1972–1973	3319	1808
1973–1974	3457	1704
1974–1975	3680	2252
1975–1976	3339	1936
1976–1977	3128	1550
1977–1978	3584	1650
1978–1979	3544	1860
1979–1980	3470	1428
1980–1981	4113	2304
1981–1982	4399	2382
1982–1983	3827	2207
1983–1984	3874	2608

Year	Area	Production
	(In '000 Hectare)	(In '000 Tonne)
1984–1985	3987	3073
1985–1986	3982	2680
1986–1987	3719	2605
1987–1988	4508	3370
1988–1989	4832	4877
1989–1990	4967	4125
1990–1991	5782	5229
1991–1992	6553	5863
1992–1993	6193	4803
1993–1994	6289	5328
1994–1995	6058	5758
1995–1996	6547	6000
1996–1997	6545	6658
1997–1998	7041	4703
1998–1999	6513	5664
1999–2000	6027	5790
2000–2001	4477	4190
2001–2002	5073	5083
2002–2003	4544	3880
2003–2004	5428	6291
2004–2005	7316	7593
2005–2006	7276	8131
2006–2007	6790	7438
2007–2008	5826	5834
2008–2009	6298	7201
2009–2010	5588	6608
2010–2011	6901	8179
2011–2012	5894	6604
2012–2013	6363	8029
2013–2014	6646	7877
2014–2015	5799	6282

Step 1: Showing the data entered for analysis in the excel. We need import this into *R* for performing analysis. File should be in csv format to import it into R. Following code will import the file to *R* software.

COR_EXAMPLE = read.csv("file:///C:/Users/Lolipop/Documents/correlation.csv")

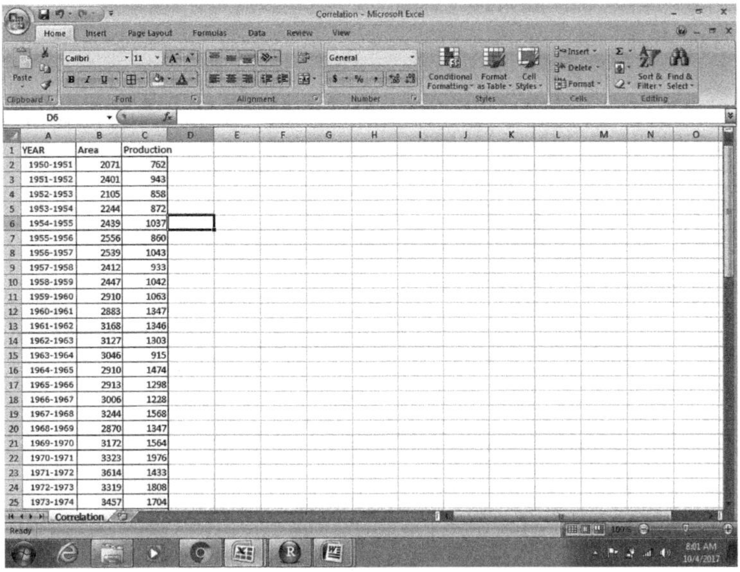

To look at the data imported, we can use type "COR_EXAMPLE" and press enter in the *R* console as shown below.

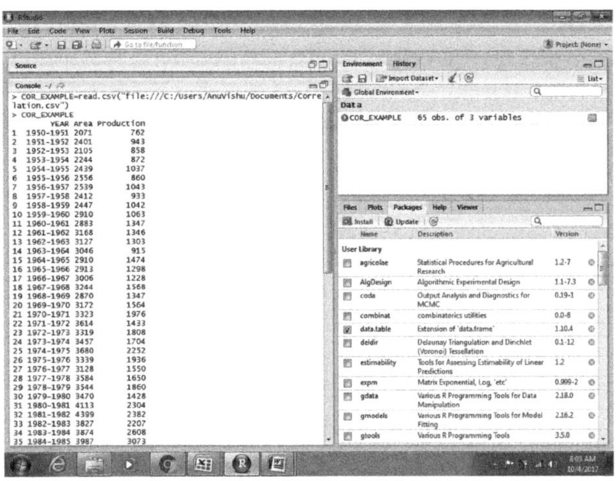

Step 2: Performing correlation analysis. Following code will use be used to perform analysis.

attach (COR_EXAMPLE): This code facilitate to use the objects in the database by simply giving their names.

cor(Area, Production): This code will give us the correlation coefficients between Area and Production.

cor. Test(Area, Production): This code is used to perform the significance test of the correlation coefficients.

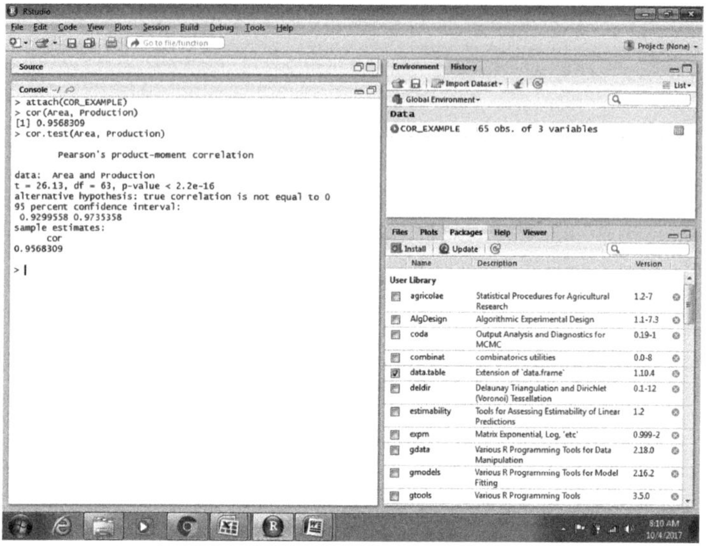

Above example shows how to deal when you need the correlation between only two variables. But in the real situation, we may need to have correlation between no. of variable. We can get the correlation between any number of variable at a time using the same code above. Following steps are to be followed to get the correlation matrix between no. of variables.

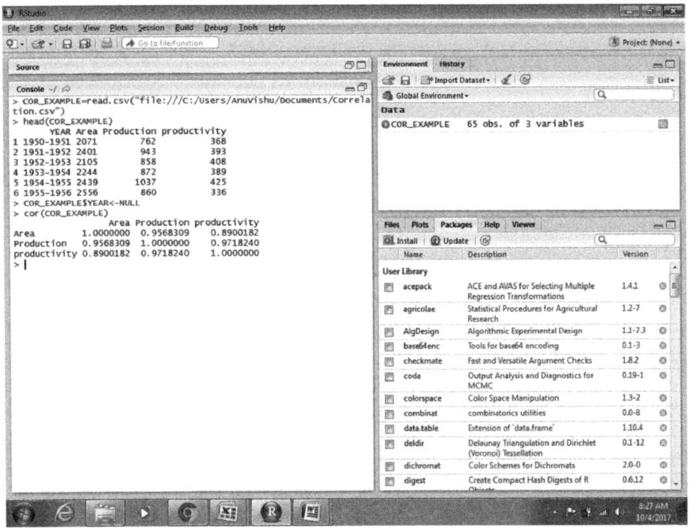

Note: COR_EXAMPLE$YEAR<-NULL deletes the variable YEAR, we need to perform this because, YEAR is a string variable, cor function in cannot take string variable/ we cannot calculate the correlation for string variable/s.

6.4.2.1. Example 2A: Find the ***Correlation Matrix*** for the with data given in the "Example 1" in the section 6.4.1.

Solution:

Step 1: Importing the data

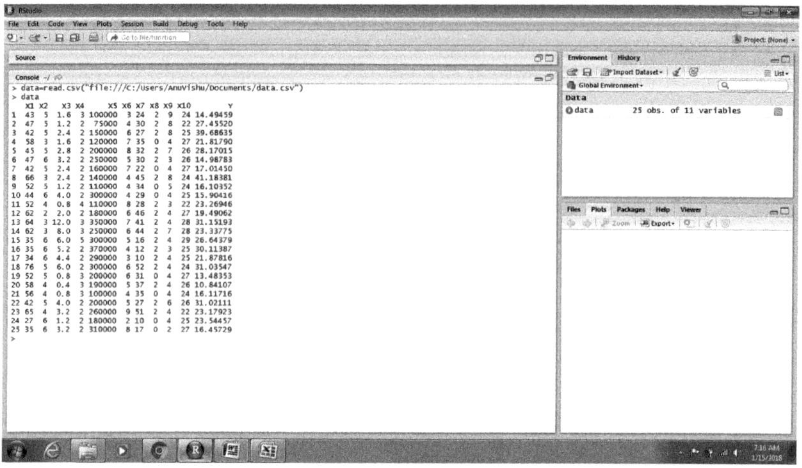

Step 2: Getting the correlation matrix:

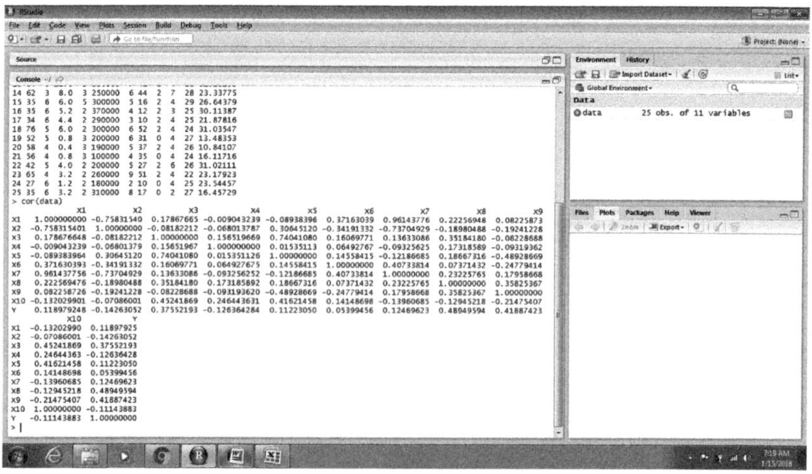

Step 3: Exporting the correlation matrix into csv

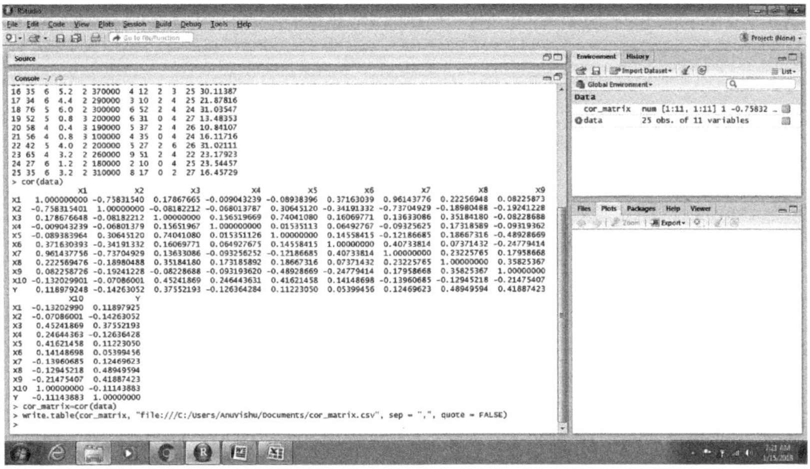

Exported matrix will be as follows:

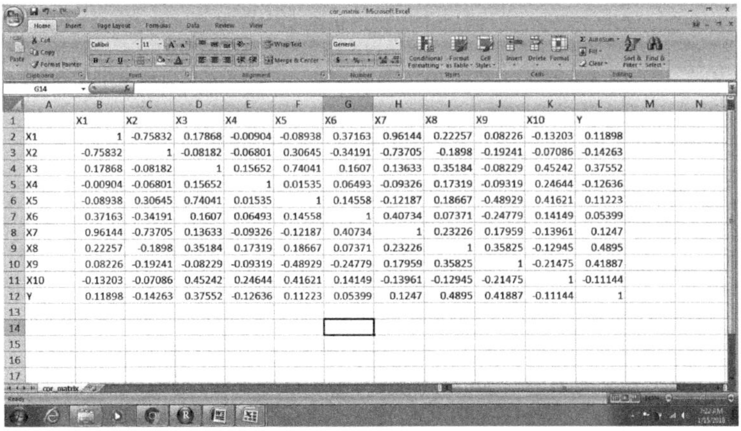

6.4.3 Example 3: Regression.

For the data given in Example 2 in the Section 6.4.2, populate the regression equation (i.e., production is a function of area).

Solution:

Step 1: Import the data as described in the correlation example.

Step 2: Use the below code to get the regression equation.

attach(COR_EXAMPLE)

model = lm(Production ~ Area): To fit the regression equation

summary(model)

Above code and regression analysis result will be as shown in figure below.

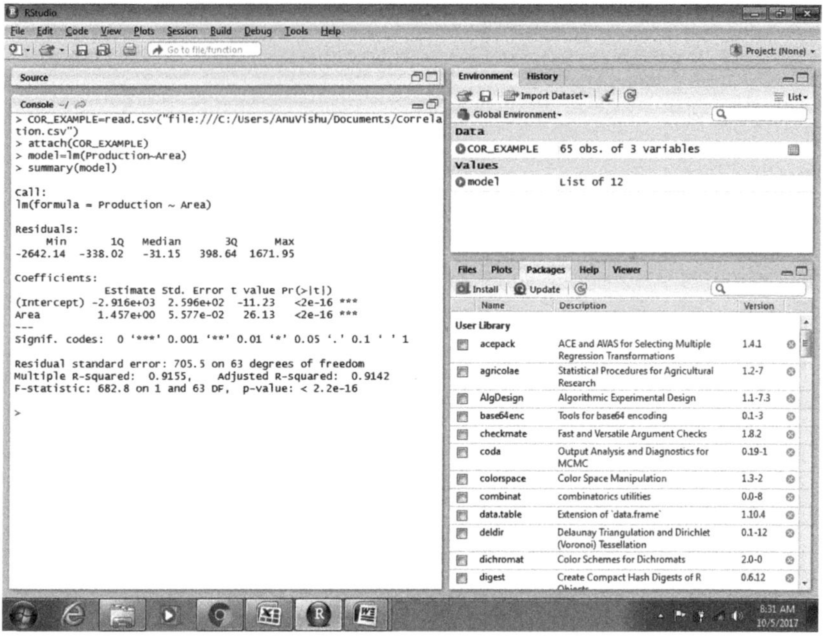

Note: We can fit the multiple regression equation using same above code, for example, if you want to fit the regression equation as Production is a function of Area and Productivity then you have to define the model as model = lm(Production ~ Area + Productivity) and then use summary(model) to get detailed result.

6.4.3.1 Example 2: For the data given in example 1 in the section 6.4.1, populate the regression equation *y* on x1, x2, ..., x10

Solution: Import the data as describe in correlation example.

Performing regression analysis:

reg = lm(Y~X1+X2+X3+X4+X5+X6+X7+X8+X9+X10, data = data): Defining the regression equation

summary(reg): to get the summary of defined equation.

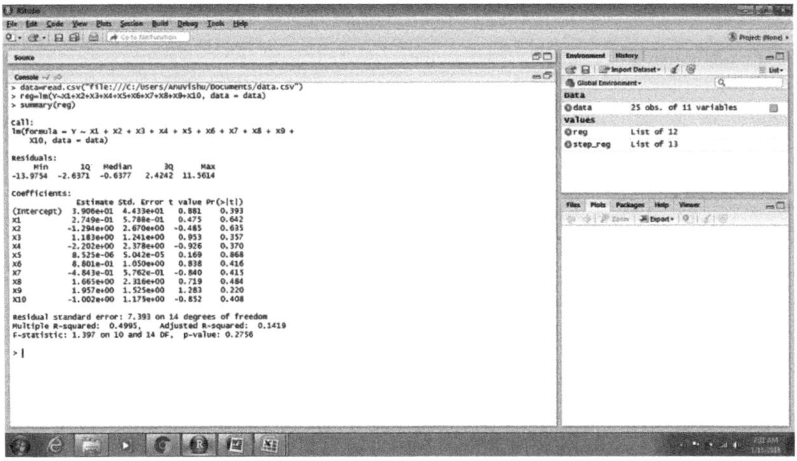

Note: This example is to show how to perform the multiple regression and hence significance of equation/variables are not considered.

6.4.4. Example 4: CRD with Unequal Replication: A rice trial under CRD design conducted with six treatments T1, T2, T3, T4, T5 and T6 with unequal replication. The yield of all treatments in Kg/plot gives below in table.

Treatment	R1	R2	R3
T1	1.12	2.49	3.43
T2	3.55	3.39	
T3	3.16	4.21	3.78
T4	2.66	1.18	
T5	3.88	2.00	1.38
T6	0.12	0.88	

Solution:

Step 1: Showing the data entered for analysis in the excel. We need import this into *R* for performing analysis. File should be in csv format to import it into R. Following code will import the file to *R* software.

CRD_EXAMPLE = read.csv("file:///C:/Users/Lolipop/Documents/CRD_EXAMPLE.csv")

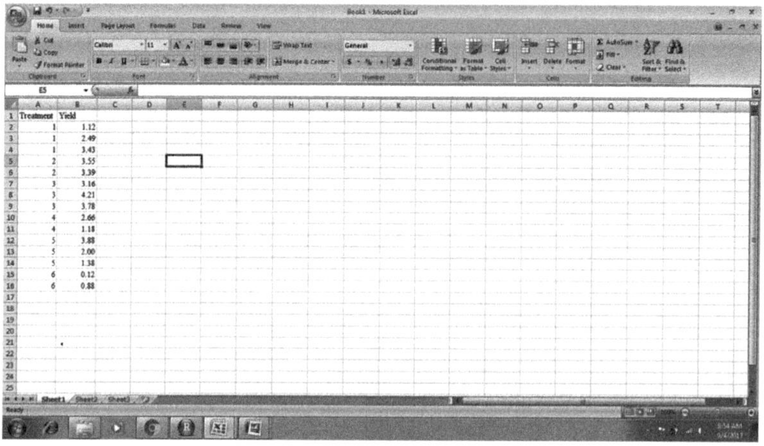

To look at the data imported, we can use type "CRD_EXAMPLE" and press enter in the *R* console as shown below.

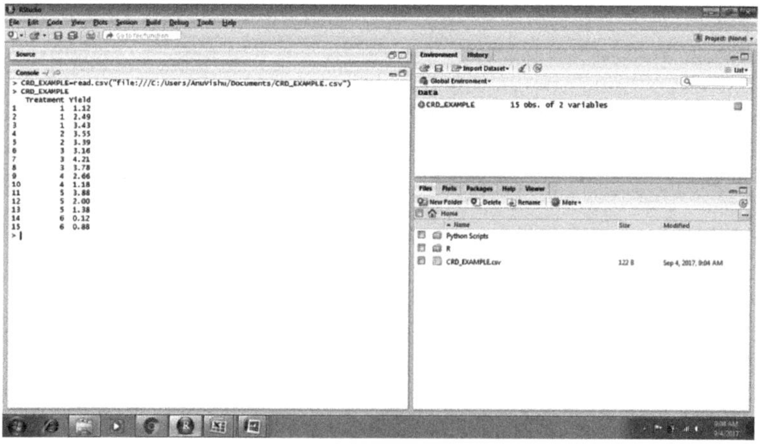

Step 2: Performing CRD analysis with unequal replication. Following code will use be used to perform analysis.

attach (CRD_EXAMPLE): This code facilitate to use the objects in the database by simply giving their names.

TREATMENT = factor(Treatment): This code convert the treatment into factor.

MODEL = lm(Yield~TREATMENT): This code is used to fit the model.

ANOVA (MODEL): This code is used to get ANOVA for the model.

Code and output in the *R* environment is as shown below.

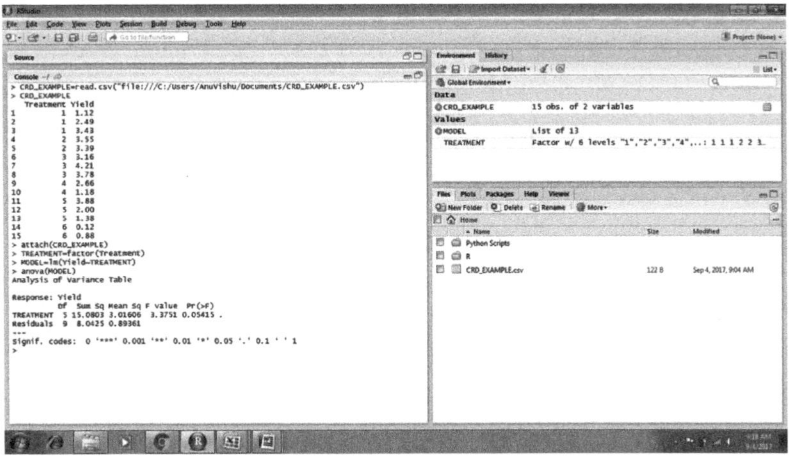

Step 3: For performing the multiple comparison test, we need to load the package "lsmeans" to perform multiple comparison test as below:

library("lsmeans")

Multiple_Comp_Test = lsmeans(MODEL, "TREATMENT")

pairs(Multiple_Comp_Test): To get pair-wise comparisons.

library(multcompView): Package to get letters for groups.

cld(Multiple_Comp_Test, Letters = "ABCDEF"): To get the letters for the groups, the treatment with same letter are onpar with each other. All the procedure are as shown below.

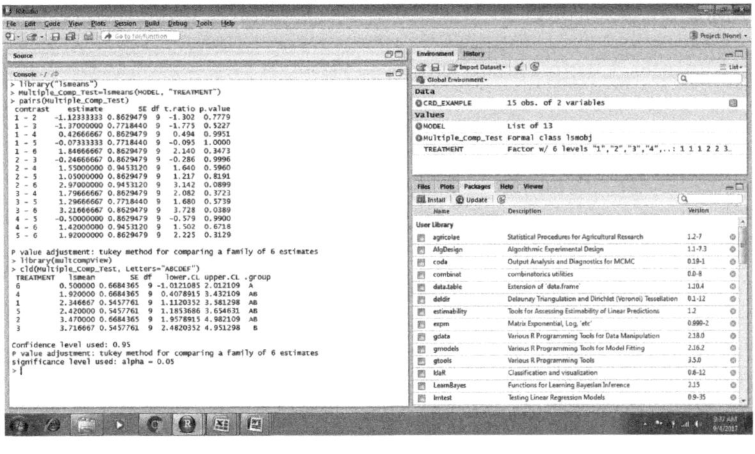

6.4.5 Example 5: *CRD with Equal Replication:* A rice trial under CRD design conducted with six treatments T1, T2, T3, T4, T5 and T6 with equal replication. The yield of all treatments in Kg/plot given below:

Treatment	R1	R2	R3
T1	1.12	2.49	3.43
T2	3.55	3.39	4.12
T3	3.16	4.21	3.78
T4	2.66	1.18	2.37
T5	3.88	2.00	1.38
T6	0.12	0.88	0.92

Solution: All the steps are same as the previous example (CRD with unequal replication). Hence we will provide the code below.

```
CRD_EXAMPLE = read.csv("file:///C:/Users/AnuVishu/Documents/
CRD_EXAMPLE.csv")
attach(CRD_EXAMPLE)
TREATMENT = factor(Treatment)
MODEL = lm(Yield~TREATMENT)
library("lsmeans")
Multiple_Comp_Test = lsmeans(MODEL, "Treatment")
pairs(Multiple_Comp_Test)
library(multcompView)
cld(Multiple_Comp_Test, Letters = "ABCDEF")
```

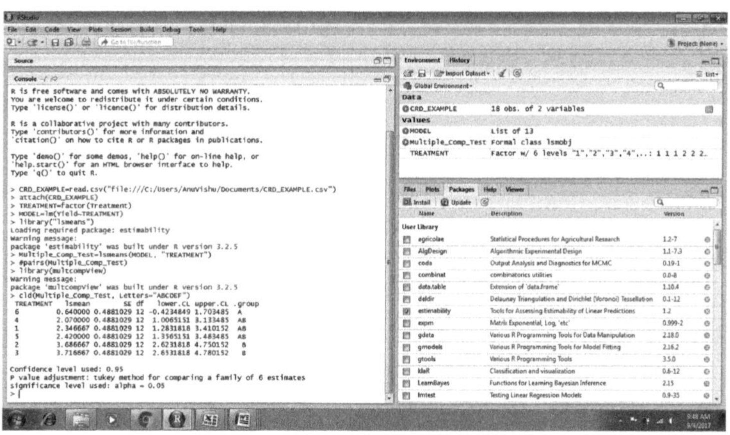

6.4.6 Example 6: RBD Design: A field trial conduct in soybean in seven different varieties and each variety is having four replications. The data are given below in table:

Treatment	R1	R2	R3	R4
T1	34.12	32.32	31.21	30.12
T2	36.44	28.00	29.32	29.55
T3	27.77	25.60	27.80	29.70
T4	30.90	30.01	34.00	32.07
T5	28.75	31.49	29.03	22.32
T6	20.39	22.99	25.78	27.40
T7	27.23	29.58	30.48	26.50

Solution:

Step 1: We have to arrange and save the data in csv format as mentioned below. Importing data into *R* is already explained in example 1.

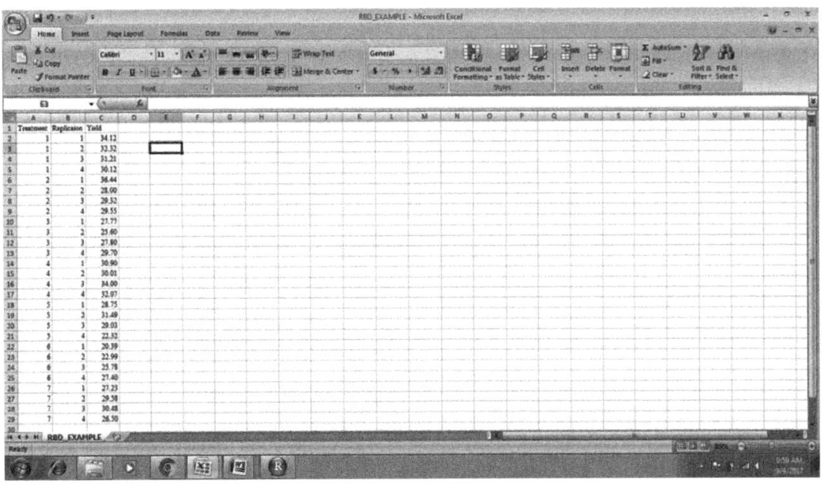

Step 2: ANOVA for RBD can be performed by using following code. Same has been shown in below image.

RBD_EXAMPLE = read.csv("file:///C:/Users/AnuVishu/Documents/RBD_EXAMPLE.csv")

attach(RBD_EXAMPLE)

TREATMENT = factor(Treatment)

REPLICATION = factor(Replication)

MODEL = lm(Yield~TREATMENT+REPLICATION)
ANOVA(MODEL)

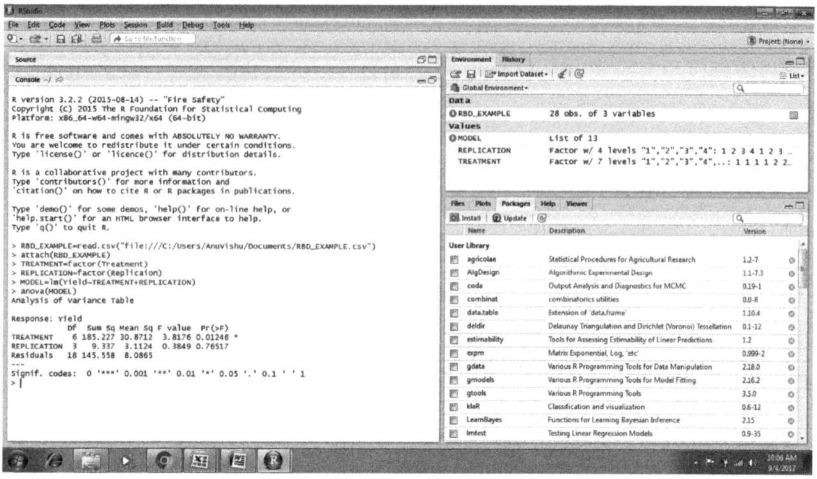

Step 3: For performing the multiple comparison test, we need to load the package "lsmeans" to perform multiple comparison test as below:

library("lsmeans")
Multiple_Comp_Test = lsmeans(MODEL, "TREATMENT")
#pairs(Multiple_Comp_Test)
library(multcompView)
cld(Multiple_Comp_Test, Letters = "ABCDEF")

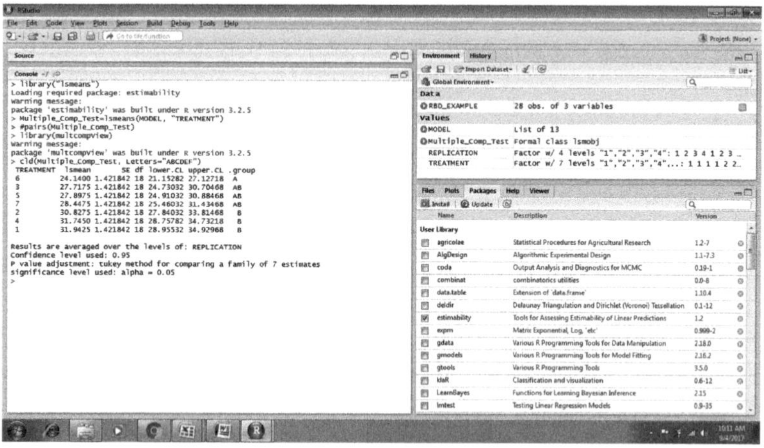

6.4.7 Example 7: LSD: A field trial conduct in rice in Rewa (M.P.) and five different level of Nitrogen applied with having five replications. To see the effect of different level of Nitrogen.

Treatment	R1	R2	R3	R4	R5
T1	20.11	14.14	18.45	17.31	17.67
T2	32.66	29.54	33.41	31.32	29.13
T3	26.43	28.11	25.25	26.43	30.22
T4	42.20	39.00	27.80	37.70	44.75
T5	32.22	29.60	38.54	29.32	34.32

Solution:

Step 1: We have to arrange and save the data in csv format as mentioned below. Importing data into *R* is already explained in example 1.

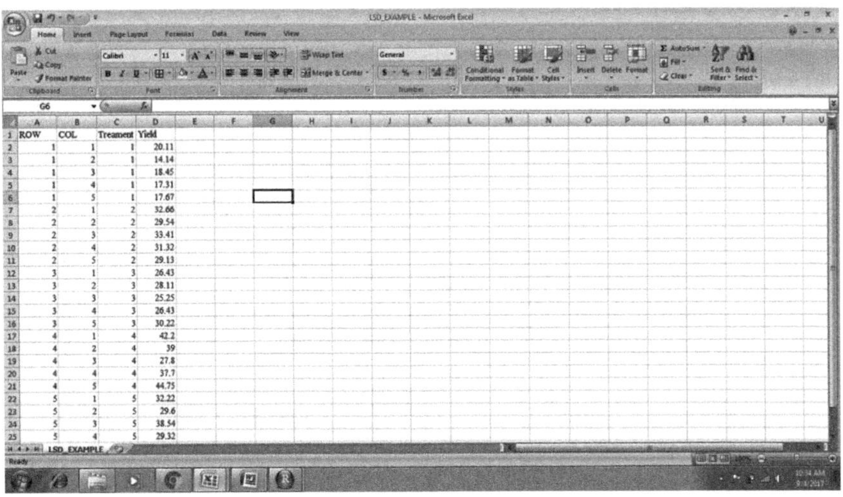

Step 2: Performing ANOVA, following are the code to perform ANOVA.

```
LSD_EXAMPLE = read.csv("file:///C:/Users/AnuVishu/Documents/LSD_EXAMPLE.csv")
attach(LSD_EXAMPLE)
ROW = factor(ROW)
COL = factor(COL)
TREATMENT = factor(Treament)
MODEL = lm(Yield~TREATMENT+ROW+COL)
ANOVA(MODEL)
```

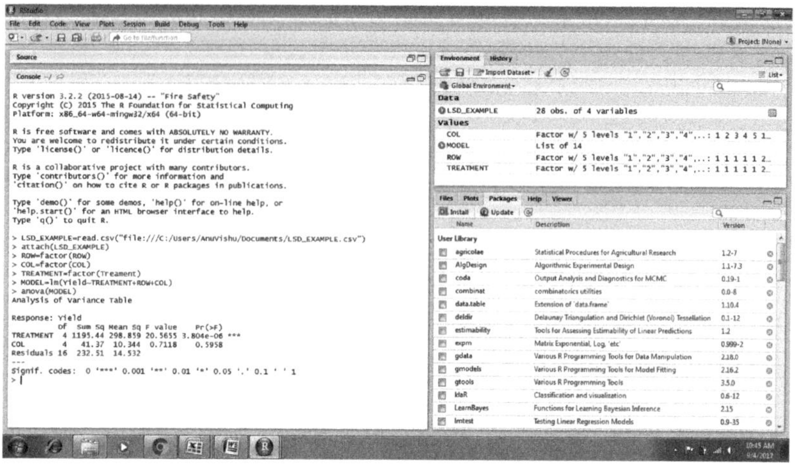

Step 3: Multiple comparison test: Following are the codes to perform the analysis.

```
library("lsmeans")
Multiple_Comp_Test = lsmeans(MODEL, "TREATMENT")
pairs(Multiple_Comp_Test)
```

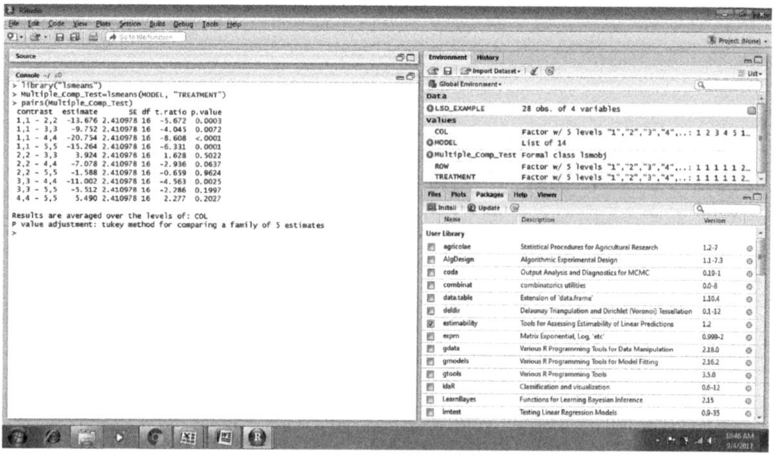

6.4.8 Example 8: Modeling and Forecasting

ARIMA: Example: Perform the stationary test and fit the ARIMA model for the area of rapeseed and mustered given the correlation analysis example.

Solution:

Step 1: We need to load the following package to perform the stationarity analysis and fitting ARIMA model using the code below

library(tseries)
library(forecast)

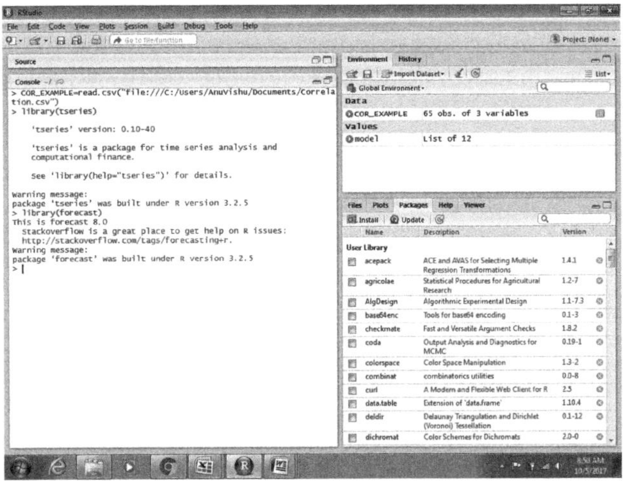

Step 2: Define the data series as time series data using the code below

Area = ts(COR_EXAMPLE$Area, start = 1950, frequency = 1)

Step 3: Performing the stationary test, use the following code to perform the analysis

adf = adf.test(Area) and then press adf to get result as shown in below image.

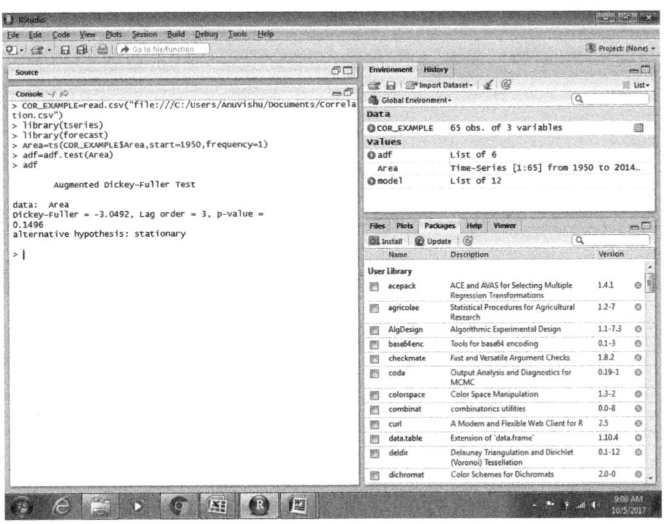

In the above image, p-value is more than 0.05 which signifies that the data is non-stationary and differencing is needed to make it stationary

Note: There is in-build function called "ndiff" in *r* to know the order of differencing to make the data stationary, if it is non-stationary.

Step 5: Fitting the ARIMA model. Following code describe how to fit and forecast the future using ARIMA in R. Please note, on very huge analysis we are showing steps to fit the best-fitted model. [It is important to note that this model is best fitted for this data set only]

Model_213 = Arima(Area, order = c(2,1,3), include.constant = TRUE)

Summary(Model_213)

Detailed code and results are as shown in below image.

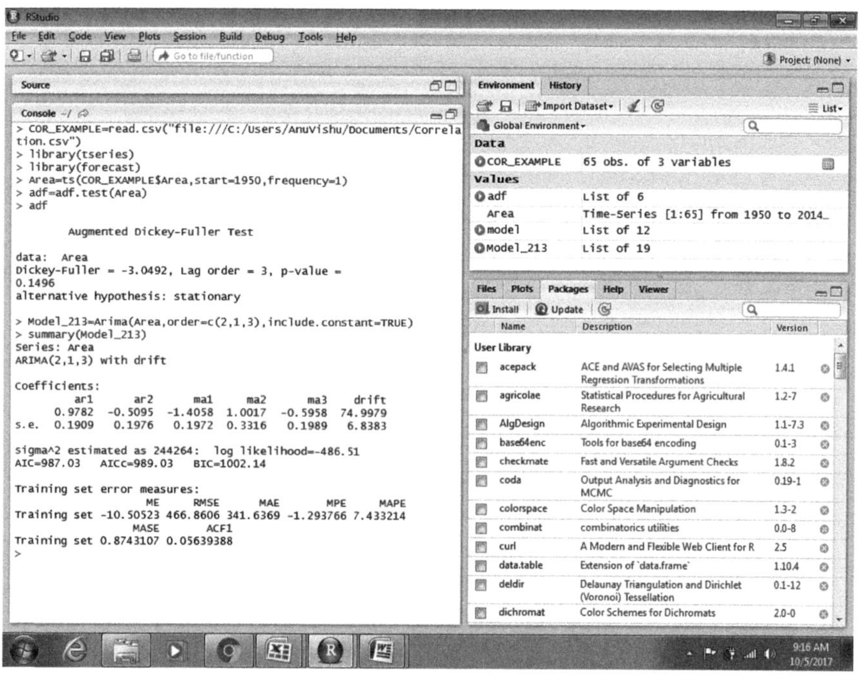

Model diagnostic: We can perform Ljung-Box test for residuals using following code.

Box. Test(resid(Model_213), type = "Ljung," lag = 20, fitdf = 1)

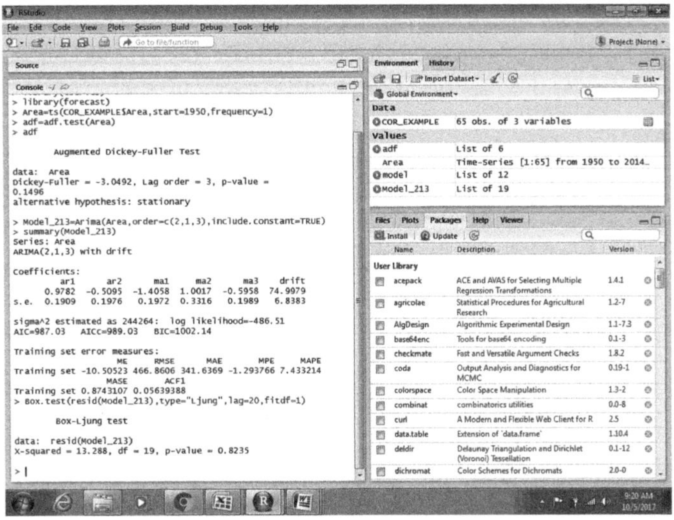

As the p-value of Ljung-Box test is more than 0.05, it signifies that there is no presence of significant auto-correlation in the residuals and hence we can use this model for forecasting. Another method to perform this test is to get ACF and PACF graph for residuals of the fitted model. If the there is no significant spikelet in the graph, we can use the model for forecasting purpose. So for the fitted model graph looks like below also code for getting above graph is give in image format.

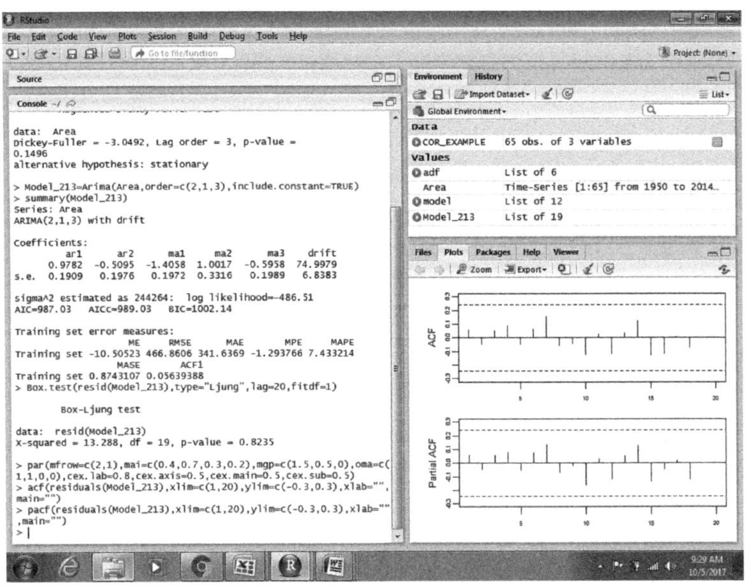

Step 5: To get the forecasted value we use the following code.

Forecast (Model_213, h = 5), h = 5 signifies forecast next 5 data point. Forecasted value and codes are given in the below image:

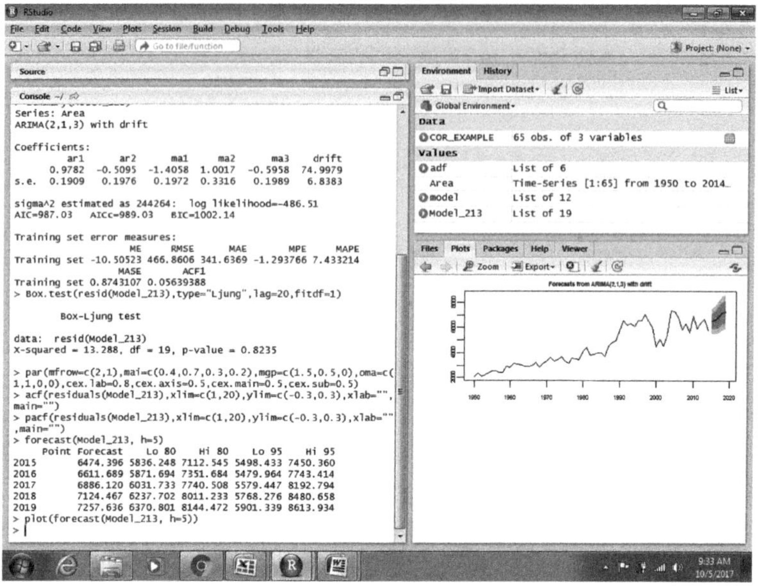

6.4.9 ***Example 9:*** Following data given (**Section** ***6.4.1 – Example 1***) the yield (Y) corresponding to 10 yield components (X_1, X_2, X_3, X_4, X_5, X_6, X_7, X_8, X_9, X_{10}) for certain agricultural crop. Work out the ***PCA Analysis***.

Solution:

Step 1: Showing the data entered for analysis in the excel. We need import this into *R* for performing analysis. File should be in csv format to import it into R. Following code will import the file to *R* software.

PCA_EXAMPLE = read.csv("file:///C:/Users/AnuVishu/Documents/PCA_EXAMPLE.csv")

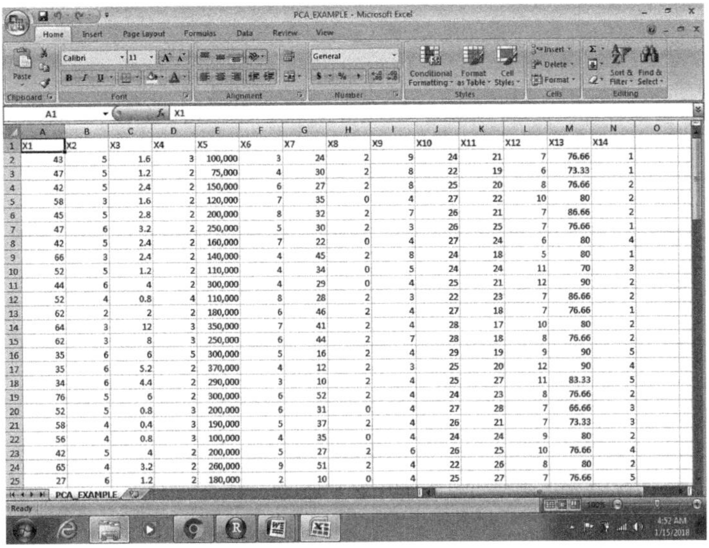

To look at the data imported, we can use type "PCA_EXAMPLE" and press enter in the *R* console as shown below.

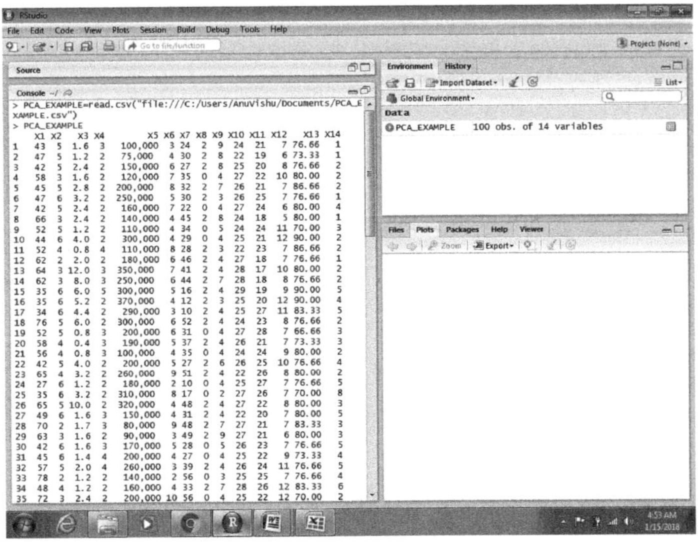

Step 2: Performing PCA analysis. Following code will use be used to perform analysis.

library(FactoMineR): To load required package, if package is not installed we need to install the package first using install.packages("FactoMineR")

PCA = PCA(PCA_EXAMPLE)

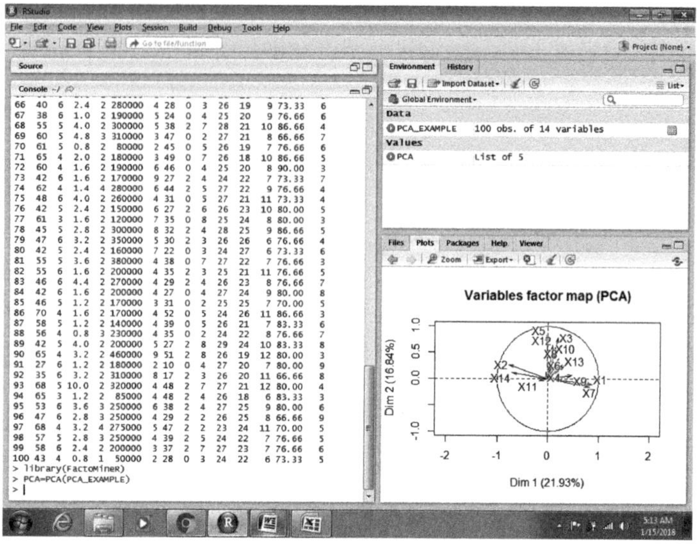

Following code can be used to get information like Eigenvalues, correlation between variables and PCs, etc.

PCA$eig: To print matrix with Eigenvalue

PCAvarcoord: To print matrix of correlation between variables and PCs.

PC_SCORES = PCAindcoord: To save PCs with name "PC_SCORES"

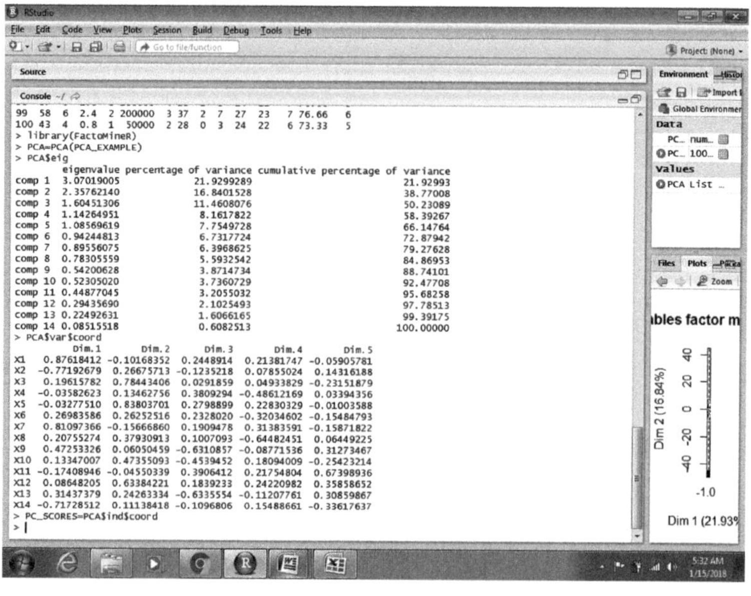

Note: Above code will produce only 5 dimensions only as 5 is the default value. If you need more dimensions, then you have to define it PCA as PCA(PCA_EXAMPLE, ncp = 10)

6.4.10 Example 10: Following data given (**Section *6.4.1 – Example 1***) the yield (Y) corresponding to 10 yield components (X_1, X_2, X_3, X_4, X_5, X_6, X_7, X_8, X_9, X_{10}) for certain agricultural crop. Work out the ***Cluster Analysis.***

Solution:

Step 1: Showing the data entered for analysis in the excel. We need import this into *R* for performing analysis. File should be in csv format to import it into R. Following code will import the file to *R* software.

The # tag is used to provide comments in *R* either individually or in conjunction with functions/objects.

#Cluster Analysis

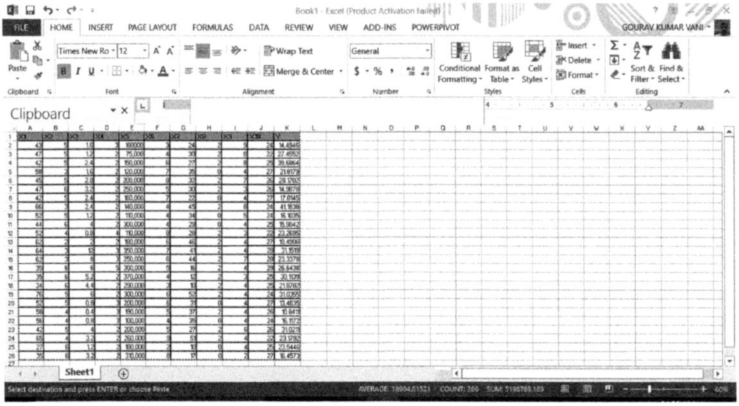

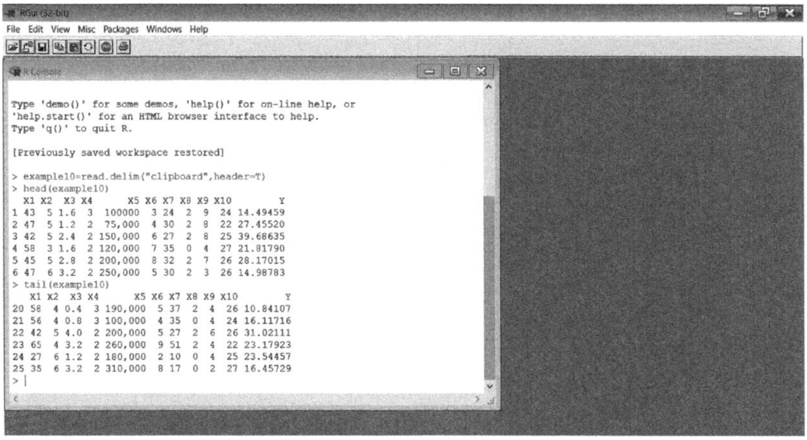

example10 = read.delim("clipboard,"header = T) #Pasting data from clipboard into R

head(example10) # view first 10 rows of data

tail(example10) # view last 10 rows of data

library(cluster) # call package cluster to service

dist = daisy (example10) #calculate distance matrix using daisy function, #for more details help(daisy)

clust = hclust(dist,method = "ward.D") # hierarchal cluster analysis with Ward Method

plot(clust,labels = example10$Names,main = "Hierarchical Clustering for Example10")

plotting dendogram

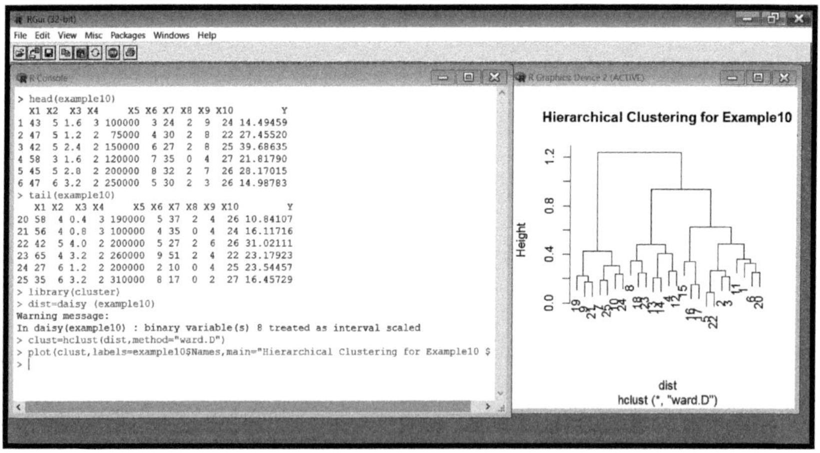

cutree(clust,k = 5) #cutting dendogram into five sections

rect.hclust(clust,k = 5,cutree(clust,k = 5)) # creating rectangular formations around five cuts

clustkmeans = kmeans(matrix,5) # performing k-means clustering on data

clustkmeans # printing output of k-means clustering

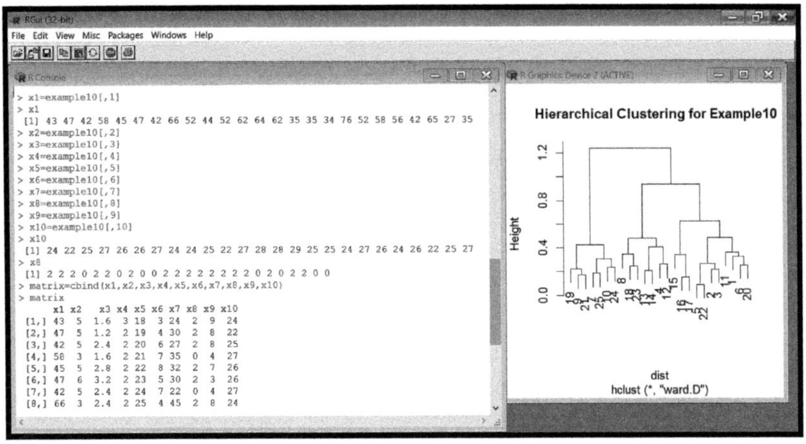

6.4.11 Example 11: Following data given (**Section *6.4.1 – Example 1***) in the yield (Y) corresponding to 10 yield components (X_1, X_2, X_3, X_4, X_5, X_6, X_7, X_8, X_9, X_{10}) for certain agricultural crop. Work out the ***Factor Analysis***.

Solution:

#Factor Analysis

matrix.1 = matrix[,1:10] # excluding *Y* variable from matrix

library(psych) # calling package "psych" from library

example10.cor = cor(matrix.1) # computing correlation matrix

example10_principal = principal(example10.cor,nfactors = 5,rotate = "varimax,"scores = T,oblique.scores = T,method = "factor")

performing factor analysis with varimax rotation and option to retain five factors

for more details on analysis use help(principal)

example10_principal # printing output of factor analysis

example10_principal$communality # Communality estimates for each itemexample10_principal$values # Eigenvalues of all components

example10_principal$loadings #A standard loading matrix of class "loadings"

example10_principal$fit #Fit of the model to the correlation matrix

example10_principal$fit.off #how well are the off-diagonal elements reproduced?

example10_principal$uniquenesses # uniqueness of items

example10_principal$weights #The beta weights to find the PCA s from the data example10_principal$complexity #Hoffman's index of complexity for each item

example10_principal$rot.mat #The rotation matrix used to produce the rotated component loadings

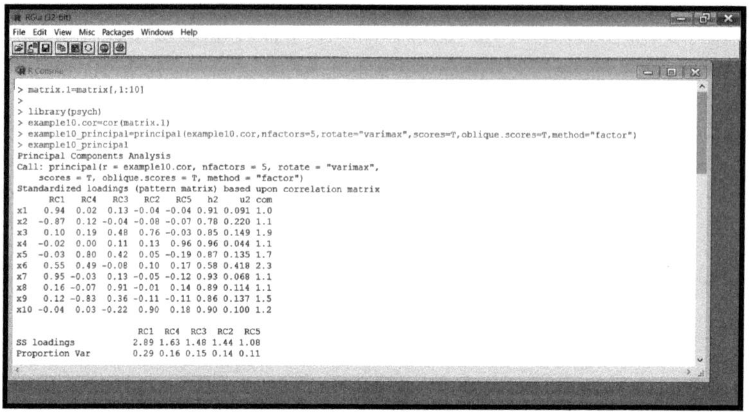

```
> matrix.1=matrix[,1:10]
>
> library(psych)
> example10.cor=cor(matrix.1)
> example10_principal=principal(example10.cor,nfactors=5,rotate="varimax",scores=T,oblique.scores=T,method="factor")
> example10_principal
Principal Components Analysis
Call: principal(r = example10.cor, nfactors = 5, rotate = "varimax",
    scores = T, oblique.scores = T, method = "factor")
Standardized loadings (pattern matrix) based upon correlation matrix
      RC1   RC4   RC3   RC2   RC5   h2    u2 com
x1   0.94  0.02  0.13 -0.04 -0.04 0.91 0.091 1.0
x2  -0.87  0.12 -0.04 -0.08 -0.07 0.78 0.220 1.1
x3   0.10  0.19  0.48  0.76 -0.03 0.85 0.149 1.9
x4  -0.02  0.00  0.11  0.13  0.96 0.96 0.044 1.1
x5  -0.03  0.80  0.42  0.05 -0.19 0.87 0.135 1.7
x6   0.55  0.49 -0.08  0.10  0.17 0.58 0.418 2.3
x7   0.95 -0.03  0.13 -0.05 -0.12 0.93 0.068 1.1
x8   0.16 -0.07  0.91 -0.01  0.14 0.89 0.114 1.1
x9   0.12 -0.83  0.36 -0.11 -0.11 0.86 0.137 1.5
x10 -0.04  0.03 -0.22  0.90  0.18 0.90 0.100 1.2

                       RC1  RC4  RC3  RC2  RC5
SS loadings           2.89 1.63 1.48 1.44 1.08
Proportion Var        0.29 0.16 0.15 0.14 0.11
```

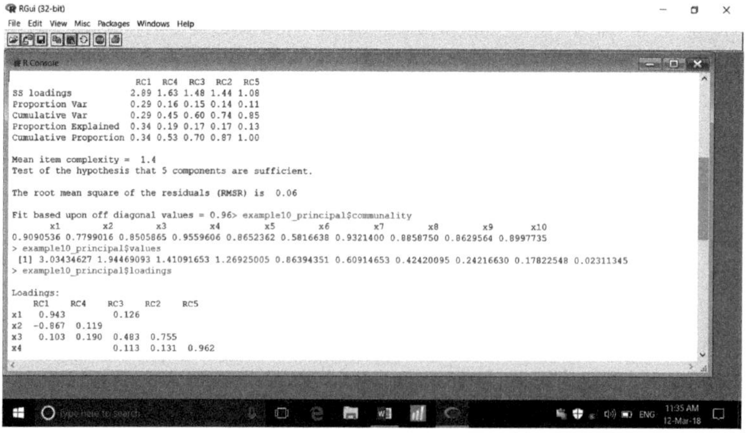

```
                       RC1  RC4  RC3  RC2  RC5
SS loadings           2.89 1.63 1.48 1.44 1.08
Proportion Var        0.29 0.16 0.15 0.14 0.11
Cumulative Var        0.29 0.45 0.60 0.74 0.85
Proportion Explained  0.34 0.19 0.17 0.17 0.13
Cumulative Proportion 0.34 0.53 0.70 0.87 1.00

Mean item complexity =  1.4
Test of the hypothesis that 5 components are sufficient.

The root mean square of the residuals (RMSR) is  0.06

Fit based upon off diagonal values = 0.96> example10_principal$communality
       x1        x2        x3        x4        x5        x6        x7        x8        x9       x10
0.9090536 0.7799016 0.8505865 0.9559606 0.8652362 0.5816638 0.9321400 0.8858750 0.8629564 0.8997735
> example10_principal$values
 [1] 3.03434627 1.94469093 1.41091653 1.26925005 0.86394351 0.60914653 0.42420095 0.24216630 0.17822548 0.02311345
> example10_principal$loadings

Loadings:
    RC1    RC4    RC3    RC2    RC5
x1   0.943         0.126
x2  -0.867  0.119
x3   0.103  0.190  0.483  0.755
x4                 0.113  0.131  0.962
```

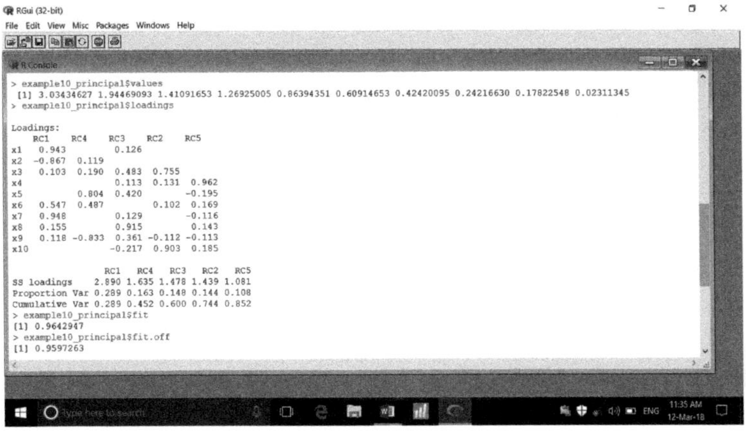

6.4.12 Example 12: Following data given (**Section *6.4.1 – Example 1***) the yield (Y) corresponding to 10 yield components (X_1, X_2, X_3, X_4, X_5, X_6, X_7, X_8, X_9, X_{10}) for certain agricultural crop. Work out the ***Canonical Correlation Analysis.***

Solution:

```
x1 = example10[, 1] #creating vector for x1
x2 = example10[, 2]
x3 = example10[, 3]
x4 = example10[, 4]
x5 = example10[, 5]
x6 = example10[, 6]
x7 = example10[, 7]
x8 = example10[, 8]
x9 = example10[, 9]
x10 = example10[, 10]
matrix = cbind(x1,x2,x3,x4,x5,x6,x7,x8,x9,x10) # column wise binding of vectors
Y = example10$Y
matrix = cbind(matrix,Y)
#Canonical correlation between Y and rest of variables
cancor(example10$Y,matrix[,1:10]) # using cancor function available in base packages [::stats]
```

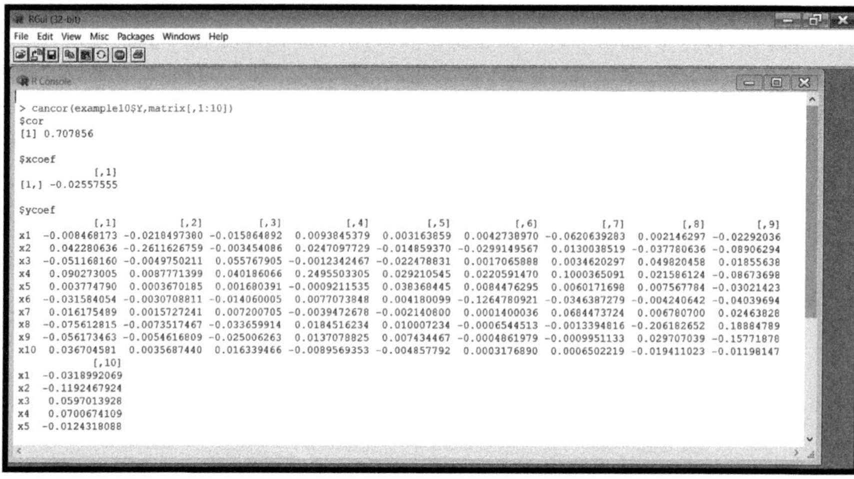

```
> cancor(example10$Y,matrix[,1:10])
$cor
[1] 0.707856

$xcoef
            [,1]
[1,] -0.02557555

$ycoef
            [,1]          [,2]         [,3]          [,4]         [,5]          [,6]          [,7]         [,8]        [,9]
x1  -0.008468173 -0.0218497380 -0.015864892  0.0093845379  0.003163859  0.0042738970 -0.0620639283  0.002146297 -0.02292036
x2   0.042280636 -0.2611626759 -0.003454086  0.0247097729 -0.014859370 -0.0299149567  0.0130038519 -0.037780636 -0.08906294
x3  -0.051168160 -0.0049750211  0.055767905 -0.0012342467 -0.022478831  0.0017065888  0.0034620297  0.049820458  0.01855638
x4   0.090273005  0.0087771399  0.040186066  0.2495503305  0.029210545  0.0220591470  0.1000365091  0.021586124 -0.08673698
x5   0.003774790  0.0003670185  0.001680391 -0.0009211535  0.038368445  0.0084476295  0.0060171698  0.007567784 -0.03021423
x6  -0.031584054 -0.0030708811 -0.014060005  0.0077073848  0.004180099 -0.1264780921 -0.0346387279 -0.004240642 -0.04039694
x7   0.016175489  0.0015727241  0.007200705 -0.0039472678 -0.002140800  0.0001400036  0.0684473724  0.006780700  0.02463828
x8  -0.075612815 -0.0073517467 -0.033659914  0.0184516234  0.010007234 -0.0006544513 -0.0013394816 -0.206182652  0.18884789
x9  -0.056173463 -0.0054616809 -0.025006263  0.0137078825  0.007434467 -0.0004861979 -0.0009951133  0.029707039 -0.15771878
x10  0.036704581  0.0035687440  0.016339466 -0.0089569353 -0.004857792  0.0003176890  0.0006502219 -0.019411023 -0.01198147
           [,10]
x1  -0.0318992069
x2  -0.1192467924
x3   0.0597013928
x4   0.0700674109
x5  -0.0124318088
```

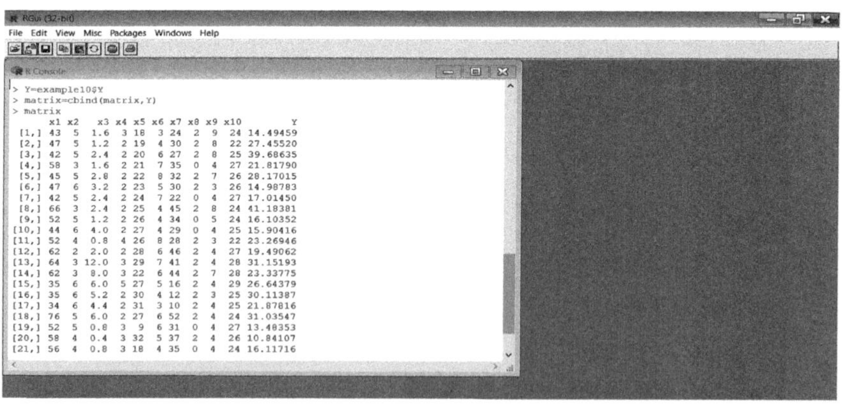

```
> Y=example10$Y
> matrix=cbind(matrix,Y)
> matrix
      x1 x2   x3 x4 x5 x6 x7 x8 x9 x10        Y
 [1,] 43  5  1.6  3 18  3 24  2  9  24 14.49459
 [2,] 47  5  1.2  2 19  4 30  2  8  22 27.45520
 [3,] 42  5  2.4  2 20  6 27  2  8  25 39.68635
 [4,] 58  3  1.6  2 21  7 35  0  4  27 21.81790
 [5,] 45  5  2.8  2 22  8 32  2  7  26 28.17015
 [6,] 47  6  3.2  2 23  5 30  2  3  26 14.98783
 [7,] 42  5  2.4  2 24  7 22  0  4  27 17.01450
 [8,] 66  3  2.4  2 25  4 45  2  8  24 41.18381
 [9,] 52  5  1.2  2 26  4 34  0  5  24 16.10352
[10,] 44  6  4.0  2 27  4 29  0  4  25 15.90416
[11,] 52  4  0.8  4 26  8 28  2  3  22 23.26946
[12,] 62  2  2.0  2 28  6 46  2  4  27 19.49062
[13,] 64  3 12.0  3 29  7 41  2  4  28 31.15193
[14,] 62  3  8.0  3 22  6 44  2  7  28 23.33775
[15,] 35  6  6.0  5 27  5 16  2  4  29 26.64379
[16,] 35  6  5.2  2 30  4 12  2  3  25 30.11387
[17,] 34  6  4.4  2 31  3 10  2  4  25 21.87816
[18,] 76  5  6.0  2 27  6 52  2  4  24 31.03547
[19,] 52  5  0.8  3  9  6 31  0  4  27 13.48353
[20,] 58  4  0.4  3 32  5 37  2  4  26 10.84107
[21,] 56  4  0.8  3 18  4 35  0  4  24 16.11716
```

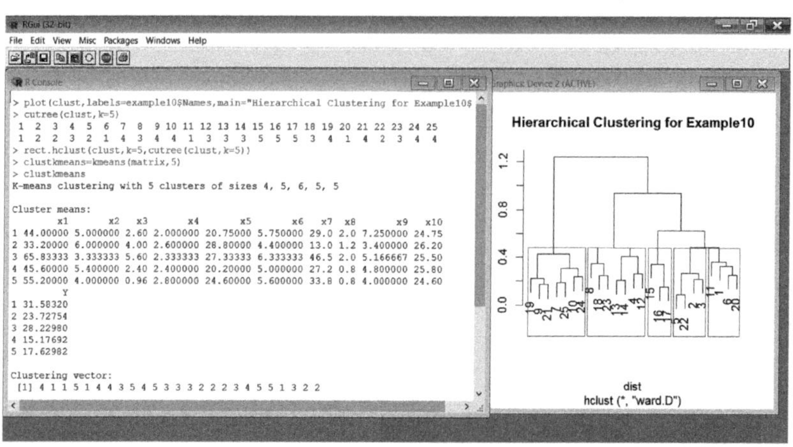

```
> plot(clust,labels=example10$Names,main="Hierarchical Clustering for Example10$
> cutree(clust,k=5)
 1  2  3  4  5  6  7  8  9 10 11 12 13 14 15 16 17 18 19 20 21 22 23 24 25
 1  2  2  3  2  1  4  3  4  4  1  3  3  3  5  5  5  3  4  1  4  2  3  4  4
> rect.hclust(clust,k=5,cutree(clust,k=5))
> clustkmeans=kmeans(matrix,5)
> clustkmeans
K-means clustering with 5 clusters of sizes 4, 5, 6, 5, 5

Cluster means:
        x1       x2   x3       x4       x5       x6   x7  x8       x9   x10
1 44.00000 5.000000 2.60 2.000000 20.75000 5.750000 29.0 2.0 7.250000 24.75
2 33.20000 6.000000 4.00 2.600000 28.80000 4.400000 13.0 1.2 3.400000 26.20
3 65.83333 3.333333 5.60 2.333333 27.33333 6.333333 46.5 2.0 5.166667 25.50
4 45.60000 5.400000 2.40 2.400000 20.20000 5.000000 27.2 0.8 4.800000 25.80
5 55.20000 4.000000 0.96 2.800000 24.60000 5.600000 33.8 0.8 4.000000 24.60
         Y
1 31.58320
2 23.72754
3 28.22980
4 15.17692
5 17.62982

Clustering vector:
 [1] 4 1 1 5 1 4 4 3 5 4 5 3 3 3 2 2 2 3 4 5 5 1 3 2 2
```

Plotting and Graph Preparation in R

The # tag is used to provide comments in *R* either individually or in conjunction with functions/objects.

Before jumping on to plotting data in R, it is necessary to import data in R. To do so, one can chose to go with following quick options:

1. Use read.delim("clipboard,"header = T/F) # to copy and paste the data from spreadsheet
2. Use read.table("FilePath,"header = T/F,sep = ,"") # to directly read data from either ASCII TXT file
3. Use read.csv("FilePath,"header = T/F,sep,"") # to directly read data from a csv file
4. Use XLConnect or xlsx library if file type is an MS-EXCEL file.

The argument header is to let the software know if there are any column names in the data being imported. T/F denotes True or False, whichever being applicable should be selected. The sep argument is to tell software what separates the data lines from each other like tab, comma, dot, semi-colon, etc.

If file is in any other format, do see the *R* Introduction Manual on importing datasets from different data formats. There are other useful arguments that can be possibly used with first three listed options for data importing such as *row.names* which can be used to specify the column number in the file which will be used in providing row names. For more details, use help(functionName) function like help(read.csv) to know more about using read.csv to import data. In case you wish to use the inbuilt data of *R* for learning purposes then use data(datasetname) to get that dataset activated in R.

We have used here inbuilt data available in *R* package "datasets" by name "faithful." For more details on this dataset use following command:

help(faithful)

data(faithful) # to use this dataset we need to call it by this command to service

duration = faithful$eruptions # to extract data on duration of eruptions from faithful dataset into a vector

Once you import data into R, then first step is to see first few lines from top and bottom of the dataset. This can be done with head() and tail() function such as:

head(x,n) # *x* is a numeric vector/data.frame and *n* is the number of first "*n*" lines that you wish to see.

tail(x,n) # here *n* is the last "*n*" data lines that you wish to see.

Upon completing summary view of data, it is useful to get a summary of the data with summary () function as follows:

summary(x) # this provides details on Min. value, 1st Quartile, Median, Mean, 3rd Quartile, Max. value

The statistics obtained using summary() function can be obtained using following commands individually:

mean(x) #mean of dataset x

median(x) # median for dataset x

max(x) # maximum value in dataset x

min(x) # minimum value in dataset x

quantile(x, number) # *x* dataset and number; for example, 0.25 & 0.75 for 1st & 3rd quartile

In addition remaining summary statistics like standard deviation, skewness, and kurtosis can be obtained as follows:

sd(x) # standard deviation for *x* dataset

install.l.packages("moments")

library(moments) # calling moments package to work from library of packages in R

skewness(x) #Skewness for dataset *x* in R, available in moments package

kurtosis(x) # Kurtosis for dataset *x* in R, available in moments package

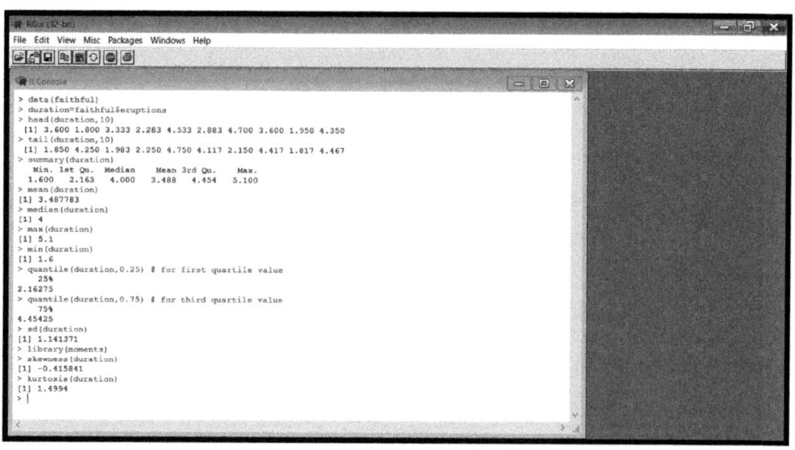

Once summary is obtained then to enhance our understanding of summary statistics, we can plot data to create graphs/charts as follows:

#**Ogive Curve** provides less than and more than ogive curve, the intersection of which is at median

breaks = seq(from = 1.5,to = 5.5,by = 0.5) # seq function creates sequence from 1.5 to 5.5 by a distance of 0.5

duration.cut = cut(duration,breaks,right = F) # to divide vector "duration" into intervals

duration.freq = table(duration.cut) # table function creates a count for each interval in 'duration.cut'

cumfreq0 = c(0,cumsum(duration.freq)) # cumsum is a function to calculate cumulative sum

cumfreq1 = length(duration)-cumfreq0 # length function provides the length of vector duration

plot(breaks,cumfreq0,main = "Old Faithful Eruptions,"xlab = "Duration Minutes,"

ylab = "Cumulative Eruptions")

for more details on plot function, use help(plot)

lines(breaks,cumfreq0) # this will add less than ogive curve

lines(breaks,cumfreq1) # this will add more than ogive curve

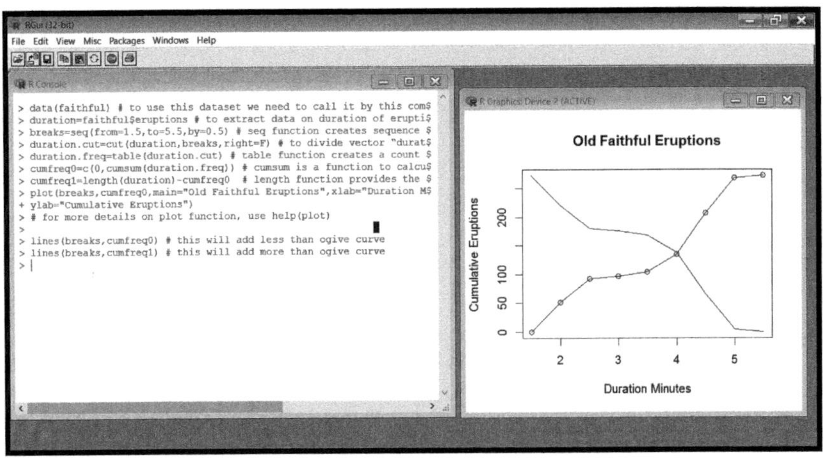

#Histogram

h = hist(duration,breaks = breaks,main = "Histogram,"xlab = "Class,"ylab = "frequency,"col = "violet,"prob = T)

hist function creates histogram, for more details on this function use help(hist)

lines(density(duration),bw = 0.1) # this function will add density lines to histogram

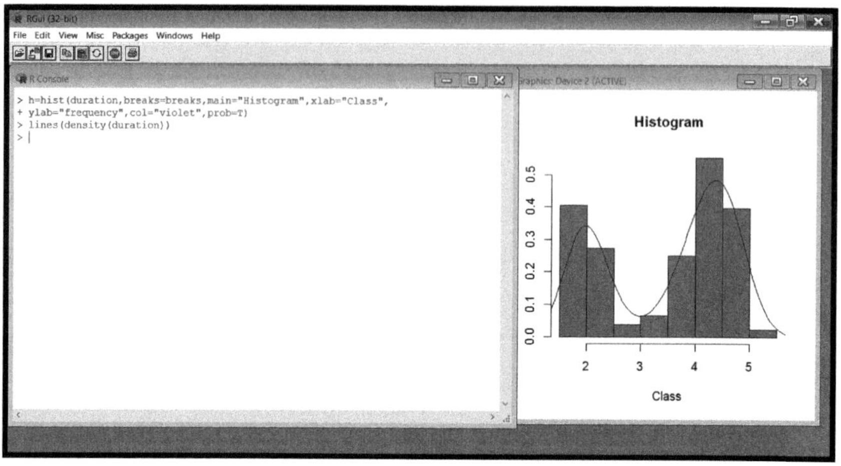

#Frequency Polygone

mp = c(min(h$mids)-(h$mids[2]-h$mids[1]),h$mids,max(h$mids)+(h$ mids[2]-h$mids[1]))

mp

freq = c(0,h$counts,0)

freq

lines(mp,freq,type = “b,”pch = 20,col = “red,”lwd = 3)

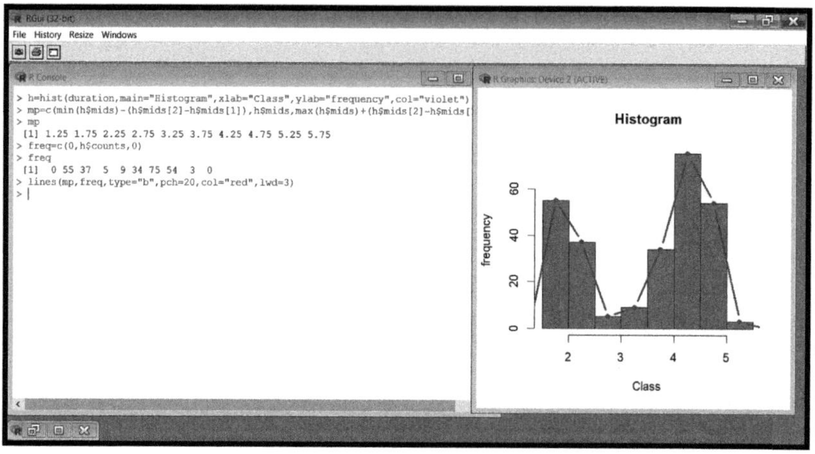

#bar chart

barplot(duration,width = 10)

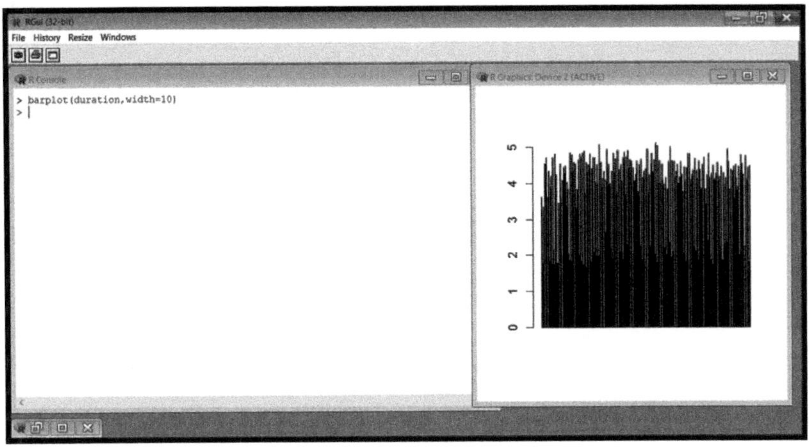

#Polygone

tt = table(duration)

plot(tt,ylab = "frequency") #creating polygone)

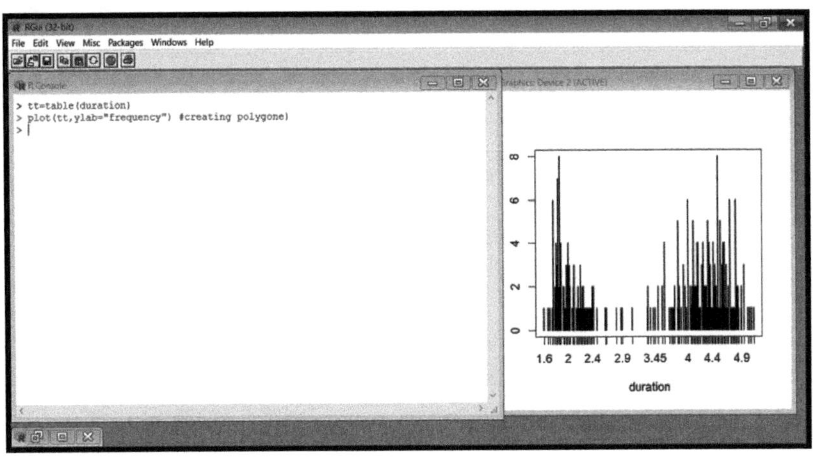

#Plotting Empicircal Cumulative Density Function

plot(ecdf(duration), do.points = FALSE, verticals = TRUE)

x = seq(2,5,0.5)

lines(x, pnorm(x, mean = mean(duration), sd = sd(duration)),lty = 3)#fit a normal distribution and overlay the fitted CDF.

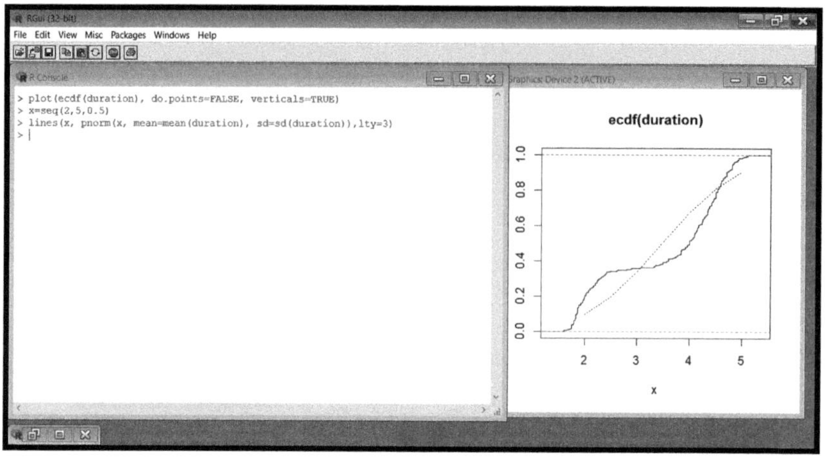

#Q-Q plot for normality

qqnorm(duration)

qqline(duration)

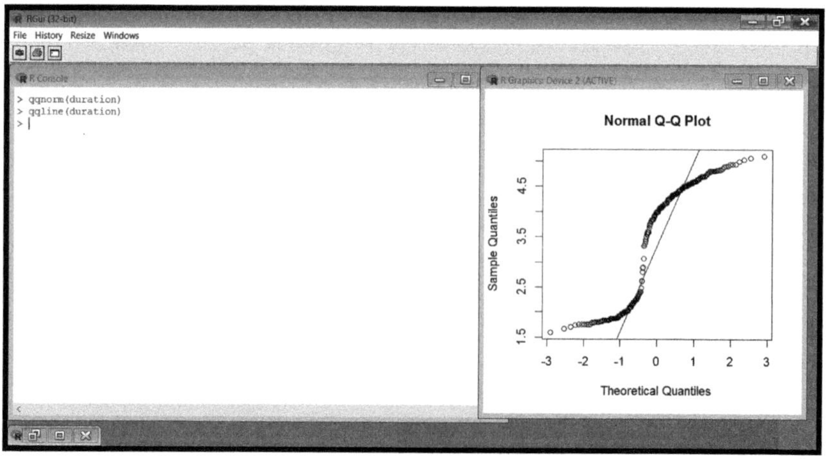

#boxplot

boxplot(duration)

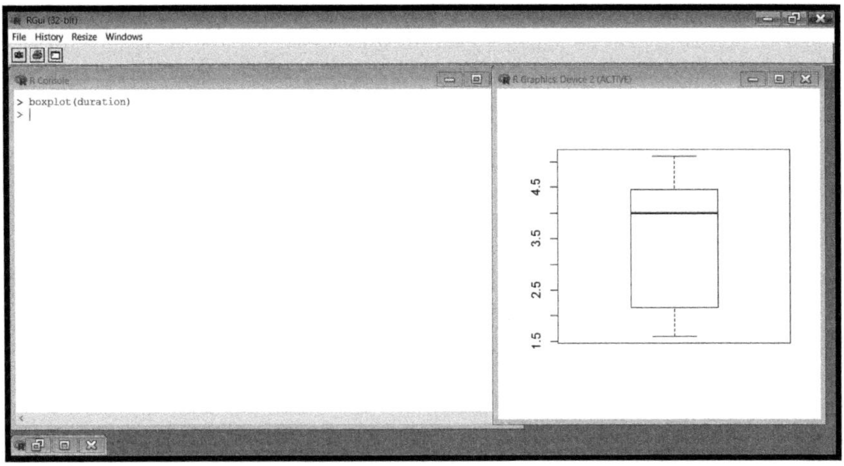

#PP Plot

library(circular)
pp.plot(duration,main = "PP Plot")

KEYWORDS

- **MS Excel**
- **principal component**
- **tick (√) on treatments**

REFERENCES

Applied Spatial Data Analysis with R: http://www.amazon.com/Applied-Spatial-Data-Analysis-Use/dp/0387781706 (accessed on 13 January 2019).

Deming, W. E., (1982). *Out of the Crisis*. Cambridge, MA: M.I.T. Center for Advanced Engineering Study.

Deming, W. E., & Edwards, D. W. (1982). *Quality, Productivity, and Competitive Position* (Vol. 183). Cambridge, MA: Massachusetts Institute of Technology, Center for Advanced Engineering Study.

Donoho, D., & Ramos, E., (1982). *PRIMDATA: Data Sets for Use With PRIM-H (DRAFT)*. FTP stat Library at Carnegie Mellon University.

Guerrero, H. (2019). Modeling and Simulation: Part 2. *In Excel Data Analysis* (pp. 265–310). Springer, Cham.

Sahu, P. K. (2013). Research methodology: A guide for researchers in agricultural science, social science and other related fields. New Delhi: Springer.

CHAPTER 7

Overview of JMP and SAS

PRADEEP MISHRA[1], D. RAMESH[2], R. B. SINGH[3], SUPRIYA[4], and A. K. TAILOR[5]

[1]*Assistant Professor (Statistics), College of Agriculture, JNKVV, Powarkheda, (M.P.), 461110, India, E-mail: pradeepjnkvv@gmail.com*

[2]*Department of Statistics and Computer Applications Acharya N. G. Ranga Agricultural University, Agricultural College, Bapatla, Andhra Pradesh–522101, India*

[3]*Professor, Department of Mathematics and Statistic College of Agriculture, JNKVV, Jabalpur (M.P.), 482004, India*

[4]*Assistant Professor, College of Agriculture Campus, Kotwa, Azamgarh, NDUAT, Faizabad, India*

[5]*Technical Officer (Statistics) National Horticultural Research and Development Foundation, Regional Research Station, Nashik–422 003, (M.H.), India*

7.1 JMP STEPS FOR ANALYSIS

7.1.1 ***Example 1:*** Following data given the yield (Y) corresponding to 10 yield components (X_1, X_2, X_3, X_4, X_5, X_6, X_7, X_8, X_9, X_{10}) for certain agricultural crop. Work out the ***Descriptive Statistics*** like mean, standard deviation, standard error, ranger minimum, and maximum value.

X1	X2	X3	X4	X5	X6	X7	X8	X9	X10	Y
43	5	1.6	3	1,00,000	3	24	2	9	24	14.49459
47	5	1.2	2	75,000	4	30	2	8	22	27.4552
42	5	2.4	2	1,50,000	6	27	2	8	25	39.68635
58	3	1.6	2	1,20,000	7	35	0	4	27	21.8179
45	5	2.8	2	2,00,000	8	32	2	7	26	28.17015
47	6	3.2	2	2,50,000	5	30	2	3	26	14.98783

X1	X2	X3	X4	X5	X6	X7	X8	X9	X10	Y
42	5	2.4	2	1,60,000	7	22	0	4	27	17.0145
66	3	2.4	2	1,40,000	4	45	2	8	24	41.18381
52	5	1.2	2	1,10,000	4	34	0	5	24	16.10352
44	6	4	2	3,00,000	4	29	0	4	25	15.90416
52	4	0.8	4	1,10,000	8	28	2	3	22	23.26946
62	2	2	2	1,80,000	6	46	2	4	27	19.49062
64	3	12	3	3,50,000	7	41	2	4	28	31.15193
62	3	8	3	2,50,000	6	44	2	7	28	23.33775
35	6	6	5	3,00,000	5	16	2	4	29	26.64379
35	6	5.2	2	3,70,000	4	12	2	3	25	30.11387
34	6	4.4	2	2,90,000	3	10	2	4	25	21.87816
76	5	6	2	3,00,000	6	52	2	4	24	31.03547
52	5	0.8	3	2,00,000	6	31	0	4	27	13.48353
58	4	0.4	3	1,90,000	5	37	2	4	26	10.84107
56	4	0.8	3	1,00,000	4	35	0	4	24	16.11716
42	5	4	2	2,00,000	5	27	2	6	26	31.02111
65	4	3.2	2	2,60,000	9	51	2	4	22	23.17923
27	6	1.2	2	1,80,000	2	10	0	4	25	23.54457
35	6	3.2	2	3,10,000	8	17	0	2	27	16.45729

Solution: For getting descriptive statistics

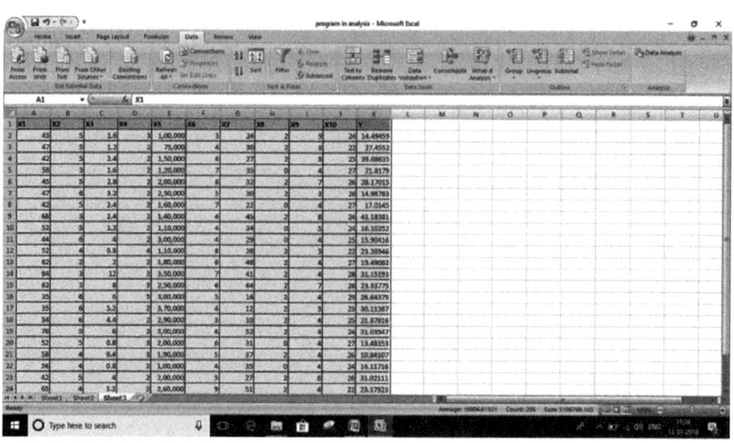

First, transfer the data into JMP data sheet.

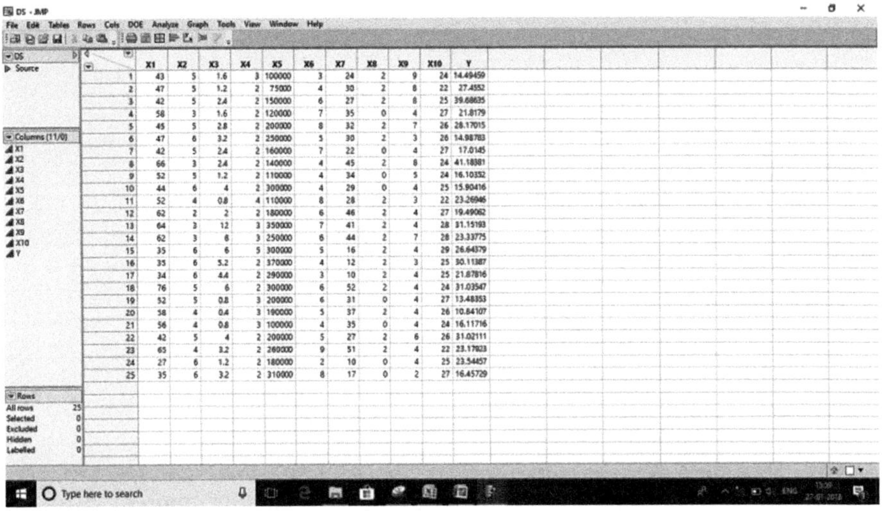

For getting summary statistics, click on Tables → Tabulate.

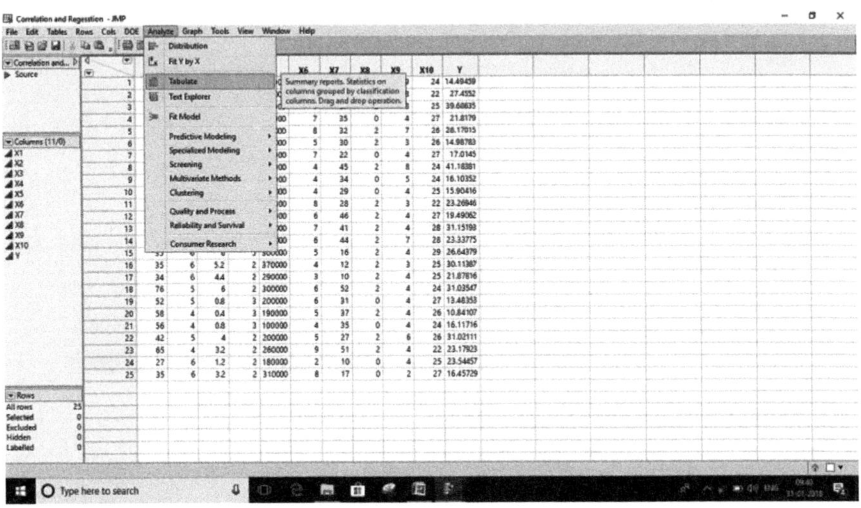

After clicking on tabulate, following dialog box appear on the screen.

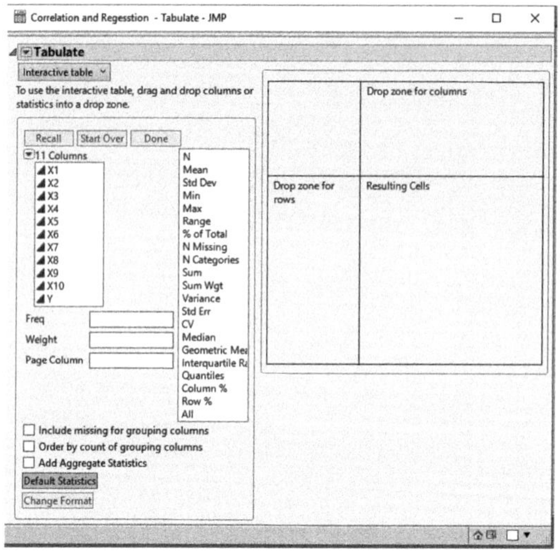

Just drag the variable and required descriptive statistics option.

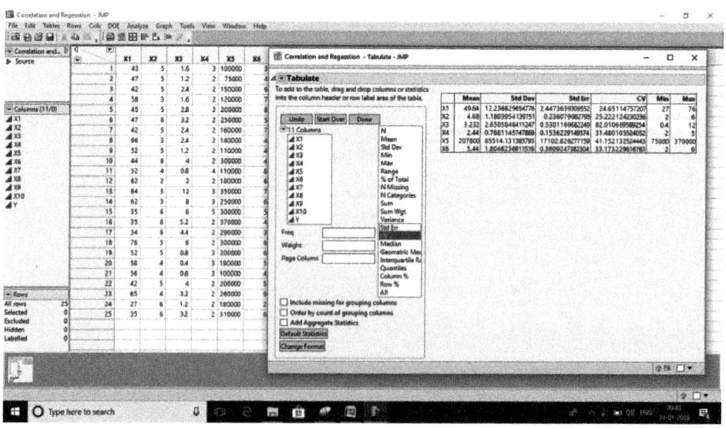

Like that we can get for all variables.

	Mean	Std Dev	Std Err	CV	Min	Max
X1	49.64	12.236829654776	2.4473659309552	24.65114757207	27	76
X2	4.68	1.1803954139751	0.236079082795	25.222124230236	2	6
X3	3.232	2.6505848411247	0.5301169682249	82.010669589254	0.4	12
X4	2.44	0.7681145747869	0.1536229149574	31.480105524052	2	5
X5	207800	85514.131385793	17102.826277159	41.152132524443	75000	370000
X6	5.44	1.8046236911519	0.3609247382304	33.173229616763	2	9
X7	30.6	11.982626312012	2.3965252624025	39.158909516381	10	52
X8	1.36	0.952190457139	0.1904380914278	70.014004201401	0	2
X9	4.84	1.8859126879754	0.3771825375951	38.96513818131	2	9
X10	25.4	1.8929694486001	0.37859388972	7.4526356244098	22	29
Y	23.1353208	7.9812221362019	1.5962444272404	34.497996397793	10.84107	41.18381

7.1.2 Example 2: Following data given the yield (Y) corresponding to 10 yield components (X_1, X_2, X_3, X_4, X_5, X_6, X_7, X_8, X_9, X_{10}) for certain agricultural crop. Work out the ***Correlation Analysis***.

X1	X2	X3	X4	X5	X6	X7	X8	X9	X10	Y
43	5	1.6	3	1,00,000	3	24	2	9	24	14.49459
47	5	1.2	2	75,000	4	30	2	8	22	27.4552
42	5	2.4	2	1,50,000	6	27	2	8	25	39.68635
58	3	1.6	2	1,20,000	7	35	0	4	27	21.8179
45	5	2.8	2	2,00,000	8	32	2	7	26	28.17015
47	6	3.2	2	2,50,000	5	30	2	3	26	14.98783
42	5	2.4	2	1,60,000	7	22	0	4	27	17.0145
66	3	2.4	2	1,40,000	4	45	2	8	24	41.18381
52	5	1.2	2	1,10,000	4	34	0	5	24	16.10352
44	6	4	2	3,00,000	4	29	0	4	25	15.90416
52	4	0.8	4	1,10,000	8	28	2	3	22	23.26946
62	2	2	2	1,80,000	6	46	2	4	27	19.49062
64	3	12	3	3,50,000	7	41	2	4	28	31.15193
62	3	8	3	2,50,000	6	44	2	7	28	23.33775
35	6	6	5	3,00,000	5	16	2	4	29	26.64379
35	6	5.2	2	3,70,000	4	12	2	3	25	30.11387
34	6	4.4	2	2,90,000	3	10	2	4	25	21.87816
76	5	6	2	3,00,000	6	52	2	4	24	31.03547
52	5	0.8	3	2,00,000	6	31	0	4	27	13.48353
58	4	0.4	3	1,90,000	5	37	2	4	26	10.84107
56	4	0.8	3	1,00,000	4	35	0	4	24	16.11716
42	5	4	2	2,00,000	5	27	2	6	26	31.02111
65	4	3.2	2	2,60,000	9	51	2	4	22	23.17923
27	6	1.2	2	1,80,000	2	10	0	4	25	23.54457
35	6	3.2	2	3,10,000	8	17	0	2	27	16.45729

Solution: For getting Correlation

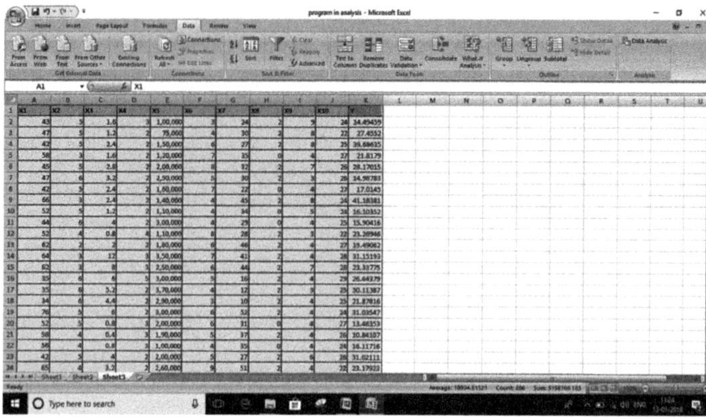

First, transfer the data into JMP data sheet.

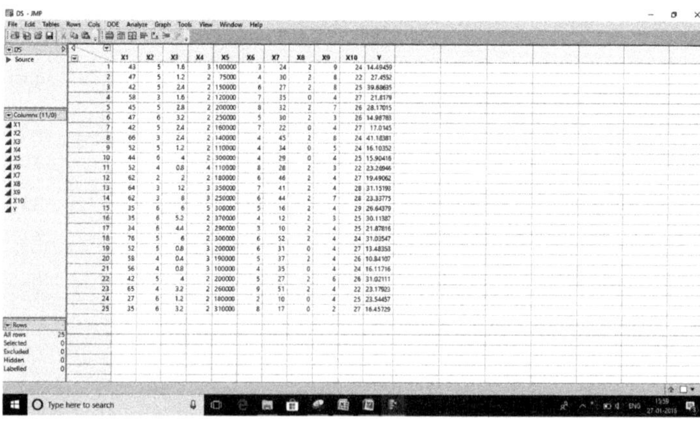

For getting Correlation, Multivariate Methods→ Multivariate. After clicking on multivariate dialog box appear on screen.

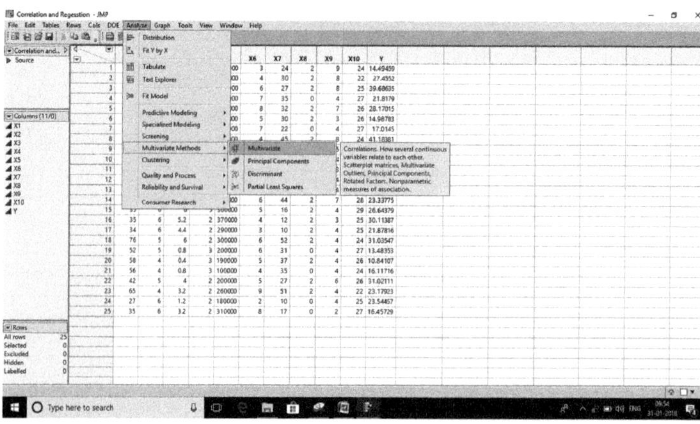

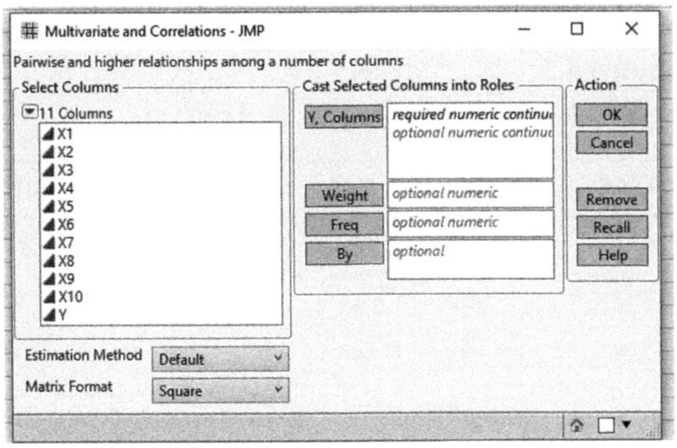

Put variables in *Y,* columns

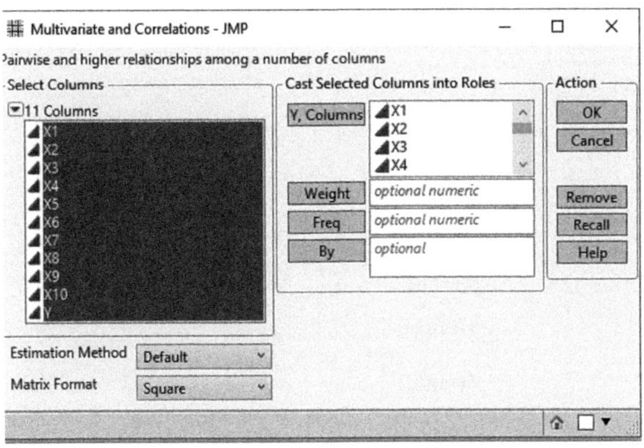

After clicking on OK, the result will appear on the screen.

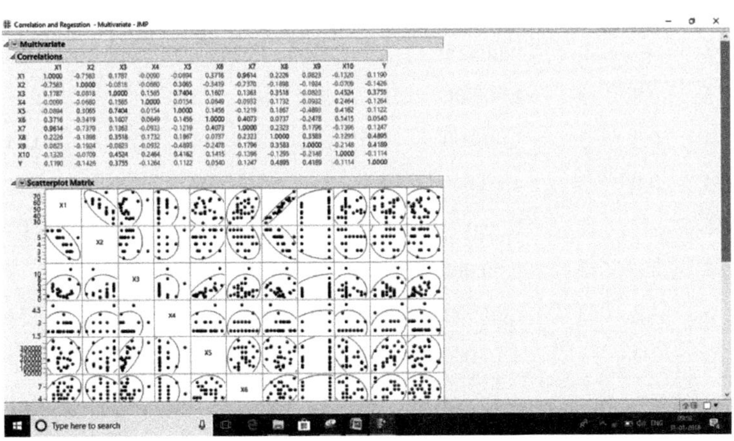

7.1.3 ***Example 3:*** Following data given the yield (Y) corresponding to 10 yield components (X_1, X_2, X_3, X_4, X_5, X_6, X_7, X_8, X_9, X_{10}) for certain agricultural crop. Work out the ***Regression Analysis***.

X1	X2	X3	X4	X5	X6	X7	X8	X9	X10	Y
43	5	1.6	3	1,00,000	3	24	2	9	24	14.49459
47	5	1.2	2	75,000	4	30	2	8	22	27.4552
42	5	2.4	2	1,50,000	6	27	2	8	25	39.68635
58	3	1.6	2	1,20,000	7	35	0	4	27	21.8179
45	5	2.8	2	2,00,000	8	32	2	7	26	28.17015
47	6	3.2	2	2,50,000	5	30	2	3	26	14.98783
42	5	2.4	2	1,60,000	7	22	0	4	27	17.0145
66	3	2.4	2	1,40,000	4	45	2	8	24	41.18381
52	5	1.2	2	1,10,000	4	34	0	5	24	16.10352
44	6	4	2	3,00,000	4	29	0	4	25	15.90416
52	4	0.8	4	1,10,000	8	28	2	3	22	23.26946
62	2	2	2	1,80,000	6	46	2	4	27	19.49062
64	3	12	3	3,50,000	7	41	2	4	28	31.15193
62	3	8	3	2,50,000	6	44	2	7	28	23.33775
35	6	6	5	3,00,000	5	16	2	4	29	26.64379
35	6	5.2	2	3,70,000	4	12	2	3	25	30.11387
34	6	4.4	2	2,90,000	3	10	2	4	25	21.87816
76	5	6	2	3,00,000	6	52	2	4	24	31.03547
52	5	0.8	3	2,00,000	6	31	0	4	27	13.48353
58	4	0.4	3	1,90,000	5	37	2	4	26	10.84107
56	4	0.8	3	1,00,000	4	35	0	4	24	16.11716
42	5	4	2	2,00,000	5	27	2	6	26	31.02111
65	4	3.2	2	2,60,000	9	51	2	4	22	23.17923
27	6	1.2	2	1,80,000	2	10	0	4	25	23.54457
35	6	3.2	2	3,10,000	8	17	0	2	27	16.45729

Solution: For getting Regression

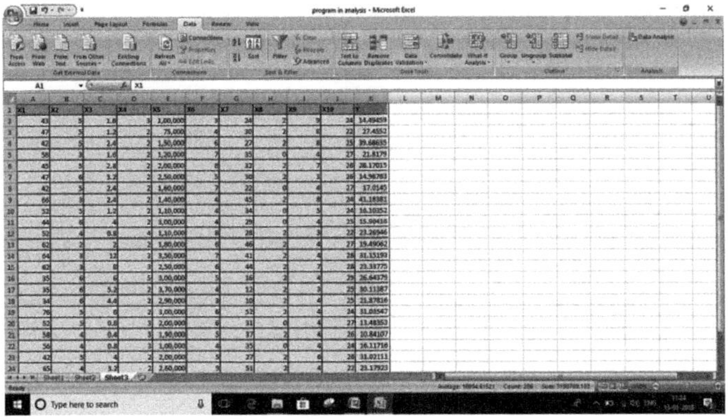

First, transfer the data into JMP data sheet.

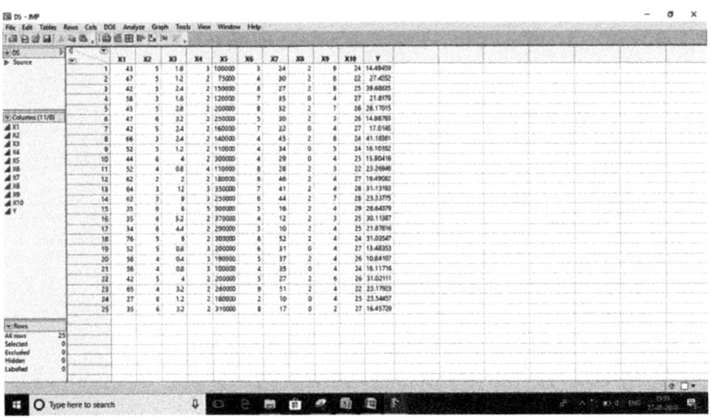

For getting Regression, after transferring the data into JMP sheet. Go to Analysis → Fit Model

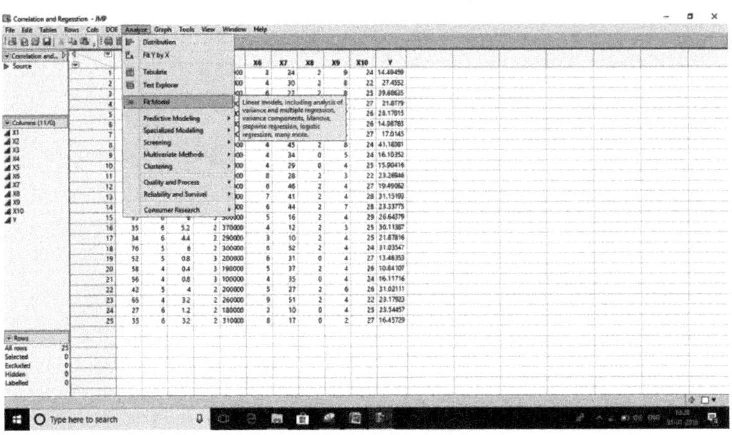

After clicking on Fit model, dialog box appear on screen.

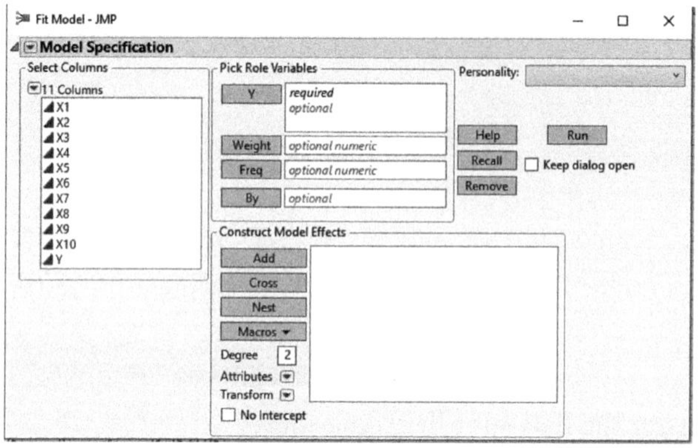

Put dependent variable on *Y* and rest in construct Model effects

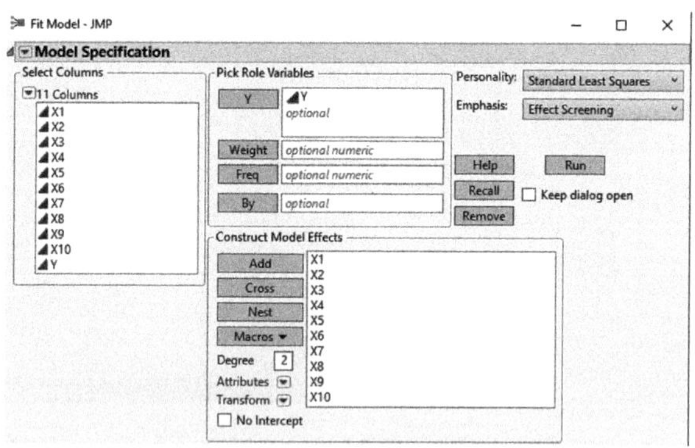

Then click on OK button and result will appear on screen.

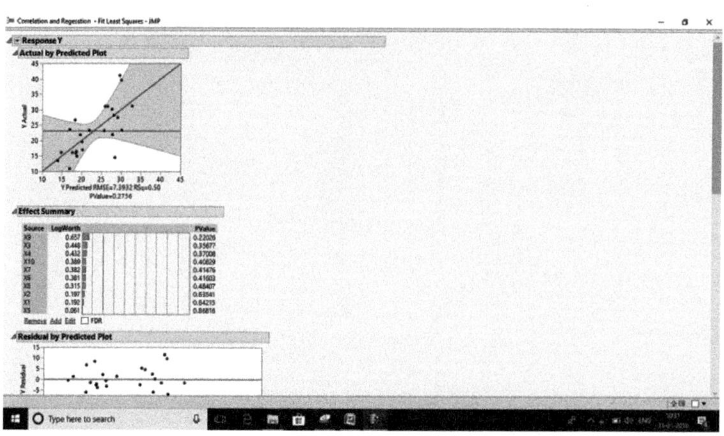

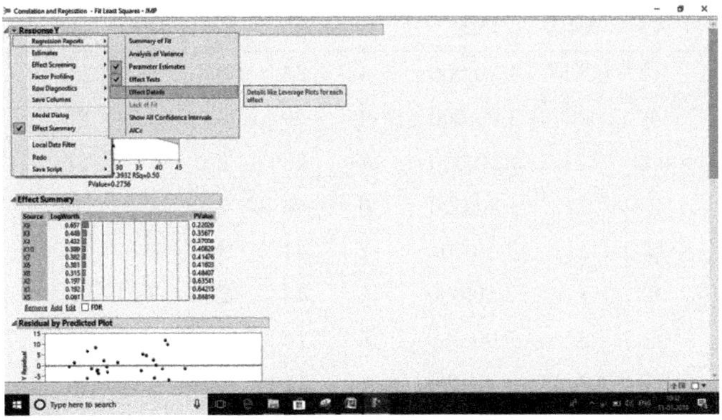

Then click on Parameter Estimates, effect test and effect details

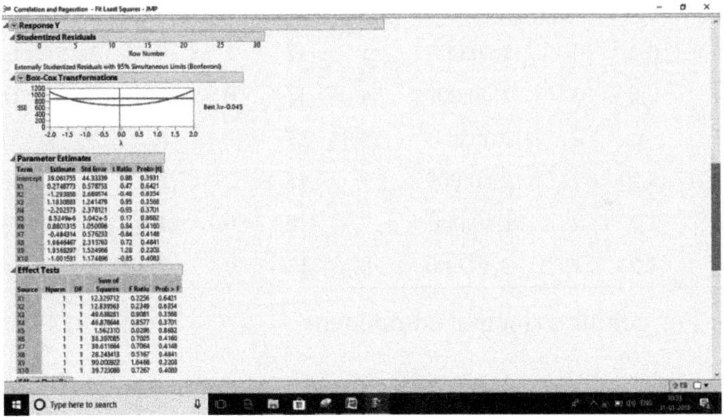

7.1.4 ***Example 4:*** Following data given the yield (Y) corresponding to 10 yield components (X_1, X_2, X_3, X_4, X_5, X_6, X_7, X_8, X_9, X_{10}) for certain agricultural crop. Work out the ***Principal Component Analysis***.

X1	X2	X3	X4	X5	X6	X7	X8	X9	X10	Y
43	5	1.6	3	1,00,000	3	24	2	9	24	14.49459
47	5	1.2	2	75,000	4	30	2	8	22	27.4552
42	5	2.4	2	1,50,000	6	27	2	8	25	39.68635
58	3	1.6	2	1,20,000	7	35	0	4	27	21.8179
45	5	2.8	2	2,00,000	8	32	2	7	26	28.17015
47	6	3.2	2	2,50,000	5	30	2	3	26	14.98783
42	5	2.4	2	1,60,000	7	22	0	4	27	17.0145
66	3	2.4	2	1,40,000	4	45	2	8	24	41.18381

X1	X2	X3	X4	X5	X6	X7	X8	X9	X10	Y
52	5	1.2	2	1,10,000	4	34	0	5	24	16.10352
44	6	4	2	3,00,000	4	29	0	4	25	15.90416
52	4	0.8	4	1,10,000	8	28	2	3	22	23.26946
62	2	2	2	1,80,000	6	46	2	4	27	19.49062
64	3	12	3	3,50,000	7	41	2	4	28	31.15193
62	3	8	3	2,50,000	6	44	2	7	28	23.33775
35	6	6	5	3,00,000	5	16	2	4	29	26.64379
35	6	5.2	2	3,70,000	4	12	2	3	25	30.11387
34	6	4.4	2	2,90,000	3	10	2	4	25	21.87816
76	5	6	2	3,00,000	6	52	2	4	24	31.03547
52	5	0.8	3	2,00,000	6	31	0	4	27	13.48353
58	4	0.4	3	1,90,000	5	37	2	4	26	10.84107
56	4	0.8	3	1,00,000	4	35	0	4	24	16.11716
42	5	4	2	2,00,000	5	27	2	6	26	31.02111
65	4	3.2	2	2,60,000	9	51	2	4	22	23.17923
27	6	1.2	2	1,80,000	2	10	0	4	25	23.54457
35	6	3.2	2	3,10,000	8	17	0	2	27	16.45729

Solution: For getting Principal component

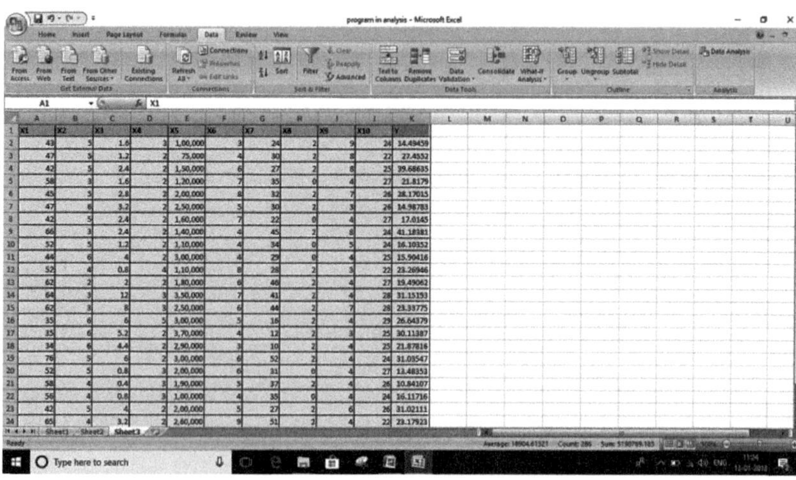

First, transfer the data into JMP data sheet.

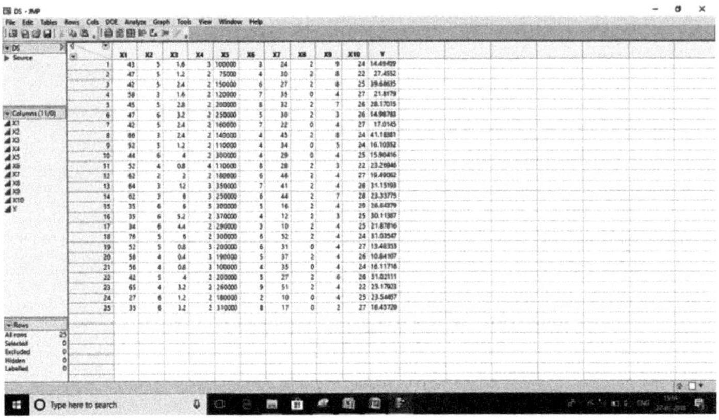

For getting principal component analysis go to Analyze → Multivariate methods→ Principal component

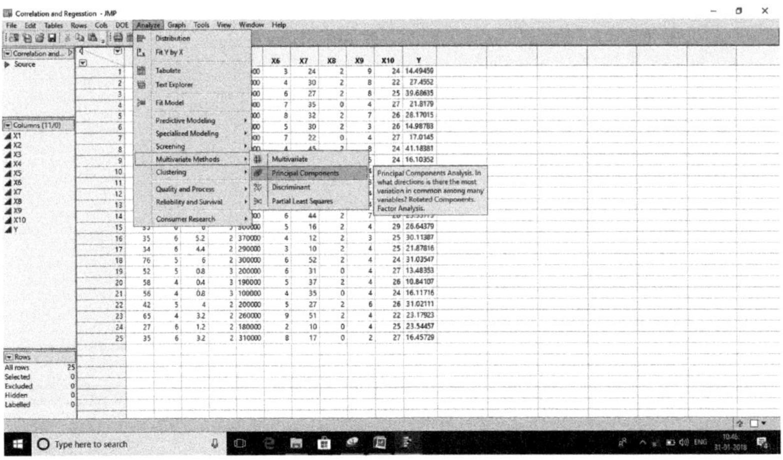

After clicking on principal component dialog box appear on screen.

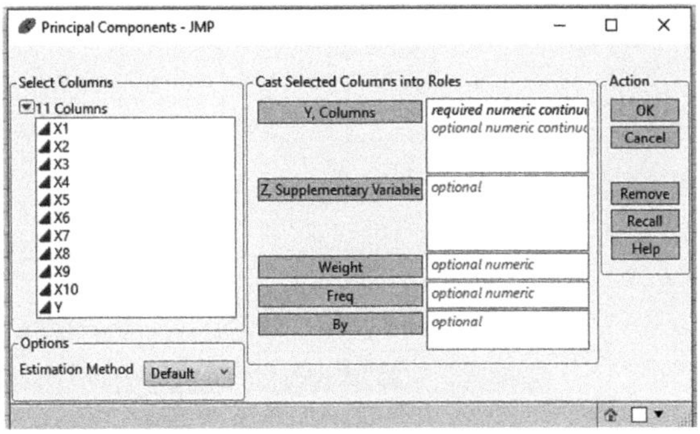

Put variables on Supplementary variable

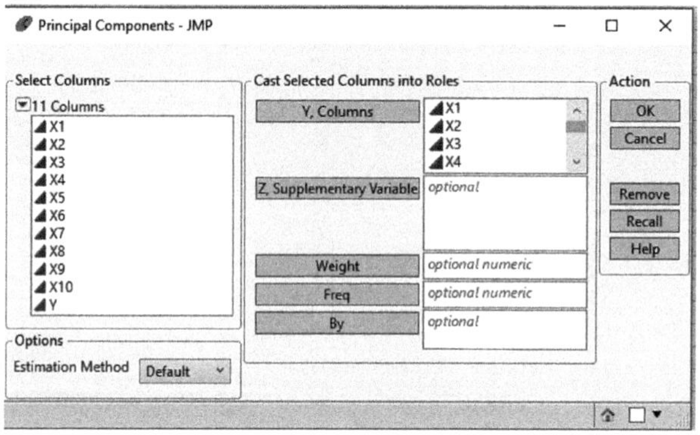

After clicking on OK, the result will appear on the screen.

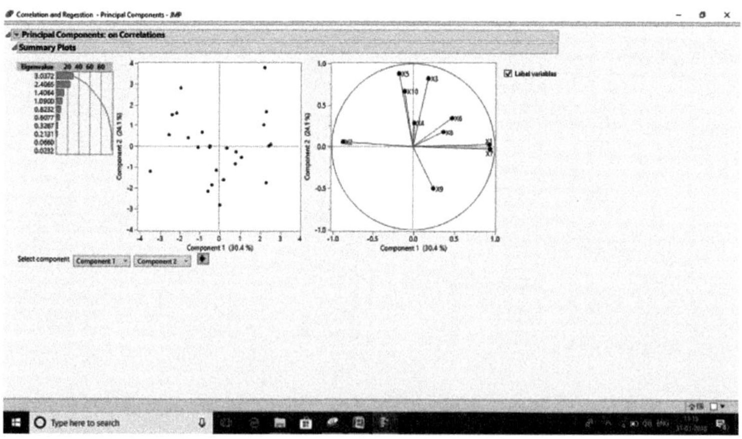

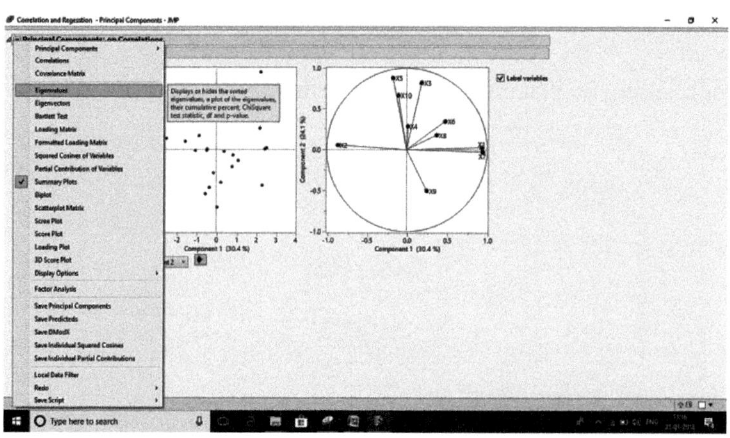

Eigenvalues

Number	Eigenvalue	Percent	Cum Percent
1	3.0372	30.372	30.372
2	2.4065	24.065	54.437
3	1.4064	14.064	68.500
4	1.0900	10.900	79.400
5	0.8232	8.232	87.632
6	0.6077	6.077	93.709
7	0.3267	3.267	96.976
8	0.2131	2.131	99.108
9	0.0660	0.660	99.768
10	0.0232	0.232	100.000

7.1.5 Example 5: RBD Design:

A field trial conduct in soybean in seven different varieties and each variety is having four replications. The data are given below in table:

Treatment	R1	R2	R3	R4
T1	34.12	32.32	31.21	30.12
T2	36.44	28.00	29.32	29.55
T3	27.77	25.60	27.80	29.70
T4	30.90	30.01	34.00	32.07
T5	28.75	31.49	29.03	22.32
T6	20.39	22.99	25.78	27.40
T7	27.23	29.58	30.48	26.50

Solution:

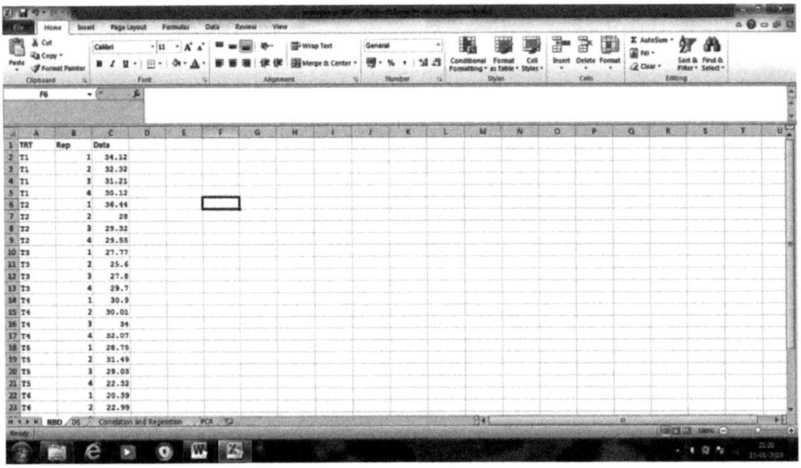

Then transfer data into JMP sheet.

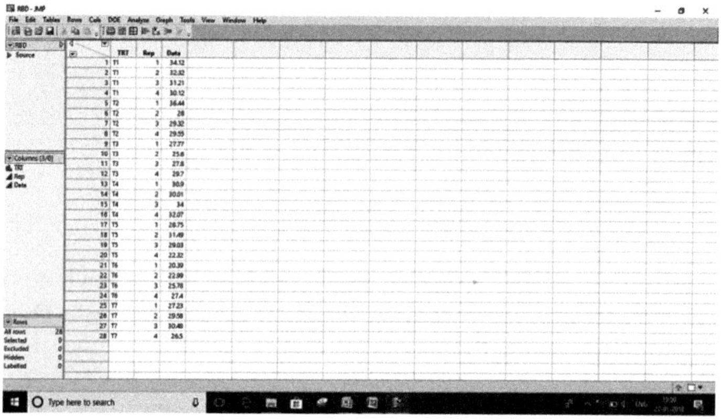

Then go to Analyze → Go to Fit *Y* by *X* and after clicking this following dialog box appear on the screen.

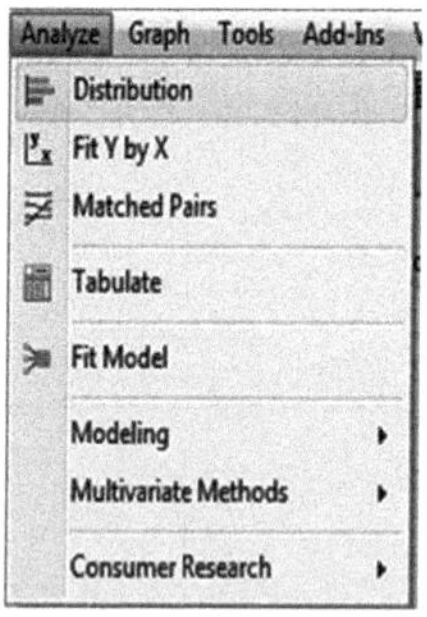

Put data in *Y*, Response, and *X*, Factor put Treatment (TRT) and Block our Replication and click on OK Button. So that result will appear on the screen.

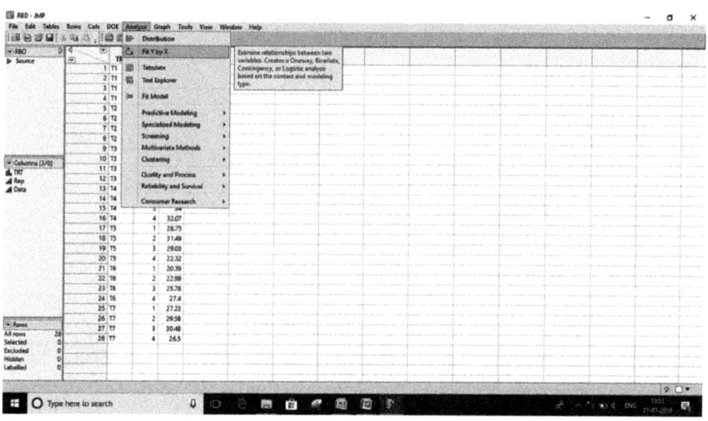

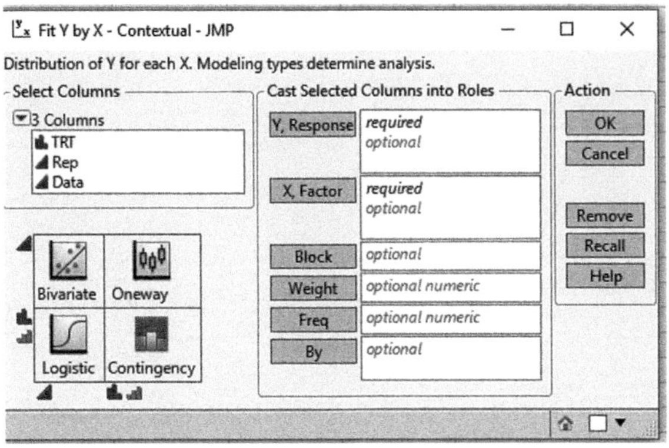

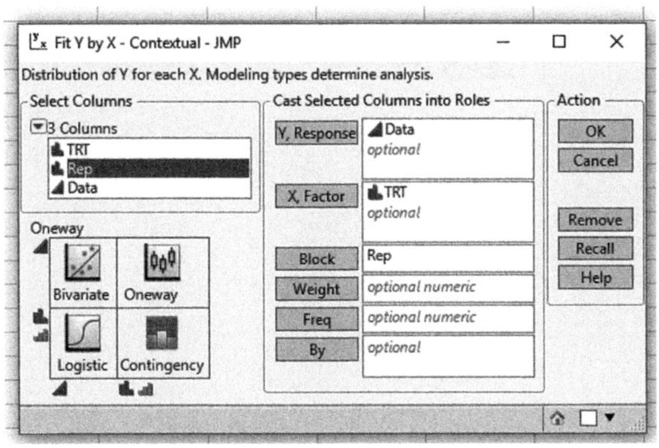

After clicking OK button, results will appear on the screen.

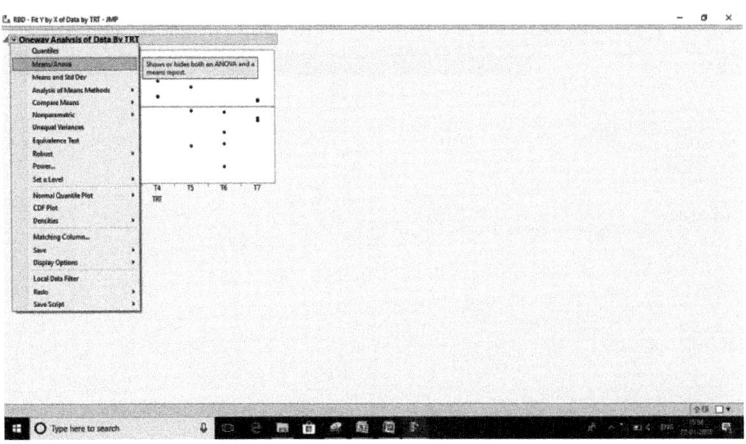

For getting ANOVA use below step.

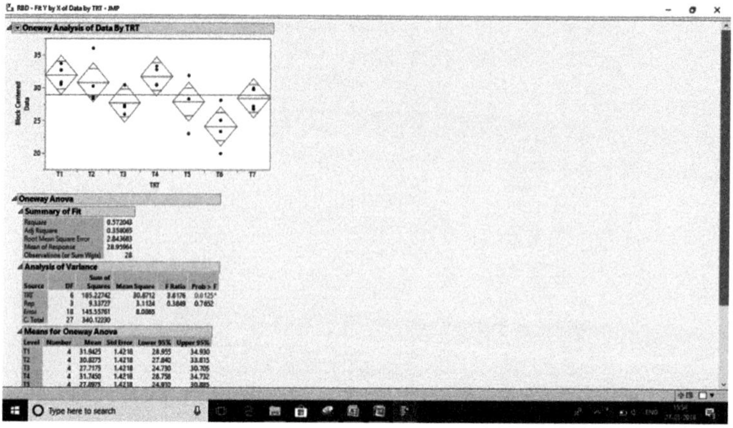

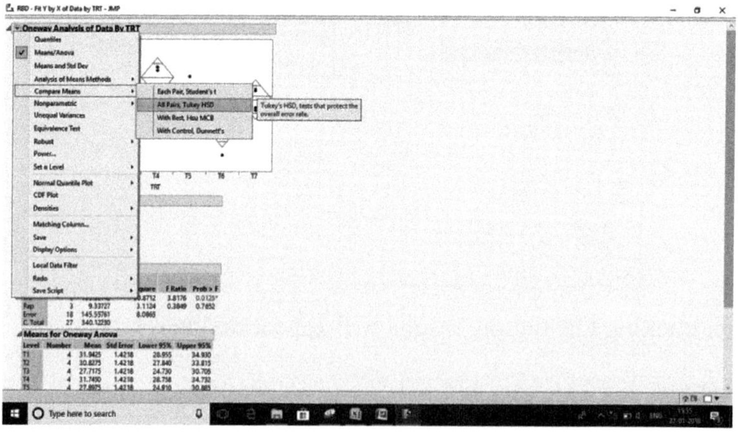

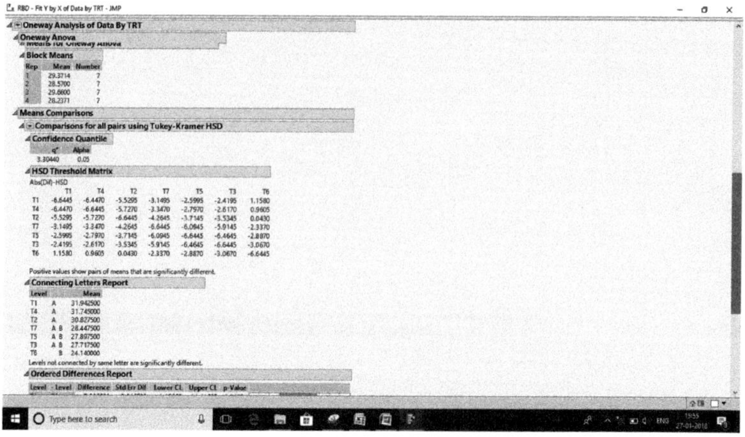

7.1.6 *Example 6:* Analysis the ***Multi-Location Analysis*** trials for below data.

Location	Treat No.	Rep	Yield	Location	Treat No.	Rep	Yield
1	2	1	2.40	2	2	1	2.00
1	2	2	2.40	2	2	2	1.60
1	2	3	3.00	2	2	3	2.80
1	3	1	2.40	2	3	1	3.00
1	3	2	2.00	2	3	2	1.60
1	3	3	2.00	2	3	3	2.00
1	4	1	2.00	2	4	1	2.00
1	4	2	2.00	2	4	2	2.00
1	4	3	1.80	2	4	3	2.40
1	5	1	1.80	2	5	1	1.80
1	5	2	1.60	2	5	2	1.60
1	5	3	1.00	2	5	3	1.60
1	6	1	2.60	2	6	1	3.80
1	6	2	1.60	2	6	2	2.40
1	6	3	1.20	2	6	3	2.20
1	7	1	2.00	2	7	1	2.20
1	7	2	2.00	2	7	2	2.20
1	7	3	1.60	2	7	3	2.20
1	8	1	1.20	2	8	1	1.80
1	8	2	1.00	2	8	2	1.60
1	8	3	1.20	2	8	3	1.00
1	9	1	2.80	2	9	1	3.20
1	9	2	2.60	2	9	2	2.60
1	9	3	2.60	2	9	3	3.00
1	10	1	0.80	2	10	1	1.20
1	10	2	1.20	2	10	2	1.20
1	10	3	1.00	2	10	3	1.20
1	11	1	0.80	2	11	1	1.00
1	11	2	0.40	2	11	2	0.20
1	11	3	1.00	2	11	3	1.00
1	UTC	1	3.60	2	UTC	1	3.20
1	UTC	2	2.80	2	UTC	2	2.00
1	UTC	3	3.20	2	UTC	3	3.00

Solution:

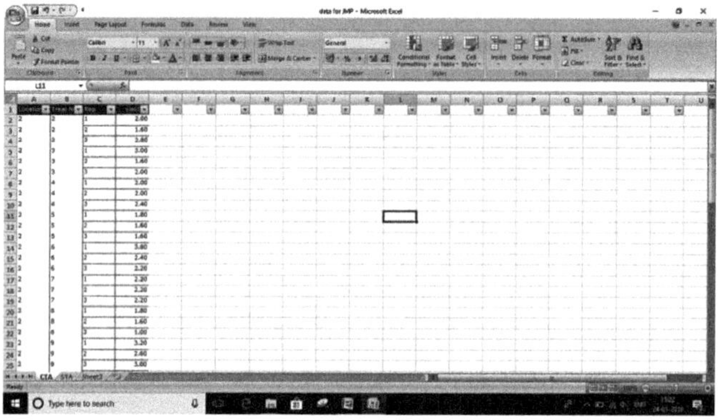

First, transfer the data into JMP sheet.

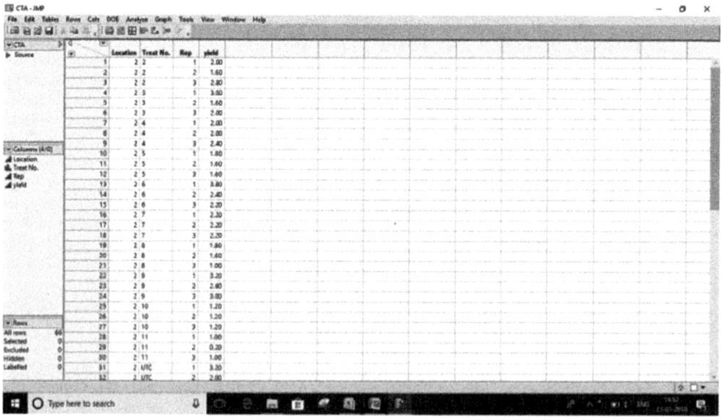

After transferring the data into JMP sheet. Go to Analysis → Fit Model

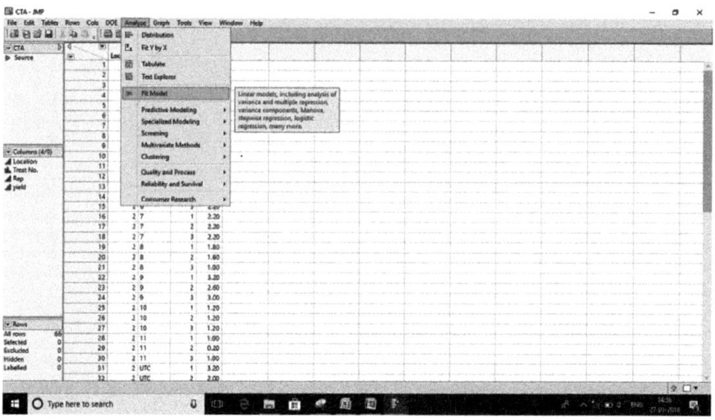

After clicking on Fit model, dialog box appear on the screen.

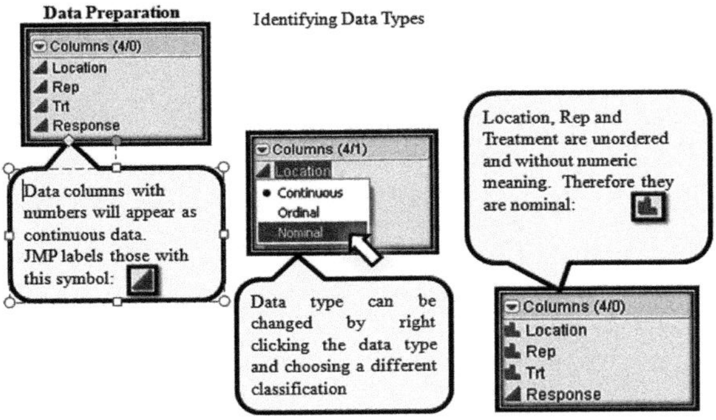

Pick role variable section; put variable data, which needs to analysis. And construct model effect select, Treatment (TRT), Location/Year and rep as well as cross trial model required statistical model.

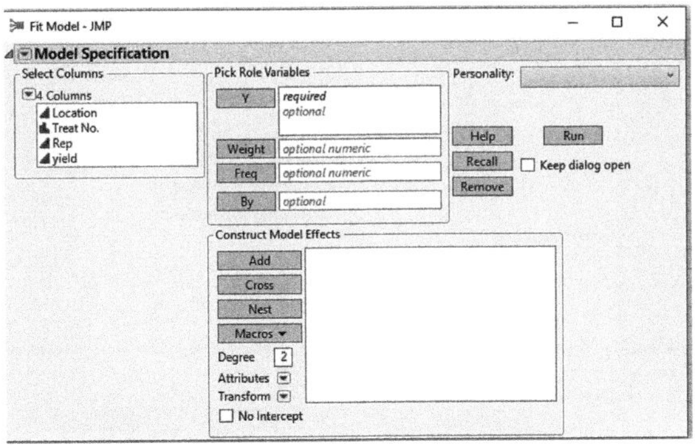

Then take interaction between location and Treatment. Then need nested effect of rep [Location].

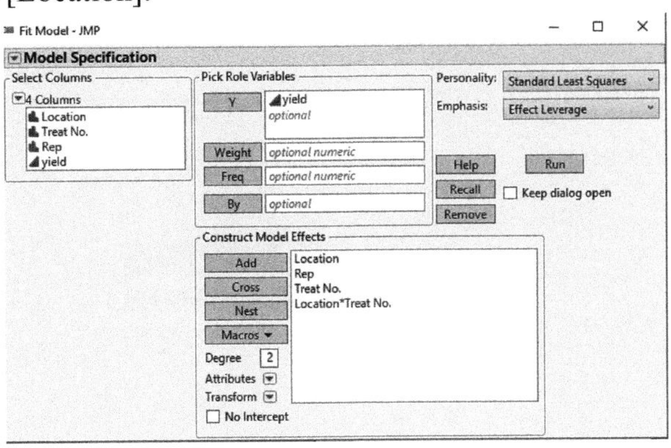

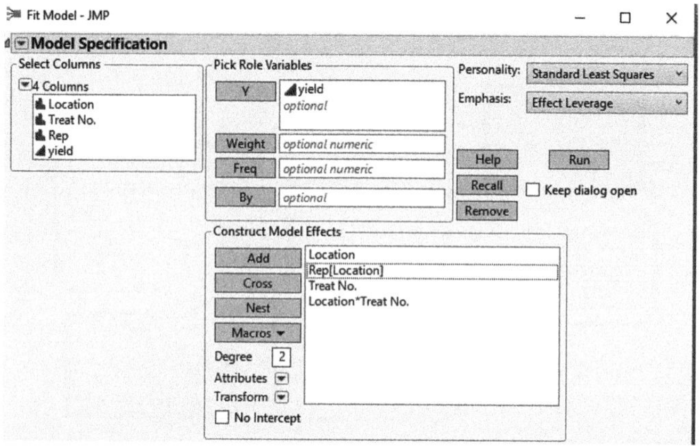

Take Location as a random effect. And click on run and results will appear on screen.

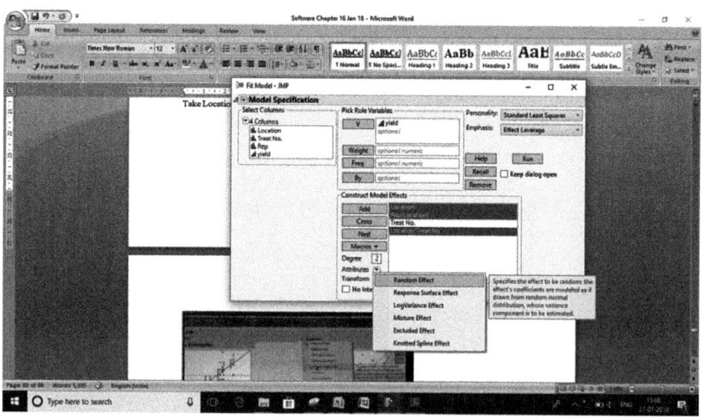

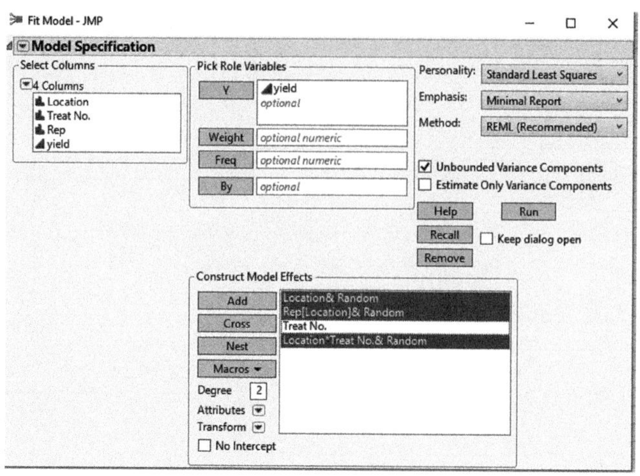

After clicking on run, the result will appear on the screen.

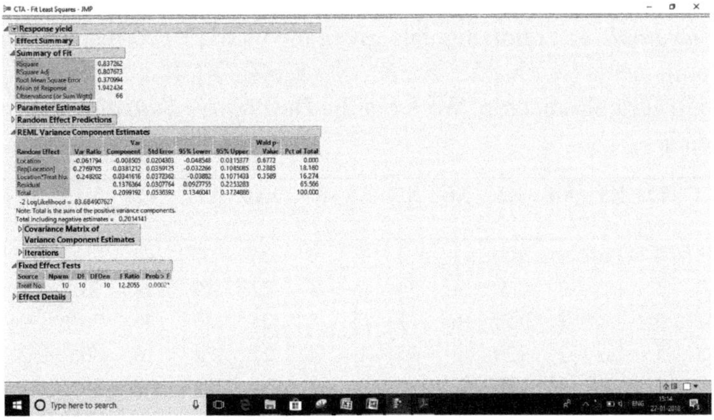

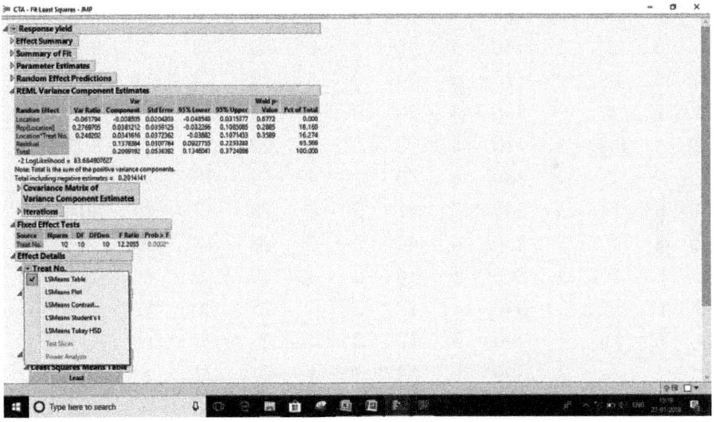

For getting mean separation select on LS Means Tukey's HSD

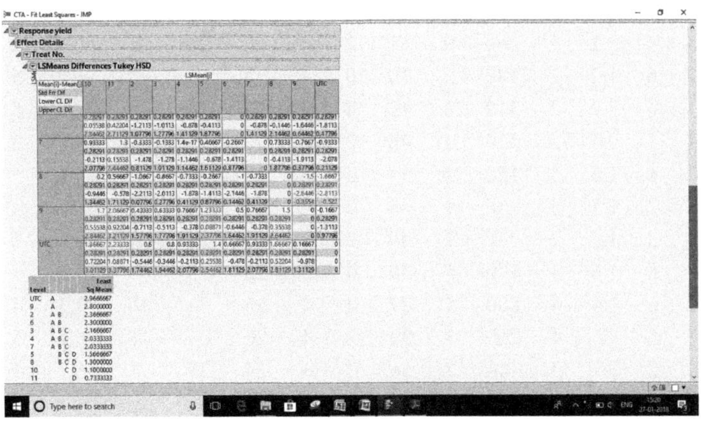

7.2 SAS STEPS FOR ANALYSIS

7.2.1 ***Example 1:*** Following data given the yield (Y) corresponding to 14 yield components (X_1, X_2, X_3, X_4, X_5, X_6, X_7, X_8, X_9, X_{10}, X_{11}, X_{12}, X_{13}, X_{14}) for certain agricultural crop. Work out the ***Descriptive Statistics Analysis*** for below data.

S. No.	X1	X2	X3	X4	X5	X6	X7	X8	X9	X10	X11	X12	X13	X14	Y
1	43	5	1.6	3	100	3	24	2	9	24	21	7	76.7	1	14.5
2	47	5	1.2	2	75	4	30	2	8	22	19	6	73.3	1	27.5
3	42	5	2.4	2	150	6	27	2	8	25	20	8	76.7	2	39.7
4	58	3	1.6	2	120	7	35	0	4	27	22	10	80.0	2	21.8
5	45	5	2.8	2	200	8	32	2	7	26	21	7	86.7	2	28.2
6	47	6	3.2	2	250	5	30	2	3	26	25	7	76.7	1	15.0
7	42	5	2.4	2	160	7	22	0	4	27	24	6	80.0	4	17.0
8	66	3	2.4	2	140	4	45	2	8	24	18	5	80.0	1	41.2
9	52	5	1.2	2	110	4	34	0	5	24	24	11	70.0	3	16.1
10	44	6	4	2	300	4	29	0	4	25	21	12	90.0	2	15.9
11	52	4	0.8	4	110	8	28	2	3	22	23	7	86.7	2	23.3
12	62	2	2	2	180	6	46	2	4	27	18	7	76.7	1	19.5
13	64	3	12	3	350	7	41	2	4	28	17	10	80.0	2	31.2
14	62	3	8	3	250	6	44	2	7	28	18	8	76.7	2	23.3
15	35	6	6	5	300	5	16	2	4	29	19	9	90.0	5	26.6
16	35	6	5.2	2	370	4	12	2	3	25	20	12	90.0	4	30.1
17	34	6	4.4	2	290	3	10	2	4	25	27	11	83.3	5	21.9
18	76	5	6	2	300	6	52	2	4	24	23	8	76.7	2	31.0
19	52	5	0.8	3	200	6	31	0	4	27	28	7	66.7	3	13.5
20	58	4	0.4	3	190	5	37	2	4	26	21	7	73.3	3	10.8
21	56	4	0.8	3	100	4	35	0	4	24	24	9	80.0	2	16.1
22	42	5	4	2	200	5	27	2	6	26	25	10	76.7	4	31.0
23	65	4	3.2	2	260	9	51	2	4	22	26	8	80.0	2	23.2
24	27	6	1.2	2	180	2	10	0	4	25	27	7	76.7	5	23.5
25	35	6	3.2	2	310	8	17	0	2	27	26	7	70.0	8	16.5
26	65	5	10	2	320	4	48	2	4	27	22	8	80.0	3	16.2
27	49	6	1.6	3	150	4	31	2	4	22	20	7	80.0	5	27.6
28	70	2	1.7	3	80	9	48	2	7	27	21	7	83.3	3	21.4
29	63	3	1.6	2	90	3	49	2	9	27	21	6	80.0	3	22.8
30	42	6	1.6	3	170	5	28	0	5	26	23	7	76.7	5	15.9
31	45	6	1.4	4	200	4	27	0	4	25	22	9	73.3	4	14.1
32	57	5	2	4	260	3	39	2	4	26	24	11	76.7	5	21.9
33	78	2	1.2	2	140	2	56	0	3	25	25	7	76.7	4	24.8

S. No.	X1	X2	X3	X4	X5	X6	X7	X8	X9	X10	X11	X12	X13	X14	Y
34	48	4	1.2	2	160	4	33	2	7	28	26	12	83.3	6	14.4
35	72	3	2.4	2	200	10	56	0	4	25	22	12	70.0	2	23.3
36	56	5	2.4	4	230	6	34	0	7	27	21	11	80.0	4	20.8
37	52	5	2	3	270	4	45	2	4	24	25	11	80.0	3	27.7
38	45	5	4	2	340	5	28	2	6	24	26	12	80.0	4	30.9
39	69	5	3.2	2	370	2	49	2	8	27	27	10	86.7	1	32.8
40	56	4	0.8	3	200	3	39	0	2	25	28	6	73.3	4	17.8
41	55	5	3.6	2	380	4	38	0	6	26	25	11	76.7	4	19.5
42	55	6	1.6	2	200	4	35	0	6	24	26	8	76.7	3	8.7
43	46	6	4.4	2	270	4	29	2	4	25	21	9	76.7	6	25.3
44	42	6	1.6	2	200	4	27	0	4	22	23	10	70.0	5	9.3
45	46	5	1.2	2	170	3	31	0	2	25	21	7	66.7	4	10.7
46	64	4	1.6	2	170	4	52	0	7	26	24	6	83.3	3	11.4
47	56	5	1.2	2	140	4	39	0	7	26	27	8	80.0	2	13.1
48	56	5	0.8	3	170	3	35	0	7	28	21	11	86.7	3	12.7
49	55	6	1.6	2	140	4	30	2	6	26	25	9	80.0	4	15.6
50	65	4	3.2	4	80	6	2	2	7	27	26	12	83.3	3	21.9
51	51	5	2	4	50	4	29	0	3	24	23	11	73.3	5	17.9
52	75	4	3.6	2	350	12	59	0	6	26	21	12	76.7	2	16.2
53	66	4	0.6	2	90	5	51	0	8	28	21	7	90.0	4	7.1
54	62	5	1.2	4	210	6	49	2	4	24	27	11	73.3	3	21.1
55	72	4	1.6	2	160	4	54	0	5	26	25	6	76.7	2	9.8
56	68	4	2.4	2	280	3	52	2	3	25	23	12	66.7	2	27.0
57	36	6	2.8	2	280	5	21	0	6	26	21	11	90.0	7	13.6
58	46	6	1.2	3	220	4	28	0	6	24	22	7	76.7	7	14.8
59	60	5	1.2	3	240	3	47	2	5	25	26	10	80.0	2	28.7
60	62	5	5.2	2	300	5	45	0	7	27	27	11	86.7	3	12.8
61	52	5	5.2	2	350	5	36	0	7	27	22	12	83.3	5	17.0
62	68	6	3.2	2	280	4	52	0	5	26	18	11	83.3	4	15.7
63	38	6	2	3	270	5	23	0	3	26	17	7	73.3	8	11.3
64	55	6	0.8	2	50	3	42	0	8	25	17	6	83.3	5	5.9
65	60	5	2	2	160	5	48	2	5	26	18	6	86.7	5	1.6
66	40	6	2.4	2	280	4	28	0	3	26	19	9	73.3	6	13.1
67	38	6	1	2	190	5	24	0	4	25	20	9	76.7	6	7.6
68	55	5	4	2	300	5	38	2	7	28	21	10	86.7	4	14.4
69	60	5	4.8	3	310	3	47	0	2	27	21	8	66.7	7	18.3
70	61	5	0.8	2	80	2	45	0	5	26	19	7	76.7	6	4.7
71	65	4	2	2	180	3	49	0	7	26	18	10	86.7	5	11.1
72	60	4	1.6	2	190	6	46	0	4	25	20	8	90.0	3	9.9
73	42	6	1.6	2	170	9	27	2	4	24	22	7	73.3	7	18.8

S. No.	X1	X2	X3	X4	X5	X6	X7	X8	X9	X10	X11	X12	X13	X14	Y
74	62	4	1.4	4	280	6	44	2	5	27	22	9	76.7	4	29.6
75	48	6	4	2	260	4	31	0	5	27	21	11	73.3	4	7.8
76	42	5	2.4	2	150	6	27	2	6	26	23	10	80.0	5	27.5
77	61	3	1.6	2	120	7	35	0	8	25	24	8	80.0	3	5.7
78	45	5	2.8	2	300	8	32	2	4	28	25	9	86.7	5	25.1
79	47	6	3.2	2	350	5	30	2	3	26	26	6	76.7	4	27.4
80	42	5	2.4	2	160	7	22	0	3	24	27	6	73.3	6	13.0
81	55	5	3.6	2	380	4	38	0	7	27	22	7	76.7	3	9.8
82	55	6	1.6	2	200	4	35	2	3	25	21	11	76.7	5	11.9
83	46	6	4.4	2	270	4	29	2	4	26	23	8	76.7	7	23.0
84	42	6	1.6	2	200	4	27	0	4	27	24	9	80.0	8	9.3
85	46	5	1.2	2	170	3	31	0	2	25	25	7	70.0	5	10.0
86	70	4	1.6	2	170	4	52	0	5	24	26	11	86.7	3	14.0
87	58	5	1.2	2	140	4	39	0	5	26	21	7	83.3	6	8.0
88	56	4	0.8	3	230	4	35	0	2	24	22	8	76.7	7	11.7
89	42	5	4	2	200	5	27	2	8	29	24	10	83.3	8	33.1
90	65	4	3.2	2	460	9	51	2	8	26	19	12	80.0	3	23.9
91	27	6	1.2	2	180	2	10	0	4	27	20	7	80.0	9	12.1
92	35	6	3.2	2	310	8	17	2	3	26	20	11	66.7	8	6.5
93	68	5	10	2	320	4	48	2	7	27	21	12	80.0	4	32.4
94	65	3	1.2	2	85	4	48	2	4	26	18	6	83.3	3	7.7
95	53	6	3.6	3	250	6	38	2	4	27	25	9	80.0	6	22.5
96	47	6	2.8	3	250	4	29	2	2	26	25	8	66.7	9	26.1
97	68	4	3.2	4	275	5	47	2	2	23	24	11	70.0	5	29.2
98	57	5	2.8	3	250	4	39	2	5	24	22	7	76.7	5	20.5
99	58	6	2.4	2	200	3	37	2	7	27	23	7	76.7	6	18.3
100	43	4	0.8	1	50	2	28	0	3	24	22	6	73.3	5	4.5

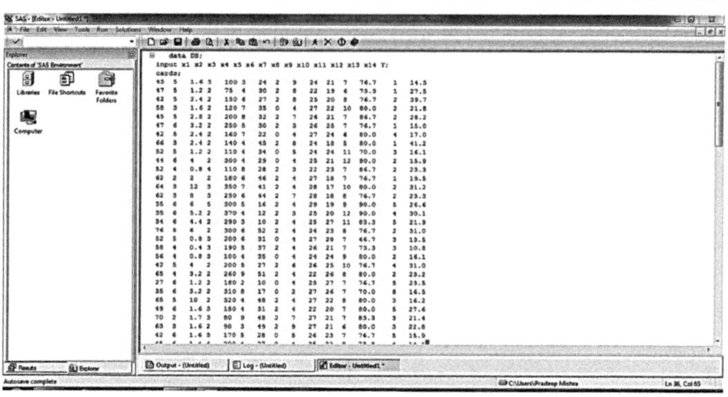

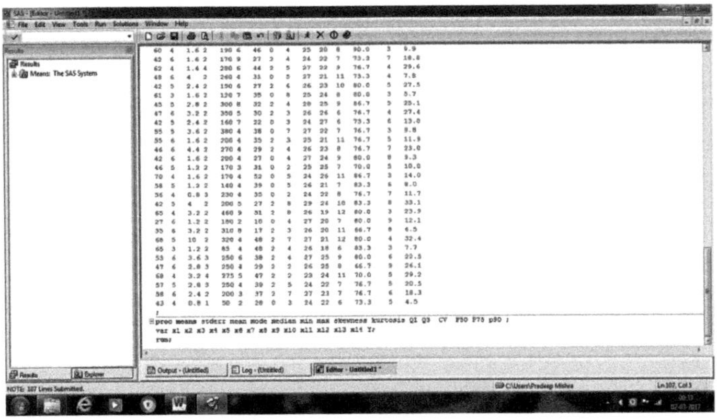

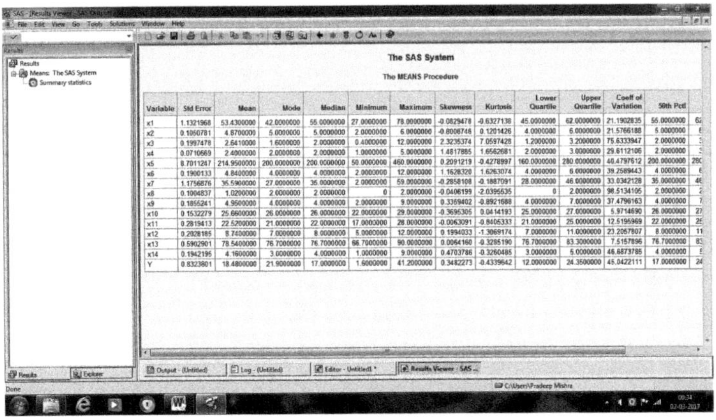


```
data DS;
input x1 x2 x3 x4 x5 x6 x7 x8 x9 x10 x11 x12 x13 x14 Y;
cards;
```

43	5	1.6	3	100	3	24	2	9	24	21	7	76.7	1	14.5
47	5	1.2	2	75	4	30	2	8	22	19	6	73.3	1	27.5
42	5	2.4	2	150	6	27	2	8	25	20	8	76.7	2	39.7
58	3	1.6	2	120	7	35	0	4	27	22	10	80.0	2	21.8
45	5	2.8	2	200	8	32	2	7	26	21	7	86.7	2	28.2
47	6	3.2	2	250	5	30	2	3	26	25	7	76.7	1	15.0
42	5	2.4	2	160	7	22	0	4	27	24	6	80.0	4	17.0
66	3	2.4	2	140	4	45	2	8	24	18	5	80.0	1	41.2
52	5	1.2	2	110	4	34	0	5	24	24	11	70.0	3	16.1
44	6	4	2	300	4	29	0	4	25	21	12	90.0	2	15.9

52	4	0.8	4	110	8	28	2	3	22	23	7	86.7	2	23.3
62	2	2	2	180	6	46	2	4	27	18	7	76.7	1	19.5
64	3	12	3	350	7	41	2	4	28	17	10	80.0	2	31.2
62	3	8	3	250	6	44	2	7	28	18	8	76.7	2	23.3
35	6	6	5	300	5	16	2	4	29	19	9	90.0	5	26.6
35	6	5.2	2	370	4	12	2	3	25	20	12	90.0	4	30.1
34	6	4.4	2	290	3	10	2	4	25	27	11	83.3	5	21.9
76	5	6	2	300	6	52	2	4	24	23	8	76.7	2	31.0
52	5	0.8	3	200	6	31	0	4	27	28	7	66.7	3	13.5
58	4	0.4	3	190	5	37	2	4	26	21	7	73.3	3	10.8
56	4	0.8	3	100	4	35	0	4	24	24	9	80.0	2	16.1
42	5	4	2	200	5	27	2	6	26	25	10	76.7	4	31.0
65	4	3.2	2	260	9	51	2	4	22	26	8	80.0	2	23.2
27	6	1.2	2	180	2	10	0	4	25	27	7	76.7	5	23.5
35	6	3.2	2	310	8	17	0	2	27	26	7	70.0	8	16.5
65	5	10	2	320	4	48	2	4	27	22	8	80.0	3	16.2
49	6	1.6	3	150	4	31	2	4	22	20	7	80.0	5	27.6
70	2	1.7	3	80	9	48	2	7	27	21	7	83.3	3	21.4
63	3	1.6	2	90	3	49	2	9	27	21	6	80.0	3	22.8
42	6	1.6	3	170	5	28	0	5	26	23	7	76.7	5	15.9
45	6	1.4	4	200	4	27	0	4	25	22	9	73.3	4	14.1
57	5	2	4	260	3	39	2	4	26	24	11	76.7	5	21.9
78	2	1.2	2	140	2	56	0	3	25	25	7	76.7	4	24.8
48	4	1.2	2	160	4	33	2	7	28	26	12	83.3	6	14.4
72	3	2.4	2	200	10	56	0	4	25	22	12	70.0	2	23.3
56	5	2.4	4	230	6	34	0	7	27	21	11	80.0	4	20.8
52	5	2	3	270	4	45	2	4	24	25	11	80.0	3	27.7
45	5	4	2	340	5	28	2	6	24	26	12	80.0	4	30.9
69	5	3.2	2	370	2	49	2	8	27	27	10	86.7	1	32.8
56	4	0.8	3	200	3	39	0	2	25	28	6	73.3	4	17.8
55	5	3.6	2	380	4	38	0	6	26	25	11	76.7	4	19.5
55	6	1.6	2	200	4	35	0	6	24	26	8	76.7	3	8.7
46	6	4.4	2	270	4	29	2	4	25	21	9	76.7	6	25.3
42	6	1.6	2	200	4	27	0	4	22	23	10	70.0	5	9.3
46	5	1.2	2	170	3	31	0	2	25	21	7	66.7	4	10.7
64	4	1.6	2	170	4	52	0	7	26	24	6	83.3	3	11.4

56	5	1.2	2	140	4	39	0	7	26	27	8	80.0	2	13.1
56	5	0.8	3	170	3	35	0	7	28	21	11	86.7	3	12.7
55	6	1.6	2	140	4	30	2	6	26	25	9	80.0	4	15.6
65	4	3.2	4	80	6	2	2	7	27	26	12	83.3	3	21.9
51	5	2	4	50	4	29	0	3	24	23	11	73.3	5	17.9
75	4	3.6	2	350	12	59	0	6	26	21	12	76.7	2	16.2
66	4	0.6	2	90	5	51	0	8	28	21	7	90.0	4	7.1
62	5	1.2	4	210	6	49	2	4	24	27	11	73.3	3	21.1
72	4	1.6	2	160	4	54	0	5	26	25	6	76.7	2	9.8
68	4	2.4	2	280	3	52	2	3	25	23	12	66.7	2	27.0
36	6	2.8	2	280	5	21	0	6	26	21	11	90.0	7	13.6
46	6	1.2	3	220	4	28	0	6	24	22	7	76.7	7	14.8
60	5	1.2	3	240	3	47	2	5	25	26	10	80.0	2	28.7
62	5	5.2	2	300	5	45	0	7	27	27	11	86.7	3	12.8
52	5	5.2	2	350	5	36	0	7	27	22	12	83.3	5	17.0
68	6	3.2	2	280	4	52	0	5	26	18	11	83.3	4	15.7
38	6	2	3	270	5	23	0	3	26	17	7	73.3	8	11.3
55	6	0.8	2	50	3	42	0	8	25	17	6	83.3	5	5.9
60	5	2	2	160	5	48	2	5	26	18	6	86.7	5	1.6
40	6	2.4	2	280	4	28	0	3	26	19	9	73.3	6	13.1
38	6	1	2	190	5	24	0	4	25	20	9	76.7	6	7.6
55	5	4	2	300	5	38	2	7	28	21	10	86.7	4	14.4
60	5	4.8	3	310	3	47	0	2	27	21	8	66.7	7	18.3
61	5	0.8	2	80	2	45	0	5	26	19	7	76.7	6	4.7
65	4	2	2	180	3	49	0	7	26	18	10	86.7	5	11.1
60	4	1.6	2	190	6	46	0	4	25	20	8	90.0	3	9.9
42	6	1.6	2	170	9	27	2	4	24	22	7	73.3	7	18.8
62	4	1.4	4	280	6	44	2	5	27	22	9	76.7	4	29.6
48	6	4	2	260	4	31	0	5	27	21	11	73.3	4	7.8
42	5	2.4	2	150	6	27	2	6	26	23	10	80.0	5	27.5
61	3	1.6	2	120	7	35	0	8	25	24	8	80.0	3	5.7
45	5	2.8	2	300	8	32	2	4	28	25	9	86.7	5	25.1
47	6	3.2	2	350	5	30	2	3	26	26	6	76.7	4	27.4
42	5	2.4	2	160	7	22	0	3	24	27	6	73.3	6	13.0
55	5	3.6	2	380	4	38	0	7	27	22	7	76.7	3	9.8
55	6	1.6	2	200	4	35	2	3	25	21	11	76.7	5	11.9

46	6	4.4	2	270	4	29	2	4	26	23	8	76.7	7	23.0
42	6	1.6	2	200	4	27	0	4	27	24	9	80.0	8	9.3
46	5	1.2	2	170	3	31	0	2	25	25	7	70.0	5	10.0
70	4	1.6	2	170	4	52	0	5	24	26	11	86.7	3	14.0
58	5	1.2	2	140	4	39	0	5	26	21	7	83.3	6	8.0
56	4	0.8	3	230	4	35	0	2	24	22	8	76.7	7	11.7
42	5	4	2	200	5	27	2	8	29	24	10	83.3	8	33.1
65	4	3.2	2	460	9	51	2	8	26	19	12	80.0	3	23.9
27	6	1.2	2	180	2	10	0	4	27	20	7	80.0	9	12.1
35	6	3.2	2	310	8	17	2	3	26	20	11	66.7	8	6.5
68	5	10	2	320	4	48	2	7	27	21	12	80.0	4	32.4
65	3	1.2	2	85	4	48	2	4	26	18	6	83.3	3	7.7
53	6	3.6	3	250	6	38	2	4	27	25	9	80.0	6	22.5
47	6	2.8	3	250	4	29	2	2	26	25	8	66.7	9	26.1
68	4	3.2	4	275	5	47	2	2	23	24	11	70.0	5	29.2
57	5	2.8	3	250	4	39	2	5	24	22	7	76.7	5	20.5
58	6	2.4	2	200	3	37	2	7	27	23	7	76.7	6	18.3
43	4	0.8	1	50	2	28	0	3	24	22	6	73.3	5	4.5

```
Proc means std err mean mode median minmax skewness kurtosis Q1 Q3
CV P50 P75 P90;
    var x1 x2 x3 x4 x5 x6 x7 x8 x9 x10 x11 x12 x13 x14 Y;
    run;
```

7.2.2 ***Example 2:*** Following data given the yield (Y) corresponding to 14 yield components (X_1, X_2, X_3, X_4, X_5, X_6, X_7, X_8, X_9, X_{10}, X_{11}, X_{12}, X_{13}, X_{14}) for certain agricultural crop. Work out the ***Correlation Analysis*** for below data.

Data:

S. no.	X1	X2	X3	X4	X5	X6	X7	X8	X9	X10	X11	X12	X13	X14	Y
1	43	5	1.6	3	100	3	24	2	9	24	21	7	76.7	1	14.5
2	47	5	1.2	2	75	4	30	2	8	22	19	6	73.3	1	27.5
3	42	5	2.4	2	150	6	27	2	8	25	20	8	76.7	2	39.7
4	58	3	1.6	2	120	7	35	0	4	27	22	10	80.0	2	21.8
5	45	5	2.8	2	200	8	32	2	7	26	21	7	86.7	2	28.2
6	47	6	3.2	2	250	5	30	2	3	26	25	7	76.7	1	15.0

S. no.	X1	X2	X3	X4	X5	X6	X7	X8	X9	X10	X11	X12	X13	X14	Y
7	42	5	2.4	2	160	7	22	0	4	27	24	6	80.0	4	17.0
8	66	3	2.4	2	140	4	45	2	8	24	18	5	80.0	1	41.2
9	52	5	1.2	2	110	4	34	0	5	24	24	11	70.0	3	16.1
10	44	6	4	2	300	4	29	0	4	25	21	12	90.0	2	15.9
11	52	4	0.8	4	110	8	28	2	3	22	23	7	86.7	2	23.3
12	62	2	2	2	180	6	46	2	4	27	18	7	76.7	1	19.5
13	64	3	12	3	350	7	41	2	4	28	17	10	80.0	2	31.2
14	62	3	8	3	250	6	44	2	7	28	18	8	76.7	2	23.3
15	35	6	6	5	300	5	16	2	4	29	19	9	90.0	5	26.6
16	35	6	5.2	2	370	4	12	2	3	25	20	12	90.0	4	30.1
17	34	6	4.4	2	290	3	10	2	4	25	27	11	83.3	5	21.9
18	76	5	6	2	300	6	52	2	4	24	23	8	76.7	2	31.0
19	52	5	0.8	3	200	6	31	0	4	27	28	7	66.7	3	13.5
20	58	4	0.4	3	190	5	37	2	4	26	21	7	73.3	3	10.8
21	56	4	0.8	3	100	4	35	0	4	24	24	9	80.0	2	16.1
22	42	5	4	2	200	5	27	2	6	26	25	10	76.7	4	31.0
23	65	4	3.2	2	260	9	51	2	4	22	26	8	80.0	2	23.2
24	27	6	1.2	2	180	2	10	0	4	25	27	7	76.7	5	23.5
25	35	6	3.2	2	310	8	17	0	2	27	26	7	70.0	8	16.5
26	65	5	10	2	320	4	48	2	4	27	22	8	80.0	3	16.2
27	49	6	1.6	3	150	4	31	2	4	22	20	7	80.0	5	27.6
28	70	2	1.7	3	80	9	48	2	7	27	21	7	83.3	3	21.4
29	63	3	1.6	2	90	3	49	2	9	27	21	6	80.0	3	22.8
30	42	6	1.6	3	170	5	28	0	5	26	23	7	76.7	5	15.9
31	45	6	1.4	4	200	4	27	0	4	25	22	9	73.3	4	14.1
32	57	5	2	4	260	3	39	2	4	26	24	11	76.7	5	21.9
33	78	2	1.2	2	140	2	56	0	3	25	25	7	76.7	4	24.8
34	48	4	1.2	2	160	4	33	2	7	28	26	12	83.3	6	14.4
35	72	3	2.4	2	200	10	56	0	4	25	22	12	70.0	2	23.3
36	56	5	2.4	4	230	6	34	0	7	27	21	11	80.0	4	20.8
37	52	5	2	3	270	4	45	2	4	24	25	11	80.0	3	27.7
38	45	5	4	2	340	5	28	2	6	24	26	12	80.0	4	30.9
39	69	5	3.2	2	370	2	49	2	8	27	27	10	86.7	1	32.8

S. no.	X1	X2	X3	X4	X5	X6	X7	X8	X9	X10	X11	X12	X13	X14	Y
40	56	4	0.8	3	200	3	39	0	2	25	28	6	73.3	4	17.8
41	55	5	3.6	2	380	4	38	0	6	26	25	11	76.7	4	19.5
42	55	6	1.6	2	200	4	35	0	6	24	26	8	76.7	3	8.7
43	46	6	4.4	2	270	4	29	2	4	25	21	9	76.7	6	25.3
44	42	6	1.6	2	200	4	27	0	4	22	23	10	70.0	5	9.3
45	46	5	1.2	2	170	3	31	0	2	25	21	7	66.7	4	10.7
46	64	4	1.6	2	170	4	52	0	7	26	24	6	83.3	3	11.4
47	56	5	1.2	2	140	4	39	0	7	26	27	8	80.0	2	13.1
48	56	5	0.8	3	170	3	35	0	7	28	21	11	86.7	3	12.7
49	55	6	1.6	2	140	4	30	2	6	26	25	9	80.0	4	15.6
50	65	4	3.2	4	80	6	2	2	7	27	26	12	83.3	3	21.9
51	51	5	2	4	50	4	29	0	3	24	23	11	73.3	5	17.9
52	75	4	3.6	2	350	12	59	0	6	26	21	12	76.7	2	16.2
53	66	4	0.6	2	90	5	51	0	8	28	21	7	90.0	4	7.1
54	62	5	1.2	4	210	6	49	2	4	24	27	11	73.3	3	21.1
55	72	4	1.6	2	160	4	54	0	5	26	25	6	76.7	2	9.8
56	68	4	2.4	2	280	3	52	2	3	25	23	12	66.7	2	27.0
57	36	6	2.8	2	280	5	21	0	6	26	21	11	90.0	7	13.6
58	46	6	1.2	3	220	4	28	0	6	24	22	7	76.7	7	14.8
59	60	5	1.2	3	240	3	47	2	5	25	26	10	80.0	2	28.7
60	62	5	5.2	2	300	5	45	0	7	27	27	11	86.7	3	12.8
61	52	5	5.2	2	350	5	36	0	7	27	22	12	83.3	5	17.0
62	68	6	3.2	2	280	4	52	0	5	26	18	11	83.3	4	15.7
63	38	6	2	3	270	5	23	0	3	26	17	7	73.3	8	11.3
64	55	6	0.8	2	50	3	42	0	8	25	17	6	83.3	5	5.9
65	60	5	2	2	160	5	48	2	5	26	18	6	86.7	5	1.6
66	40	6	2.4	2	280	4	28	0	3	26	19	9	73.3	6	13.1
67	38	6	1	2	190	5	24	0	4	25	20	9	76.7	6	7.6
68	55	5	4	2	300	5	38	2	7	28	21	10	86.7	4	14.4
69	60	5	4.8	3	310	3	47	0	2	27	21	8	66.7	7	18.3
70	61	5	0.8	2	80	2	45	0	5	26	19	7	76.7	6	4.7
71	65	4	2	2	180	3	49	0	7	26	18	10	86.7	5	11.1
72	60	4	1.6	2	190	6	46	0	4	25	20	8	90.0	3	9.9

S. no.	X1	X2	X3	X4	X5	X6	X7	X8	X9	X10	X11	X12	X13	X14	Y
73	42	6	1.6	2	170	9	27	2	4	24	22	7	73.3	7	18.8
74	62	4	1.4	4	280	6	44	2	5	27	22	9	76.7	4	29.6
75	48	6	4	2	260	4	31	0	5	27	21	11	73.3	4	7.8
76	42	5	2.4	2	150	6	27	2	6	26	23	10	80.0	5	27.5
77	61	3	1.6	2	120	7	35	0	8	25	24	8	80.0	3	5.7
78	45	5	2.8	2	300	8	32	2	4	28	25	9	86.7	5	25.1
79	47	6	3.2	2	350	5	30	2	3	26	26	6	76.7	4	27.4
80	42	5	2.4	2	160	7	22	0	3	24	27	6	73.3	6	13.0
81	55	5	3.6	2	380	4	38	0	7	27	22	7	76.7	3	9.8
82	55	6	1.6	2	200	4	35	2	3	25	21	11	76.7	5	11.9
83	46	6	4.4	2	270	4	29	2	4	26	23	8	76.7	7	23.0
84	42	6	1.6	2	200	4	27	0	4	27	24	9	80.0	8	9.3
85	46	5	1.2	2	170	3	31	0	2	25	25	7	70.0	5	10.0
86	70	4	1.6	2	170	4	52	0	5	24	26	11	86.7	3	14.0
87	58	5	1.2	2	140	4	39	0	5	26	21	7	83.3	6	8.0
88	56	4	0.8	3	230	4	35	0	2	24	22	8	76.7	7	11.7
89	42	5	4	2	200	5	27	2	8	29	24	10	83.3	8	33.1
90	65	4	3.2	2	460	9	51	2	8	26	19	12	80.0	3	23.9
91	27	6	1.2	2	180	2	10	0	4	27	20	7	80.0	9	12.1
92	35	6	3.2	2	310	8	17	2	3	26	20	11	66.7	8	6.5
93	68	5	10	2	320	4	48	2	7	27	21	12	80.0	4	32.4
94	65	3	1.2	2	85	4	48	2	4	26	18	6	83.3	3	7.7
95	53	6	3.6	3	250	6	38	2	4	27	25	9	80.0	6	22.5
96	47	6	2.8	3	250	4	29	2	2	26	25	8	66.7	9	26.1
97	68	4	3.2	4	275	5	47	2	2	23	24	11	70.0	5	29.2
98	57	5	2.8	3	250	4	39	2	5	24	22	7	76.7	5	20.5
99	58	6	2.4	2	200	3	37	2	7	27	23	7	76.7	6	18.3
100	43	4	0.8	1	50	2	28	0	3	24	22	6	73.3	5	4.5

Code:

```
Data corr; /*one can enter any other name for data*/
inputs x1 x2 x3 x4 x5 x6 x7 x8 x9 x10 x11 x12 x13 x14 yld;
cards;
```

1	43	5	1.6	3	100	3	24	2	9	24	21	7	76.7	1	14.5
2	47	5	1.2	2	75	4	30	2	8	22	19	6	73.3	1	27.5
99	58	6	2.4	2	200	3	37	2	7	27	23	7	76.7	6	18.3
100	43	4	0.8	1	50	2	28	0	3	24	22	6	73.3	5	4.5

```
;
/* Obtain correlation coefficient */
proc corr;
var x1 x2 x3 x4 x5 x6 x7 x8 x9 x10 x11 x12 x13 x14 yld;
run;
```

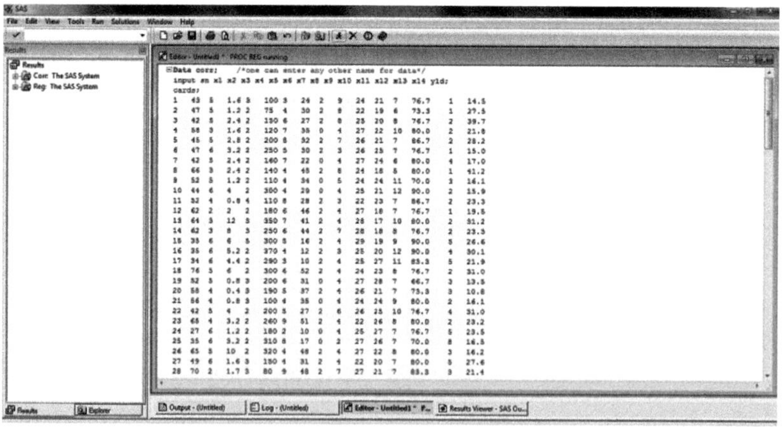

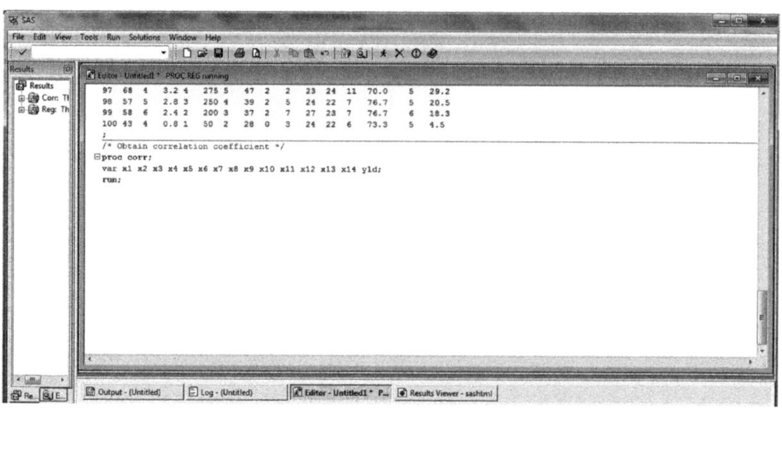

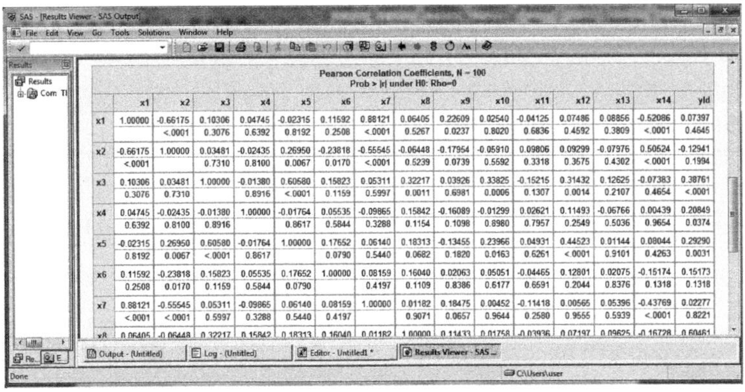

Pearson Correlation Coefficients, N = 100
Prob > |r| under H0: Rho=0

	x1	x2	x3	x4	x5	x6	x7	x8	x9	x10	x11	x12	x13	x14	yld
x1	1.00000	-0.66175	0.10306	0.04745	-0.02315	0.11592	0.88121	0.06405	0.22609	0.02540	-0.04125	0.07486	0.08856	-0.52086	0.07397
		<.0001	0.3076	0.6392	0.8192	0.2508	<.0001	0.5267	0.0237	0.8020	0.6836	0.4592	0.3809	<.0001	0.4645
x2	-0.66175	1.00000	0.03481	-0.02435	0.26950	-0.23818	-0.55545	-0.06448	-0.17954	-0.05910	0.09806	0.09299	-0.07976	0.50624	-0.12941
	<.0001		0.7310	0.8100	0.0067	0.0170	<.0001	0.5239	0.0739	0.5592	0.3318	0.3675	0.4302	<.0001	0.1994
x3	0.10306	0.03481	1.00000	-0.01380	0.60580	0.15823	0.05311	0.32217	0.03926	0.33825	-0.15215	0.31432	0.12625	-0.07383	0.38761
	0.3076	0.7310		0.8916	<.0001	0.1159	0.5997	0.0011	0.6981	0.0006	0.1307	0.0014	0.2107	0.4654	<.0001
x4	0.04745	-0.02435	-0.01380	1.00000	-0.01764	0.05535	-0.09865	0.15842	-0.16089	-0.01299	0.02621	0.11493	-0.06766	0.00439	0.20849
	0.6392	0.8100	0.8916		0.8617	0.5844	0.3288	0.1154	0.1098	0.8980	0.7957	0.2549	0.5036	0.9654	0.0374
x5	-0.02315	0.26950	0.60580	-0.01764	1.00000	0.17652	0.06140	0.18313	-0.13455	0.23966	0.04931	0.44523	0.01144	0.08044	0.29290
	0.8192	0.0067	<.0001	0.8617		0.0790	0.5440	0.0682	0.1820	0.0163	0.6261	<.0001	0.9101	0.4263	0.0031
x6	0.11592	-0.23818	0.15823	0.05535	0.17652	1.00000	0.08159	0.16040	0.02063	0.05051	-0.04465	0.12801	0.02075	-0.15174	0.15173
	0.2508	0.0170	0.1159	0.5844	0.0790		0.4197	0.1109	0.8386	0.6177	0.6591	0.2044	0.8376	0.1318	0.1318
x7	0.88121	-0.55545	0.05311	-0.09865	0.06140	0.08159	1.00000	0.01182	0.18475	0.00452	-0.11418	0.00565	0.05396	-0.43769	0.02277
	<.0001	<.0001	0.5997	0.3288	0.5440	0.4197		0.9071	0.0657	0.9644	0.2580	0.9555	0.5939	<.0001	0.8221
x8	0.06405	-0.06448	0.32217	0.15842	0.18313	0.16040	0.01182	1.00000	0.11433	0.01758	-0.03936	0.07197	0.09625	-0.16728	0.60461

7.2.3 *Example 3:* Following data given the yield (Y) corresponding to 14 yield components (X_1, X_2, X_3, X_4, X_5, X_6, X_7, X_8, X_9, X_{10}, X_{11}, X_{12}, X_{13}, X_{14}) for certain agricultural crop. Work out the ***regression analysis*** for below data.

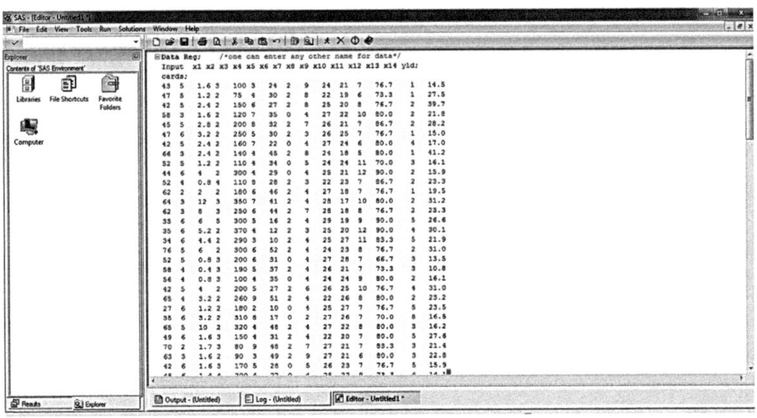

```
Data Reg;     /*one can enter any other name for data*/
Input  x1 x2 x3 x4 x5 x6 x7 x8 x9 x10 x11 x12 x13 x14 yld;
cards;
43  5  1.6 3  100 3  24  2  9  24  21  7   76.7  1  14.5
47  5  1.2 2  75  4  30  2  8  22  19  6   73.3  1  27.5
42  5  2.4 2  150 6  27  2  8  25  20  8   76.7  2  39.7
58  3  1.6 2  120 7  35  0  4  27  22  10  80.0  2  21.8
45  5  2.8 2  200 8  32  2  7  26  21  7   86.7  2  28.2
47  6  3.2 2  250 5  30  2  3  26  25  7   76.7  1  15.0
42  5  2.4 2  160 7  22  0  4  27  24  6   80.0  4  17.0
66  3  2.4 2  140 4  45  2  8  24  18  5   80.0  1  41.2
52  5  1.2 2  110 4  34  0  5  24  24  11  70.0  3  16.1
44  6  4   2  300 4  29  0  4  25  21  12  90.0  2  15.9
52  4  0.8 4  110 8  28  2  3  22  23  7   86.7  2  23.3
62  2  2   2  180 6  46  2  4  27  18  7   76.7  1  19.5
64  3  12  3  350 7  41  2  4  28  17  10  80.0  2  31.2
62  3  8   3  250 6  44  2  7  28  18  8   76.7  2  23.3
35  6  6   5  300 5  16  2  4  29  19  9   90.0  5  26.6
35  6  5.2 2  370 4  12  2  3  25  20  12  90.0  4  30.1
34  6  4.4 2  290 3  10  2  4  25  27  11  83.3  5  21.9
76  5  6   2  300 6  52  2  4  24  23  8   76.7  2  31.0
52  5  0.8 3  200 6  31  0  4  27  28  7   66.7  3  13.5
58  4  0.4 3  190 5  37  2  4  26  21  7   73.3  3  10.8
56  4  0.8 3  100 4  35  0  4  24  24  9   80.0  2  16.1
42  5  4   2  200 5  27  2  6  26  25  10  76.7  4  31.0
65  4  3.2 2  260 9  51  2  4  22  26  8   80.0  2  23.2
27  6  1.2 2  180 2  10  0  4  25  27  7   76.7  5  23.5
35  6  3.2 2  310 8  17  0  2  27  26  7   70.0  8  16.5
65  5  10  2  320 4  48  2  4  27  22  8   80.0  3  16.2
49  6  1.6 3  150 4  31  2  4  22  20  7   80.0  5  27.6
70  2  1.7 3  80  9  48  2  7  27  21  7   83.3  3  21.4
63  3  1.6 2  90  3  49  2  9  27  21  6   80.0  3  22.8
42  6  1.6 3  170 5  28  0  5  26  23  7   76.7  5  15.9
```

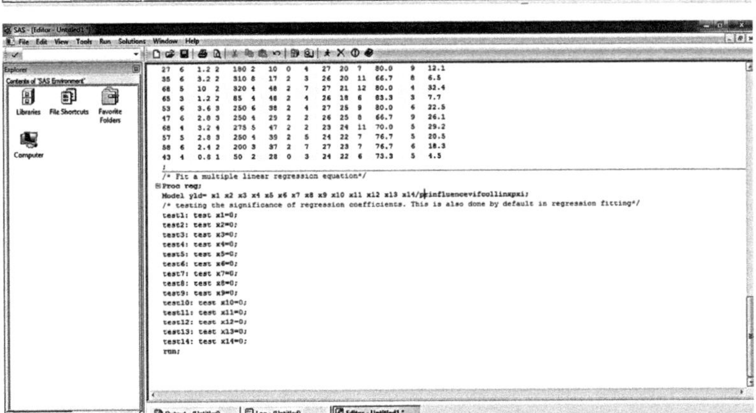

```
27  6  1.2 2  180 2  10  0  4  27  20  7   80.0  9  12.1
35  6  3.2 2  310 8  17  2  3  26  20  11  66.7  8  6.5
68  5  10  2  320 4  48  2  7  27  21  12  80.0  4  32.4
65  3  1.2 2  85  4  48  2  4  26  18  6   83.3  3  7.7
53  6  3.6 3  250 6  38  2  4  27  25  9   80.0  6  22.5
47  6  2.8 3  250 4  29  2  2  26  25  8   66.7  9  26.1
68  4  3.2 4  275 5  47  2  2  23  24  11  70.0  5  29.2
57  5  2.8 3  250 4  39  2  5  24  22  7   76.7  5  20.5
58  6  2.4 2  200 3  37  2  7  27  23  7   76.7  6  18.3
43  4  0.8 1  50  2  28  0  3  24  22  6   73.3  5  4.5
;
/* Fit a multiple linear regression equation*/
Proc reg;
Model yld= x1 x2 x3 x4 x5 x6 x7 x8 x9 x10 x11 x12 x13 x14/prinfluencevifcollinxpxi;
/* testing the significance of regression coefficients. This is also done by default in regression fitting*/
test1: test x1=0;
test2: test x2=0;
test3: test x3=0;
test4: test x4=0;
test5: test x5=0;
test6: test x6=0;
test7: test x7=0;
test8: test x8=0;
test9: test x9=0;
test10: test x10=0;
test11: test x11=0;
test12: test x12=0;
test13: test x13=0;
test14: test x14=0;
run;
```

Code:
Data Reg; /*one can enter any other name for data*/
inputs x1 x2 x3 x4 x5 x6 x7 x8 x9 x10 x11 x12 x13 x14 yld;
cards;

1	43	5	1.6	3	100	3	24	2	9	24	21	7	76.7	1	14.5
2	47	5	1.2	2	75	4	30	2	8	22	19	6	73.3	1	27.5
99	58	6	2.4	2	200	3	37	2	7	27	23	7	76.7	6	18.3
100	43	4	0.8	1	50	2	28	0	3	24	22	6	73.3	5	4.5

;

/* Fit a multiple linear regression equation*/

proc reg;

model yld = x1 x2 x3 x4 x5 x6 x7 x8 x9 x10 x11 x12 x13 x14/p *r* influence vif Collin xpx i;

/* testing the significance of regression coefficients. This is also done by default in regression fitting*/

test1: test x1 = 0;

test2: test x2 = 0;

test3: test x3 = 0;

test4: test x4 = 0;

test5: test x5 = 0;

test6: test x6 = 0;

test7: test x7 = 0;

test8: test x8 = 0;

test9: test x9 = 0;

test10: test x10 = 0;

test11: test x11 = 0;

test12: test x12 = 0;

test13: test x13 = 0;

test14: test x14 = 0;

run;

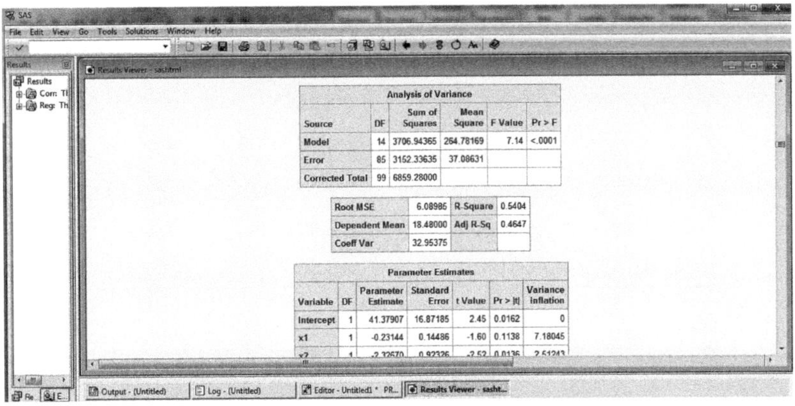

Analysis of Variance					
Source	DF	Sum of Squares	Mean Square	F Value	Pr > F
Model	14	3706.94365	264.78169	7.14	<.0001
Error	85	3152.33635	37.08631		
Corrected Total	99	6859.28000			

Root MSE	6.08985	R-Square	0.5404
Dependent Mean	18.48000	Adj R-Sq	0.4647
Coeff Var	32.95375		

Parameter Estimates						
Variable	DF	Parameter Estimate	Standard Error	t Value	Pr > |t|	Variance Inflation
Intercept	1	41.37907	16.87185	2.45	0.0162	0
x1	1	-0.23144	0.14486	-1.60	0.1138	7.18045
x2	1	-2.32670	0.92326	-2.52	0.0136	2.51243
x3	1	1.03332	0.44809	2.31	0.0235	2.13851
x4	1	2.05376	0.95094	2.16	0.0336	1.21916
x5	1	0.01741	0.01124	1.55	0.1251	2.55233
x6	1	-0.19512	0.34971	-0.56	0.5783	1.17873
x7	1	0.06265	0.12761	0.49	0.6247	6.00828
x8	1	3.73017	0.67801	5.50	<.0001	1.23905
x9	1	0.78008	0.42409	1.84	0.0693	1.65248
x10	1	-1.03268	0.49117	-2.10	0.0385	1.51202
x11	1	0.50274	0.23774	2.11	0.0374	1.19934
x12	1	0.20521	0.35824	0.57	0.5683	1.40927
x13	1	-0.06294	0.12032	-0.52	0.6023	1.34658
x14	1	-0.29487	0.42958	-0.69	0.4943	1.85821

7.2.4 ***Example 4:*** CRD with unequal replication: A Rice trial under Completely Randomized ***Design (CRD)*** conducted with six treatments T1, T2, T3, T4, T5 and T6 unequal replication. The yield of all treatments in Kg/ plot, give below in table.

Treatment	R1	R2	R3
T1	1.12	2.49	3.43
T2	3.55	3.39	
T3	3.16	4.21	3.78
T4	2.66	1.18	
T5	3.88	2.00	1.38
T6	0.12	0.88	

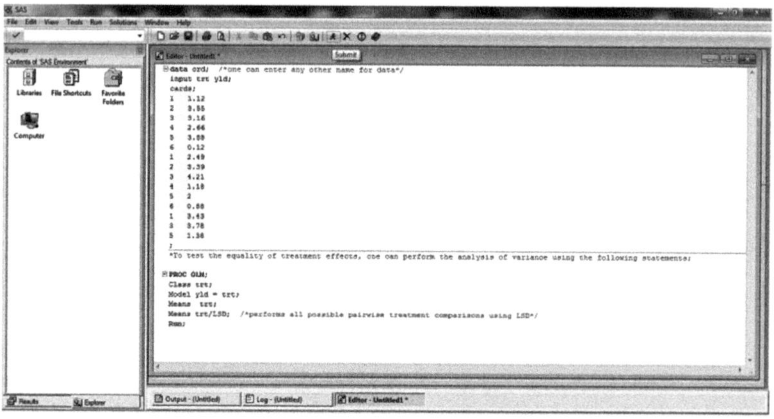

Code:

```
data crd; /*one can enter any other name for data*/
input trt yld;
cards;
1    1.12
2    3.55
3    3.16
4    2.66
5    3.88
6    0.12
1    2.49
2    3.39
3    4.21
4    1.18
5    2
6    0.88
1    3.43
```

3 3.78
5 1.38
;
*To test the equality of treatment effects, one can perform the analysis of variance using the following statements;
PROC GLM;
Class trt;
Model yld = trt;
Means trt;
Means trt/LSD; /*performs all possible pairwise treatment comparisons using LSD*/
Run;

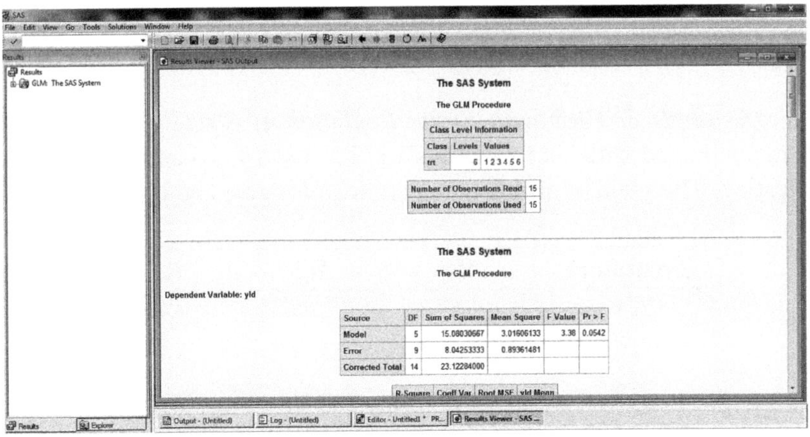

Results:

The SAS System

The GLM Procedure

Class Level Information		
Class	Levels	Values
trt	6	1 2 3 4 5 6

Number of Observations Read	15
Number of Observations Used	15

Source	DF	Sum of Squares	Mean Square	F Value	Pr > F
Model	5	15.08030667	3.01606133	3.38	0.0542
Error	9	8.04253333	0.89361481		
Corrected Total	14	23.12284000			

R-Square	Coeff Var	Root MSE	yld Mean
0.652182	38.08671	0.945312	2.482000

The SAS System

The GLM Procedure

t Tests (LSD) for yld

Note: This test controls the Type I comparisonwise error rate, not the experimentwise error rate.

Alpha	0.05
Error Degrees of Freedom	9
Error Mean Square	0.893615
Critical Value of t	2.26216

7.2.5 Example 5: CRD with Equal Replication: A rice trial under CRD design conducted with six treatments T1, T2, T3, T4, T5 and T6 with equal replication. The yield of all treatments in Kg/plot gives below in

Treatment	R1	R2	R3
T1	1.12	2.49	3.43
T2	3.55	3.39	4.12
T3	3.16	4.21	3.78
T4	2.66	1.18	2.37
T5	3.88	2.00	1.38
T6	0.12	0.88	0.92

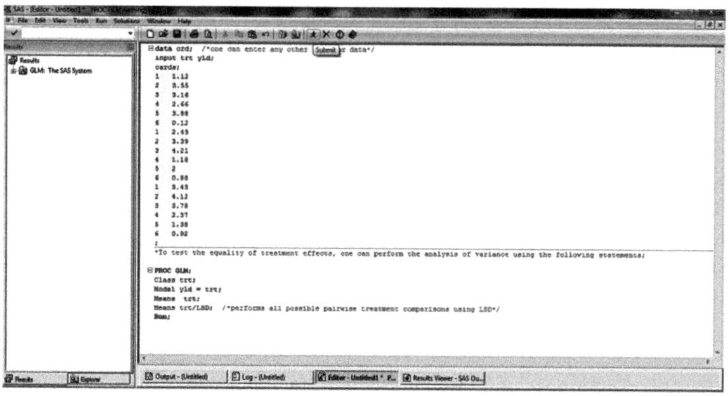

Here, code will be same for Unequal and equal replications (except inclusion of data).

The SAS System

The GLM Procedure

Class Level Information		
Class	Levels	Values
trt	6	1 2 3 4 5 6

Number of Observations Read	18
Number of Observations Used	18

Source	DF	Sum of Squares	Mean Square	F Value	Pr > F
Model	5	19.68140000	3.93628000	5.51	0.0073
Error	12	8.57680000	0.71473333		
Corrected Total	17	28.25820000			

R-Square	Coeff Var	Root MSE	yld Mean
0.696485	34.08948	0.845419	2.480000

Alpha	0.05
Error Degrees of Freedom	12
Error Mean Square	0.714733
Critical Value of t	2.17881
Least Significant Difference	1.504

7.2.6 Example 6: RBD Design: A field trial conduct in soybean in seven different varieties and each variety is having four replications. The data are given below in table:

Treatment	R1	R2	R3	R4
T1	34.12	32.32	31.21	30.12
T2	36.44	28.00	29.32	29.55
T3	27.77	25.60	27.80	29.70
T4	30.90	30.01	34.00	32.07
T5	28.75	31.49	29.03	22.32
T6	20.39	22.99	25.78	27.40
T7	27.23	29.58	30.48	26.50

Solution:

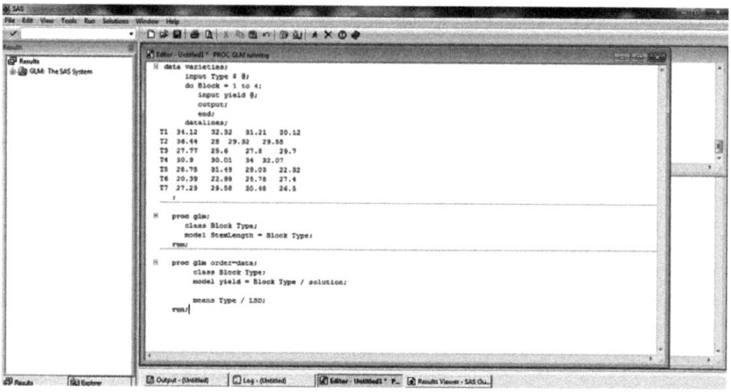

Code:

```
data varieties;
input Type $ @;
do Block = 1to4;
input yield @;
output;
end;
datalines;
T1   34.12   32.32   31.21   30.12
T2   36.44   28      29.32   29.55
T3   27.77   25.6    27.8    29.7
T4   30.9    30.01   34      32.07
T5   28.75   31.49   29.03   22.32
T6   20.39   22.99   25.78   27.4
T7   27.23   29.58   30.48   26.5
;
Proc glm;
class Block Type;
model StemLength = Block Type;
run;
proc glm order = data;
class Block Type;
model yield = Block Type / solution;
means Type / LSD;
run;
```

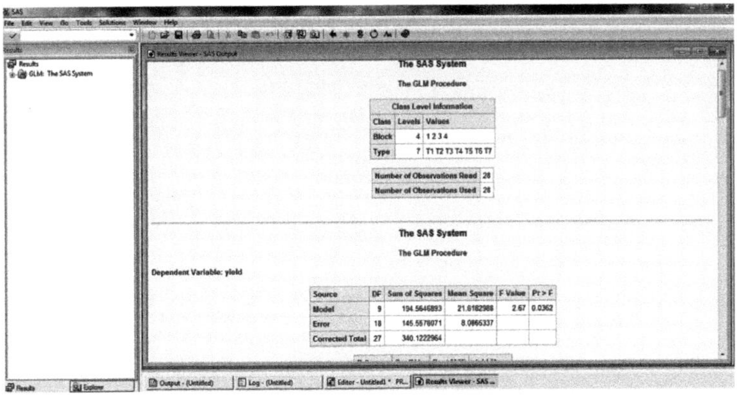

Results:

The GLM Procedure

Class Level Information		
Class	Levels	Values
Block	4	1 2 3 4
Type	7	T1 T2 T3 T4 T5 T6 T7

Number of Observations Read	28
Number of Observations Used	28

Source	DF	Sum of Squares	Mean Square	F Value	Pr > F
Model	9	194.5646893	21.6182988	2.67	0.0362
Error	18	145.5576071	8.0865337		
Corrected Total	27	340.1222964			

R-Square	Coeff Var	Root MSE	yield Mean
0.572043	9.819469	2.843683	28.95964

Note: This test controls the Type I comparisonwise error rate, not the experimentwise error rate.

Alpha	0.05
Error Degrees of Freedom	18
Error Mean Square	8.086534
Critical Value of t	2.10092
Least Significant Difference	4.2245

Means with the same letter are not significantly different.					
t Grouping			Mean	N	Type
	A		31.943	4	T1
	A				
B	A		31.745	4	T4
B	A				
B	A		30.828	4	T2
B	A				
B	A		28.448	4	T7
B	A				
B	A	C	27.898	4	T5
B		C			
B		C	27.718	4	T3
		C			
		C	24.140	4	T6

7.2.7 ***Example 7: LSD:*** The following information is pertaining to the yield data (q/ha) from an experiment on six varieties of Rice in Rewa (M.P.) conducted in LSD

	C–1	C–2	C–3	C–4	C–5	C–6
R–1	(T3) 3.1	(T6) 5.95	(T1) 1.75	(T5) 6.4	(T2) 3.85	(T4) 5.3
R–2	(T2) 4.8	(T1) 2.7	(T3) 3.3	(T6) 5.95	(T4) 3.7	(T5) 5.4
R–3	(T1) 3	(T2) 2.95	(T5) 6.7	(T4) 5.95	(T6) 7.75	(T3) 7.1
R–4	(T5) 6.4	(T4) 5.8	(T2) 3.8	(T3) 6.55	(T1) 4.8	(T6) 9.4
R–5	(T6) 5.2	(T3) 4.85	(T4) 6.6	(T2) 4.6	(T5) 7	(T1) 5
R–6	(T4) 4.25	(T5) 6.65	(T6) 9.3	(T1) 4.95	(T3) 9.3	(T2) 8.4

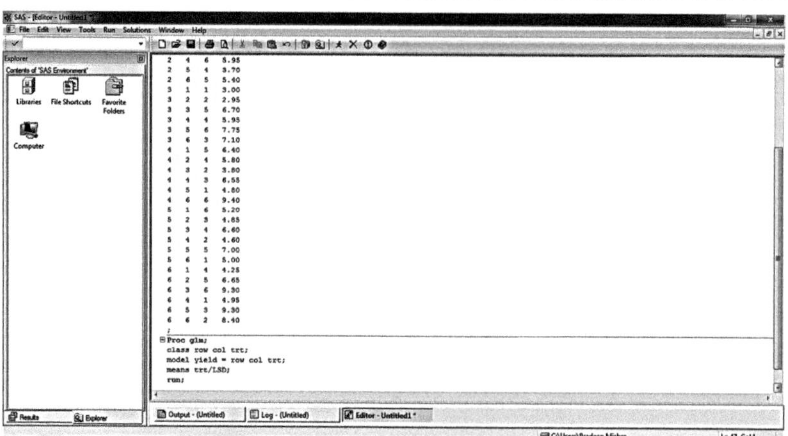

```
Code:
data latin;
input row col trt yield;
cards;
1   1   3   3.10
1   2   6   5.95
1   3   1   1.75
1   4   5   6.40
1   5   2   3.85
1   6   4   5.30
2   1   2   4.80
2   2   1   2.70
2   3   3   3.30
2   4   6   5.95
2   5   4   3.70
2   6   5   5.40
3   1   1   3.00
3   2   2   2.95
3   3   5   6.70
3   4   4   5.95
3   5   6   7.75
3   6   3   7.10
4   1   5   6.40
4   2   4   5.80
4   3   2   3.80
4   4   3   6.55
4   5   1   4.80
4   6   6   9.40
5   1   6   5.20
5   2   3   4.85
5   3   4   6.60
5   4   2   4.60
5   5   5   7.00
5   6   1   5.00
6   1   4   4.25
6   2   5   6.65
6   3   6   9.30
6   4   1   4.95
6   5   3   9.30
6   6   2   8.40
```

```
;
proc glm;
class row col trt;
model yield = row col trt;
meanstrt/LSD;
run;
```

Result:

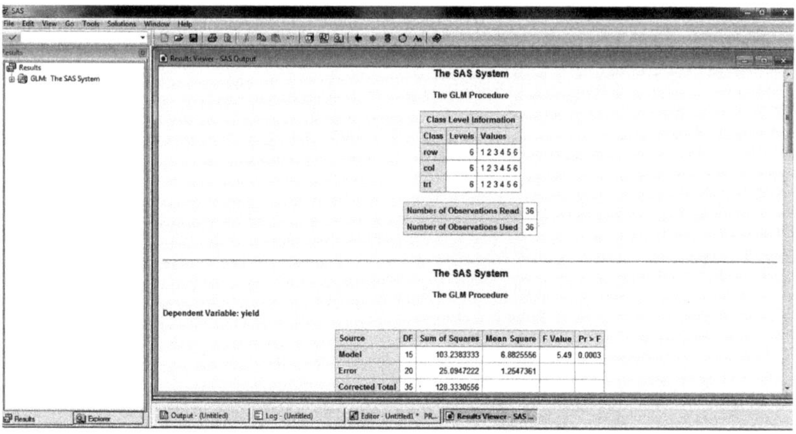

Source	DF	Sum of Squares	Mean Square	F Value	Pr > F
Model	15	103.2383333	6.8825556	5.49	0.0003
Error	20	25.0947222	1.2547361		
Corrected Total	35	128.3330556			

R-Square	Coeff Var	Root MSE	yield Mean
0.804456	20.31506	1.120150	5.513889

Source	DF	Type I SS	Mean Square	F Value	Pr > F
row	5	34.44222222	6.88844444	5.49	0.0024
col	5	21.58555556	4.31711111	3.44	0.0210
trt	5	47.21055556	9.44211111	7.53	0.0004

Alpha	0.05
Error Degrees of Freedom	20
Error Mean Square	1.254736
Critical Value of t	2.08596
Least Significant Difference	1.349

Means with the same letter are not significantly different.				
t Grouping		Mean	N	trt
	A	7.2583	6	6
	A			
B	A	6.4250	6	5
B				
B	C	5.7000	6	3
B	C			
B	C	5.2667	6	4
	C			
D	C	4.7333	6	2
D				
D		3.7000	6	1

7.2.8 Example 8: LSD: The following information is pertaining to the yield data (q/ha) from an experiment on five varieties of Rice in Rewa (M.P.) conducted in LSD.

D–44.0	A–29.1	E–31.1	B–42.0	C–47.2
E–26.2	B–43.1	A–29.0	C–44.3	D–38.1
C–40.6	E–38.5	B–43.1	D–45.8	A–29.2
A–35.8	C–36.1	D–51.7	E–33.7	B–49.9
B–49.3	D–34.6	C–46.1	A–31.3	E–29.4

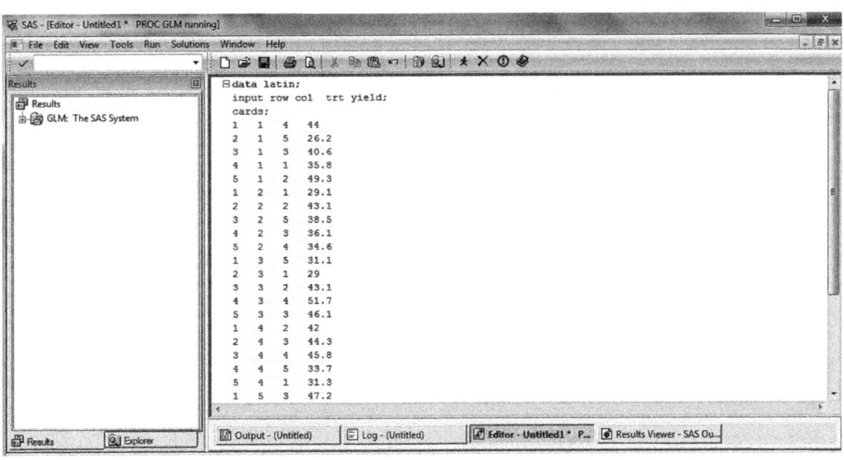

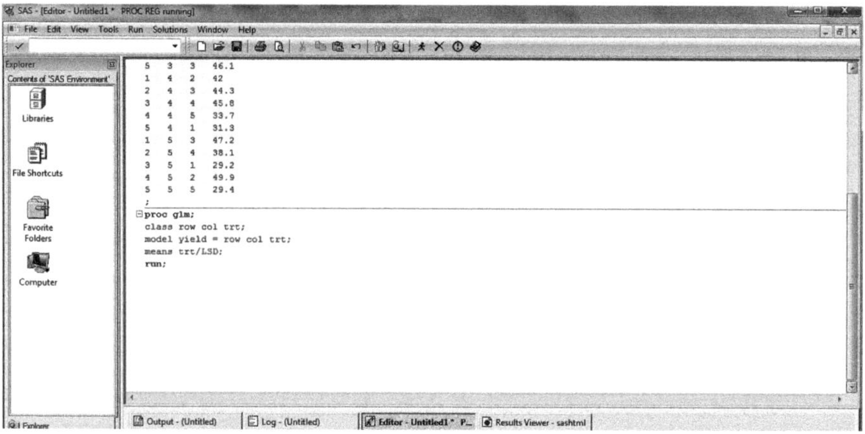


```
Code:
data latin;
input row col trt yield;
cards;
1    1    4    44
2    1    5    26.2
3    1    3    40.6
4    1    1    35.8
5    1    2    49.3
1    2    1    29.1
2    2    2    43.1
3    2    5    38.5
4    2    3    36.1
5    2    4    34.6
1    3    5    31.1
2    3    1    29
3    3    2    43.1
4    3    4    51.7
5    3    3    46.1
1    4    2    42
2    4    3    44.3
3    4    4    45.8
4    4    5    33.7
5    4    1    31.3
1    5    3    47.2
2    5    4    38.1
```

```
3    5    1    29.2
4    5    2    49.9
5    5    5    29.4
;
proc glm;
class row col trt;
model yield = row col trt;
meanstrt/LSD;
run;
```

Result:

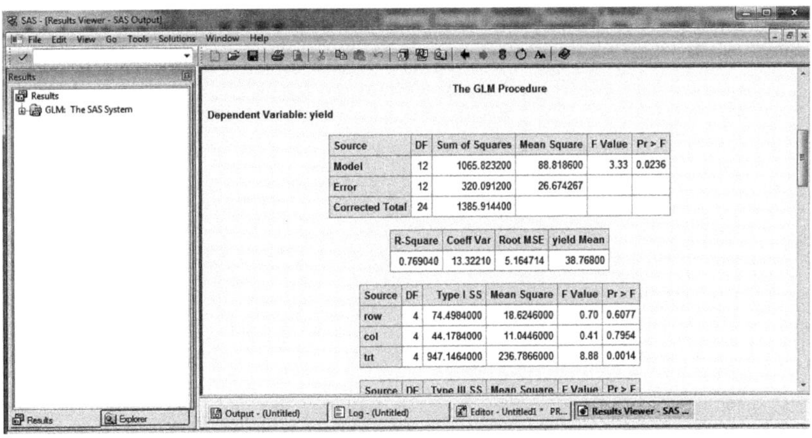

Dependent Variable: yield

Source	DF	Sum of Squares	Mean Square	F Value	Pr > F
Model	12	1065.823200	88.818600	3.33	0.0236
Error	12	320.091200	26.674267		
Corrected Total	24	1385.914400			

R-Square	Coeff Var	Root MSE	yield Mean
0.769040	13.32210	5.164714	38.76800

Source	DF	Type I SS	Mean Square	F Value	Pr > F
row	4	74.4984000	18.6246000	0.70	0.6077
col	4	44.1784000	11.0446000	0.41	0.7954
trt	4	947.1464000	236.7866000	8.88	0.0014

7.2.9 Example 9: Calculate the *Factorial RBD* for below data.

Subsample	Rep	Treatment	Yield
1	1	1	19.5
1	1	2	12.8
1	1	3	16
1	1	4	11.7
2	1	1	14
2	1	2	16.5
2	1	3	12.2
2	1	4	12.8
3	1	1	15.2
3	1	2	15.8
3	1	3	13.8
3	1	4	15
1	2	1	18
1	2	2	17.2
1	2	3	10.8
1	2	4	12.5
2	2	1	15
2	2	2	14.8
2	2	3	15.5
2	2	4	15.7
3	2	1	18
3	2	2	19.2
3	2	3	14.3
3	2	4	12.8
1	3	1	11.5
1	3	2	16
1	3	3	13.5
1	3	4	15
2	3	1	12.5
2	3	2	16.2
2	3	3	14.8

Subsample	Rep	Treatment	Yield
2	3	4	15.2
3	3	1	14.8
3	3	2	15.5
3	3	3	14.5
3	3	4	14.7

Codes:

Data factrbd;

input y	s	rep	n;
datalines;			
19.5	1	1	1
12.8	1	1	2
16	1	1	3
11.7	1	1	4
14	2	1	1
16.5	2	1	2
12.2	2	1	3
12.8	2	1	4
15.2	3	1	1
15.8	3	1	2
13.8	3	1	3
15	3	1	4
18	1	2	1
17.2	1	2	2
10.8	1	2	3
12.5	1	2	4
15	2	2	1
14.8	2	2	2
15.5	2	2	3
15.7	2	2	4
18	3	2	1
19.2	3	2	2

14.3	3	2	3
12.8	3	2	4
11.5	1	3	1
16	1	3	2
13.5	1	3	3
15	1	3	4
12.5	2	3	1
16.2	2	3	2
14.8	2	3	3
15.2	2	3	4
14.8	3	3	1
15.5	3	3	2
14.5	3	3	3
14.7	3	3	4

```
;
proc glm;
class s       rep     n;
model y  =  rep s             n s*n/ss3;
means s n s*n/lsd;
run;
```

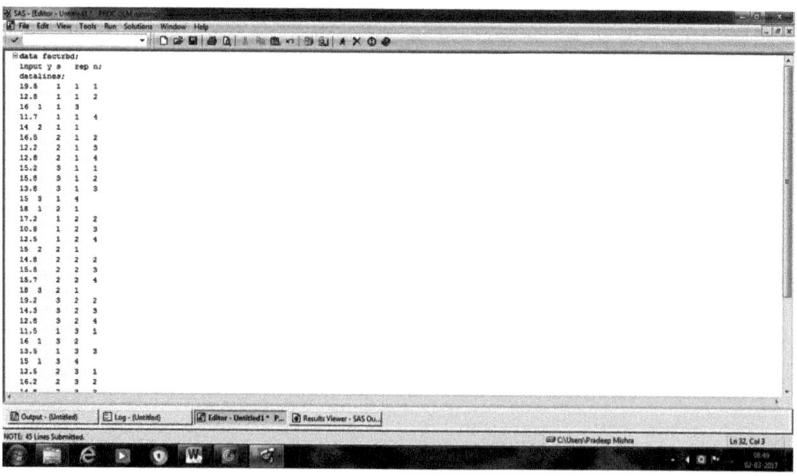

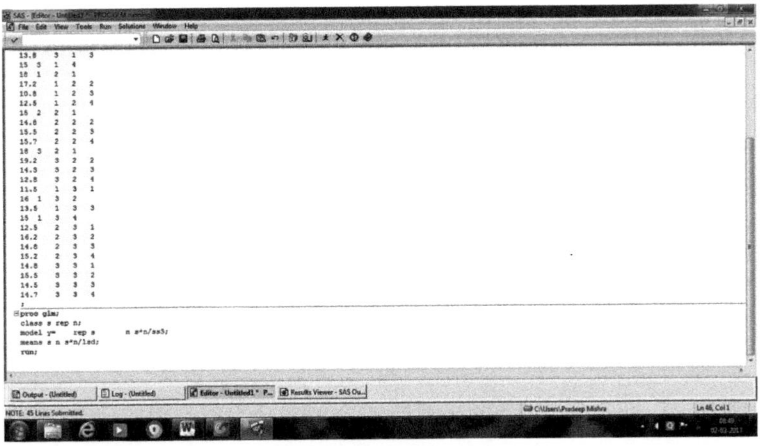
13.8 3 1 3
15 3 1 4
18 1 2 1
17.2 1 2 2
10.8 1 2 3
12.5 1 2 4
15 2 2 1
14.8 2 2 2
15.5 2 2 3
15.7 2 2 4
18 3 2 1
19.2 3 2 2
14.3 3 2 3
12.8 3 2 4
11.5 1 3 1
16 1 3 2
13.5 1 3 3
15 1 3 4
12.5 2 3 1
16.2 2 3 2
14.6 2 3 3
15.2 2 3 4
14.8 3 3 1
15.5 3 3 2
14.5 3 3 3
14.7 3 3 4
;
proc glm;
class s rep n;
model y= rep s n s*n/ss3;
means s n s*n/lsd;
run;
Output - (Untitled)
Log - (Untitled)
NOTE: 45 Lines Submitted.

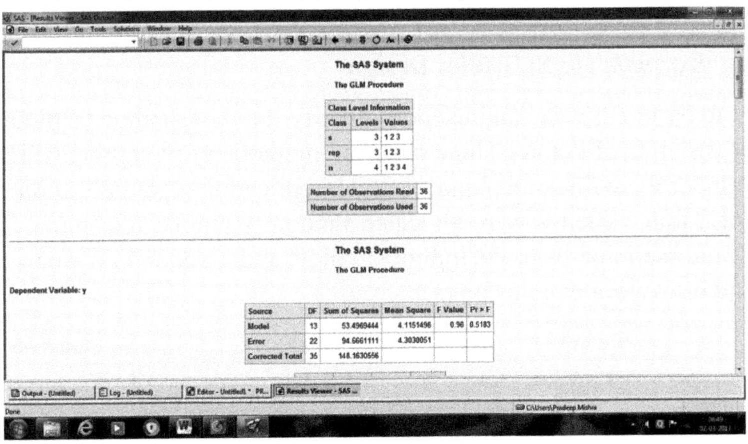
The SAS System
The GLM Procedure
Class Level Information
Class Levels Values
s 3 1 2 3
rep 3 1 2 3
n 4 1 2 3 4
Number of Observations Read 36
Number of Observations Used 36
The SAS System
The GLM Procedure
Dependent Variable: y
Source DF Sum of Squares Mean Square F Value Pr > F
Model 13 53.4969444 4.1151496 0.96 0.5183
Error 22 94.6661111 4.3030051
Corrected Total 35 148.1630556

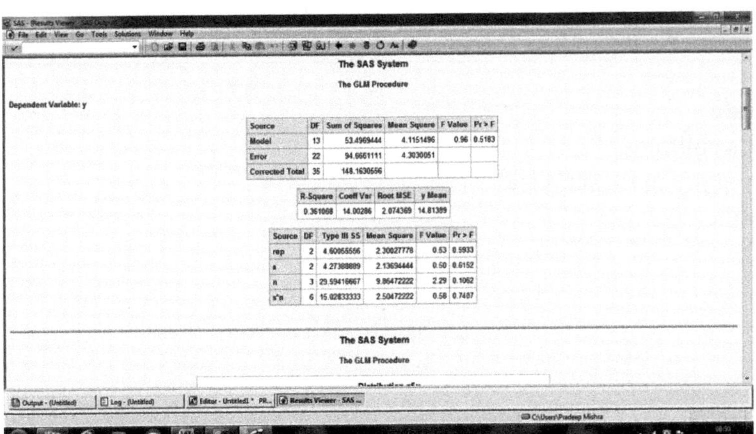
The SAS System
The GLM Procedure
Dependent Variable: y
Source DF Sum of Squares Mean Square F Value Pr > F
Model 13 53.4969444 4.1151496 0.96 0.5183
Error 22 94.6661111 4.3030051
Corrected Total 35 148.1630556
R-Square Coeff Var Root MSE y Mean
0.361068 14.00286 2.074369 14.81389
Source DF Type III SS Mean Square F Value Pr > F
rep 2 4.60055556 2.30027778 0.53 0.5933
s 2 4.27388889 2.13694444 0.50 0.6152
n 3 29.59416667 9.86472222 2.29 0.1062
s*n 6 15.02833333 2.50472222 0.58 0.7407
The SAS System
The GLM Procedure

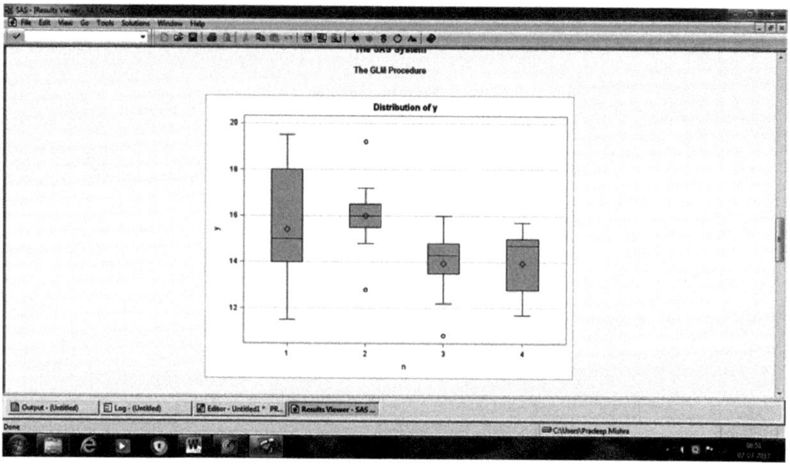

7.2.10 *Example 10:* Split-Plot Design

A field experiment was conducted to identify the best method of ploughing in four varieties of Potato. Three different methods of ploughing were allocated randomly to three main plots in each replication separately and in each main plot four varieties were allocated randomly among four plots. Yields (t/ha) were recorded from the individual plots and given below. Analyze the data and draw your conclusions?

	Plough–1				Plough–2				Plough–3			
	V1	**V2**	**V3**	**V4**	**V1**	**V2**	**V3**	**V4**	**V1**	**V2**	**V3**	**V4**
Rep–1	8.7	9.1	7.8	7.2	9.5	12.6	11.2	9.8	7.5	9.5	8.2	7.9
Rep–2	8.6	9.2	7.9	7.3	9.4	12.5	11	9.6	7.6	9.8	8.4	8
Rep–3	8.5	9.3	8.2	7.4	9.6	12.3	10.9	10	7.4	9.7	8.5	8.1

Code:

```
data splitplot;
input rep main sub yield;
cards;
1 1 1 8.7
2 1 1 8.6
3 1 1 8.5
1 1 2 9.1
2 1 2 9.2
3 1 2 9.3
1 1 3 7.8
```

```
2	1	3	7.9
3	1	3	8.2
1	1	4	7.2
2	1	4	7.3
3	1	4	7.4
1	2	1	9.5
2	2	1	9.4
3	2	1	9.6
1	2	2	12.6
2	2	2	12.5
3	2	2	12.3
1	2	3	11.2
2	2	3	11
3	2	3	10.9
1	2	4	9.8
2	2	4	9.6
3	2	4	10
1	3	1	7.5
2	3	1	7.6
3	3	1	7.4
1	3	2	9.5
2	3	2	9.8
3	3	2	9.7
1	3	3	8.2
2	3	3	8.4
3	3	3	8.5
1	3	4	7.9
2	3	4	8
3	3	4	8.1
;
proc glm;
class rep main sub;
model yield = rep main rep*main sub main*sub;
testh =  main e = rep*main;
means main /lsde = rep*main;
means sub main*sub/lsd;
lsmeans main*sub/pdiff;
```

run;

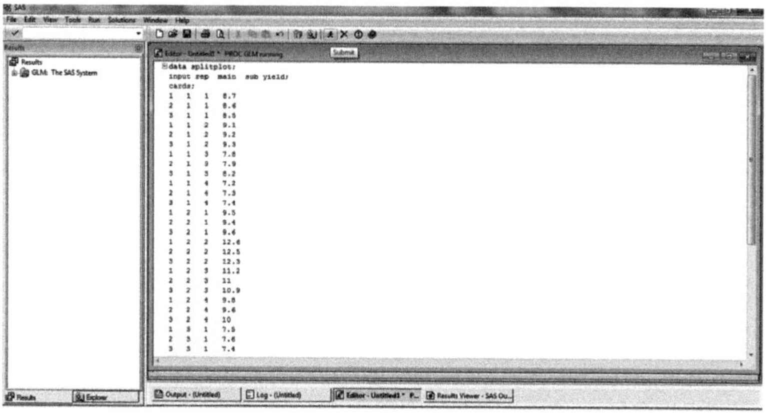
Editor - Untitled1 * PROC GLM running
data splitplot;
input rep main sub yield;
cards;
1 1 1 8.7
2 1 1 8.6
3 1 1 8.5
1 1 2 9.1
2 1 2 9.2
3 1 2 9.3
1 1 3 7.8
2 1 3 7.9
3 1 3 8.2
1 1 4 7.2
2 1 4 7.3
3 1 4 7.4
1 2 1 9.5
2 2 1 9.4
3 2 1 9.6
1 2 2 12.6
2 2 2 12.5
3 2 2 12.3
1 2 3 11.2
2 2 3 11
3 2 3 10.9
1 2 4 9.8
2 2 4 9.6
3 2 4 10
1 3 1 7.5
2 3 1 7.6
3 3 1 7.4
Output - (Untitled)
Log - (Untitled)

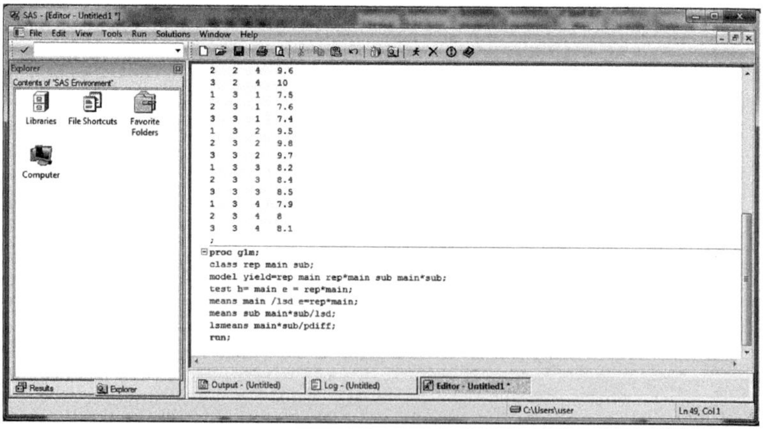
SAS - [Editor - Untitled1 *]
File Edit View Tools Run Solutions Window Help
Explorer
Contents of 'SAS Environment'
Libraries
File Shortcuts
Favorite Folders
Computer
2 2 4 9.6
3 2 4 10
1 3 1 7.5
2 3 1 7.6
3 3 1 7.4
1 3 2 9.5
2 3 2 9.8
3 3 2 9.7
1 3 3 8.2
2 3 3 8.4
3 3 3 8.5
1 3 4 7.9
2 3 4 8
3 3 4 8.1
;
proc glm;
class rep main sub;
model yield=rep main rep*main sub main*sub;
test h= main e = rep*main;
means main /lsd e=rep*main;
means sub main*sub/lsd;
lsmeans main*sub/pdiff;
run;
Results
Explorer
Output - (Untitled)
Log - (Untitled)
Editor - Untitled1 *
C:\Users\user
Ln 49, Col 1

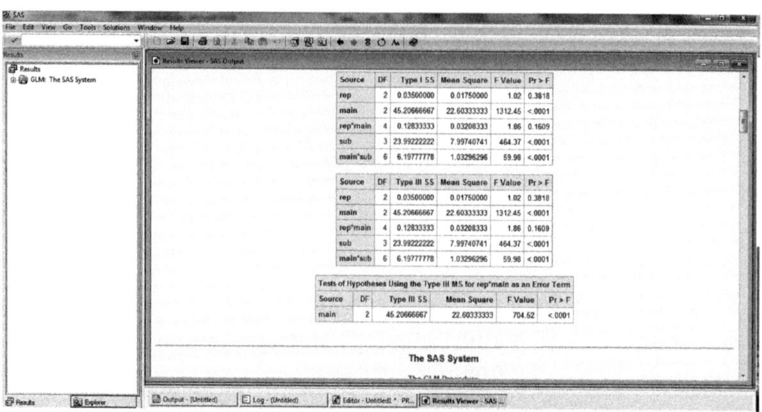
Results Viewer - SAS Output
Source DF Type I SS Mean Square F Value Pr > F
rep 2 0.03500000 0.01750000 1.02 0.3818
main 2 45.20666667 22.60333333 1312.45 <.0001
rep*main 4 0.12833333 0.03208333 1.86 0.1609
sub 3 23.99222222 7.99740741 464.37 <.0001
main*sub 6 6.19777778 1.03296296 59.98 <.0001
Source DF Type III SS Mean Square F Value Pr > F
rep 2 0.03500000 0.01750000 1.02 0.3818
main 2 45.20666667 22.60333333 1312.45 <.0001
rep*main 4 0.12833333 0.03208333 1.86 0.1609
sub 3 23.99222222 7.99740741 464.37 <.0001
main*sub 6 6.19777778 1.03296296 59.98 <.0001
Tests of Hypotheses Using the Type III MS for rep*main as an Error Term
Source DF Type III SS Mean Square F Value Pr > F
main 2 45.20666667 22.60333333 704.52 <.0001
The SAS System

Source	DF	Type III SS	Mean Square	F Value	Pr > F
rep	2	0.03500000	0.01750000	1.02	0.3818
main	2	45.20666667	22.60333333	1312.45	<.0001
rep*main	4	0.12833333	0.03208333	1.86	0.1609
sub	3	23.99222222	7.99740741	464.37	<.0001
main*sub	6	6.19777778	1.03296296	59.98	<.0001

Alpha	0.05
Error Degrees of Freedom	4
Error Mean Square	0.032083
Critical Value of t	2.77645
Least Significant Difference	0.203

Means with the same letter are not significantly different.			
t Grouping	Mean	N	main
A	10.70000	12	2
B	8.38333	12	3
B			
B	8.26667	12	1

7.2.11. Example 11: Strip Plot Design

A field experiment was conducted to find the efficacy of three different sources of organic manure and four irrigation schedules was conducted in strip plot design with three replications? Given below are yield (t/ha) for different treatments? Identify the best organic manure, irrigation schedule along with their combination?

	Vertical Plot											
Horizontal Plot	**M–1**				**M–2**				**M–3**			
	I–1	I–2	I–3	I–4	I–1	I–2	I–3	I–4	I–1	I–2	I–3	I–4
Rep–1	11.2	10.2	14.5	12.3	11.8	10.9	16.2	13.5	11	9.5	12.5	10.6
Rep–2	11.3	10.5	14.6	12.5	11.9	10.8	16.5	13.8	10.9	9.8	12	10.8
Rep–3	11.5	10.4	14.2	12.8	11.6	10.7	16.6	13.9	10.5	9.7	12.4	10.7

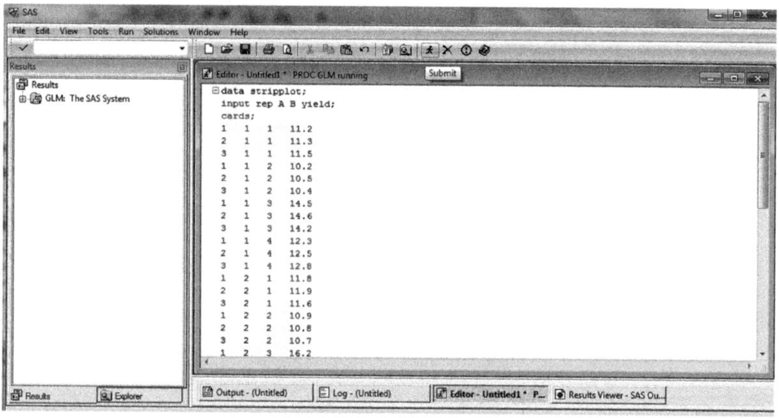

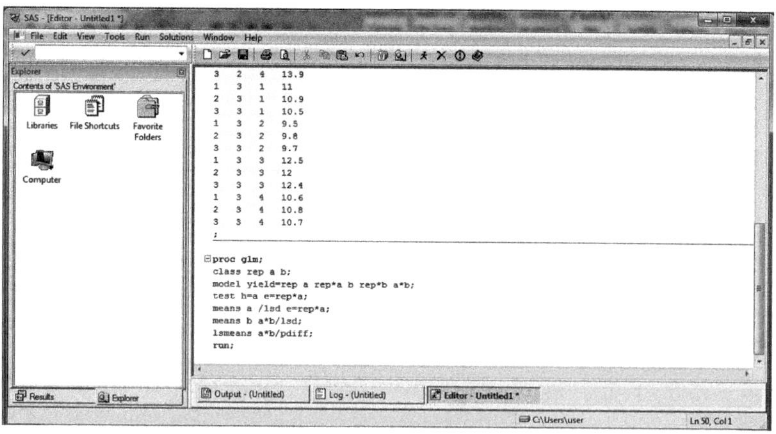

Code:

```
data strip plot;
input rep A B yield;
cards;
1    1    1    11.2
2    1    1    11.3
3    1    1    11.5
1    1    2    10.2
2    1    2    10.5
3    1    2    10.4
1    1    3    14.5
2    1    3    14.6
```

```
3    1    3    14.2
1    1    4    12.3
2    1    4    12.5
3    1    4    12.8
1    2    1    11.8
2    2    1    11.9
3    2    1    11.6
1    2    2    10.9
2    2    2    10.8
3    2    2    10.7
1    2    3    16.2
2    2    3    16.5
3    2    3    16.6
1    2    4    13.5
2    2    4    13.8
3    2    4    13.9
1    3    1    11
2    3    1    10.9
3    3    1    10.5
1    3    2    9.5
2    3    2    9.8
3    3    2    9.7
1    3    3    12.5
2    3    3    12
3    3    3    12.4
1    3    4    10.6
2    3    4    10.8
3    3    4    10.7
;
proc glm;
class rep a b;
model yield = rep a rep*a b rep*b a*b;
testh = a e = rep*a;
means a /lsde = rep*a;
```

```
means b a*b/lsd;
lsmeans a*b/pdiff;
run;
```

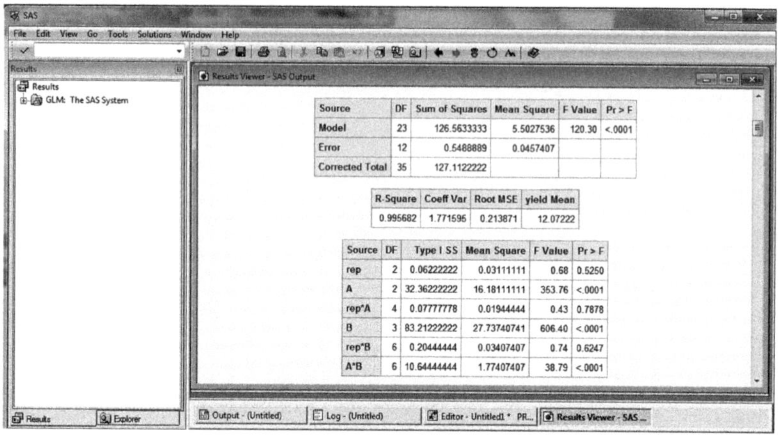

Source	DF	Sum of Squares	Mean Square	F Value	Pr > F
Model	23	126.5633333	5.5027536	120.30	<.0001
Error	12	0.5488889	0.0457407		
Corrected Total	35	127.1122222			

R-Square	Coeff Var	Root MSE	yield Mean
0.995682	1.771595	0.213871	12.07222

Source	DF	Type I SS	Mean Square	F Value	Pr > F
rep	2	0.06222222	0.03111111	0.68	0.5250
A	2	32.36222222	16.18111111	353.76	<.0001
rep*A	4	0.07777778	0.01944444	0.43	0.7878
B	3	83.21222222	27.73740741	606.40	<.0001
rep*B	6	0.20444444	0.03407407	0.74	0.6247
A*B	6	10.64444444	1.77407407	38.79	<.0001

Alpha	0.05
Error Degrees of Freedom	4
Error Mean Square	0.019444
Critical Value of t	2.77645
Least Significant Difference	0.1581

Means with the same letter are not significantly different.			
t Grouping	Mean	N	A
A	13.18333	12	2
B	12.16667	12	1
C	10.86667	12	3

7.2.12 Example 12: Following data give the yield (Y) corresponding to 14 yield components (X_1, X_2, X_3, X_4, X_5, X_6, X_7, X_8, X_9, X_{10}, X_{11}, X_{12}, X_{13}, X_{14}) for certain agricultural crop. Work out the ***Principal Component Analysis*** for below data.

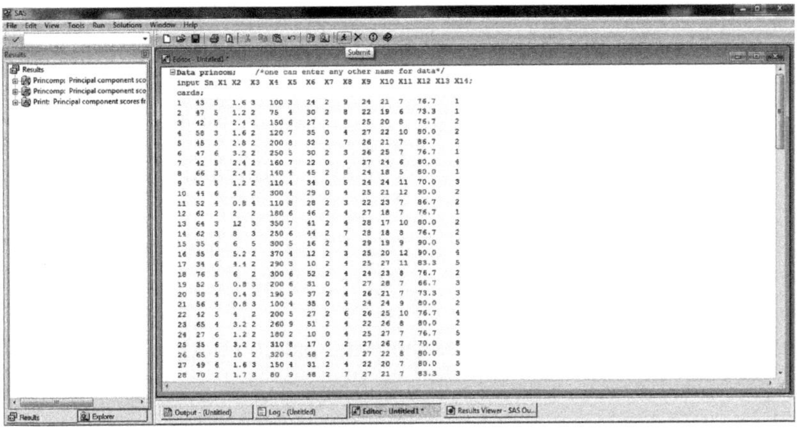

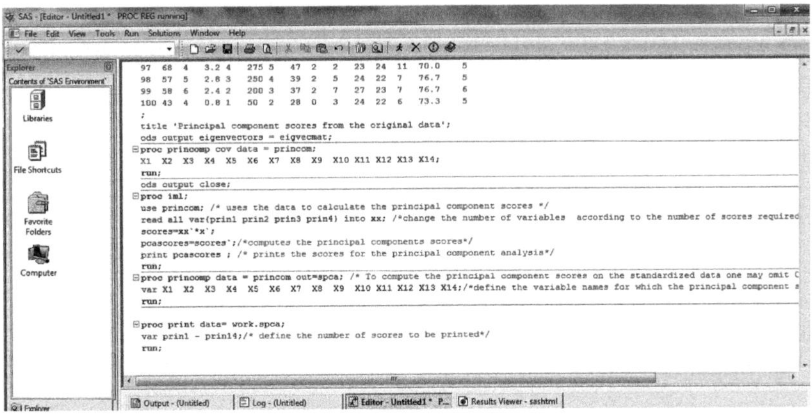

Codes:

Data princom; /*one can enter any other name for data*/

Input X1 X2 X3 X4 X5 X6 X7 X8 X9 X10 X11 X12 X13 X14;

cards;

1	43	5	1.6	3	100	3	24	2	9	24	21	7	76.7	1
2	47	5	1.2	2	75	4	30	2	8	22	19	6	73.3	1
3	42	5	2.4	2	150	6	27	2	8	25	20	8	76.7	2
97	68	4	3.2	4	275	5	47	2	2	23	24	11	70.0	5
98	57	5	2.8	3	250	4	39	2	5	24	22	7	76.7	5
99	58	6	2.4	2	200	3	37	2	7	27	23	7	76.7	6
100	43	4	0.8	1	50	2	28	0	3	24	22	6	73.3	5

```
;
title 'Principal component scores from the original data';
odsoutput eigenvectors = eigvecmat;
procprincompcovdata = princom;
X1	X2	X3	X4	X5	X6	X7	X8	X9	X10
X11	X12	X13	X14;
run;
odsoutput close;
proc iml;
useprincom; /* uses the data to calculate the principal component scores
*/
read all var{prin1 prin2 prin3 prin4} into xx; /*change the number of
variables according to the number of scores required*/
scores = xx`*x`;
pcascores = scores`;/*computes the principal components scores*/
printpcascores;/* prints the scores for the principal component analysis*/
run;
procprincompdata = princomout = spca; /* To compute the principal
component scores on the standardized data one may omit COV option*/
var X1	X2	X3	X4	X5	X6	X7	X8	X9	X10
X11	X12	X13	X14;/*define the variable names for which the principal
component scores are to be computed*/
run;
procprintdata = work.spca;
var prin1 – prin14;/* define the number of scores to be printed*/
run;
```

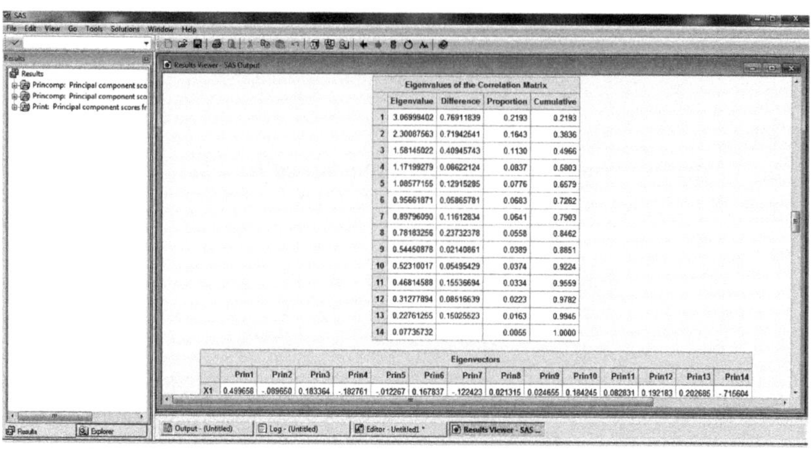

Eigenvalues of the Correlation Matrix				
	Eigenvalue	Difference	Proportion	Cumulative
1	3.06999402	0.76911839	0.2193	0.2193
2	2.30087563	0.71942541	0.1643	0.3836
3	1.58145022	0.40945743	0.1130	0.4966
4	1.17199279	0.08622124	0.0837	0.5803
5	1.08577155	0.12915285	0.0776	0.6579
6	0.95661871	0.05865781	0.0683	0.7262
7	0.89796090	0.11612834	0.0641	0.7903
8	0.78183256	0.23732378	0.0558	0.8462
9	0.54450878	0.02140861	0.0389	0.8851
10	0.52310017	0.05495429	0.0374	0.9224
11	0.46814588	0.15536694	0.0334	0.9559
12	0.31277894	0.08516639	0.0223	0.9782
13	0.22761255	0.15025523	0.0163	0.9945
14	0.07736732		0.0055	1.0000

Eigenvectors														
	Prin1	Prin2	Prin3	Prin4	Prin5	Prin6	Prin7	Prin8	Prin9	Prin10	Prin11	Prin12	Prin13	Prin14
X1	0.499658	-.089650	0.183364	-.182761	-.012267	0.167837	-.122423	0.021315	0.024655	0.184245	0.082831	0.192183	0.202686	-.715604

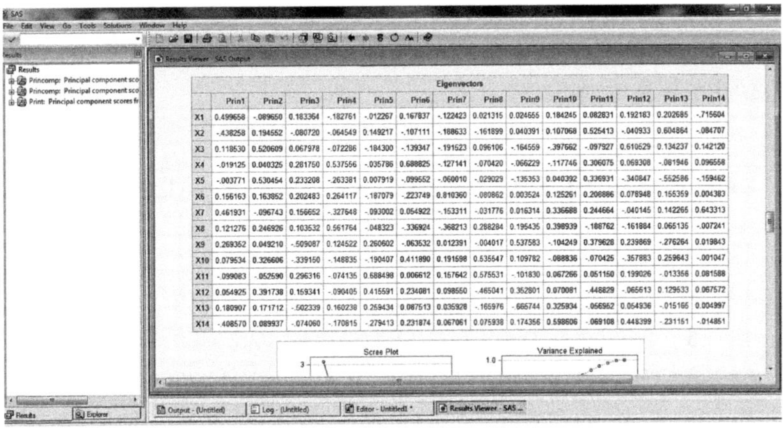
Eigenvectors
Scree Plot
Variance Explained

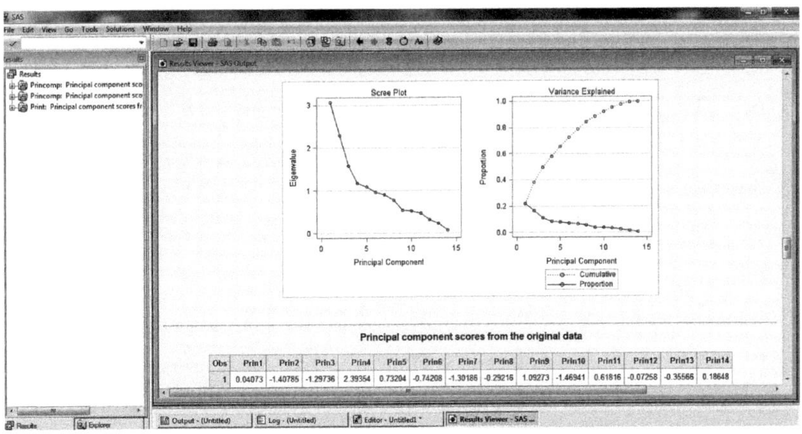
Scree Plot
Variance Explained
Eigenvalue
Proportion
Principal Component
Cumulative
Proportion
Principal component scores from the original data

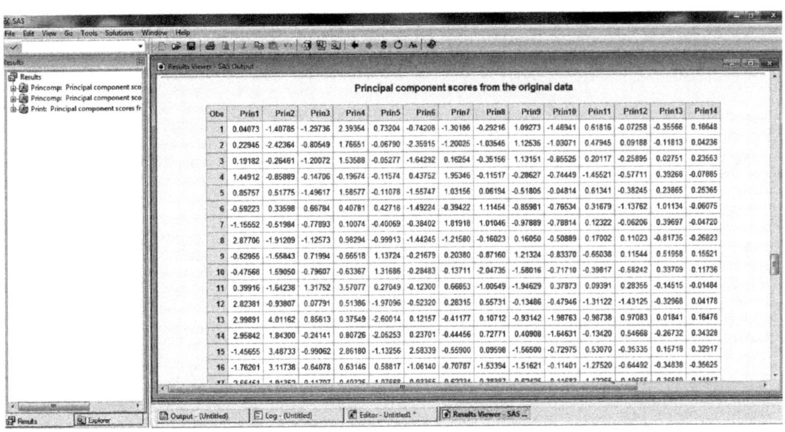
Principal component scores from the original data

7.2.13 ***Example 13:*** Following data (X_1, X_2, X_3, X_4, X_5, X_6, X_7, X_8, X_9, X_{10}, X_{11}, X_{12}, X_{13}, X_{14}) for certain agricultural crop. Work out the ***Cluster Analysis*** for below data.

Data:

S. No.	X1	X2	X3	X4	X5	X6	X7	X8	X9	X10	X11	X12	X13	X14
1	43	5	1.6	3	100	3	24	2	9	24	21	7	76.7	1
2	47	5	1.2	2	75	4	30	2	8	22	19	6	73.3	1
3	42	5	2.4	2	150	6	27	2	8	25	20	8	76.7	2
4	58	3	1.6	2	120	7	35	0	4	27	22	10	80.0	2
5	45	5	2.8	2	200	8	32	2	7	26	21	7	86.7	2
6	47	6	3.2	2	250	5	30	2	3	26	25	7	76.7	1
7	42	5	2.4	2	160	7	22	0	4	27	24	6	80.0	4
8	66	3	2.4	2	140	4	45	2	8	24	18	5	80.0	1
9	52	5	1.2	2	110	4	34	0	5	24	24	11	70.0	3
10	44	6	4	2	300	4	29	0	4	25	21	12	90.0	2
11	52	4	0.8	4	110	8	28	2	3	22	23	7	86.7	2
12	62	2	2	2	180	6	46	2	4	27	18	7	76.7	1
13	64	3	12	3	350	7	41	2	4	28	17	10	80.0	2
14	62	3	8	3	250	6	44	2	7	28	18	8	76.7	2
15	35	6	6	5	300	5	16	2	4	29	19	9	90.0	5
16	35	6	5.2	2	370	4	12	2	3	25	20	12	90.0	4
17	34	6	4.4	2	290	3	10	2	4	25	27	11	83.3	5
18	76	5	6	2	300	6	52	2	4	24	23	8	76.7	2
19	52	5	0.8	3	200	6	31	0	4	27	28	7	66.7	3
20	58	4	0.4	3	190	5	37	2	4	26	21	7	73.3	3
21	56	4	0.8	3	100	4	35	0	4	24	24	9	80.0	2
22	42	5	4	2	200	5	27	2	6	26	25	10	76.7	4
23	65	4	3.2	2	260	9	51	2	4	22	26	8	80.0	2
24	27	6	1.2	2	180	2	10	0	4	25	27	7	76.7	5
25	35	6	3.2	2	310	8	17	0	2	27	26	7	70.0	8
26	65	5	10	2	320	4	48	2	4	27	22	8	80.0	3
27	49	6	1.6	3	150	4	31	2	4	22	20	7	80.0	5

S. No.	X1	X2	X3	X4	X5	X6	X7	X8	X9	X10	X11	X12	X13	X14
28	70	2	1.7	3	80	9	48	2	7	27	21	7	83.3	3
29	63	3	1.6	2	90	3	49	2	9	27	21	6	80.0	3
30	42	6	1.6	3	170	5	28	0	5	26	23	7	76.7	5
31	45	6	1.4	4	200	4	27	0	4	25	22	9	73.3	4
32	57	5	2	4	260	3	39	2	4	26	24	11	76.7	5
33	78	2	1.2	2	140	2	56	0	3	25	25	7	76.7	4
34	48	4	1.2	2	160	4	33	2	7	28	26	12	83.3	6
35	72	3	2.4	2	200	10	56	0	4	25	22	12	70.0	2
36	56	5	2.4	4	230	6	34	0	7	27	21	11	80.0	4
37	52	5	2	3	270	4	45	2	4	24	25	11	80.0	3
38	45	5	4	2	340	5	28	2	6	24	26	12	80.0	4
39	69	5	3.2	2	370	2	49	2	8	27	27	10	86.7	1
40	56	4	0.8	3	200	3	39	0	2	25	28	6	73.3	4
41	55	5	3.6	2	380	4	38	0	6	26	25	11	76.7	4
42	55	6	1.6	2	200	4	35	0	6	24	26	8	76.7	3
43	46	6	4.4	2	270	4	29	2	4	25	21	9	76.7	6
44	42	6	1.6	2	200	4	27	0	4	22	23	10	70.0	5
45	46	5	1.2	2	170	3	31	0	2	25	21	7	66.7	4
46	64	4	1.6	2	170	4	52	0	7	26	24	6	83.3	3
47	56	5	1.2	2	140	4	39	0	7	26	27	8	80.0	2
48	56	5	0.8	3	170	3	35	0	7	28	21	11	86.7	3
49	55	6	1.6	2	140	4	30	2	6	26	25	9	80.0	4
50	65	4	3.2	4	80	6	2	2	7	27	26	12	83.3	3
51	51	5	2	4	50	4	29	0	3	24	23	11	73.3	5
52	75	4	3.6	2	350	12	59	0	6	26	21	12	76.7	2
53	66	4	0.6	2	90	5	51	0	8	28	21	7	90.0	4
54	62	5	1.2	4	210	6	49	2	4	24	27	11	73.3	3
55	72	4	1.6	2	160	4	54	0	5	26	25	6	76.7	2
56	68	4	2.4	2	280	3	52	2	3	25	23	12	66.7	2
57	36	6	2.8	2	280	5	21	0	6	26	21	11	90.0	7
58	46	6	1.2	3	220	4	28	0	6	24	22	7	76.7	7

S. No.	X1	X2	X3	X4	X5	X6	X7	X8	X9	X10	X11	X12	X13	X14
59	60	5	1.2	3	240	3	47	2	5	25	26	10	80.0	2
60	62	5	5.2	2	300	5	45	0	7	27	27	11	86.7	3
61	52	5	5.2	2	350	5	36	0	7	27	22	12	83.3	5
62	68	6	3.2	2	280	4	52	0	5	26	18	11	83.3	4
63	38	6	2	3	270	5	23	0	3	26	17	7	73.3	8
64	55	6	0.8	2	50	3	42	0	8	25	17	6	83.3	5
65	60	5	2	2	160	5	48	2	5	26	18	6	86.7	5
66	40	6	2.4	2	280	4	28	0	3	26	19	9	73.3	6
67	38	6	1	2	190	5	24	0	4	25	20	9	76.7	6
68	55	5	4	2	300	5	38	2	7	28	21	10	86.7	4
69	60	5	4.8	3	310	3	47	0	2	27	21	8	66.7	7
70	61	5	0.8	2	80	2	45	0	5	26	19	7	76.7	6
71	65	4	2	2	180	3	49	0	7	26	18	10	86.7	5
72	60	4	1.6	2	190	6	46	0	4	25	20	8	90.0	3
73	42	6	1.6	2	170	9	27	2	4	24	22	7	73.3	7
74	62	4	1.4	4	280	6	44	2	5	27	22	9	76.7	4
75	48	6	4	2	260	4	31	0	5	27	21	11	73.3	4
76	42	5	2.4	2	150	6	27	2	6	26	23	10	80.0	5
77	61	3	1.6	2	120	7	35	0	8	25	24	8	80.0	3
78	45	5	2.8	2	300	8	32	2	4	28	25	9	86.7	5
79	47	6	3.2	2	350	5	30	2	3	26	26	6	76.7	4
80	42	5	2.4	2	160	7	22	0	3	24	27	6	73.3	6
81	55	5	3.6	2	380	4	38	0	7	27	22	7	76.7	3
82	55	6	1.6	2	200	4	35	2	3	25	21	11	76.7	5
83	46	6	4.4	2	270	4	29	2	4	26	23	8	76.7	7
84	42	6	1.6	2	200	4	27	0	4	27	24	9	80.0	8
85	46	5	1.2	2	170	3	31	0	2	25	25	7	70.0	5
86	70	4	1.6	2	170	4	52	0	5	24	26	11	86.7	3
87	58	5	1.2	2	140	4	39	0	5	26	21	7	83.3	6
88	56	4	0.8	3	230	4	35	0	2	24	22	8	76.7	7
89	42	5	4	2	200	5	27	2	8	29	24	10	83.3	8

S. No.	X1	X2	X3	X4	X5	X6	X7	X8	X9	X10	X11	X12	X13	X14
90	65	4	3.2	2	460	9	51	2	8	26	19	12	80.0	3
91	27	6	1.2	2	180	2	10	0	4	27	20	7	80.0	9
92	35	6	3.2	2	310	8	17	2	3	26	20	11	66.7	8
93	68	5	10	2	320	4	48	2	7	27	21	12	80.0	4
94	65	3	1.2	2	85	4	48	2	4	26	18	6	83.3	3
95	53	6	3.6	3	250	6	38	2	4	27	25	9	80.0	6
96	47	6	2.8	3	250	4	29	2	2	26	25	8	66.7	9
97	68	4	3.2	4	275	5	47	2	2	23	24	11	70.0	5
98	57	5	2.8	3	250	4	39	2	5	24	22	7	76.7	5
99	58	6	2.4	2	200	3	37	2	7	27	23	7	76.7	6
100	43	4	0.8	1	50	2	28	0	3	24	22	6	73.3	5

Code:

```
title 'cluster analysis using plant data';
data cluster;
inputtrt $ X1 X2 X3 X4 X5 X6 X7 X8 X9 X10 X11
X12 X13 X14;
cards;
```

X1	X2	X3	X4	X5	X6	X7	X8	X9	X10	X11	X12	X13	X14
43	5	1.6	3	100	3	24	2	9	24	21	7	76.7	1
47	5	1.2	2	75	4	30	2	8	22	19	6	73.3	1
42	5	2.4	2	150	6	27	2	8	25	20	8	76.7	2
58	3	1.6	2	120	7	35	0	4	27	22	10	80.0	2
45	5	2.8	2	200	8	32	2	7	26	21	7	86.7	2
47	6	3.2	2	250	5	30	2	3	26	25	7	76.7	1
42	5	2.4	2	160	7	22	0	4	27	24	6	80.0	4
66	3	2.4	2	140	4	45	2	8	24	18	5	80.0	1
52	5	1.2	2	110	4	34	0	5	24	24	11	70.0	3
44	6	4	2	300	4	29	0	4	25	21	12	90.0	2
52	4	0.8	4	110	8	28	2	3	22	23	7	86.7	2
62	2	2	2	180	6	46	2	4	27	18	7	76.7	1
64	3	12	3	350	7	41	2	4	28	17	10	80.0	2
62	3	8	3	250	6	44	2	7	28	18	8	76.7	2
35	6	6	5	300	5	16	2	4	29	19	9	90.0	5

X1	X2	X3	X4	X5	X6	X7	X8	X9	X10	X11	X12	X13	X14
35	6	5.2	2	370	4	12	2	3	25	20	12	90.0	4
34	6	4.4	2	290	3	10	2	4	25	27	11	83.3	5
76	5	6	2	300	6	52	2	4	24	23	8	76.7	2
52	5	0.8	3	200	6	31	0	4	27	28	7	66.7	3
58	4	0.4	3	190	5	37	2	4	26	21	7	73.3	3
56	4	0.8	3	100	4	35	0	4	24	24	9	80.0	2
42	5	4	2	200	5	27	2	6	26	25	10	76.7	4
65	4	3.2	2	260	9	51	2	4	22	26	8	80.0	2
27	6	1.2	2	180	2	10	0	4	25	27	7	76.7	5
35	6	3.2	2	310	8	17	0	2	27	26	7	70.0	8
65	5	10	2	320	4	48	2	4	27	22	8	80.0	3
49	6	1.6	3	150	4	31	2	4	22	20	7	80.0	5
70	2	1.7	3	80	9	48	2	7	27	21	7	83.3	3
63	3	1.6	2	90	3	49	2	9	27	21	6	80.0	3
42	6	1.6	3	170	5	28	0	5	26	23	7	76.7	5
45	6	1.4	4	200	4	27	0	4	25	22	9	73.3	4
57	5	2	4	260	3	39	2	4	26	24	11	76.7	5
78	2	1.2	2	140	2	56	0	3	25	25	7	76.7	4
48	4	1.2	2	160	4	33	2	7	28	26	12	83.3	6
72	3	2.4	2	200	10	56	0	4	25	22	12	70.0	2
56	5	2.4	4	230	6	34	0	7	27	21	11	80.0	4
52	5	2	3	270	4	45	2	4	24	25	11	80.0	3
45	5	4	2	340	5	28	2	6	24	26	12	80.0	4
69	5	3.2	2	370	2	49	2	8	27	27	10	86.7	1
56	4	0.8	3	200	3	39	0	2	25	28	6	73.3	4
55	5	3.6	2	380	4	38	0	6	26	25	11	76.7	4
55	6	1.6	2	200	4	35	0	6	24	26	8	76.7	3
46	6	4.4	2	270	4	29	2	4	25	21	9	76.7	6
42	6	1.6	2	200	4	27	0	4	22	23	10	70.0	5
46	5	1.2	2	170	3	31	0	2	25	21	7	66.7	4
64	4	1.6	2	170	4	52	0	7	26	24	6	83.3	3
56	5	1.2	2	140	4	39	0	7	26	27	8	80.0	2
56	5	0.8	3	170	3	35	0	7	28	21	11	86.7	3
55	6	1.6	2	140	4	30	2	6	26	25	9	80.0	4
65	4	3.2	4	80	6	2	2	7	27	26	12	83.3	3

X1	X2	X3	X4	X5	X6	X7	X8	X9	X10	X11	X12	X13	X14
51	5	2	4	50	4	29	0	3	24	23	11	73.3	5
75	4	3.6	2	350	12	59	0	6	26	21	12	76.7	2
66	4	0.6	2	90	5	51	0	8	28	21	7	90.0	4
62	5	1.2	4	210	6	49	2	4	24	27	11	73.3	3
72	4	1.6	2	160	4	54	0	5	26	25	6	76.7	2
68	4	2.4	2	280	3	52	2	3	25	23	12	66.7	2
36	6	2.8	2	280	5	21	0	6	26	21	11	90.0	7
46	6	1.2	3	220	4	28	0	6	24	22	7	76.7	7
60	5	1.2	3	240	3	47	2	5	25	26	10	80.0	2
62	5	5.2	2	300	5	45	0	7	27	27	11	86.7	3
52	5	5.2	2	350	5	36	0	7	27	22	12	83.3	5
68	6	3.2	2	280	4	52	0	5	26	18	11	83.3	4
38	6	2	3	270	5	23	0	3	26	17	7	73.3	8
55	6	0.8	2	50	3	42	0	8	25	17	6	83.3	5
60	5	2	2	160	5	48	2	5	26	18	6	86.7	5
40	6	2.4	2	280	4	28	0	3	26	19	9	73.3	6
38	6	1	2	190	5	24	0	4	25	20	9	76.7	6
55	5	4	2	300	5	38	2	7	28	21	10	86.7	4
60	5	4.8	3	310	3	47	0	2	27	21	8	66.7	7
61	5	0.8	2	80	2	45	0	5	26	19	7	76.7	6
65	4	2	2	180	3	49	0	7	26	18	10	86.7	5
60	4	1.6	2	190	6	46	0	4	25	20	8	90.0	3
42	6	1.6	2	170	9	27	2	4	24	22	7	73.3	7
62	4	1.4	4	280	6	44	2	5	27	22	9	76.7	4
48	6	4	2	260	4	31	0	5	27	21	11	73.3	4
42	5	2.4	2	150	6	27	2	6	26	23	10	80.0	5
61	3	1.6	2	120	7	35	0	8	25	24	8	80.0	3
45	5	2.8	2	300	8	32	2	4	28	25	9	86.7	5
47	6	3.2	2	350	5	30	2	3	26	26	6	76.7	4
42	5	2.4	2	160	7	22	0	3	24	27	6	73.3	6
55	5	3.6	2	380	4	38	0	7	27	22	7	76.7	3
55	6	1.6	2	200	4	35	2	3	25	21	11	76.7	5
46	6	4.4	2	270	4	29	2	4	26	23	8	76.7	7
42	6	1.6	2	200	4	27	0	4	27	24	9	80.0	8
46	5	1.2	2	170	3	31	0	2	25	25	7	70.0	5

X1	X2	X3	X4	X5	X6	X7	X8	X9	X10	X11	X12	X13	X14
70	4	1.6	2	170	4	52	0	5	24	26	11	86.7	3
58	5	1.2	2	140	4	39	0	5	26	21	7	83.3	6
56	4	0.8	3	230	4	35	0	2	24	22	8	76.7	7
42	5	4	2	200	5	27	2	8	29	24	10	83.3	8
65	4	3.2	2	460	9	51	2	8	26	19	12	80.0	3
27	6	1.2	2	180	2	10	0	4	27	20	7	80.0	9
35	6	3.2	2	310	8	17	2	3	26	20	11	66.7	8
68	5	10	2	320	4	48	2	7	27	21	12	80.0	4
65	3	1.2	2	85	4	48	2	4	26	18	6	83.3	3
53	6	3.6	3	250	6	38	2	4	27	25	9	80.0	6
47	6	2.8	3	250	4	29	2	2	26	25	8	66.7	9
68	4	3.2	4	275	5	47	2	2	23	24	11	70.0	5
57	5	2.8	3	250	4	39	2	5	24	22	7	76.7	5
58	6	2.4	2	200	3	37	2	7	27	23	7	76.7	6
43	4	0.8	1	50	2	28	0	3	24	22	6	73.3	5

```
;
PROC cluster data = distmat method = average outtree = tree;
idtrt;
run;
PROC print data = distmat;
idtrt;
run;
PROC tree data = Tree nclusters = 3horizontalhor display = right lines =
(color = blue) out = out;
idtrt;
run;
PROC distance data = cluster method = EUCLID out = distmat;
var interval(x1-x14);
idtrt;
run;
```

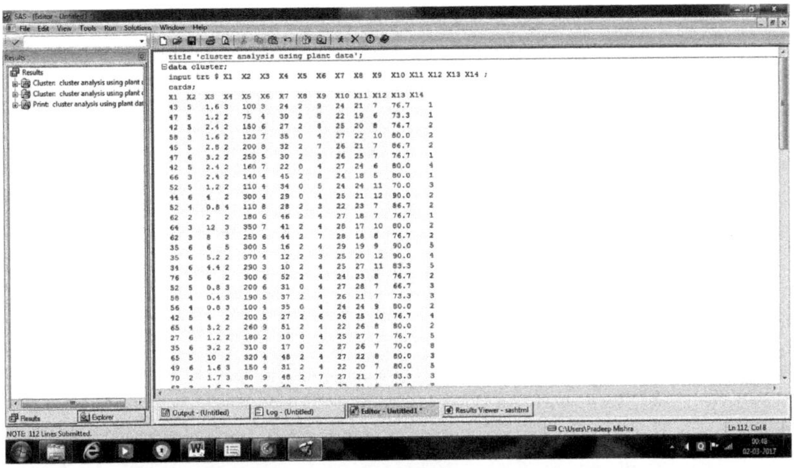
title 'cluster analysis using plant data';
data cluster;
input trt $ X1 X2 X3 X4 X5 X6 X7 X8 X9 X10 X11 X12 X13 X14 ;
cards;

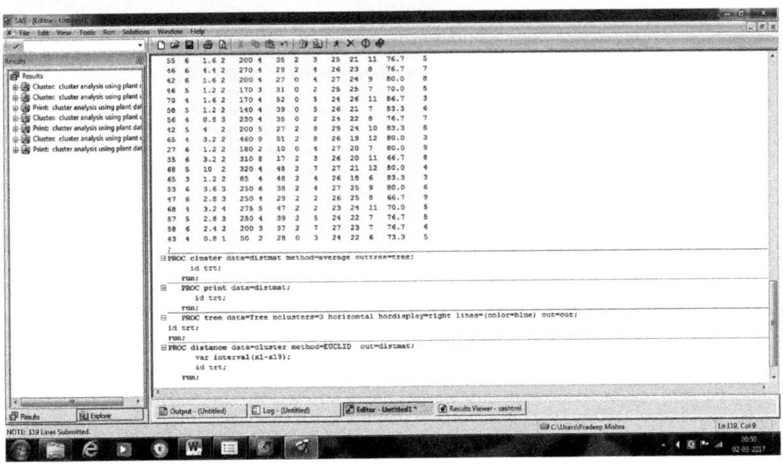
PROC cluster data=distmat method=average outtree=tree;
id trt;
run;
PROC print data=distmat;
id trt;
run;
PROC tree data=Tree nclusters=3 horizontal hordisplay=right lines=(color=blue) out=out;
id trt;
run;
PROC distance data=cluster method=EUCLID out=distmat;
var interval(x1-x19);
id trt;
run;

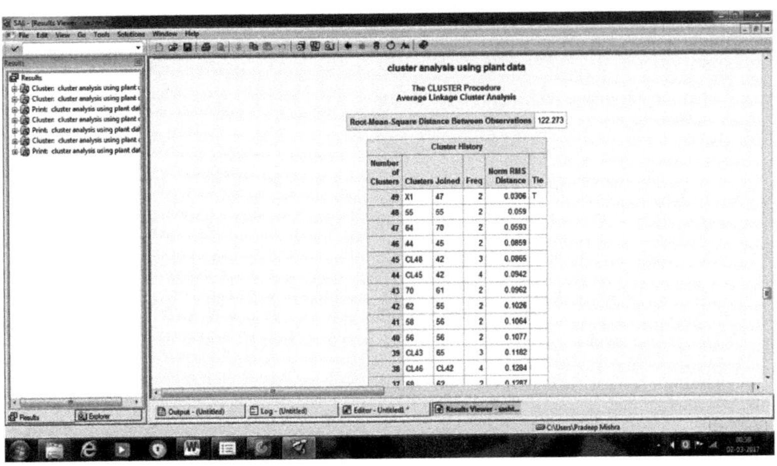
cluster analysis using plant data
The CLUSTER Procedure
Average Linkage Cluster Analysis
Root-Mean-Square Distance Between Observations 122.273
Cluster History
Number of Clusters | Clusters Joined | Freq | Norm RMS Distance | Tie
49 X1 47 2 0.0306 T
48 55 55 2 0.059
47 64 70 2 0.0593
46 44 45 2 0.0859
45 CL48 42 3 0.0865
44 CL45 42 4 0.0942
43 70 61 2 0.0962
42 62 55 2 0.1026
41 58 56 2 0.1064
40 56 56 2 0.1077
39 CL43 65 3 0.1182
38 CL46 CL42 4 0.1284

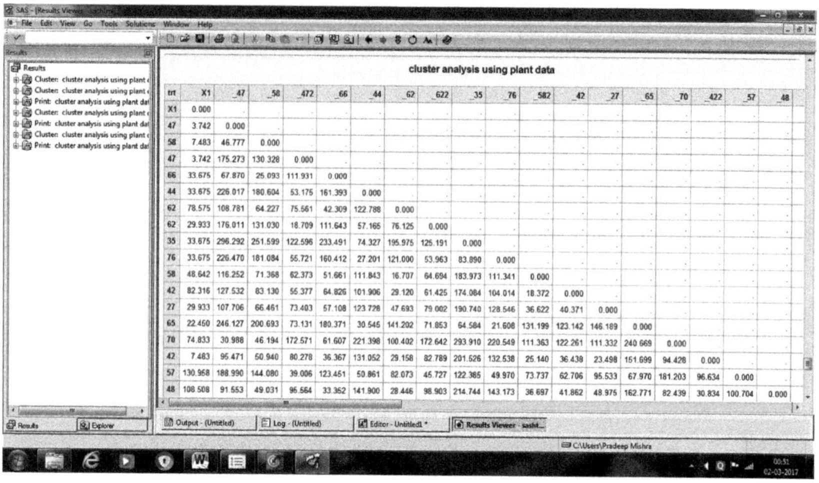

cluster analysis using plant data

trt	X1	_47	_58	_472	_66	_44	_62	_622	_35	_76	_582	_42	_27	_65	_70	_422	_57	_48
X1	0.000	.	.	.	.	.	.	.	.	.	.	.	.	.	.	.	.	.
47	3.742	0.000	.	.	.	.	.	.	.	.	.	.	.	.	.	.	.	.
58	7.483	46.777	0.000	.	.	.	.	.	.	.	.	.	.	.	.	.	.	.
47	3.742	175.273	130.328	0.000	.	.	.	.	.	.	.	.	.	.	.	.	.	.
66	33.675	67.870	25.093	111.931	0.000	.	.	.	.	.	.	.	.	.	.	.	.	.
44	33.675	226.017	180.604	53.175	161.393	0.000	.	.	.	.	.	.	.	.	.	.	.	.
62	78.575	108.781	64.227	75.561	42.309	122.788	0.000	.	.	.	.	.	.	.	.	.	.	.
62	29.933	176.011	131.030	18.709	111.643	57.165	76.125	0.000	.	.	.	.	.	.	.	.	.	.
35	33.675	296.292	251.599	122.596	233.491	74.327	195.975	125.191	0.000	.	.	.	.	.	.	.	.	.
76	33.675	226.470	181.084	55.721	160.412	27.201	121.000	53.963	83.890	0.000	.	.	.	.	.	.	.	.
58	48.642	116.252	71.368	62.373	51.661	111.843	16.707	64.694	183.973	111.341	0.000	.	.	.	.	.	.	.
42	82.316	127.532	83.130	55.377	64.826	101.906	29.120	61.425	174.084	104.014	18.372	0.000	.	.	.	.	.	.
27	29.933	107.706	66.461	73.403	57.108	123.728	47.693	79.002	190.740	128.546	36.622	40.371	0.000	.	.	.	.	.
65	22.460	246.127	200.693	73.131	180.371	30.546	141.202	71.853	64.584	21.608	131.199	123.142	146.189	0.000	.	.	.	.
70	74.833	30.988	46.194	172.571	61.607	221.398	100.402	172.642	293.910	220.549	111.363	122.261	111.332	240.669	0.000	.	.	.
42	7.483	95.471	50.940	80.278	36.367	131.052	29.158	82.789	201.526	132.538	25.140	36.438	23.498	151.699	94.428	0.000	.	.
57	130.958	188.990	144.080	39.006	123.451	50.861	82.073	45.727	122.385	49.970	73.737	62.706	95.533	67.970	181.203	96.634	0.000	.
48	108.508	91.553	49.031	96.564	33.362	141.900	28.446	98.903	214.744	143.173	36.697	41.862	48.975	162.771	82.439	30.834	100.704	0.000

7.2.14 *Example 14:* Calculate the ***Canonical Correlation*** for below data.

S. No.	X1	X2	X3	X4	X5	X6	X7	X8	X9	X10	X11	X12	X13	X14
1	43	5	1.6	3	100	3	24	2	9	24	21	7	76.7	1
2	47	5	1.2	2	75	4	30	2	8	22	19	6	73.3	1
3	42	5	2.4	2	150	6	27	2	8	25	20	8	76.7	2
4	58	3	1.6	2	120	7	35	0	4	27	22	10	80.0	2
5	45	5	2.8	2	200	8	32	2	7	26	21	7	86.7	2
6	47	6	3.2	2	250	5	30	2	3	26	25	7	76.7	1
7	42	5	2.4	2	160	7	22	0	4	27	24	6	80.0	4
8	66	3	2.4	2	140	4	45	2	8	24	18	5	80.0	1
9	52	5	1.2	2	110	4	34	0	5	24	24	11	70.0	3
10	44	6	4	2	300	4	29	0	4	25	21	12	90.0	2
11	52	4	0.8	4	110	8	28	2	3	22	23	7	86.7	2
12	62	2	2	2	180	6	46	2	4	27	18	7	76.7	1
13	64	3	12	3	350	7	41	2	4	28	17	10	80.0	2
14	62	3	8	3	250	6	44	2	7	28	18	8	76.7	2
15	35	6	6	5	300	5	16	2	4	29	19	9	90.0	5
16	35	6	5.2	2	370	4	12	2	3	25	20	12	90.0	4
17	34	6	4.4	2	290	3	10	2	4	25	27	11	83.3	5
18	76	5	6	2	300	6	52	2	4	24	23	8	76.7	2
19	52	5	0.8	3	200	6	31	0	4	27	28	7	66.7	3

S. No.	X1	X2	X3	X4	X5	X6	X7	X8	X9	X10	X11	X12	X13	X14
20	58	4	0.4	3	190	5	37	2	4	26	21	7	73.3	3
21	56	4	0.8	3	100	4	35	0	4	24	24	9	80.0	2
22	42	5	4	2	200	5	27	2	6	26	25	10	76.7	4
23	65	4	3.2	2	260	9	51	2	4	22	26	8	80.0	2
24	27	6	1.2	2	180	2	10	0	4	25	27	7	76.7	5
25	35	6	3.2	2	310	8	17	0	2	27	26	7	70.0	8
26	65	5	10	2	320	4	48	2	4	27	22	8	80.0	3
27	49	6	1.6	3	150	4	31	2	4	22	20	7	80.0	5
28	70	2	1.7	3	80	9	48	2	7	27	21	7	83.3	3
29	63	3	1.6	2	90	3	49	2	9	27	21	6	80.0	3
30	42	6	1.6	3	170	5	28	0	5	26	23	7	76.7	5
31	45	6	1.4	4	200	4	27	0	4	25	22	9	73.3	4
32	57	5	2	4	260	3	39	2	4	26	24	11	76.7	5
33	78	2	1.2	2	140	2	56	0	3	25	25	7	76.7	4
34	48	4	1.2	2	160	4	33	2	7	28	26	12	83.3	6
35	72	3	2.4	2	200	10	56	0	4	25	22	12	70.0	2
36	56	5	2.4	4	230	6	34	0	7	27	21	11	80.0	4
37	52	5	2	3	270	4	45	2	4	24	25	11	80.0	3
38	45	5	4	2	340	5	28	2	6	24	26	12	80.0	4
39	69	5	3.2	2	370	2	49	2	8	27	27	10	86.7	1
40	56	4	0.8	3	200	3	39	0	2	25	28	6	73.3	4
41	55	5	3.6	2	380	4	38	0	6	26	25	11	76.7	4
42	55	6	1.6	2	200	4	35	0	6	24	26	8	76.7	3
43	46	6	4.4	2	270	4	29	2	4	25	21	9	76.7	6
44	42	6	1.6	2	200	4	27	0	4	22	23	10	70.0	5
45	46	5	1.2	2	170	3	31	0	2	25	21	7	66.7	4
46	64	4	1.6	2	170	4	52	0	7	26	24	6	83.3	3
47	56	5	1.2	2	140	4	39	0	7	26	27	8	80.0	2
48	56	5	0.8	3	170	3	35	0	7	28	21	11	86.7	3
49	55	6	1.6	2	140	4	30	2	6	26	25	9	80.0	4
50	65	4	3.2	4	80	6	2	2	7	27	26	12	83.3	3
51	51	5	2	4	50	4	29	0	3	24	23	11	73.3	5
52	75	4	3.6	2	350	12	59	0	6	26	21	12	76.7	2

S. No.	X1	X2	X3	X4	X5	X6	X7	X8	X9	X10	X11	X12	X13	X14
53	66	4	0.6	2	90	5	51	0	8	28	21	7	90.0	4
54	62	5	1.2	4	210	6	49	2	4	24	27	11	73.3	3
55	72	4	1.6	2	160	4	54	0	5	26	25	6	76.7	2
56	68	4	2.4	2	280	3	52	2	3	25	23	12	66.7	2
57	36	6	2.8	2	280	5	21	0	6	26	21	11	90.0	7
58	46	6	1.2	3	220	4	28	0	6	24	22	7	76.7	7
59	60	5	1.2	3	240	3	47	2	5	25	26	10	80.0	2
60	62	5	5.2	2	300	5	45	0	7	27	27	11	86.7	3
61	52	5	5.2	2	350	5	36	0	7	27	22	12	83.3	5
62	68	6	3.2	2	280	4	52	0	5	26	18	11	83.3	4
63	38	6	2	3	270	5	23	0	3	26	17	7	73.3	8
64	55	6	0.8	2	50	3	42	0	8	25	17	6	83.3	5
65	60	5	2	2	160	5	48	2	5	26	18	6	86.7	5
66	40	6	2.4	2	280	4	28	0	3	26	19	9	73.3	6
67	38	6	1	2	190	5	24	0	4	25	20	9	76.7	6
68	55	5	4	2	300	5	38	2	7	28	21	10	86.7	4
69	60	5	4.8	3	310	3	47	0	2	27	21	8	66.7	7
70	61	5	0.8	2	80	2	45	0	5	26	19	7	76.7	6
71	65	4	2	2	180	3	49	0	7	26	18	10	86.7	5
72	60	4	1.6	2	190	6	46	0	4	25	20	8	90.0	3
73	42	6	1.6	2	170	9	27	2	4	24	22	7	73.3	7
74	62	4	1.4	4	280	6	44	2	5	27	22	9	76.7	4
75	48	6	4	2	260	4	31	0	5	27	21	11	73.3	4
76	42	5	2.4	2	150	6	27	2	6	26	23	10	80.0	5
77	61	3	1.6	2	120	7	35	0	8	25	24	8	80.0	3
78	45	5	2.8	2	300	8	32	2	4	28	25	9	86.7	5
79	47	6	3.2	2	350	5	30	2	3	26	26	6	76.7	4
80	42	5	2.4	2	160	7	22	0	3	24	27	6	73.3	6
81	55	5	3.6	2	380	4	38	0	7	27	22	7	76.7	3
82	55	6	1.6	2	200	4	35	2	3	25	21	11	76.7	5
83	46	6	4.4	2	270	4	29	2	4	26	23	8	76.7	7
84	42	6	1.6	2	200	4	27	0	4	27	24	9	80.0	8
85	46	5	1.2	2	170	3	31	0	2	25	25	7	70.0	5

S. No.	X1	X2	X3	X4	X5	X6	X7	X8	X9	X10	X11	X12	X13	X14
86	70	4	1.6	2	170	4	52	0	5	24	26	11	86.7	3
87	58	5	1.2	2	140	4	39	0	5	26	21	7	83.3	6
88	56	4	0.8	3	230	4	35	0	2	24	22	8	76.7	7
89	42	5	4	2	200	5	27	2	8	29	24	10	83.3	8
90	65	4	3.2	2	460	9	51	2	8	26	19	12	80.0	3
91	27	6	1.2	2	180	2	10	0	4	27	20	7	80.0	9
92	35	6	3.2	2	310	8	17	2	3	26	20	11	66.7	8
93	68	5	10	2	320	4	48	2	7	27	21	12	80.0	4
94	65	3	1.2	2	85	4	48	2	4	26	18	6	83.3	3
95	53	6	3.6	3	250	6	38	2	4	27	25	9	80.0	6
96	47	6	2.8	3	250	4	29	2	2	26	25	8	66.7	9
97	68	4	3.2	4	275	5	47	2	2	23	24	11	70.0	5
98	57	5	2.8	3	250	4	39	2	5	24	22	7	76.7	5
99	58	6	2.4	2	200	3	37	2	7	27	23	7	76.7	6
100	43	4	0.8	1	50	2	28	0	3	24	22	6	73.3	5

Code:

```
data cancor;
input x1 x2 x3 x4 x5 x6 x7 x8 x9 x10 x11 x12 x13 x14;
cards;
```

X1	X2	X3	X4	X5	X6	X7	X8	X9	X10	X11	X12	X13	X14
43	5	1.6	3	100	3	24	2	9	24	21	7	76.7	1
47	5	1.2	2	75	4	30	2	8	22	19	6	73.3	1
42	5	2.4	2	150	6	27	2	8	25	20	8	76.7	2
58	3	1.6	2	120	7	35	0	4	27	22	10	80.0	2
45	5	2.8	2	200	8	32	2	7	26	21	7	86.7	2
47	6	3.2	2	250	5	30	2	3	26	25	7	76.7	1
42	5	2.4	2	160	7	22	0	4	27	24	6	80.0	4
66	3	2.4	2	140	4	45	2	8	24	18	5	80.0	1
52	5	1.2	2	110	4	34	0	5	24	24	11	70.0	3
44	6	4	2	300	4	29	0	4	25	21	12	90.0	2
52	4	0.8	4	110	8	28	2	3	22	23	7	86.7	2
62	2	2	2	180	6	46	2	4	27	18	7	76.7	1
64	3	12	3	350	7	41	2	4	28	17	10	80.0	2
62	3	8	3	250	6	44	2	7	28	18	8	76.7	2

X1	X2	X3	X4	X5	X6	X7	X8	X9	X10	X11	X12	X13	X14
35	6	6	5	300	5	16	2	4	29	19	9	90.0	5
35	6	5.2	2	370	4	12	2	3	25	20	12	90.0	4
34	6	4.4	2	290	3	10	2	4	25	27	11	83.3	5
76	5	6	2	300	6	52	2	4	24	23	8	76.7	2
52	5	0.8	3	200	6	31	0	4	27	28	7	66.7	3
58	4	0.4	3	190	5	37	2	4	26	21	7	73.3	3
56	4	0.8	3	100	4	35	0	4	24	24	9	80.0	2
42	5	4	2	200	5	27	2	6	26	25	10	76.7	4
65	4	3.2	2	260	9	51	2	4	22	26	8	80.0	2
27	6	1.2	2	180	2	10	0	4	25	27	7	76.7	5
35	6	3.2	2	310	8	17	0	2	27	26	7	70.0	8
65	5	10	2	320	4	48	2	4	27	22	8	80.0	3
49	6	1.6	3	150	4	31	2	4	22	20	7	80.0	5
70	2	1.7	3	80	9	48	2	7	27	21	7	83.3	3
63	3	1.6	2	90	3	49	2	9	27	21	6	80.0	3
42	6	1.6	3	170	5	28	0	5	26	23	7	76.7	5
45	6	1.4	4	200	4	27	0	4	25	22	9	73.3	4
57	5	2	4	260	3	39	2	4	26	24	11	76.7	5
78	2	1.2	2	140	2	56	0	3	25	25	7	76.7	4
48	4	1.2	2	160	4	33	2	7	28	26	12	83.3	6
72	3	2.4	2	200	10	56	0	4	25	22	12	70.0	2
56	5	2.4	4	230	6	34	0	7	27	21	11	80.0	4
52	5	2	3	270	4	45	2	4	24	25	11	80.0	3
45	5	4	2	340	5	28	2	6	24	26	12	80.0	4
69	5	3.2	2	370	2	49	2	8	27	27	10	86.7	1
56	4	0.8	3	200	3	39	0	2	25	28	6	73.3	4
55	5	3.6	2	380	4	38	0	6	26	25	11	76.7	4
55	6	1.6	2	200	4	35	0	6	24	26	8	76.7	3
46	6	4.4	2	270	4	29	2	4	25	21	9	76.7	6
42	6	1.6	2	200	4	27	0	4	22	23	10	70.0	5
46	5	1.2	2	170	3	31	0	2	25	21	7	66.7	4
64	4	1.6	2	170	4	52	0	7	26	24	6	83.3	3
56	5	1.2	2	140	4	39	0	7	26	27	8	80.0	2
56	5	0.8	3	170	3	35	0	7	28	21	11	86.7	3
55	6	1.6	2	140	4	30	2	6	26	25	9	80.0	4

X1	X2	X3	X4	X5	X6	X7	X8	X9	X10	X11	X12	X13	X14
65	4	3.2	4	80	6	2	2	7	27	26	12	83.3	3
51	5	2	4	50	4	29	0	3	24	23	11	73.3	5
75	4	3.6	2	350	12	59	0	6	26	21	12	76.7	2
66	4	0.6	2	90	5	51	0	8	28	21	7	90.0	4
62	5	1.2	4	210	6	49	2	4	24	27	11	73.3	3
72	4	1.6	2	160	4	54	0	5	26	25	6	76.7	2
68	4	2.4	2	280	3	52	2	3	25	23	12	66.7	2
36	6	2.8	2	280	5	21	0	6	26	21	11	90.0	7
46	6	1.2	3	220	4	28	0	6	24	22	7	76.7	7
60	5	1.2	3	240	3	47	2	5	25	26	10	80.0	2
62	5	5.2	2	300	5	45	0	7	27	27	11	86.7	3
52	5	5.2	2	350	5	36	0	7	27	22	12	83.3	5
68	6	3.2	2	280	4	52	0	5	26	18	11	83.3	4
38	6	2	3	270	5	23	0	3	26	17	7	73.3	8
55	6	0.8	2	50	3	42	0	8	25	17	6	83.3	5
60	5	2	2	160	5	48	2	5	26	18	6	86.7	5
40	6	2.4	2	280	4	28	0	3	26	19	9	73.3	6
38	6	1	2	190	5	24	0	4	25	20	9	76.7	6
55	5	4	2	300	5	38	2	7	28	21	10	86.7	4
60	5	4.8	3	310	3	47	0	2	27	21	8	66.7	7
61	5	0.8	2	80	2	45	0	5	26	19	7	76.7	6
65	4	2	2	180	3	49	0	7	26	18	10	86.7	5
60	4	1.6	2	190	6	46	0	4	25	20	8	90.0	3
42	6	1.6	2	170	9	27	2	4	24	22	7	73.3	7
62	4	1.4	4	280	6	44	2	5	27	22	9	76.7	4
48	6	4	2	260	4	31	0	5	27	21	11	73.3	4
42	5	2.4	2	150	6	27	2	6	26	23	10	80.0	5
61	3	1.6	2	120	7	35	0	8	25	24	8	80.0	3
45	5	2.8	2	300	8	32	2	4	28	25	9	86.7	5
47	6	3.2	2	350	5	30	2	3	26	26	6	76.7	4
42	5	2.4	2	160	7	22	0	3	24	27	6	73.3	6
55	5	3.6	2	380	4	38	0	7	27	22	7	76.7	3
55	6	1.6	2	200	4	35	2	3	25	21	11	76.7	5
46	6	4.4	2	270	4	29	2	4	26	23	8	76.7	7
42	6	1.6	2	200	4	27	0	4	27	24	9	80.0	8

X1	X2	X3	X4	X5	X6	X7	X8	X9	X10	X11	X12	X13	X14
46	5	1.2	2	170	3	31	0	2	25	25	7	70.0	5
70	4	1.6	2	170	4	52	0	5	24	26	11	86.7	3
58	5	1.2	2	140	4	39	0	5	26	21	7	83.3	6
56	4	0.8	3	230	4	35	0	2	24	22	8	76.7	7
42	5	4	2	200	5	27	2	8	29	24	10	83.3	8
65	4	3.2	2	460	9	51	2	8	26	19	12	80.0	3
27	6	1.2	2	180	2	10	0	4	27	20	7	80.0	9
35	6	3.2	2	310	8	17	2	3	26	20	11	66.7	8
68	5	10	2	320	4	48	2	7	27	21	12	80.0	4
65	3	1.2	2	85	4	48	2	4	26	18	6	83.3	3
53	6	3.6	3	250	6	38	2	4	27	25	9	80.0	6
47	6	2.8	3	250	4	29	2	2	26	25	8	66.7	9
68	4	3.2	4	275	5	47	2	2	23	24	11	70.0	5
57	5	2.8	3	250	4	39	2	5	24	22	7	76.7	5
58	6	2.4	2	200	3	37	2	7	27	23	7	76.7	6
43	4	0.8	1	50	2	28	0	3	24	22	6	73.3	5

```
;
proc cancorr data = cancorv pefix = f10 wpefix = l9;
var x1 x2 x3 x4 x5 x6 x7 x8 x9 x10;with x11 x12 x13 x14 x15 x16 x17
x18 x19;
run;
```

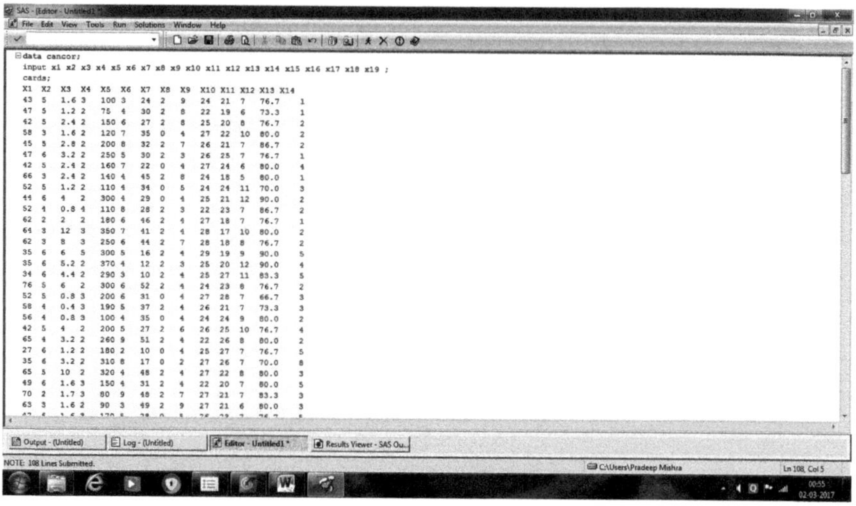

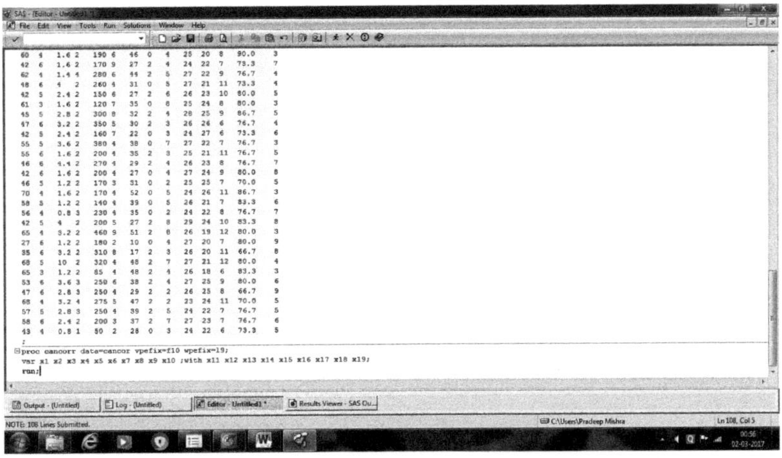

```
60  4  1.6 2  190 6  46 0  4  25 20 8  90.0  3
42  6  1.6 2  170 9  27 2  4  24 22 7  73.3  7
62  4  1.4 4  280 6  44 2  5  27 22 9  76.7  4
48  6  4   2  260 4  31 0  5  27 21 11 73.3  4
42  5  2.4 2  150 6  27 2  6  26 23 10 80.0  5
61  3  1.6 2  120 7  35 0  8  25 24 8  80.0  3
45  5  2.8 2  300 8  32 2  4  28 25 9  86.7  5
47  6  3.2 2  350 5  30 2  3  26 26 6  76.7  4
42  5  2.4 2  160 7  22 0  3  24 27 6  73.3  6
55  5  3.6 2  380 4  38 0  7  27 22 7  76.7  3
55  6  1.6 2  200 4  35 2  3  25 21 11 76.7  5
46  6  4.4 2  270 4  29 2  4  26 23 8  76.7  7
42  6  1.6 2  200 4  27 0  4  27 24 9  80.0  8
46  5  1.2 2  170 3  31 0  2  25 25 7  70.0  5
70  4  1.6 2  170 4  52 0  5  24 26 11 86.7  3
58  5  1.2 2  140 4  39 0  5  26 21 7  83.3  6
56  4  0.8 3  230 4  35 0  2  24 22 8  76.7  7
42  5  4   2  200 5  27 2  8  29 24 10 83.3  8
65  4  3.2 2  460 9  51 2  8  26 19 12 80.0  3
27  6  1.2 2  180 2  10 0  4  27 20 7  80.0  9
35  6  3.2 2  310 8  17 2  3  26 20 11 66.7  8
68  5  10  2  320 4  48 2  7  27 21 12 80.0  4
65  3  1.2 2  85  4  48 2  4  26 18 6  83.3  3
53  6  3.6 3  250 6  38 2  4  27 25 9  80.0  6
47  6  2.8 3  250 4  29 2  2  26 25 8  66.7  9
68  4  3.2 4  275 5  47 2  2  23 24 11 70.0  5
57  5  2.8 3  250 4  39 2  5  24 22 7  76.7  5
58  6  2.4 2  200 3  37 2  7  27 23 7  76.7  6
43  4  0.8 1  50  2  28 0  3  24 22 6  73.3  5
;
proc cancorr data=cancor vpefix=f10 wpefix=19;
var x1 x2 x3 x4 x5 x6 x7 x8 x9 x10 ;with x11 x12 x13 x14 x15 x16 x17 x18 x19;
run;
```


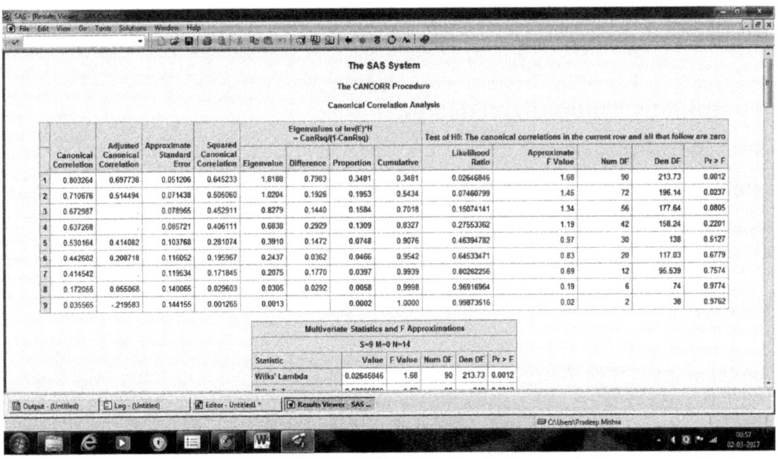

The SAS System

The CANCORR Procedure

Canonical Correlation Analysis

	Canonical Correlation	Adjusted Canonical Correlation	Approximate Standard Error	Squared Canonical Correlation	Eigenvalues of Inv(E)*H = CanRsq/(1-CanRsq): Eigenvalue	Difference	Proportion	Cumulative	Test of H0: The canonical correlations in the current row and all that follow are zero: Likelihood Ratio	Approximate F Value	Num DF	Den DF	Pr > F
1	0.803264	0.697738	0.051206	0.645233	1.8188	0.7983	0.3481	0.3481	0.02646846	1.68	90	213.73	0.0012
2	0.710676	0.514494	0.071438	0.505060	1.0204	0.1926	0.1953	0.5434	0.07460799	1.45	72	196.14	0.0237
3	0.672987	.	0.078965	0.452911	0.8279	0.1440	0.1584	0.7018	0.15074141	1.34	56	177.64	0.0805
4	0.637268	.	0.085721	0.406111	0.6838	0.2929	0.1309	0.8327	0.27553362	1.19	42	158.24	0.2201
5	0.530164	0.414082	0.103768	0.281074	0.3910	0.1472	0.0748	0.9076	0.46394782	0.97	30	138	0.5127
6	0.442682	0.208718	0.116052	0.195967	0.2437	0.0362	0.0466	0.9542	0.64533471	0.83	20	117.03	0.6779
7	0.414542	.	0.119534	0.171845	0.2075	0.1770	0.0397	0.9939	0.80262256	0.69	12	95.639	0.7574
8	0.172055	0.055068	0.140065	0.029603	0.0305	0.0292	0.0058	0.9998	0.96916964	0.19	6	74	0.9774
9	0.035565	-.219583	0.144155	0.001265	0.0013		0.0002	1.0000	0.99873516	0.02	2	38	0.9762

Multivariate Statistics and F Approximations					
S=9 M=0 N=14					
Statistic	Value	F Value	Num DF	Den DF	Pr > F
Wilks' Lambda	0.02646846	1.68	90	213.73	0.0012

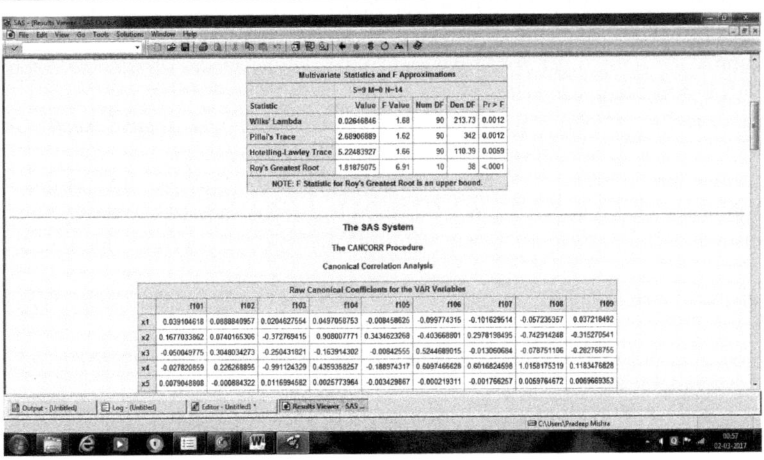

Multivariate Statistics and F Approximations					
S=9 M=0 N=14					
Statistic	Value	F Value	Num DF	Den DF	Pr > F
Wilks' Lambda	0.02646846	1.68	90	213.73	0.0012
Pillai's Trace	2.68906889	1.62	90	342	0.0012
Hotelling-Lawley Trace	5.22483927	1.66	90	110.39	0.0059
Roy's Greatest Root	1.81875075	6.91	10	38	<.0001
NOTE: F Statistic for Roy's Greatest Root is an upper bound.					

The SAS System

The CANCORR Procedure

Canonical Correlation Analysis

Raw Canonical Coefficients for the VAR Variables	f101	f102	f103	f104	f105	f106	f107	f108	f109
x1	0.039104618	0.0888840957	0.0204627554	0.0497058753	-0.008458625	-0.099774315	-0.101629514	-0.057235357	0.037218492
x2	0.1677033862	0.0740165306	-0.372769415	0.908007771	0.3434623268	-0.403668801	0.2978198495	-0.742914248	-0.315270541
x3	-0.050049775	0.3048034273	-0.250431821	-0.163914302	-0.00842555	0.5244689015	-0.013060684	-0.078751106	-0.282768755
x4	-0.027820859	0.226268895	-0.991124329	0.4359358257	-0.188974317	0.6097466628	0.6016824598	1.0158175319	0.1183476828
x5	0.0079048808	-0.000884322	0.0116994582	0.0025773964	-0.003429867	-0.000219311	-0.001766257	0.0059764672	0.0069669353

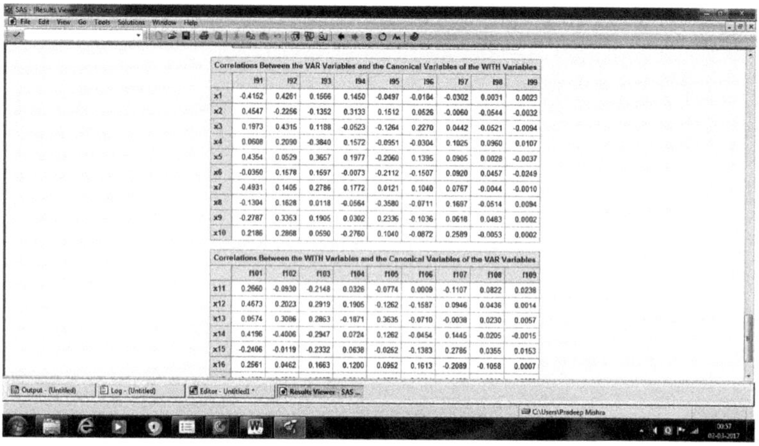

Correlations Between the VAR Variables and the Canonical Variables of the WITH Variables									
	I91	I92	I93	I94	I95	I96	I97	I98	I99
x1	-0.4152	0.4261	0.1566	0.1450	-0.0497	-0.0184	-0.0302	0.0031	0.0023
x2	0.4547	-0.2256	-0.1352	0.3133	0.1512	0.0526	-0.0060	-0.0544	-0.0032
x3	0.1973	0.4315	0.1188	-0.0523	-0.1264	0.2270	0.0442	-0.0521	-0.0094
x4	0.0608	0.2090	-0.3840	0.1572	-0.0951	-0.0304	0.1025	0.0960	0.0107
x5	0.4354	0.0529	0.3657	0.1977	-0.2060	0.1395	0.0905	0.0028	-0.0037
x6	-0.0350	0.1678	0.1597	-0.0073	-0.2112	-0.1507	0.0920	0.0457	-0.0249
x7	-0.4931	0.1405	0.2786	0.1772	0.0121	0.1040	0.0767	-0.0044	-0.0010
x8	-0.1304	0.1628	0.0118	-0.0564	-0.3580	-0.0711	0.1697	-0.0514	0.0094
x9	-0.2787	0.3353	0.1905	0.0302	0.2336	-0.1036	0.0618	0.0483	0.0002
x10	0.2186	0.2868	0.0590	-0.2760	0.1040	-0.0872	0.2589	-0.0053	0.0002

Correlations Between the WITH Variables and the Canonical Variables of the VAR Variables									
	f101	f102	f103	f104	f105	f106	f107	f108	f109
x11	0.2660	-0.0930	-0.2148	0.0326	-0.0774	0.0009	-0.1107	0.0822	0.0238
x12	0.4673	0.2023	0.2919	0.1905	-0.1262	-0.1587	0.0946	0.0436	0.0014
x13	0.0574	0.3086	0.2863	-0.1871	0.3635	-0.0710	-0.0038	0.0230	0.0057
x14	0.4196	-0.4006	-0.2947	0.0724	0.1262	-0.0454	0.1445	-0.0205	-0.0015
x15	-0.2406	-0.0119	-0.2332	0.0638	-0.0252	-0.1383	0.2786	0.0355	0.0153
x16	0.2561	0.0462	0.1663	0.1200	0.0952	0.1613	-0.2089	-0.1058	0.0007

KEYWORDS

- **completely randomized design**
- **factor put treatment**
- **JMP**
- **SAS**

REFERENCES

SAS Institute Inc., (1986). *SAS/QC User's Guide* (Version 5). SAS Institute Inc., Cary, NC.

SAS Institute Inc., (1987). *SAS/STAT Guide for Personal Computers* (Version 6). Cary NC: SAS, Institute Inc.

SAS Institute Inc., (1989). *"SAS/Technical Report P–188: SAS/QC Software Examples* (Version 6)." SAS Institute Inc., Cary, NC.

SAS Institute Inc., (1995). *SAS/QC Software: Usage and References* (Version 6, 1st edn., Vol. 1). SAS Institute Inc., Cary, NC.

SAS Institute Inc., (1996). "*SAS/STAT Software, Changes, and Enhancements Through Version 6.11*. The Mixed Procedure, Cary, NC: SAS Institute Inc.

SAS Institute Inc., (1999). *SAS/ETS User's Guide* (Version 8). Cary NC: SAS Institute Inc.

SAS Institute, (1990). *SAS/STAT User's Guide* (Vols. 1 & 2). Cary, NC: Author.

SAS Institute, (2002). *Getting Started With SAS Enterprise Guide* (2nd edn.). Cary, NC: Author.

Schlotzhauer, S., & Littell, R., (1997). *SAS System for Elementary Statistical Analysis* (2nd edn). Cary, NC: SAS Institute.

Slaughter, S. J., & Delwiche, L. D., (2006). *The Little SAS Book for Enterprise Guide 4.1*. Cary, NC: SAS Institute.

Snedecor, G. W., (1934). *Analysis of Variance and Covariance*. Ames, IA: Collegiate Press.

Index

D

E

M

P

Q

R

S

T

9781771887526